This book deals with the relationship between land use and land cover: between human activities and the transformation of the earth's surface. It describes the recent changes in the world's forests, grasslands and settlements, and the impacts of these changes on soil, water resources, and the atmosphere. It explores what is known about the importance of various underlying human sources of land transformation: population growth, technological change, political–economic institutions, and attitudes and beliefs.

Three working group reports outline important avenues for future research: the construction of a global land model, the division of the world into regional situations of land transformation, and a wiring diagram to structure the division of research among fields of study.

Changes in Land Use and Land Cover:
A Global Perspective

Changes in Land Use and Land Cover:

A Global Perspective

Edited by
WILLIAM B. MEYER and B. L. TURNER II

CAMBRIDGE
UNIVERSITY PRESS

Published by the Press Syndicate of the University of Cambridge
The Pitt Building, Trumpington Street, Cambridge CB2 1RP
40 West 20th Street, New York, NY 10011-4211, USA
10 Stamford Road, Oakleigh, Melbourne 3166, Australia

First published 1994

Printed in Great Britain at the University Press, Cambridge

A catalogue record for this book is available from the British Library

Library of Congress cataloguing in publication data

OIES Global Change Institute (1991: Snowmass Village, Colo.)
Changes in land use and land cover: a global perspective: papers arising from the 1991 OIES Global Change Institute/William B. Meyer and B. L. Turner II, editors.
    p. cm.
 Includes index.
 ISBN 0 521 47085 4
 1. Human ecology–Congresses. 2. Land use–Environmental aspects–Congresses. 3. Land use–Mathematical models–Congresses. 4. Land use–Research–Congresses. I. Meyer, William B. II. Turner, B. L. (Billie Lee), 1925– . III. University Corporation of America.
Office for Interdisciplinary Earth Studies. IV. Title.
 GF3.033 1994
 333.73'13–dc20 94-1257 CIP

Office for Interdisciplinary Earth Studies
Global Change Institute Volume 4
Series Editors: Tom M.L. Wigley and Carol Rasmussen

Papers arising from the 1991 OIES Global Change Institute

  Produced through support from the Committee on Earth and Environmental Sciences Subcommittee on Global Change. Additional support for the color illustrations was provided by the U.S. Geological Survey.

  Any opinions, findings, conclusions, or recommendations expressed in this publication are those of the authors and do not necessarily reflect the views of the sponsors.

ISBN 0 521 47085 4 hardback

Contents

Contents

The color plates will be found between pages 12 and 13.

Preface

Global change science represents a new interdisciplinary approach to understanding the complex systems, such as energy flows and biogeochemical cycles, that underlie the climate and other conditions that support the millions of life forms on the planet. The field is also concerned with how the actions of one of those life forms, namely, human beings, are altering processes, changing material and energy flows, transforming ecosystems, eliminating and rearranging species, and introducing artificial chemicals and species into the environment.

In 1986 the Office for Interdisciplinary Earth Studies (OIES) was formed at the University Corporation for Atmospheric Research to stimulate research in global change science and to provide a way for scientists from many disciplines and institutions to work together with government officials and the international community to shape this new field. Under its founding director, John A. Eddy, OIES organized a series of two-week summer workshops held at Snowmass, Colorado. The first three of these Global Change Institutes addressed different aspects of global change from an interdisciplinary perspective: greenhouse gases, past climate changes, and earth system modeling. The papers from each of these institutes have been published by OIES.

In 1991, a new and expanded approach to the Global Change Institutes was attempted; global change scientists were joined by social scientists to examine both the human causes and global consequences of altered land-use patterns on the planet. This volume comprises the collected papers and findings of that workshop. In order to achieve a wider distribution, this book and its companion from the 1992 workshop, *Industrial Ecology and Global Change*, have been published by Cambridge University Press.

Driving the development of the U.S. Global Change Research Program and the International Geosphere–Biosphere Program has been concern over increasing evidence of global environmental change due to basic human activities. For several decades, natural scientists have been observing increasing concentrations of greenhouse gases in the atmosphere, plausibly leading to changes in climate; acid precipitation and fertilizer runoff that have substantially altered the chemical environment for natural as well as managed ecosystems in many parts of the world; and continuing widespread deforestation and soil degradation with apparent disregard for species diversity and the long term sustainability. Understanding the implications of such trends and their interconnections for our global environment is a major challenge to earth scientists from many different disciplines, and was the primary focus of the earlier institutes in this series.

Seemingly underlying these trends are pressures from growing world population and increasing per capita resource use. Yet as one tries to isolate specific examples and develop a quantitative description of these pressures, it becomes apparent how little is known about how human societies actually interact with their environment. Social scientists have extensively studied structure and function in individual communities, but it is not easy to generalize this experience in ways to which natural scientists can relate.

The 1991 institute was intended to foster a dialog among representatives of various communities of scholars with insights into this issue, taking as focus the relationship between changes in land use and land cover. By land cover is meant the physical, chemical, or biological categorization of the terrestrial surface, e.g. grassland, forest, or concrete, whereas land use refers to the human purposes that are associated with that cover, e.g. raising cattle, recreation, or urban living. The group of some 40 researchers and 5 graduate students ranged from anthropologists and political scientists and economists on the one hand to agronomists and ecologists and chemists on the other. They had to establish a common language as well as address an agenda that included such diverse topics as the human drivers of land-use change and models of nutrient cycles in ecosystems. The two weeks of intense discussion marked only the beginning of a process of mutual education which will have results reaching far beyond the formal report that is presented here.

It sounds like a tall order, but hopefully this book will stimulate others to lend their disciplines and insights to addressing a critical global challenge.

Acknowledgments

The 1991 Global Change Institute and the publication of this volume were made possible by the continuing and generous support from the interagency Committee on Earth and Environmental Sciences Subcommittee on Global Change Research. The committee is supported by the agency members of the U.S. Global Change Research Program. The color section was made possible by support from the U.S. Geological Survey.

We are proud to give credit to John A. Eddy as the originator and motivator, not only of the 1991 institute but of the entire series. Additional thanks are due to the many individuals who gave a special impetus to this project: GCI director B. L. Turner II, series co-editors Carol Rasmussen and Tom M.L. Wigley, and the members of the Organizing Committee for the institute: Barbara Anderson, Lourdes Arizpe, James Broadus, John Eddy, Thomas Graedel, Jerry Melillo, John Richards, William Riebsame, Peter Rogers, and Vernon Ruttan. As always, we thank the OIES staff, whose support makes the institutes possible: Lisa Butler,

Sarah Danaher, Diane Ehret, Paula Robinson, and Pamela Witter.

Finally, we are indebted to all who came to Snowmass, especially to the rapporteurs for spending countless hours pulling together the material for the working group reports, and to the moderators of the various sessions for guiding the discussions toward the purpose assigned. In addition, special thanks are due to those scientists who have taken time to review the manuscripts contained in the 1991 GCI volume. Without their efforts, the scientific worth of this volume would have been lessened.

William R. Moomaw
Chair, OIES Steering Committee
Tufts University

Francis P. Bretherton
University of Wisconsin, Madison

I

INTRODUCTION

1

Global Land-Use and Land-Cover Change: An Overview

B. L. Turner II and William B. Meyer

Human activities have in recent years become recognized as a major force shaping the biosphere. The early insights of some prescient scholars are now common understanding: those of George Perkins Marsh in the mid-19th century documenting the pervasive modification of the land surface by humankind, and of V. I. Vernadsky early in the 20th century analyzing the growing human impact on the major biogeochemical cycles. Human actions rather than natural forces are the source of most contemporary change in the states and flows of the biosphere. Understanding these actions and the social forces that drive them is thus of crucial importance for understanding, modeling, and predicting global environmental change and for managing and responding to such change.

The need to divide the vast field of human-induced environmental transformation into more manageable areas of study led the U.S. Committee on Global Change to distinguish between two principal human sources of global environmental change: industrial metabolism and land transformation. The former category includes 'energy and materials flows' through the systems of production and consumption of an industrialized world. The scope of the latter covers 'not only activities like forest clearing that change land cover type and physical properties, but also those like fertilizer use that change chemical flows' (Committee on Global Change, 1990: 112, 117–18).

This volume from the 1991 Global Change Institute of the Office for Interdisciplinary Earth Studies (OIES) examines land transformation; its companion volume from the 1992 institute studies industrial ecology. The topic of land transformation invites the longer-term perspective. Of the two broad sources of human impact, it has by far the more extensive history, dating back to antiquity (Thomas, 1956; Wolman and Fournier, 1987). For most of human time, the modification of the earth by human action mainly involved impacts on the soil and biotic resources central to the agricultural base. Land transformation did not abate, but rather accelerated and diversified with the onset of the Industrial Revolution, the globalization of the world economy, and the expansion of population and technological capacity. Species have been thinned, exterminated, domesticated, or transplanted across continents; forests cleared; grasslands plowed or grazed; cropland and cities expanded; and wetlands drained for millennia, yet never so rapidly worldwide as at present. Almost all of the world's lands are now used and managed, albeit in widely varying degrees of intensity (Richards, 1990). Land transformations, though localized in their immediate occurrence, contribute to wider-reaching processes that include the globally systemic (Turner *et al.*, 1990) ones of atmospheric trace gas accumulation, climatic change, and stratospheric ozone depletion. Not until after the middle of the 20th century did annual carbon releases from fossil fuel combustion substantially exceed those from land transformation, and the latter remain today a large fraction of net anthropogenic emissions (Houghton and Skole, 1990).

Global demand for the products of the land is likely to continue accelerating for the foreseeable future. The capacity of the land—and of the environment more generally—to sustain that demand will remain an issue of fundamental importance. The level of concern that current trajectories of change have elicited reflects the possibility that much land transformation in some sense constitutes land degradation, whether that is defined as a decrease in the capacity of the land to meet the demands placed upon it (Blaikie and Brookfield, 1987) or is given some other meaning. Environmental change in the aggregate or in particular cases is not necessarily environmental degradation. Changes in the land have often been seen as improvements by some or all land users. Many of the world's most productive lands have unusually long histories of continuous settlement and increased output. Alteration is nearly inseparable from human occupation and use; the goal of distinguishing improvement from degradation is to encourage the former and to counter forces that encourage the latter. To assess which of the two a particular land transformation and its consequences represent is, of course, to engage difficult issues of forecasting (future resource demands and opportunities as affected by technical and socioeconomic change) and evaluation (distribution questions; the rights of future generations; the rights of other species). It also raises serious questions about ecocentric versus anthropocentric views of nature. Regardless, better scientific knowledge than we now possess of the physical extent, character, and consequences of land transformation is an indispensable foundation for any such assessment.

Purpose

The mission of the 1991 OIES Global Change Institute was 'to improve understanding of the links among human activity, land-use/land-cover change, and environmental change to improve projections of environmental change.' Toward that end, the goals of this volume are two: a broad review of what we know about the extent, character, causes, and consequences of global land-use and land-cover change and an identification of the important next steps in research. It offers a tutorial survey of the main issues in global land transformation, a critical inventory of what is known and is not known about the various facets of the topic, and, through the working group reports but also through the tutorial papers, contributions to the elaboration of a sound and promising interdisciplinary research agenda.

In accordance with the structure of the OIES Global Change Institutes, the papers were commissioned as tutorials and presented as such at the meeting in Snowmass, Colorado. Their major function was to ensure a common basis of knowledge among participants from widely varying backgrounds and to inform the consideration of the questions posed to the working groups. Revised, they serve the same function in the book.

Definitions

The topic of land transformation divides conveniently into two linked components (as reflected in the mission statement and in the title of this volume): those of land-use and land-cover change (Turner and Meyer, 1991). The two terms denote areas of study that have historically been separate. *Land use* has been a concern primarily of social scientists: economists, geographers, anthropologists, planners, and others. The term denotes the human employment of the land. Land uses include settlement, cultivation, pasture, rangeland, recreation, and so on. Land-use change at any location may involve either a shift to a different use or an intensification of the existing one. *Land cover*, a concern principally of the natural sciences, denotes the physical state of the land. It embraces, for example, the quantity and type of surface vegetation, water, and earth materials. Land-cover changes fall into two ideal types, *conversion* and *modification*. The former is a change from one class of land-cover to another: from grassland to cropland, for example. The latter is a change of condition within a land-cover category, such as the thinning of a forest or a change in its composition.

A single land use may correspond fairly well to a single class of land cover: pastoralism to unimproved grassland, for example. On the other hand, a single class of cover may support multiple uses (forest used for combinations of timbering, slash-and-burn agriculture, hunting/gathering, fuelwood collection, recreation, wildlife preserve, and watershed and soil protection), and a single system of use may involve the maintenance of several distinct covers (as certain farming systems combine cultivated land, woodlots, improved pasture, and settlements). Land-use change is likely to cause some land-cover change, but land cover may change even if the land use remains unaltered. A forest will steadily shrink if a constant rate of timber extraction or shifting cultivation exceeding regrowth is maintained.

The realms of land use and land cover are connected, in the terminology we adopt, by the *proximate sources* of change: those human actions that directly alter the physical environment. It is through the proximate sources that the human goals of land use are translated into changed physical states of land cover. Examples of proximate sources are biomass burning, fertilizer application, species transfer, plowing, irrigation, drainage, livestock pasturing, and pasture improvement. The proximate sources represent the point of intersection between the core concerns of the natural and the social sciences, between physical processes and human behavior. On one side, these proximate sources produce *land-cover changes*, or alterations of the properties of the land surface. They may take the form of either conversion or modification, and they may lead as well to *secondary environmental impacts*; trace gas emissions, biodiversity loss, soil erosion and degradation, albedo alteration and microclimatic change, and water flow and water quality changes are examples. On the human side of the chain, the proximate sources reflect human goals mirrored in land uses, and the *land-use change* that drives land-cover change, in turn, is shaped by *human driving forces*, the range of social, economic, political, and cultural attributes of humankind that shape the direction and intensity of land use.

This series of interactions is not unidirectional. Environmental changes may have feedback effects on land covers, land uses, and human driving forces. These effects, real or perceived, have a further set of human dimensions to the extent that they provoke societal responses intended to manage or mitigate harmful changes. (This last aspect lies outside the focus of the present volume except insofar as responses alter the proximate sources or the driving forces.)

Content

This framework governed the selection of topics for the tutorial papers presented at the 1991 institute. As revised for this volume, they examine several major areas of inquiry in land transformation. The first set focuses on global and world-regional trajectories of change in three hybrid classes of land use/land cover (forest/tree cover, grassland/pasture, and settlement). The second set deals with three forms of environmental impacts produced by land-cover changes (on atmosphere and air quality, soils, and water flow and quality). The third set reviews what is asserted and what is known about the roles of four categories of human driving forces (population/income change, technological change, political/economic institutions, and culture). A final set of papers explores and illustrates issues in the modeling and monitoring of land-use/cover change. We discuss each group of papers in more detail in the introductions to these sections of the volume.

The material presented in the tutorial papers informed the discussion of the questions dealt with in the second week of the institute. A brief survey of some of the key background issues is appropriate here as a preface to the working group reports.

Land-cover changes are of great importance in their own right. Much of the current research interest in them, though, is driven by their connection to other environmental changes. Global environmental models dealing with a variety of processes in which land transformations and their consequences play an important part require projections of land-cover changes. Yet the widely variable character of land transformation across the earth's surface significantly complicates the development of meaningful global-scale assessments and projections. Today as in the past, not only the rate but even the direction of change varies widely from place to place. The past few decades have been a time of afforestation and cropland abandonment in parts of the world even as rapid clearance for agriculture has been the norm in others. Though a global view is required for some purposes, a globally aggregate one is insufficient for answering many pressing questions. The net worldwide trajectories of land-cover change are rarely duplicated in any region or locality. Consequently, explanations, forecasts, and prescriptions developed only from global aggregate data are likely to be worse than useless when applied in subglobal units. Nor can adequate global projections be developed from global aggregate data alone because the global totals represent aggregations of quite dissimilar regional dynamics.

Projections, if they are not to be simple extrapolations, must be developed from models of the human driving forces of land transformation. It is clear that the patterns of environmental change across the earth's surface do not simply mirror the spatial pattern of one basic human cause, be it population growth or poverty or affluence or political-economic structure. Rather, the evidence suggests that the human driving forces of land transformation are systematically interacting clusters of variables and that their operation is strongly influenced by social and environmental context. The fragility or the robustness of the physical environment mediates the impact upon it of human activities; similar levels of human pressure may affect different environments to different degrees. The ways in which social context influences cause-impact relationships in land-cover change are also clearly important but are little understood.

Yet the opposite extreme from the global aggregate approach, a plethora of microstudies highly attentive to local context and singularity, is equally unsatisfactory given the needs and constraints of the global change research program. A large literature of small-area studies does exist, and it offers many insights into the complexities of nature–society interactions. Practical considerations, though, prohibit the separate study for global modeling purposes of every piece in the world's mosaic of environmental and socioeconomic conditions. Nor could the results, even if collected in a systematic and comparable way, necessarily be aggregated unproblematically for higher scales of analysis. The most appealing middle route toward a meaningful global assessment is the identification of a manageable number of world–regional 'situations' (which need not be spatially contiguous areas) of land-cover/land-use transformation. Such situations are defined as ones whose trends of land-cover change, secondary impacts, and driving forces are broadly similar.

It is a further complication that land-cover changes are increasingly experienced far from some of their ultimate human sources. This problem occurs both in the physical realm—transfer of impacts through the atmosphere or the hydrologic system (as in acid rain, atmospheric composition changes, or river basin flows)—and in the social realm, as in developed world demand for timber and agricultural products leading to tropical deforestation. The opportunity to export the costs of change through either of these means—or to the future through the discounting of costs and benefits—may encourage activities that otherwise would not be undertaken. Such spatial and temporal separation poses difficult problems in tracing and modeling the associations between driving forces and their environmental consequences, which may be masked if associations are sought only within spatially and temporally bounded data units. The regional situations of change are constantly interacting with one another.

With so many key issues unresolved, Bretherton's 'wiring diagram' of global change in the earth system (ESSC, 1988) could do no more than simply label one box 'human activities.' Mapping the contents and the structure of that box will be, first of all, a delicate exercise in locating and integrating the contributions of the various social sciences, many of which have developed separate disciplinary identities by defining their approaches in opposition to those of their neighbors. Posing the same challenges at a larger scale is the long-deferred need in studying human/environment interactions to begin to draw upon and merge the insights of both the social and the natural sciences. It is well recognized that the task will not be an easy one (Turner, 1991). The substantially different methods and philosophies that often characterize the two realms pose difficulties whose resolution will require innovative approaches.

The challenges given to the three working groups at Snowmass were:

Group A: to assess the feasibility of producing a flow or 'wiring' diagram of the processes of land-use/land-cover change (akin to the Bretherton diagram of earth system science) and of distinguishing the connections and boxes of the diagram that are relatively well understood from those that particularly require attention.

Group B: to devise a scheme for demarcating world regions or 'situations' characterized by social and environmental conditions sufficiently homogeneous to make them suitable midlevel units for characterizing trajectories of change and for identifying the significant clusters of human driving forces.

Group C: to assess (1) the prospects for projecting land-use/land-cover change over 10- and 50-year periods using existing data and frameworks and possibly employing historical 'retrojections,' and (2) the requirements for better data and conceptualizations that might improve such projections in the future.

Working Group A responded by devising a wiring diagram that represents the flows of information among fields and subfields of study concerned with the explanation of land-use/land-cover change. Group B developed a three-dimensional matrix (the 'cube') blending considerations of prevalent land-cover type, proximate sources of change, and social driving forces as a basis for developing a global typology of regional situations of land transformation. Group C reviewed modeling approaches to identify those best suited to the creation of a global land model capable of development and elaboration through iterative processes and improvements in the existing data base.

These reports outline promising directions for research. Indeed, the International Geosphere–Biosphere Program (International Council of Scientific Unions) and the Human Dimensions of Global Environmental Change Program (International Social Science Council) have already built on the proceedings and working group reports of the 1991 institute to formulate an international and interdisciplinary research project on global land-use/land-cover change (Turner *et al.*, 1993). Central activities of the project are the creation and refinement of a typology of regional situations, drawing on the cubic framework developed at Snowmass; the organization of detailed case studies in areas representing key situations; and the development of a regionally sensitive global model of land transformation, drawing on the case study findings.

Beyond the projects and areas indicated by the working groups and the tutorial papers, an extensive range of important research topics exists where interested natural and social scientists can contribute. It is clear, for example, that the data base on most aspects of global land use and land cover is wanting. The problems include inadequate and uncoordinated collection activities and a lack of standard definitions and measures. Our understanding would benefit significantly from the improvement of data sets both historical and contemporary. A recent report by a National Research Council committee (Stern *et al.*, 1992) outlines a large number of critical research tasks in the human dimensions of global change that await interested social scientists. They emphasize in particular the need for geographically comparative studies at varying time and space scales rather than detailed studies of individual locales. In addition to the framework used in both that report and this introduction, a perspective focusing on the social construction of global change concerns and knowledge has also been widely adopted (e.g., Rayner, 1992).

Color Plates (facing pp. 12–13)

Plates 1 to 9 were chosen from among many available ones to illustrate two features of data gathered by remote sensing: the power and utility of such imagery for observing land-use and land-cover changes, such as the image of forest cutting in Rondônia, Brazil, and the availability of large-scale data sets or data bases, such as the vegetation index maps of North America and Niger.

References

Blaikie, P., and H. C. Brookfield. 1987. *Land Degradation and Society*. Methuen, London.

Committee on Global Change. 1990. *Research Strategies for the U.S. Global Change Research Program*. National Academy Press, Washington, D.C.

ESSC (Earth System Science Committee), NASA Advisory Council. 1988. *Earth System Science: A Closer View*. National Aeronautics and Space Administration, Washington, D.C.

Houghton, R. A., and D. L. Skole. 1990. Carbon. In *The Earth as Transformed by Human Action* (B. L. Turner II, W. C. Clark, R. W. Kates, J. F. Richards, J. T. Mathews, and W. B. Meyer, eds.), Cambridge University Press, Cambridge, U.K., 393–408.

Rayner, S. 1992. Review of *Global Environmental Change: Understanding the Human Dimensions*. *Environment 34(7)*, 25–28.

Richards, J. F. 1990. Land transformation. In *The Earth as Transformed by Human Action* (B. L.

Turner, W. C. Clark, R. W. Kates, J. F. Richards, J. T. Mathews, and W. B. Meyer, eds.), Cambridge University Press, Cambridge, U.K., 163–178.

Stern, P. C., D. Druckman, and O. R. Young, eds. 1992. *Global Environmental Change: Understanding the Human Dimensions*. National Academy Press, Washington, D.C.

Thomas, W. L., Jr., ed. 1956. *Man's Role in Changing the Face of the Earth*. University of Chicago Press, Chicago, Illinois.

Turner, B. L. 1991. Thoughts on linking the physical and human sciences in the study of global environmental change. *Research and Exploration (Spring)*, 133–35.

Turner, B. L., II, and W. B. Meyer. 1991. Land use and land cover in global environmental change: Considerations for study. *International Social Science Journal 130*, 669–79.

Turner, B. L., II, R. E. Kasperson, W. B. Meyer, K. M. Dow, D. Golding, J. X. Kasperson, R. C. Mitchell, and S. J. Ratick. 1990. Two types of global environmental change: Definitional and spatial-scale issues in their human dimensions. *Global Environmental Change 1*, 14–22.

Turner, B. L., II, R. H. Moss, and D. L. Skole, eds. 1993. *Relating Land Use and Global Land-Cover Change*, International Geosphere–Biosphere Program Report No. 24/HDP Report No. 5, Stockholm, Sweden.

Wolman, M. G., and F. G. A. Fournier, eds. 1987. *Land Transformation in Agriculture*. John Wiley and Sons, Chichester, U.K.

II

WORKING GROUP REPORTS

Residential (11)
All other Urban or Built-up Land (12-17)
Agricultural Land (21-24)
Herbaceous Rangeland (31)
Shrub and Brush Rangeland (32)
Mixed Rangeland (33)
Deciduous Forest Land (41)
Evergreen Forest Land (42)
Mixed Forest Land (43)
Water (51-53)
Wetland (61-62)
Barren Land (71-77)
Tundra (81-85)
Perennial Snow or Ice (91-92)

Plate 1. Colorado land use/land cover. Over 20 years ago, the Land Use Data Analysis (LUDA) program resulted in a 1970s snapshot of Anderson Level 2 land-use classes that were photointerpreted from and mapped at a scale of 1:250,000. This data base has recently been digitized for the 48 contiguous states, and is now available as 1° × 2° blocks. Because of the potential value of this historical information for many applications, the U.S. Geological Survey's (USGS) Earth Resources Observation Systems (EROS) Data Center undertook an experiment to assemble 16 of these data sets into a Colorado LUDA data base which covers all of Colorado and a part of eastern Utah. A plot of the data base was generated at a scale of 1:600,000 in the UTM projection. For clarity, the map is color-coded (as shown in the legend).

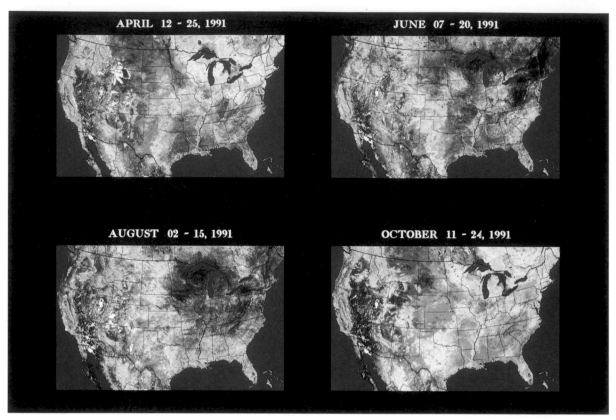

Plate 2. The conterminous U.S. advanced very high resolution radiometer (AVHRR) data set. The 1991 AVHRR data set is a comprehensive collection of calibrated, georegistered, biweekly maximum normalized difference vegetation index (NDVI) composites for January through December. The NDVI, computed from the visible and near-infrared channels of the AVHRR sensor, has been shown to be representative of vegetation condition and density. The maps depict the NDVI values in ten classes, with vigorous vegetation shown in shades of green; sparse or stressed vegetation in yellow to brown; clouds, snow and other bright surfaces in white; and water in blue.

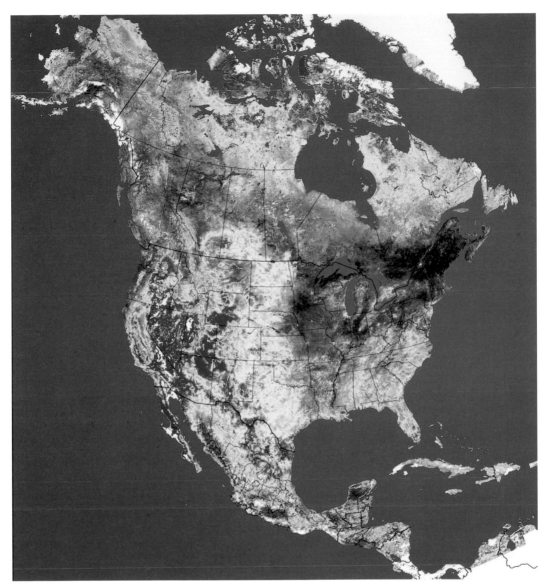

Plate 3. The North American vegetation greenness index map, produced from the AVHRR data set of North
America. As in plate 2, colors show NDVI values.

The EDC and the Canada Centre for Remote Sensing (CCRS) have developed programs to prepare
periodic 1-km AVHRR data sets to support global climate change research. The major emphasis of
this joint activity is the routine production of 1-km AVHRR data sets of the North American
continent that provide calibrated and georeferenced spectral data and a vegetation greenness index.

This map was prepared from a data set of approximately 30 daily observations using the National
Oceanic and Atmospheric Administration (NOAA)-11 satellite, acquired over the 10-day period of
August 11–20, 1990. The data for Canada, Alaska, and Greenland were processed by the CCRS, and
the data for the conterminous United States, Mexico, and Central America were processed by the
EDC. The daily observations were merged or composited using the maximum NDVI value of all
observations. This process effectively selects the most green and cloud-free observation for each
1 km in the study area. The data process by the CCRS and EDC were merged in the same manner to
complete the North American data set.

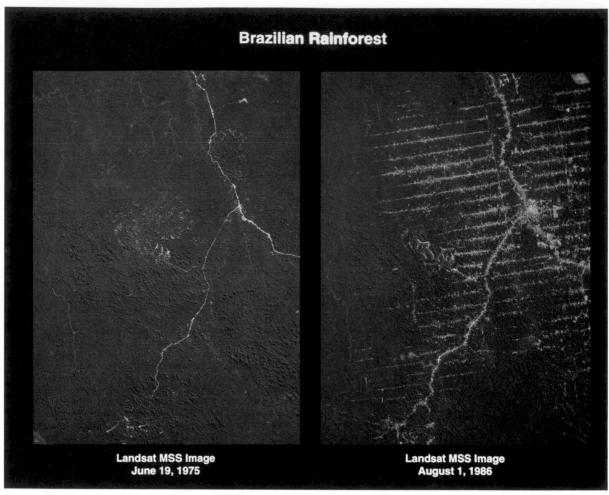

Brazilian Rainforest

**Landsat MSS Image
June 19, 1975**

**Landsat MSS Image
August 1, 1986**

Plate 4. Forest cutting in Rondônia, Brazil, as shown on Landsat multispectral scanner (MSS) images: June 19, 1975, and August 1, 1986. Systematic cutting of the forest vegetation starts along roads and then fans out to create the 'feather' pattern shown in the eastern half of the August 1, 1986, image. The cutover land and urban areas appear in light green and blue, whereas healthy vegetation is red.

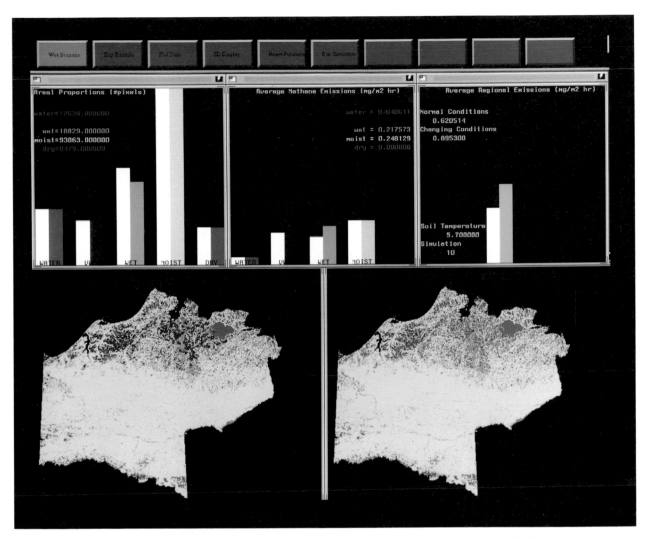

Plate 5. Modeling methane emissions. An interactive modeling environment under development at the USGS and National Aeronautics and Space Administration (NASA) Ames Research Center allows scientists to study how climate changes may affect the land. The modeling tools are used in this example to predict methane emission rates for arctic Alaska in response to changing environmental conditions. The lower right image depicts a color-coded map of the current land-cover classes for the National Petroleum Reserve in Alaska, and the white bars on the bar graphs above show current conditions. The simulation illustrated here used a warmer and wetter scenario (one possible outcome of climate change). The lower left image and colored bars on the bar graph show the results of the simulation. In the image, some of the wet tundra (light green) categories are changed to very wet (dark green). The three sets of bar graphs show (left) areal proportions of the change, (center) changes in average methane emissions by each vegetation type and its contribution to the total emission, and (right) the cumulative impact on the regional methane flux rate. Results of any model can be instantly displayed on the screen in a map and analyzed or correlated with other data layers in a geographic information system.

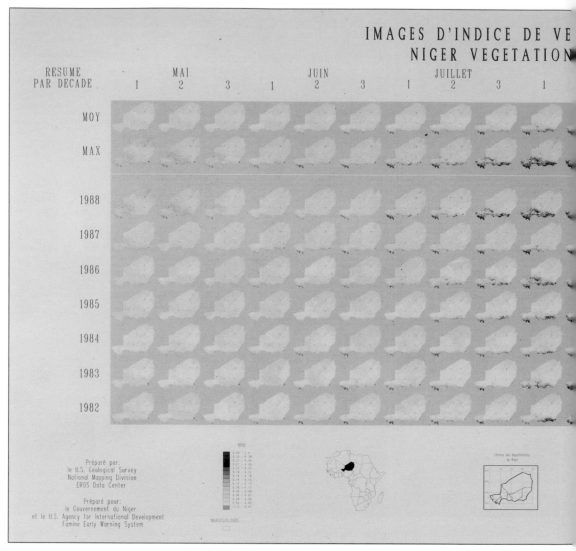

Plate 6. Vegetation index time series images of Niger. A growing number of operational programs in Africa use AVHRR data for monitoring vegetation. Though increasing attention is being paid to the land use of Local Area Coverage (LAC) AVHRR image data, which have a nominal ground resolution of 1.1 km, the coarser-resolution Global Area Coverage (GAC) data sets are valuable because they represent a source of long-term, uninterrupted data from 1982 to the present.

This figure is a graphic representation of the time series image data for Niger spanning the years 1982–88. The images are based on a special African (7.4-km resolution) NDVI image data base prepared by the NASA Goddard Space Flight Center, with image processing and graphic representation by the EDC. The images were prepared for vegetation monitoring by the Famine Early Warning Systems (FEWS) of the U.S. Agency for International Development. These NDVI image data are created by resampling maximum value weekly NDVI composite GAC images (with approximately 4 km resolution) to generate images based on a 7.4-km pixel, composited over a ten-day period.

The NDVI range is represented by a logical color sequence. Class intervals were chosen

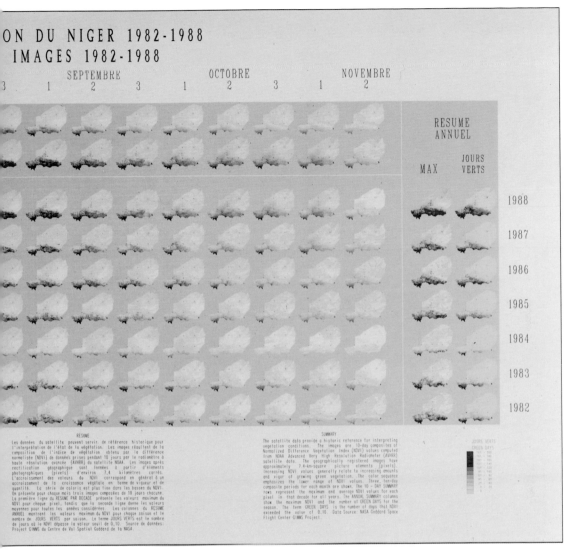

for increased sensitivity to the lower range of green vegetation cover. The summary rows (top two rows) represent the average and maximum NDVI values for each pixel in each ten-day period. The annual summary colums (right) show the maximum and average NDVI values for the season for each year. Cloud-contaminated pixels are represented in white. They were identified by using AVHRR thermal infrared low temperature values.

EDC scientists and FEWS analysts use the images for qualitative comparisons of vegetation conditions throughout the year. Seasonal initiation of growth, length of the growing season, and the amount of aboveground photosynthetically active biomass are easily observed. The drought years of 1983 and 1984, and the relatively wet year of 1988, are easily contrasted. The timing of the start of the growing season is markedly different from year to year, as is the intensity of green-up. These indicators contain considerable information that can be related to crop and rangeland growing conditions, and therefore to food production assessments. Since 1986, these data have played a critical role for monitoring environmental conditions in Africa that can lead to food shortages and famine.

ONSET OF GREENNESS

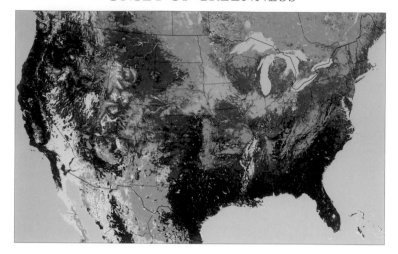

PEAK GREENNESS

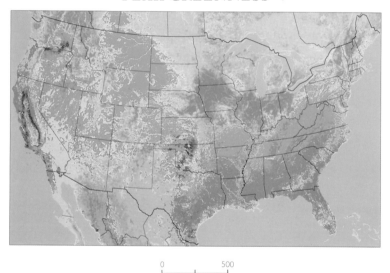

PERIOD

BEFORE
MARCH

MARCH

APRIL

MAY

JUNE

JULY

AUGUST

SEPTEMBER

OCTOBER

BARREN

WATER

0   500
STATUS MILES

Plate 7. Four interpretive maps developed at the EDC from the 1990 NDVI biweekly
data set to portray the seasonal behavior of vegetation in the conterminous
United States, These images portray four seasonal parameters: (1) the month
in which the NDVI first rose above a threshold value (onset of greenness), (2)
the month in which maximum NDVI occurred (peak of greenness), (3) the
number of days in which the NDVI reached or exceeded a threshold value
(duration of greenness), and (4) the cumulative value of the NDVI (total
NDVI) for March through October.

DURATION OF GREENNESS

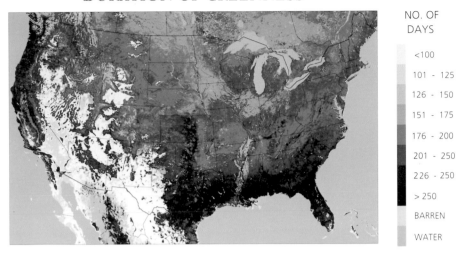

NO. OF DAYS

	<100
	101 - 125
	126 - 150
	151 - 175
	176 - 200
	201 - 250
	226 - 250
	> 250
	BARREN
	WATER

TOTAL NDVI

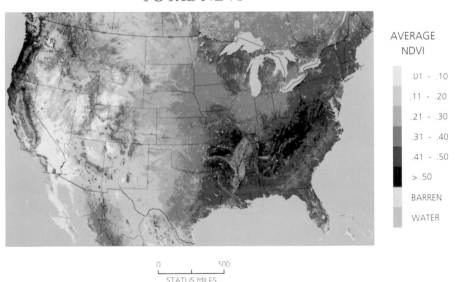

AVERAGE NDVI

	.01 - .10
	.11 - .20
	.21 - .30
	.31 - .40
	.41 - .50
	> .50
	BARREN
	WATER

0   500

STATUS MILES

The four factors are strongly related to the phenologic cycle of vegetation. The month in which the NDVI increases dramatically corresponds to the beginning of the growing season. The month of maximum NDVI coincides with the time of maximum photosynthetic activity. The length of time during which the NDVI exceeds a certain threshold value is analogous to the length of the growing season. The cumulative NDVI measured over the growing season generally reflects total photosynthetic activity or annual net primary production.

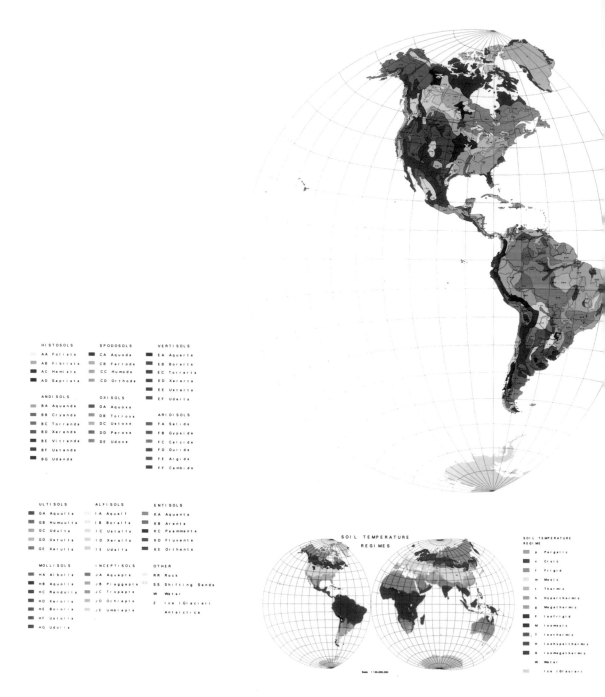

Plate 8. Major soil regions of the world, classified at the suborder level of the U.S. soil taxonomy system. This map was developed so that interpretations of soil properties such as soil carbon and water holding capacity can be quickly produced for global change studies. The interpretations are developed by aggregating data from soil samples (the pedon data base) to the units of soil

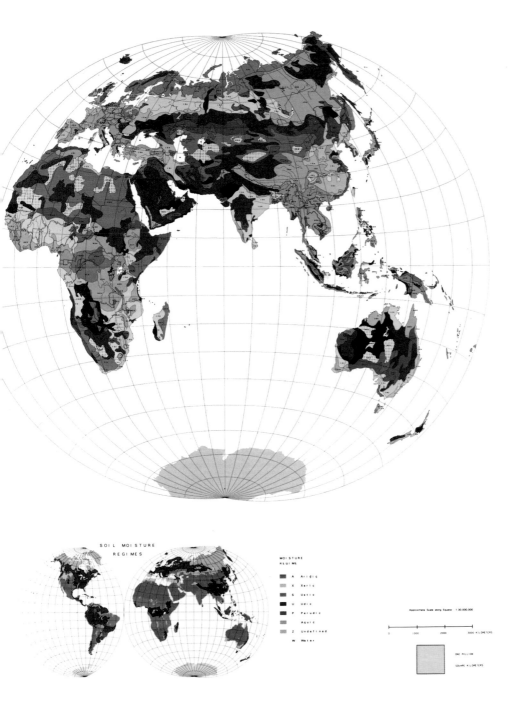

SOIL MOISTURE
REGIMES

MOISTURE
REGIMES

A   Aridic
X   Xeric
S   Ustic
U   Udic
P   Perudic
     Aquic
Z   Undefined
W   Water

Approximate Scale along Equator   1:30,000,000

0   1000   2000   3000 KILOMETERS

ONE MILLION
SQUARE KILOMETERS

taxonomy shown on the map. The analyses will be repeated using more detailed soil maps when additional data are available. The World Soil Resources map of the U.N. Food and Agriculture Organization was adapted as a basis for this map. Scientists from the EDC, the U.S. Department of Agriculture Soil Conservation Service, and the U.S. Forest Service participated in developing the map.

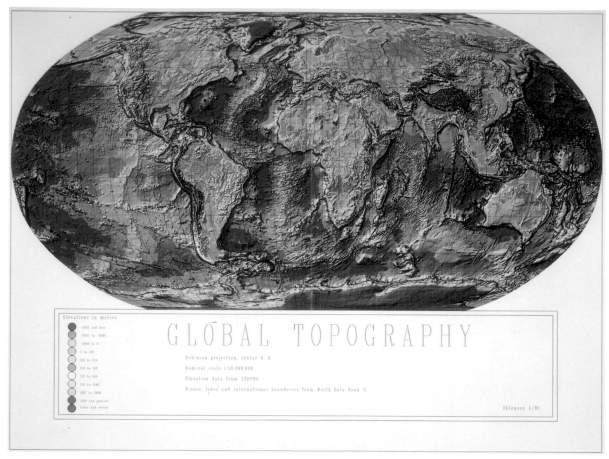

Plate 9. Global digital topographic data. The need among earth scientists for digital topographic data at many scales
continues to increase for many applications, including geometric and radiometric correction of satellite data, slope
computations for surface water modeling, and studies of plate tectonics. Global coverage, however, is limited to
NOAA's ETOPO5 data set illustrated here. The rivers, lakes, and international boundaries also shown are taken
from the World Data Bank II vector data set. ETOPO5 has a five-minute (approximately 10-km) resolution that is
too large for many applications. The USGS is identifying, acquiring, and generating higher-resolution data sets to
provide these essential data for global change science.

(a)

(b)

Plate 10. (a) Temperate, tropical, and arid grasslands have a ground story in which grasses are the dominant vegetation life form; this temperate example is from the Mongolian steppe.
(b) Arid and tropical grasslands grade into savannas as the very sparse overstory of trees increases; this arid example is from Australia.

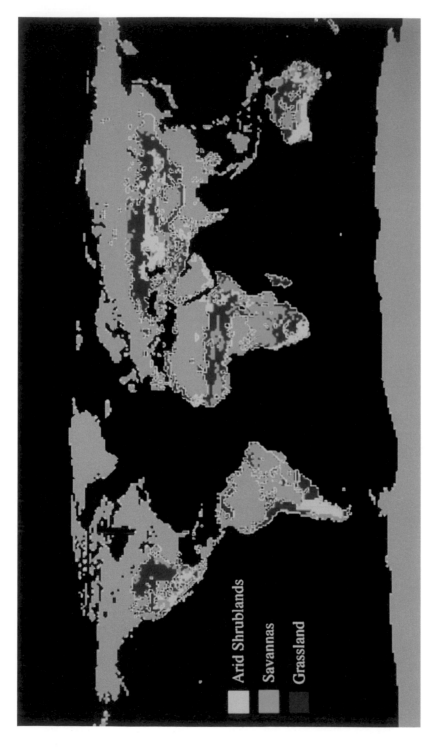

Plate 11. The distribution of three major vegetation types (biomes) that are collectively called grasslands in chapter 6 plotted at 1°×1° spatial resolution (from Matthews, 1983). The first category (green) is the extensive, treeless grasslands, Matthews classes 24–28. The second category (midgreen) is the savannas, Matthews classes 12, 13, 15, and 23. The last category is the arid shrubby grasslands (light green), Matthews class 21.

Plate 12. The temperate grasslands of Mongolia. For millennia this land-cover type has been subject only to the land use of nomadic pastoralism. Since World War II, however, the population has increased tenfold, from approximately 2 to 20 million people, almost entirely by migration. Sections of the grasslands are now being converted to croplands in a manner reminiscent of the cultivation of the steppes in the USSR and the prairies of North America.

Plate 13. A NOAA advanced very high resolution radiometer (AVHRR) image of the top end of Australia on June 28, 1990 (the dry season). The image has been enhanced to show fires, by either their smoke plumes or their footprints. There are scores of individual fires in this image. This area (~250,000 km^2) of tropical tall-grass savannas is subject to pastoral and traditional aboriginal use. Both of these land users use fire to transform the land cover for their purposes. Even though this has been done for decades, there is evidence to suggest that the intensity and frequency of this burning is increasing. As a consequence, the impact of fire on these savannas may no longer be temporary. Satellite images such as this have demonstrated that this is an increasingly common situation, globally.

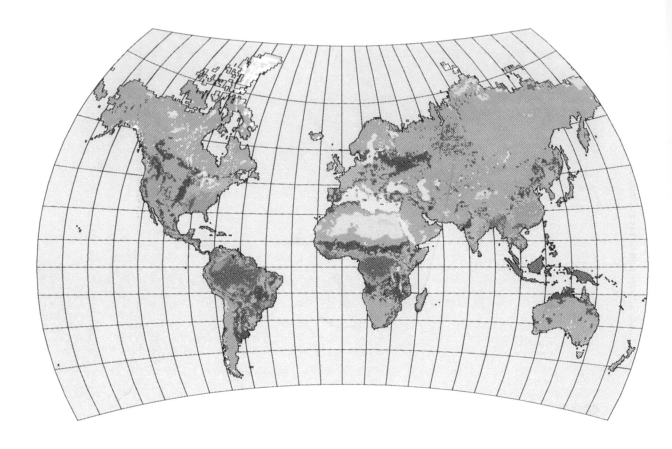

Annual Net Primary Production
gC -m-2-yr-1

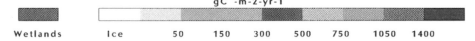

| Wetlands | Ice | 50 | 150 | 300 | 500 | 750 | 1050 | 1400 |

Plate 14. The global pattern of terrestrial net primary production for an 'undisturbed' or 'potential' vegetation under an 'average' contemporary climate (from Melillo *et al.*, 1993; printed with permission from *Nature*, copyright by Macmillan Magazines Limited).

2

A Wiring Diagram for the Study of Land-Use/Cover Change: Report of Working Group A

Steve Rayner (rapporteur), Francis Bretherton, Stanley Buol, Michael Fosberg, Wolf Grossman, Richard Houghton, Rattan Lal, Jeffrey Lee, Stephen Lonergan, Jennifer Olson, Richard Rockwell, Colin Sage, Evert van Imhoff

Introduction

Land use/cover has been recognized by a variety of national and international bodies as a critical factor mediating between socioeconomic, political, and cultural behavior and global environmental changes, especially changes in atmospheric chemistry and potential climate change (IGBP, 1988; NRC, 1990; ISSC, 1990). However, conceptual elucidation of these linkages seldom has proceeded beyond the simple assertion of interdependence among human behaviors, land cover, and the state of the global environment. The task of this working group was to begin to spell out how these interdependencies might be modeled and studied.

Scientists and policy-makers agree that humankind now possesses the capability to change the environment on a global scale. There is considerable speculation and much assertion about the importance of various human factors as drivers of land-use change. Significant uncertainty also remains about some of the biogeophysical processes involved. Furthermore, the important relationships among human and physical processes have not been identified systematically.

Achieving a better understanding of these relationships requires interdisciplinary collaboration, but this is currently not prevalent. This chapter attempts to facilitate collaboration by:

- Identifying interdisciplinary linkages required to solve problems that cannot be explored adequately within a single discipline
- Pinpointing areas of needed research that are in good shape and in good hands, areas that are not yet in good shape but are in good hands, and areas that are not yet being addressed
- Assisting researchers in identifying what information and models they need and can expect from colleagues focusing on specific aspects of the system.

After evaluating several approaches for achieving these objectives, the working group chose to develop a diagram of fields of study and the necessary information flows among them that contribute to an understanding of land-use processes. It was agreed that the framework should:

- Be capable of linking our understanding of processes at different scales—local, regional, and global
- Specify, where possible, information that is susceptible to measurement and modeling rather than merely identifying topics for research
- Assist in establishing concrete data needs
- Place strong emphasis on decision-making
- Take a modular approach
- Be parsimonious.

Developing the diagram required making certain assumptions. First, we postulated that the effects of land-use practices on the environment and the impacts of environmental change upon human societies may best be analyzed by examining the production systems and management incentives at the local level for a set of commodities that involve the use of land as a resource. This involves tracing the demand for these commodities back to human aspirations. It also involves tracing the effects of production forward through the modification of land cover, nutrient cycling, soil quality, and the emission of trace gases aggregated into global processes, manifest as regional changes in climate and potential ecosystem productivity. This emphasis makes the extent, technology, and intensity of the associated production and resource management systems the primary linking variables between the social science and natural science aspects of global change. Hence, we are able to sidestep the many problems that arise in attempting to reconcile existing land-use and land-cover typologies.

Assumptions on which the diagram is based include:

- Land use is the outcome of competition among potential uses.
- Political institutions and other decision-making agents operating above the level of the individual decision-maker need to be included.
- Both markets and administrative systems of decision-making (courts, government agencies, etc.) are vital carriers of information throughout the system.
- Cultural filters on information flows are significant (and may be modeled quantitatively).

Basic Structure of the Wiring Diagram

Figure 1 is a schematic framework for understanding how general human needs and wants are translated into land-use changes which in turn affect regional air quality and global climate. The large boxes in the diagram, called *modules*, represent fields of study. The smaller boxes, or *submodules,* represent subfields.

The arrows on the diagram represent information flows between modules. It is important to distinguish the flow of information about a variable from the flow of

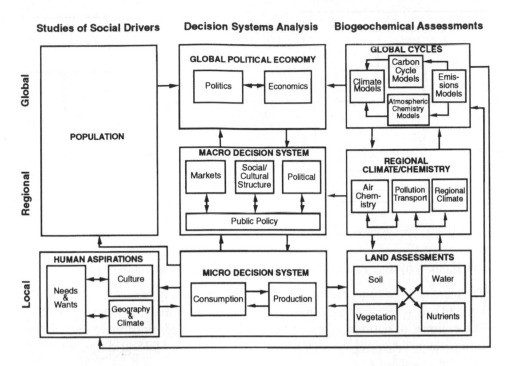

Figure 1. Land-use/cover diagram.

the physical or social quantity represented by that variable. In some cases, material flows may be in the opposite direction to the information flows.

We also emphasize that the diagram represents pathways for analysis of the phenomena involved in land-use decision-making. It is a framework for under-standing the processes, not an influence diagram or decision tree that represents the processes themselves.

The diagram encompasses three kinds of analysis. The left side shows our understanding of the basic drivers of land use represented by human aspirations and population. This is modified by our understanding of the human decision sys-tems examined in the middle of the diagram. The right-hand side of the diagram represents our understanding of natural processes and the changes wrought in these by human decision-making.

The diagram is also organized into three levels of scale for both organizations and biogeophysical impacts. These two types of scales are only approximately iso-morphic. The organizational (sociocultural and economic) scale, which dominates the left side and middle of the diagram, consists of:

- The micro decision level: the local agent (such as a household) immediately responsible for decisions about how a given plot is used
- The macro decision level: the regional decision system responsible for func-tional aggregations, which may be national or subnational
- The global decision level: the scale of international trade and politics.

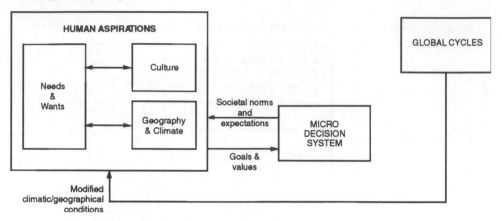

Figure 2. Human aspirations module.

The biogeophysical scale, on the right side of the diagram, consists of:

- The local land-use unit, perhaps corresponding as closely as possible to a 10×10 km grid square (the size used in some geographic information systems, or GIS)
- The regional or national scale, corresponding to the administrative unit of concern perhaps disaggregated at the level of 200×200 km GIS grid squares
- The global scale.

In the remainder of this chapter we work through the diagram module by module, explaining the contents of each box and the information flows between boxes.

The Human Aspirations Module

Individual and collective human aspirations are key variables in any land-use decision. These are the subject of the first module (Figure 2).

Module Components

The first submodule is basic needs and wants. These needs are modified by physical geography and climate and by culture—the other two submodules. We begin with six basic categories of human needs and wants: diet, shelter, prestige, security, recreation, and spiritual satisfaction.

Health, age, gender, and other differences affect individual nutritional requirements and needs for clothing and shelter. However, needs also vary because of differences in environmental conditions, the subject of the geography and climate submodule. For example, altitude and aridity may directly influence the need for water, and seasonal temperature variation (especially low winter temperatures) affects clothing needs.

Cultural and social contexts influence not only the individual's efforts to satisfy specific needs and wants but also his or her perception of their importance

(Douglas and Isherwood, 1979). Definitions of such intangibles as the 'poverty line' vary among different cultures and social groups, as well as over time. Elucidating the role of culture in translating basic needs and wants into specific goals and values is the task of the culture submodule.

As needs and wants are modified by culture and geography, we expand the category of diet beyond essential nutrition to include quantity and variety of food. Shelter includes technologies for heating and cooling, etc. Prestige needs are met in such varied ways as the production of pigs for communal feasting in Papua New Guinea (Rappaport, 1967), and the construction of large public works in developing and developed countries. Security needs call attention to the importance of territorial demarcation in control of land. Recreational aspirations may involve beaches, lakes, ski resorts, or national parks. Walks in the woods may be recreational but may also satisfy spiritual aspirations through wilderness experiences, vision quests, or even recognition of the intrinsic value of natural environments (Stankey, 1972).

Module Inputs

The most basic human needs are endogenous to this module. However, the module receives important feedback from other parts of the diagram with respect to the modification of culture, which is a direct feedback from the micro decision unit, and geography and climate, which may be modified both by regional atmospheric chemistry and by changes in the global cycles.

Module Outputs

The outputs of the human aspirations module do not yet constitute economic demands for goods and services; rather, they are the goals and values of the decision-makers described in the micro decision-making module.

Examples of Relevant Research

Attempts to identify and measure the drivers of individual behavior generally distinguish between biophysical needs and psychosocial wants, i.e., what is indispensable for survival and what makes life more comfortable and satisfying. For example, Maslow (1970) formulates a hierarchy of motivations in which, he claims, people strive to satisfy their needs at each level before they begin to be motivated by the next higher level. In ascending order of priority, these are: physiological necessities, safety, belonging and love, esteem, and self-actualization. Although this hierarchy is useful in analyzing psychological motivations, it tends to obscure the interdependence among the various needs, particularly with regard to resource consumption and land use.

In a slightly different approach to human needs and wants, McHale and McHale (1975) define three categories: (1) deficiency needs that must be satisfied to avert destitution, sickness, and death; (2) sufficiency needs or wants to attain a desired standard of living; and (3) growth needs or desires for personal development and satisfaction. Inglehart (1990) argues that in advanced industrial societies and in developing countries such as South Korea and Mexico, people increasingly assume that their elemental needs will be satisfied. In place of a materialist ideology focused on obtaining food, shelter, and clothing, people adopt a 'postmaterialist' ideology. This provides scope for development of what the McHales term 'growth needs' and Maslow, 'self-actualization.'

Potential Contributions

Clearly, understanding the most basic needs requires information from physiologists, nutritionists, and psychologists to establish the minimum daily requirements for food, shelter, rest, etc., that are necessary to avoid absolute destitution—the physical deterioration of the human organism. However, the medical disciplines may tend towards ethnocentric definitions of need that obscure the important conceptual boundary between destitution and poverty.

The shaping of specific needs by geography and climate is addressed by some subfields of human geography (e.g., cultural ecology) and by ecological anthropology. Both disciplines have generated sizeable literatures relating factors such as terrain, precipitation, and vegetation to specific aspects of economic, political, social, and ritual organization as well as the development of technology. Sophisticated exponents of this viewpoint describe not only how the environment has shaped consumption choices according to available resources but also how societies may develop complex ritual arrangements that constrain them from exceeding the carrying capacity of the local ecosystem (Rappaport, 1967). This literature has sometimes been accused of reducing society and culture to mere mechanisms of environmental response and management, as well as of seeking biophysical motivations for human behavior at the expense of other dimensions of culture (Geertz, 1963).

Cultural theorists in anthropology and sociology have invested the cultural submodule with greater autonomy than some geographers and ecological anthropologists allow. Cultural theory defines poverty (as distinct from destitution) as the restriction of social choices and recognizes that goods are necessary for projects that, like feasting, may fulfill subsistence as well as other needs or, like cut flowers, may serve purely expressive goals (Douglas, 1976). In rural societies, possession of the right sort of goods in appropriate quantities is likely to be directly related to control over land use. Thus cultural theory is an important component in understanding the transformation of basic needs into the goals and values that guide the activities described in the micro decision module. To understand this process in regard to land use, especially how land is directly valued (e.g., as a resource to be exploited or to be conserved), culture must be related explicitly to social choices about human organization as well as to ecological factors (Schwarz and Thompson, 1990).

Data Issues

Research on human aspirations begins with an understanding of humans as an animal species. Therefore, it requires not only the cognitive data that might be expected but also data on human physiology, especially data on responses to climatic variations, such as heat stress, and epidemiological data such as patterns of disease vectors. Insofar as perceptions are shaped by experience of the natural environment, microclimate records are important data inputs. These records can also be invaluable in revealing disparate perceptions of measurably identical climates (Douglas, 1962). Quantitative data on individual nutritional needs disaggregated by age and gender may be important, as well as data on patterns of nutritional deficiency that might be related to land use or location.

Understanding the structure of value systems also requires detailed qualitative data such as those most often collected through ethnographic techniques such as key-informant interviews and participant observation. Such data include information about the shaping of perceptions of risk and of both social and environmental vulnerability, as well as the formation and ordering of values. Many of these data are distributed widely across a large literature, but some are consolidated in the Human Relations Area Files (HRAF, 1992). Surveys of what people are willing and unwilling to pay for and what trade-offs they are willing to make may be helpful for understanding how people value intangibles for which no price information exists (Cummings *et al.*, 1986). Understanding the way we value unpriced goods, such as the intrinsic value of wilderness or the spiritual/mental health benefits of a wilderness experience, has been the subject of psychological survey research (Feingold, 1979; Haas *et al.*, 1980).

The cultural shaping of perceptions, beliefs, attitudes, and values has been shown to be closely related to the transitivity and closeness of social networks and experiences of equality or hierarchy (Douglas, 1978). Network data (Holland and Leinhardt, 1979) can be combined with multiple hierarchy measurements (Jeffries and Ransford, 1980) to understand and predict this cultural shaping (Gross and Rayner, 1985).

A large body of data permits the tracking of changes in sociopolitical attitudes across time and the analysis of differences in attitudes across countries. The Eurobarometer surveys, initiated in 9 member nations of the European Community in 1974, replicated attitudinal questions that had been asked of the U.S. population for many years. The Eurobarometers now total more than 30 studies and cover 13 nations. There have been similar efforts in the International Social Science Program, which began a series of studies in 1984; in the International Social Justice Program, recently formed; and in the Comparative Project on Class Structure and Class Consciousness, which includes data from 10 countries in the 1980s. With the exception of this last project, one of the limitations of these bodies of data is their restriction to the countries of the North Atlantic.

One of the few comparative international surveys to focus on values as well as attitudes is the World Values Survey. In 1981, this survey encompassed a sample of 23 nations; the 1990–91 survey had a sample of 42 societies on all six inhabited continents. It is the first survey ever to have sampled a majority of the human population. The survey provides measures of both individual attitudes and values, and it permits analyses relating these measures to economic indicators, political variables, and social indicators. It is a rich store of information, even though it, like most other surveys, still excludes people living in most of Africa and much of Asia and Latin America (Inglehart, 1990).

Psychographic profile distributions relating survey research and georeferenced satellite data could facilitate measurement of the spatial distribution of value systems (Worcester and Barnes, 1991). Other useful data can be collected through advanced market research techniques focusing on desired goods as well as through attitude surveys. Content analyses of media attention to indirect measures, such as consumption, as well as to more direct land-use and environmental issues may also be appropriate, especially where literacy is high.

The Micro Decision Module

The micro decision module (Figure 3) represents our understanding of the smallest units of decision-making whose actions directly or indirectly affect land-use/cover change. These actions may be decisions about consumption that shape demands whose satisfaction involves use of the land as well as decisions about production made by those who directly control plots of land.

Module Components

For analytic purposes, this module is divided into submodules for consumption and for production. In many cases, especially close to the subsistence level, consumption and production units are really one and the same—agricultural households producing primarily for their own consumption (Cantor *et al.*, 1992). In advanced industrial societies, the production and consumption units may be on different continents.

To understand production, we combine information about the goals and values of consumers with information about income to derive an economic demand function for goods and services. Similarly, to derive a picture of supply opportunities, information about the goals and values of producers is combined with data about the availability of suitable land, its potential sustainable productivity, and capital and labor available to the producer.

Certain internal characteristics of the decision-making unit influence its decisions. These include the politics within the group (e.g., gender and age roles), characteristics of the decision-maker, the social organization within which the group functions, and the group's physical environment. Decisions may not be made equally by all members of the unit, and this may have implications for perception of the larger social organization and the environment, as well as for how

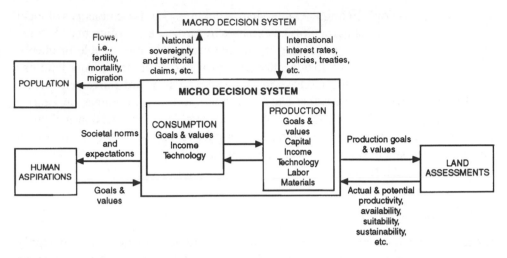

Figure 3. Micro decision system module.

the labor and other resources are allocated. For example, gender roles may dictate that women collect the fuelwood, so they may be more aware of the decline of forest resources, whereas men who sell the products may be less willing to engage in reforestation (Blaikie, 1985).

Characteristics of the members of the unit influence or directly affect land-use decisions. For example, access to capital and knowledge for production activities affects the agent's ability to gain access to and use machinery or other inputs, hire additional labor, and gain control over additional land. Capital and knowledge also influence the choice of crops and other agricultural management practices such as irrigation and terracing (Schultz, 1990).

Land-use decisions are only a subset of many decisions concerning income generation and other goals, and they are often affected by these other goals. Nonfarm activities must be considered, for example, because they affect labor availability and potential capital investment in the farm (Bohannan, 1970).

Module Inputs

Three fields of study contribute to an enhanced understanding of the micro decision module: human aspirations, land assessments, and macro decision-making. Information about goals and values comes from the human aspirations module. Data on land availability, productivity, etc., are derived from the land assessments module. Information about capital availability (especially concerning interest rates) and the labor market is derived from the macro decision module, as are other important pieces of information about technology, transportation, laws, taxes and subsidies, and national price structures.

The causes of land-use change may not always be due primarily to decisions at the smallest unit, but may lie outside the area or at a higher level. For example, the expansion of agriculture into a forested area may be due to land tenure conflicts elsewhere that force people to migrate rather than to changing land-use practices of local farmers (Hecht and Cockburn, 1989). A change in government policy may encourage agricultural expansion or logging in an area previously unavailable or uneconomic for settlement. Environmental changes may also contribute to major land-use changes. An example is the increase in rainfall and northward movement of the Sahel in the 1950s and 1960s that allowed farmers to move into areas previously dominated by pastoralists (Gritzner, 1988). Local-level land-use decision-making is thus the outcome of factors emanating from higher levels interacting with critical elements at the smallest unit. To understand this process we must weigh and interpret these external information flows from the viewpoint of the land user.

Module Outputs

The micro decision module is the source of information about demand and supply to the market submodule of the macro decision module. Also, political demands expressed through voting behavior, pressure groups, or threat of civil disorder are made available to the political submodule, and data about societal norms and expectations to the cultural submodule. The micro decision module is also the primary unit of human reproduction (the family) and provides vital information on fertility, mortality, and migration to the population assessment module.

Examples of Relevant Research

A variety of approaches has been developed that incorporate information flows, individual characteristics, and individual goals and values into land-use decision-making. Most are based on a rational model of decision-making, emphasizing positive rather than normative information. There are three basic approaches.

The first is the *multivariate analysis approach*. Researchers correlate the outcome of a decision to characteristics of the individual (e.g., age, gender, education, wealth) and place (e.g., distance to market, environment), usually using multiple regression (Evenson and Kislev, 1975). This approach provides a broad view of decision patterns and can identify associations among variables. However, the causes of the associations are not revealed, so the approach cannot generally explain the complex reasons behind a particular decision. For example, education may appear to be important where the more significant factor in decision-making is risk aversion, which may be associated with a lack of exposure to education.

The second is the *behavioral approach,* in which researchers establish a choice criterion, such as maximization of profit subject to a number of constraints, and test the hypothesis that land users make decisions consistent with the outcomes of the model. The approach can be used to identify constraints, evaluate the impact of new technologies and policy changes, and identify and evaluate management strategies. Such models are most often used by economists. Like those of the first approach, these models simplify the actual decision criteria and are susceptible to measurement problems. They rarely attempt to explain the reasons behind decision-making.

If the behavioral models applied to land use and cover have been mechanistic, they are underlain by rather well-developed constructions of farming behavior that could readily be incorporated into land-use/cover models. These constructions are of three broad types: market-based in developed economies, market-based in less developed economies, and market/subsistence–based in less developed economies (Hayami and Ruttan, 1985; Pingali *et al.*, 1987; Turner and Brush, 1987). Behavioral models have been tested at local (household and village) levels, largely through assessments of the inputs and outputs that follow from the logic of the models.

The third is the *cognitive anthropological approach,* which begins with the assumption that the decision-maker's own perspective of the world is the best starting point for model building. Intensive interviews are used to understand the criteria used in making decisions and the management strategies pursued. More emphasis is placed on normative and prescriptive information than in the other two approaches. Examples of this approach include participant observation, hierarchical decision modeling, and some diagnostic survey procedures in farming systems research. Researchers emphasize understanding the logic behind decisions rather than gathering facts about production.

Potential Contributions

The types of analyses discussed in the preceding section are seldom if ever combined. An interdisciplinary combination of these perspectives and approaches holds out the promise of a far more complete understanding of land-use decision-making at the micro level than we currently possess. Such an approach would closely resemble the emerging field of socio-economics (Etzioni, 1988).

Data Issues

Like most other modules, the micro decision module requires data that are both endogenous and exogenous. Exogenous data will be discussed under their modules of origin. They include measures of national or regional prices for agricultural inputs and outputs, and commodities purchased from outside the local economy but paid for out of income from land use. Capital availability, national and regional interest rates, currency exchange rates, tax rates, subsidies, and technologies are other important inputs to this module, as are data on laws and regulations, off-farm employment opportunities, productivity, sustainability of planned uses, etc.

Endogenous data requirements concern the institutions (families, firms) that process information from these other sources in pursuit of the goals originating in the human aspirations module. The innovativeness of the production/consumption unit and the rate of land-use change may be significantly different when the goal is maximizing profit from the land than when the goal is satisficing (in the terminology of Herbert Simon) across a range of desired benefits. The structure of authority in the family or firm also has a significant impact on rates of behavioral change. Both goals and structures have been shown to be significant to innovativeness of firms in the energy industry (Trauger *et al.*, 1986). These variables can be measured according to social network and multiple hierarchy data.

The degree of specialization of labor according to age and gender can make a fundamental difference to both productivity and sustainability. These sorts of data can best be collected ethnographically in the early stages, and the information used to structure subsequent surveys. Data on educational status are also important; not only does education directly influence employment opportunities, but literacy, urbanization, and off-farm employment have historically been elements of the demographic transition to low fertility (Watkins, 1992).

Spatial data derived from maps and remote sensing provide information about distance to markets, transportation costs, access to extension services, rates of technological diffusion, etc. These are linked to the structure of authority in land tenure rules. Written land tenure rules may be collected through archival research, although key informant interviewing and ethnographic observation often reveal discrepancies between written systems and local practices. The latter are collected ethnographically. The Land Tenure Center at the University of Wisconsin is a major repository of this information (LTC, 1992).

Many of these data provide the context for information derived from microeconomic studies and decision research in small group settings (Plott, 1986; Smith, 1986). The insights from these fields can be revealing, but their application to real decision settings must be subjected to 'ground truthing' through intensive field interviews and cognitive diagnostic surveys. Behavioral microeconomic data, for example concerning local prices, will also be important sources of insight to the input/output aspects of the firm or family. However, especially in developing countries, the web of social obligations and opportunities for exchange in the unofficial economic sphere are more attractive than those in the formal economy and are likely to distort any microeconomic analysis that does not take account of them (Pahl, 1984; Ferman and Ferman, 1973).

Macro Decision Module

The next level at which human decision-making shapes land-use/cover change is the regional or national level. The macro decision-making module (Figure 4) describes two basic functions: coordination of actions of the lower-level decision-making units, and communication with other regional decision-making units through higher-level (global) coordinating mechanisms.

Module Components

Each of the four submodules concerns an arena in which human interactions produce macro-level decisions: markets (the economic arena), political institutions (the political arena), social and cultural structures and institutions (the social

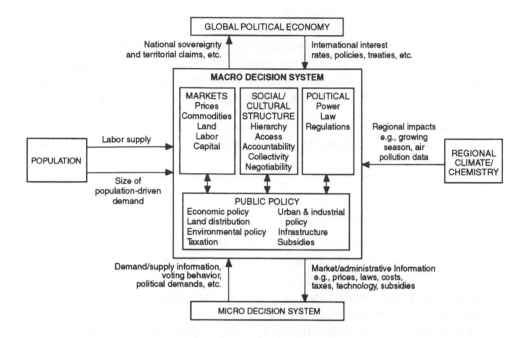

Figure 4. Macro decision system module.

arena), and public policy (the result of interaction among the other arenas). Essentially each of these represents a different social mechanism for aggregating the aspirations of the local agents discussed in the micro decision module.

Module Inputs

Data on interest rates, prices, policies, and treaties come from the global political economy module. The regional climate/atmospheric chemistry module supplies information on regional temperatures, precipitation, growing season, and air pollution. The micro decision module provides information on demand and supply to the market submodule, on voting behavior and political demand to the political submodule, and on societal norms and expectations to the cultural submodule. The population module offers information on the size, distribution, and composition of the labor supply to the market submodule.

Module Outputs

The macro decision module provides information about territorial claims to the global political economy module. It also supplies the micro decision module with information on general policy directions as well as data on the labor market, technology, taxation, regulations, and capital availability.

Examples of Relevant Research

Models for how markets function are abundant in the economic literature. For the study of land use and land cover, the markets for labor, capital, and agricultural products are especially important. In selecting a particular model and adapting it to the problem at hand, the following issues should be taken into account:

- The extent to which individual agents have access to regional markets
- The extent to which regional markets are integrated into the national or world markets
- The extent to which markets are characterized by market power or market concentration (monopolies, oligopolies, cartels)
- The influence of political institutions on the ways in which markets operate.

In economic theory, the rationale for the existence of political institutions stems primarily from considerations of *market failure*. Although markets are generally considered to be highly efficient in coordinating individual behavior, in many important cases they cannot produce certain outcomes (e.g., equity, public goods) or the outcomes are nonoptimal from a social point of view (e.g., when production is characterized by externalities like congestion or pollution). Since externalities are typically predominant in issues like land use and environmental change, the study of political decision-making is especially important here.

Models of political processes are provided by political scientists and by practitioners of public choice theory, which is basically the economic theory of collective decision-making (Mueller, 1979; Olson, 1965). Three elements of these models are crucial.

First, there are various ways in which individual agents communicate their preferences to the political process. The lower-level agents influence the higher-level agent through political behavior, e.g., voting, formation of pressure groups, mobilization of public opinion. These various forms of upward communication are called *political signals*. Generally speaking, political signals indicate the degree of support for particular policy actions.

Second, the process that transforms individual preferences into collective action must be considered. Important considerations here are the decision-making process (referendum, representative democracy, totalitarian system), the distribution of political power, and the transformation of political decisions into collective action through bureaucracies and hierarchical systems, as, for example, studied in principal-agent theory (Levinthal, 1988).

Third, collective actions affect the lower-level agents in different ways. Important categories are the following:

- Actions affecting market prices, in the form of taxes and subsidies (e.g., fuel taxes, fertilizer subsidies, water prices)
- Provision of public goods (e.g., education, roads, water works)
- Redistribution of resources (e.g., social security, income taxation, land reform)

- Legislation (e.g., bans on CFCs, afforestation requirements)
- Publicity campaigns (e.g., promoting voluntary recycling or family planning).

Recent literature on the social and cultural dimensions of market behavior argues that the sharp distinction of economic and political aggregation of preferences may obscure important characteristics of both markets and political systems of allocating goods and services (Etzioni, 1988; Cantor *et al.*, 1992). For example, activity in the so-called hidden economy may be motivated as much by human aspirations for social bonding or prestige as by demand for goods and services. Markets in the hidden economy rely on cultural rules that are not captured in conventional economic analysis. The size of this sector is difficult to estimate, but may account for 7–20% of a country's GNP (Carson, 1984).

Culture also influences political actions through *political culture* (Jasanoff, 1986). The institutional norms and values of political parties, councils, parliaments, civil bureaucracies, and nongovernmental organizations (NGOs) translate into constraints and opportunities for decision-makers at this level and that of the local agent. Similarly, the legal system responsible for upholding land tenure and zoning rules influences decision-making directly through judicial rulings and indirectly through judicial culture.

Potential Contributions

As with the micro decision module, social sciences appear to already possess the basic tools and theoretical frameworks to understand the processes studied in this module. The challenge is to combine these understandings in an interdisciplinary fashion and to focus the attention of leading practitioners on the issue of land-use change.

Data Issues

The macro decision module uses four sorts of data, corresponding to each of the four submodules.

Data on the operations of markets are plentiful. Many relevant data sets are prepared by national governments. These cover statistics on employment, income, cost of living, agricultural production, other commodity flows, gross national product/gross domestic product, public interest rates, and taxation; most are widely available. However, the principles used to organize many of these data in national income accounts are seriously flawed. Most countries fail to measure national income correctly because natural resources are treated as gifts of nature rather than as productive assets whose value should be depreciated as they are consumed (Repetto *et al.*, 1989). This failure seriously distorts economic evaluations in resource-dependent countries.

Other important economic data are published by market institutions themselves. These include banking interest rates, stocks and share prices, corporate performance data, and commodity prices. Once again, however, investigators should approach these data with a highly critical eye. Political interference in publication of economic statistics is not unknown, because the statistics are often used by the public as a measure of 'how well the government is doing.' In addition, governments have security reasons for distorting certain kinds of economic data, including the need to conceal resources or gain prestige internationally. Beyond issues of tampering with data, some governments simply do not have the capacity to take a reliable census of the population every ten years or so, much less the capacity to generate meaningful economic statistics. Lack of comparability of measures across nations continues to be a problem despite efforts by the U.N. Statistical Office to provide international standards. Finally, as noted, hidden economy activity may account for a significant sector of national production that is not included in the accounting procedures of national, public, or formal private institutions. These data must be estimated.

The availability of quantitative data on political institutions and behavior varies greatly among nations. Political opinion polling is highly institutionalized in the United States and Western Europe (usually dating back to the 1930s) and is increasingly common in Eastern Europe, the USSR, Latin America, and Asia. Such data often do not exist over long time periods for developing countries or for developed countries that have or have had single-party political systems. Content analysis of news reports may be more revealing in these cases, depending on the degree of government control over the media.

Data on social and cultural institutions vary across a wide spectrum of qualitative and quantitative sources. Qualitative information may be gleaned from interviews and the public statements of institutions such as churches, bar associations, and community and civic organizations. Interviews and ethnographic studies of NGOs active in environmental issues also provide valuable data. Quantitative data can be derived from social network measurements (Holland and Leinhardt, 1979). Other quantitative data include surveys of institutions and organizations as well as larger public opinion and attitude surveys (NSB, 1985). A number of studies of 'elites' have been conducted in several countries, including surveys of legislators, bureaucrats, and scientists.

Understanding of public policy also involves very diverse types of data regarding both the goals and motivations for policies and the effectiveness of various policy instruments in different settings. Many of the data on goals overlap with the political and cultural data already discussed. Important data on implementation instruments overlap with information on the structure and operations of institutions involved in implementation of goals and with economic data regarding the effectiveness and efficiency of instruments. There is a significant literature on policy implementation, monitoring, and compliance; this literature is currently expanding to focus on environmental policy.

Exogenous data relevant to the macro decision module include international banking interest rates, world trade prices, and treaty information from the global political economy module, as well as national and regional climate and impacts

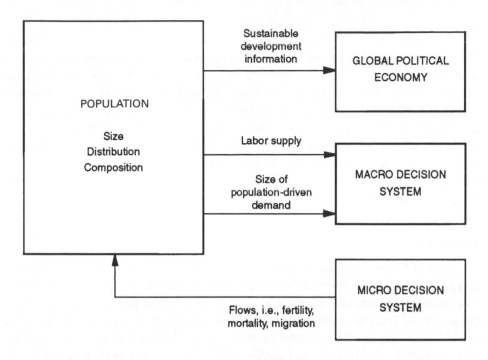

Figure 5. Population module.

data from the regional climate/chemistry module. The macro decision module also combines important demand and supply data from the micro decision module with population data to provide national statistics on labor supply, productivity, consumption, political behavior, and normative expectations.

The Population Module

The population module (Figure 5) is concerned with the study of the size, distribution, and composition of the population. These factors are vital to our understanding of the cumulative social and environmental impacts of decisions made at the micro level.

Module Components

The population module comprises the study of stocks and flows of population. The size of a population and its age and sex composition at the beginning of a period are the stocks, and the births, deaths, and migrations it experiences during that period are the flows. Flows are usually expressed in the form of rates of fertility, mortality, and migration.

29

Module Inputs

The direct inputs to the population module are from the micro decision unit, which in most societies is equivalent to the family. Families may be small nuclear units or large extended groups, and may include members who are not related by ties of blood or marriage. However it is constituted, the family is where people are born, reproduce, and die, or from which they leave to form new production/consumption units.

Among the needs and wants expressed through family behavior are preferences regarding completed family size, child spacing, and age at the initiation of child-bearing. The world's societies differ widely on these factors. It is possible that today's variation among societies is as great as has ever been seen in human history.

Module Outputs

For most of the societies in the world it is possible for demographers to produce detailed quantitative models of the size and composition of a population. For most, it is possible to produce reasonable near-term forecasts of population characteristics and size. Such information can be transmitted to other scientific fields in the form of descriptions and forecasts of the total number and characteristics of the population to be fed, housed, and employed. The population module also can supply information on the available labor force supply, but it cannot provide a labor force participation rate or an unemployment rate.

Examples of Relevant Research

Projection of population flow rates into the future permits population forecasting in both the near and long term, but long-term population forecasting is always based upon fairly strong and often questionable assumptions about individual and societal behavior. These assumptions are required because humans control their fertility to greater or lesser degrees, and they exert considerable control over morbidity and mortality. However, even given this control, it takes human societies a long time to make major changes in their population growth curves.

Demographers conventionally provide several scenarios in their forecasts of population size, so as to capture the results of differences in assumptions, rather than make a single-point estimate. For example, reasonable forecasts of world population in 2050 range from 8 billion to more than 20 billion. The study of land-use/cover changes will have to accustom itself to dealing with such varying forecasts.

Demographic forecasting rarely relies upon simple extrapolation of current trends. Instead, it relies upon compounding calculations based upon the fundamental flow rates applied to data on population stock. The results of straight-line extrapolations are often much different from the results of analytical, historically based, quantitative models.

Despite these known and unknown uncertainties, for most countries both current descriptions and near-term population forecasts are rather precise and accurate (Keilman, 1990). One of the reasons is that—with the usually minor exception of migrants—most of the people who will shape the near-term future of a population are already in place. The momentum of population change requires years to reverse. This 'inertia' of population makes confidence in near-term predictions of population size relatively reasonable (although it also makes reductions in population growth rates difficult to achieve). Interestingly, many of the forecasting errors have come when fertility rates are unexpectedly moving up; the surprise of the U.S. baby boom of the 1940s and 1950s is the classic example.

Such forecasts can, of course, be disrupted by external events that markedly alter mortality, fertility, or migration, such as plagues. But these Malthusian factors have less long-term effect than might be expected. Historically, wars have chewed away at the number of young men; but if the number of young women is not sharply reduced as well, little has occurred to change population growth. It takes few men to provide for the reproductive needs of a society. Women in the age bracket 15–45 are the people who matter demographically. After men and women finish their reproductive years, their survival is of marginal importance to the long-term size of the population.

Although technically sophisticated, demography has significant limits on what it can contribute to the understanding of land-use/cover change. The complexity of long-term forecasts has already been mentioned. Perhaps more importantly, the very elegance of the basic equations of demography constrains the field. Demographers are reluctant to venture outside their field, often seeing such things as a government's population control policy as exogenous to their models. Given a statement that a government will somehow reduce the fertility rate by 10% in five years, demographers can confidently model the effects of that reduction. But they are unlikely to study the processes by which governments make or implement such decisions. (The contiguous field of family planning is concerned with such matters.)

Potential Contributions

It is reasonable to expect that demographers will successfully collaborate in interdisciplinary research in this area. Demography itself is an interdisciplinary field, involving actuarial scientists, biologists, biostatisticians, economists, historians, psychologists, sociologists, and statisticians. Its quantitative methods are highly developed, partly because of its importance to government planning and the insurance business. To contribute to our understanding of global land-use/cover change, demographers will have to work with other researchers, particularly with the community that models the regional decision system.

Data Issues

Demography's data bases are the envy of many other fields. Except at the local level, population size is usually accurately known to an order of magnitude or better. Virtually every society collects the needed data on population stocks, usually

31

through periodic censuses or population registries. However, the quality of these data varies from society to society, and, even in such advanced industrial societies as that of the United States, there are known to be errors in census data (Clarke and Rhind, 1992). These errors generally affect the less significant digits in a measurement, although they may be socially or politically quite important.

Data on flows are less reliably collected. The developing countries often have no reliable system for registering births and deaths. Even in the industrialized countries, registration is incomplete. Migration flows across national borders are known to be monitored with a considerable degree of error (many countries have large numbers of undocumented aliens). Except for those few countries that maintain current population registries, internal migration is usually measured only through sample surveys. Forecasting the distribution of a population within a country is therefore difficult, and in some cases it is difficult to analyze current population densities.

In addition to their reliance on censuses, population registration, and vital statistics, demographers design and conduct important sample surveys (Clarke and Rhind, 1992). These surveys usually collect data beyond those needed for filling in the terms of the demographic equation; among the items frequently measured are race, socioeconomic standing, and labor force participation. The World Fertility Survey is an international collection of surveys of family planning behavior. This is one of the few surveys in the social sciences that attempt to measure the same phenomena in comparable ways in many different countries. The political sensitivity of population issues is reflected in the fact that the survey could not be conducted in several countries.

Global Political Economy Module

With the possible exception of national or ethnic territorial disputes, the impacts on the land of the human activity studied in this module (Figure 6) are rarely direct. More usually the impact of the global political economy is mediated by the actions of subglobal decision-making at the national, regional, or micro decision levels.

Module Components

For analytic purposes, the global political economy module is divided into two submodules, the global political system and the global economy. These submodules are inextricably linked because patterns of international authority shape economic decisions (particularly those decisions pertaining to international transfers of goods and byproducts that are of greatest environmental significance) and vice versa.

The political system submodule is concerned primarily with understanding the activities of the international regime of sovereign nation-states, their formal relations through enduring instruments such as treaties, and the shifting pattern of their informal alliances. This submodule also addresses the structure and functioning of international and supranational organizations.

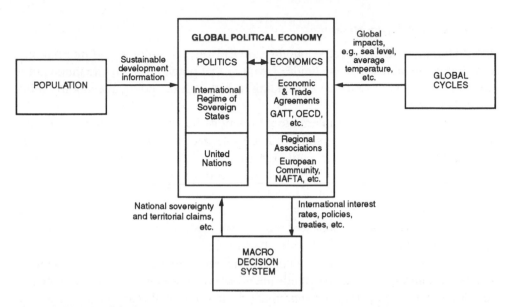

Figure 6. Global political economy module.

The economic system submodule addresses our understanding of commodities and capital flows around the world, particularly in the context of economic and trade agreements, such as the General Agreement on Tariffs and Trade, and in light of the increasing importance of regional free trade associations such as the European Community.

Module Inputs

Global biogeochemical assessments, macro decision systems analysis, and population studies all contribute to the analysis of decision systems in the global political economy. Information on global environmental conditions, such as sea level and average temperature, is derived from global biogeochemical cycle assessments. The macro decision module supplies information on national sovereignty and territorial claims, which may be resolved at the global level with significant implications for land use, as through warfare and territorial occupation. Studies of population yield relevant information for evaluating resource needs and political, military, and economic power within the global political economy.

Module Outputs

This module provides key data on international interest rates and commodity prices to the macro decision module. It also supplies information on the terms and implementation options for international policies and treaties on, for example, global trade or specific environmental goals.

Examples of Relevant Research

International relations, macroeconomics, world systems studies, and international development studies are huge fields of inquiry that potentially provide much of the context that we require for understanding how the global political economy influences land-use change.

Historically, the field of international relations has been dominated by a state-centered view of politics as the pursuit of power and national self-interest (Morgenthau, 1974). It depicts nations as frequently being in conflict because their leaders aggressively pursue short-term national interests to enhance their position in the international contest for dominance and control. The absence of an international sovereign or magistrate makes war at least as likely as cooperation. Short of war, this emphasis on national sovereignty can lead to land uses that are globally undesirable. This paradigm is described as *realist*.

The realist perspective has its analog in the dominant paradigm of international macroeconomics, which also extends the notion of individual rational self-interest to nation-states competing for supremacy in international markets. Interestingly, even radical development theorists seeking to criticize the status quo in international politics and economics have not challenged the self-interest core of these perspectives. Marxist perspectives spawned theories of economic dependency that laid all responsibility for underdevelopment at the doors of greedy developed nations (Frank, 1967). According to Marxist theorists, land use in developing countries is driven by the international economy towards intensive resource extraction (minerals, timber, etc.) that is highly degrading and unsustainable.

Growing recognition of economic and environmental interdependence among nation-states presents significant challenges to these paradigms of unadorned self-interest. Political and economic arrangements are increasingly seeking cooperation to advance common goals, not out of idealism but out of a practical concern for maintaining the sustainability of resources (Haas, 1990; Benedick, 1991). For example, the International Tropical Timber Agreement exchanges developing country acceptance of export limits on timber for guaranteed price supports for agricultural commodities and debt reduction strategies.

Concern for sustainability of economic development is also changing patterns of technology transfer. As pressures increase to transfer technologies that lower greenhouse gas emissions, more energy-efficient technologies, and more advanced pollution control methods from developed to developing nations, dramatic impacts on developed and developing country markets will occur. In some industries (e.g., cement, primary metals, commodity chemicals, automotive and aircraft manufacturing) these improvements will result in greater economic competitiveness for the developing countries and possibly a realignment of economic standing among the developed nations.

A further change in global political economy is the rapid institutionalization of the concept of equity in negotiations and agreements on global change. Exchanges between environmental and nonenvironmental criteria such as price supports, development assistance, and debt reduction strategies are now a permanent part of the developed–developing country exchange system. This is exemplified by demands of developing countries for assistance to improve national development infrastructure and to reduce the risks of global environmental change through adaptation strategies. These demands were most recently articulated at the 1992 U.N. Conference on the Environment and Development.

Thus, the nature and exercise of economic and political power are changing dramatically as *low politics,* encompassing technological, equity, and ecological concerns, is replacing *high politics* issues, involving market domination and potential military force.

Potential Contributions

New paradigms are emerging both to explain changes in the relationships among the populations of nation-states and to reestablish the conceptual linkage of international politics and economics. One such paradigm, the *inclusionist* approach, sees nations as human populations competing and cooperating within a system composed of a physical ecosphere, a structured set of practices and rules referred to as the international political economy, and an ideological realm characterized by value conflicts among supporters of various organizational alternatives (Pirages, 1989).

The attention to values and competing organizational preferences represents a new emphasis on procedural rationality rather than an exclusive focus on outcomes. This approach has also been proposed by *regime* theorists (Young, 1982) and characterized in anthropological perspectives on international cooperation and competition as *polycentric* decision processes (Rayner, 1991). In these processes, national identity provides only one dimension of affiliation for decision-makers, who also play roles in various, often competing, interest groups that communicate rapidly and effectively with organizationally compatible actors in other countries.

These approaches hold out the prospect of identifying much tighter links than are presently possible between the international political economy on the one hand, and the national and local levels of decision-making that affect land use on the other.

Data Issues

There is an extensive body of data on many aspects of the global political economy. The International Monetary Fund maintains computer-readable data sets on balance of payments statistics (1965–), imports and exports (1948–), and international and domestic finance statistics (1948–). The World Bank has a data set on international debt (1970–). Many but not all nations are represented in these data. In addition, the World Bank publishes tables of economic and social indicators (1950–). These tables consolidate some of the international economic data and also provide limited social data. A broader collection of social and political data is available in Taylor and Hudson (1973). International population data are published by both the United Nations and the International Office of the U.S. Bureau of the Census.

There is a body of data on international intergovernmental organizations, such as the United Nations. The rising importance of international NGOs is reflected in the emergence of new data sets about their organization, activities, and influence. More generally, international interactions are tracked in a variety of data sets, many of them focusing on violent conflict among nations. The source for many of these data sets is newspaper reports of conflict. Lewis Fry Richardson's pathbreaking data set on 'deadly quarrels' dates back to 1809.

Land Assessments Module

The land assessments module (Figure 7) describes those fields of knowledge and expertise necessary to determine the availability and suitability of land for alternative uses, as well as its productivity, the sustainability of those uses, and their impact on other environmental resources, especially air quality. Issues of soil, water resources, climate, and terrain require that these analyses be conducted at the local level.

Module Components

The land assessments module consists of four submodules: soil, water, nutrients, and vegetation. These components are interconnected both by *in situ* interactions among elements of an ecosystem and also by the import and export of goods and services. For example, net primary productivity depends on availability of nutrients (local or imported), pest infestation and management, and water allocation or availability. Soils provide the primary building block for agriculture and forestry. Vegetation in this module also addresses animal husbandry as expressed in grazing land and pasture. Water is an integrating theme in the land assessments module and connects the soil and vegetation components.

Module Inputs

The principal inputs to the land assessments module come from both the micro decision module and the regional climate and atmospheric chemistry module. The micro decision module provides information about goals (e.g., housing, agriculture, transportation infrastructure) for the plots or parcels in question, whose suitability for those goals must be assessed. The micro decision module also provides information about production practices such as grading for construction or tillage and biomass burning for agriculture, as well as quantities of inputs such as fertilizers, herbicides, and pesticides.

The regional climate and atmospheric chemistry module provides information on microclimatic conditions, such as temperature and precipitation, and atmospheric deposition, such as acid and airborne nutrients.

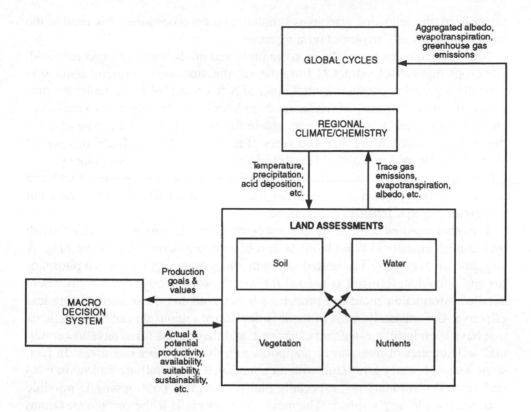

Figure 7. Land assessments module.

Module Outputs

Land available for cover change or management manipulation is constrained by the natural attributes of the area and by regulation. Many options at a local level may be constrained by infrastructure needs beyond the capacity of local action (roads, railroads, etc.). The natural attributes can be assessed via soil and other surveys and conveyed to decision-makers in forms of potential productivity for specific commodities. The spatial distribution of land availability is best evaluated on maps.

Quantitative evaluations of atmospheric emissions from land use can be made available to the regional climate/chemistry module and global cycles module. Of particular interest are carbon dioxide, methane, the oxides of nitrogen, and the sulfur aerosols.

Examples of Relevant Research

Resupply of plant nutrients extracted from the soil and exported in food and fiber products is accomplished by concentrated fertilizer in almost all commercial farming operations. Although exact procedures vary for predicting the quantity of fertilizer need, the following principles summarize most procedures now used in the most technologically advanced farming areas.

Nitrogen requirements of the crop are predicted on the basis of expected yield. For crops that cannot extract N from the air, the amount of fertilizer applied is annually adjusted to consider natural input of N from rainfall and whether the previous crop was a legume, which leaves N in its residue. In only a few cases does enough residual and rainfall nitrogen remain for an economic yield, so some additional N is required. More than 100 years of research in Illinois found that one of the most fertile soils in the world produced about 42 bushels of corn grain per acre (2600 kg/ha) with only natural N from rainfall (see Buol, this volume). Yields are approximately doubled if an N-fixing legume crop is planted in alternate years, but more than tripled if fertilizer N is applied.

The other required plant nutrients must come from minerals in the soil, although they can accumulate in plant biomass and be released upon decay or burning. A vast amount of research has been done on methods to predict how much phosphorus and potassium fertilizer is needed for specific soils. However, in areas where detailed information about soil properties is not available, these methods are less effective. Unfortunately, most of the less developed regions do not have methods that have been locally tested and calibrated, and these areas often have lower natural soil supplies of phosphorus and potassium than do temperate areas. In fact, some areas of nearly level land with an abundance of rainfall are so low in mineral-derived plant nutrients, especially phosphorus, that no cropping is possible without first adding phosphate. The most visible example is the cerrado (savanna) area of Brazil surrounding Brasilia. Only recently has research identified appropriate technology for some such areas, and farm production has been initiated on a large scale.

Potential Contributions

Requirements for food can be met either by subjecting more of the earth's land area to food production or by increasing the amount of food production per unit land area. The land area used for crop production is decreasing in both the United States and Europe, although food production has increased. Much of this land-use change can be attributed to higher soil fertility created by mineral fertilizer application.

Research to determine appropriate quantities of supplemental fertilizer not only reduces the amount of land required for food production but also controls undesirable off-site environmental pollution from excess fertilizer use. There are 16,000 different kinds of soil in the United States and an undetermined number of different kinds of soil throughout the world. Perhaps most of these are not well known to researchers. Determination of correct, environmentally safe, and economically feasible management for each individual kind of soil is perhaps the most significant potential contribution for global change concerns. Which kinds of soil produce greenhouse gases under certain management scenarios? If maximum food production is obtained on level, low-erosion-potential land, how much hillside, erosion-prone land can be left in tree cover? What are the infrastructure requirements that need to be put in place by society to encourage maximum use of relatively erosion-free land? How are these incentives transmitted to the very site-specific soil conditions within a farmer's field?

Data Issues

Local-level decisions concerning the suitability of each segment of the landscape can be assessed by comparisons to similar areas with like properties, including soil moisture regimes, soil temperature regimes, slope characteristics, etc. Information needs about soils include evaluation of the terrain characteristics, mineralogical properties, and the size and distribution of primary and secondary particles. Farmers will have this knowledge for locally grown crops but may not be familiar with cropping systems not common to their area. Thus, outside expertise may be needed to assess the suitability of management systems that are not practiced locally. Assessing nutrients at a small scale requires highly specific soil information, knowledge of past fertilizer application, and possibly soil sampling. This information may be available through local planning agencies or agricultural services that have conducted soil sampling experiments or surveys.

Water supply information is perhaps one of the most difficult forms of data to gather. The U.S. Geological Survey (USGS) keeps extensive records on water end use by county for the United States. However, this very good data collection and distribution system is highly unusual among nations. If the data cannot be obtained, it may be appropriate to use proxy methods such as calculations of river flow and precipitation or small-scale hydrographic data sets which provide surface water existence in percentage by grid cell.

Understanding the vegetation processes involves analyses of growth, growth response functions relevant to management, productivity, and canopy characteristics including canopy cover and leaf area index. Information on host-parasite relationships is also required. Quantification of the degree of biodiversity is important to the assessment of its resilience, adaptability, and sustainability. Information needs here include data on flora and fauna, genetic diversity, species succession within natural and agricultural ecosystems, ecosystem productivity, and competition. On a fine scale, vegetation and general land-cover data may be derived from several sources. Local maps and atlases may be used for general land classifications, as can aerial and satellite photographs. In some instances, local planning agencies may have fine-detail land assessments that provide many of the needed data. In other cases, a survey of current cover may be necessary.

For many studies, information on prevailing regional conditions, whether or not this information is available to the local farmer, is sufficient to understand why individuals at the local level make the choices that they do. Additionally, aggregate data from the land assessments module are essential to the regional climate and atmospheric chemistry module.

For example, data on soil temperature and soil moisture regimes exist at the 1:24,000 scale for most of the United States (USSCS, 1975). The worldwide standard in geographic identification of soil types is the UNESCO soil map of the world (1:250,000), which defines over 130 soil types. This map has been newly revised (1990) by the U.N. Food and Agriculture Organization (FAO). A digitized version is being prepared.

The National Aeronautics and Space Administration Goddard Institute for Space Studies (GISS) has used the FAO data to create a $1° \times 1°$ grid cell data set where each cell is given a value for its primary soil type. NASA also offers a data set titled 'Global Distribution of Soil Water Holding Capacity at 1×1 Degree Resolution,' which uses principles of soil science to identify water holding capacity in the FAO soil types. An additional $1° \times 1°$ data set is the Global Soil and Vegetation data set (Wilson and Henderson-Sellers, 1985), archived at the National Center for Atmospheric Research (NCAR). It lists soil type with breakdown by color, texture, and drainage. It also includes an index of the data reliability for each grid cell.

Nutrient assessment requires data on natural as well as applied nutrients. Natural nutrients must be derived from soil type information. The FAO collects and publishes statistics on fertilizer use by country, but the data are available only at the national scale. Agricultural agencies in some countries may be able to provide more detailed data.

Various climate change studies have given much attention to the question of global distribution of vegetation in the last several years. The result has been the creation of several data sets of vegetation cover. The Wilson and Henderson-Sellers set lists over 50 classes of vegetation, including crop type and natural vegetation, urban areas, and others. Another data set is the GISS 'Global Distribution of Vegetation at 1×1 Degree Resolution in 32 Vegetation Types' (Matthews, 1983). This set is primarily concerned with natural vegetation types rather than present-day conditions but does show percentages of cultivated and uncultivated land.

Another global data base of interest is designed to provide information critical for assessing carbon quantities in vegetation and natural biological carbon processes (Olson *et al.*, 1985). It identifies and ranks 44 land ecosystem complexes in seven broad groups. The accompanying map is derived from patterns of preagricultural vegetation, modern area surveys, and data from research sites.

Regional Climate and Atmospheric Chemistry Module

This module (Figure 8) performs two basic functions. The first is to translate information from atmospheric general circulation models (GCMs) into descriptions of regional climate tailored to the needs of local land assessments. The second is to describe the chemistry and transportation of regional air pollutants that emanate from and/or impact upon use of the land.

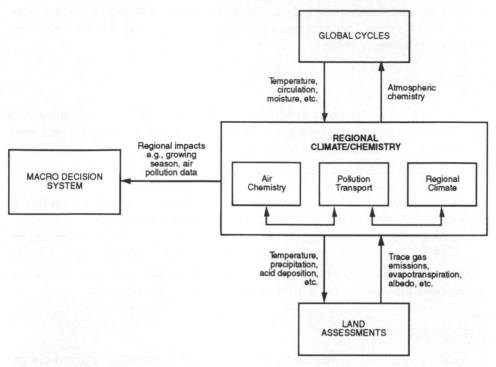

Figure 8. Regional climate/chemistry module.

Module Components

This module consists of three submodules. The regional climate submodule generates descriptions of climate tailored to the operational requirements of local ecosystem and land productivity models. The regional pollutant transport submodule describes the distribution of chemical species over source-to-receptor distances of up to 1000 km. The regional atmospheric chemistry submodule describes the behavior of short-lived trace species in the atmosphere for their influence on the lifetime of methane (a greenhouse gas), for the concentrations of tropospheric ozone and carbon monoxide (both regional air pollutants), for contributions to sulfate aerosols, and for acid deposition on the land surface.

Module Inputs

The regional climate submodule receives information about patterns of circulation, temperature, and moisture from GCMs located in the global cycles module. Global climate change and stratospheric ozone change directly affect local and regional manifestations of climate and radiation balances. The other two submodules accept information on anthropogenic and natural trace gas sources. Biogenic sources include nonmethane hydrocarbons from forests, nitrogen oxides from soils, and carbon monoxide from biomass burning, all of which affect acid deposition. Industrial sources include energy production and consumption (including automobile emissions) as well as airborne release of manufactured substances. Information on evapotranspiration, topography, and albedo is also obtained from the land assessments module.

Module Outputs

Changes in local and regional manifestations of climate and radiation balances in turn affect vegetation (including managed and unmanaged ecosystems) and water resources. Hence, the regional climate and atmospheric chemistry module passes information on insolation, temperature, and precipitation to the land assessments module. These same factors may also affect human health both directly (e.g., through heat stress) and indirectly (e.g., through water availability), leading to changed assessments of the suitability of land for human habitation.

Changes in regional air chemistry also will influence global atmospheric chemistry and climate. Hence, the regional module provides this information to the global cycles module. Assessments of regional air pollution are important inputs to the policy process described in the macro decision system module.

State of Relevant Research

Various approaches have been discussed for the development of regional climate models. These range from correlating weather observations with GCM output (a variation of the model output statistics technique) to 'embedding' output from an atmospheric mesoscale model (scale of a few km) within a GCM. To estimate changes in the frequency of extreme events like hurricanes, extreme winds, and intense precipitation, GCMs will have to be greatly improved, and process studies linking local storms to regional circulation patterns will have to be adapted.

Our current understanding of regional air chemistry and transportation is based on measurements and estimates of emissions, combined with detailed subdaily meteorological data and highly localized topographical information. The primary concern for local pollutant transport modeling is how the boundary layer is disturbed by density-driven mesoscale forcings, particularly land-sea breezes, mountain-valley wind systems, and conurbations. These forcings are often modeled independently of regional-scale pollutant transport because they require much finer grid resolution (NAPAP, 1991). Synoptic-scale transport is determined by the distribution of the pollutant relative to air masses, cyclonic systems, and associated fronts.

Air chemistry models attempt to simulate the reactions of the atmosphere with pollutants in gaseous, liquid, and particulate states. Much of the existing understanding of the complex photochemical reactions involved has been obtained in the last two decades through studies of urban air pollution (Whitten *et al.*, 1980) and the formation and destruction of stratospheric ozone (WMO, 1985). Chemical models embedded within mesoscale dynamic models (e.g., Stockwell *et al.*, 1990) have yielded important results, though significant problems remain in dealing with the wide range of spatial and temporal scales involved. They provide estimates of tropospheric ozone and the lifetime of methane, and, when coupled with suitable treatments of cloud, of acid precipitation.

Potential Contributions

The atmospheric motions that contribute to the complexity of regional transport modeling tend to be in the vertical rather than the horizontal dimension. Although several approaches have been developed to simulate these vertical motions, it appears that no single model has sufficient resolution to resolve them while also addressing regional transport applications economically. The limitation appears to be computing power (NAPAP, 1991).

Better understanding of atmospheric chemistry and improvements in emissions monitoring are required. In particular, nonmethane hydrocarbons and the products of their reactions with nitrogen species and oxidants are suspected to play important roles in the global change arena (NAPAP, 1991). Better estimates of natural and anthropogenic emissions factors are essential to determining economically efficient policy responses.

Data Issues

Because of their large quantities, the pollutant emissions needed for air transport model simulations usually must be estimated rather than measured. The number of emissions sources is so large that generalized techniques are required to develop inventories, identify trends, and estimate uncertainties. These techniques include development of representative emissions factors representing the ratio of emissions to activity levels; development of demographic, economic, and energy use data bases; construction of computerized data handling systems; and analysis of the uncertainties in emissions estimates. A pressing need is for improved techniques for estimating nonanthropogenic emissions.

The pollutant transport models require detailed meteorological information. As the existing data are too coarse to represent mesoscale motions properly, many modelers develop their own meteorological fields. However, this solution has created other problems. A comparison of four regional-scale (500 km) models (van Dop, 1986) showed that the differences in results were due primarily to differences in the ways the models processed meteorological data.

Existing GCM data are too coarse to be of great utility in determining impacts at the land assessments level (McKenney and Rosenberg, 1991). Many climate impact assessments employ biophysical models representing spatial scales of <1 to >100 km, in contrast to GCM grids of several hundred kilometers square. A daily time step would be highly desirable. Hydrologists may require both daily and annual (or longer) values. There may also be a need for daytime and nighttime data, since changes in diurnal conditions affect biological processes such as photosynthesis and respiration. Improved regional estimates of climate variability and incidence of extreme events would also be important.

Global Cycles Module

This module (Figure 9) describes those fields of knowledge necessary to understand the major global-scale biogeochemical processes. Certain interconnected regional and local phenomena also play roles.

Module Components

The global cycles module consists of four submodules: emissions, carbon cycle, atmospheric chemistry, and climate. The emissions module contains information about flow rates for key species. To understand climate change it is necessary to include flux rates of the usual suite of greenhouse gases (see Penner, this volume), and also changes in albedo caused by land-use changes. Water vapor and ozone are not included because they are not released in large quantities by anthropogenic activities, but rather are the indirect consequence of emissions of directly released gases.

The carbon cycle submodule describes our knowledge of the relationship between reservoirs and fluxes of carbon. Three classes of reservoirs can be usefully defined: atmospheric, terrestrial, and oceanic. When anthropogenic emissions of carbon enter the atmosphere, fast-moving processes lead to its relatively rapid distribution throughout the atmosphere and to secondary fluxes into and out of oceanic and terrestrial reservoirs. Process-level descriptions of oceanic and terrestrial carbon cycling have led investigators to increasingly fine spatial resolution of physical, chemical, and biological processes.

The atmospheric chemistry submodule defines our understanding of the relationship between the broad array of non-CO_2 fluxes into the atmosphere and the global distribution of the suite of relevant gases. This relationship is nontrivial. In one way, atmospheric chemistry processes are simpler than those of the carbon cycle, the ultimate sink for most gases in the atmosphere. On the other hand, the relationship between those sink relationships and other atmospheric constituents is complex and differs by altitude, for many important constituents. Even for chlorofluorocarbons (CFCs), whose destruction mechanisms are relatively uninfluenced by local lower atmosphere conditions, decay rates and subsequent ozone depletion vary with season, altitude, and latitude. These patterns, in turn, are critical to the overall assessment of CFC contributions to climate change. Processes described in the local and regional climate/chemistry module are important determinants of the overall global cycle. For example, understanding the behavior of CO, the nitrogen oxides, and sulfur dioxide requires knowledge of local conditions including weather, local emissions rates, and initial local atmospheric composition.

The climate submodule defines how changes in atmospheric composition will affect climate. If there were no feedback effects, the consequences of some changes would be relatively easy to model. In such a case, the effect on surface temperature of doubling the concentration of atmospheric CO_2 would be approxi-

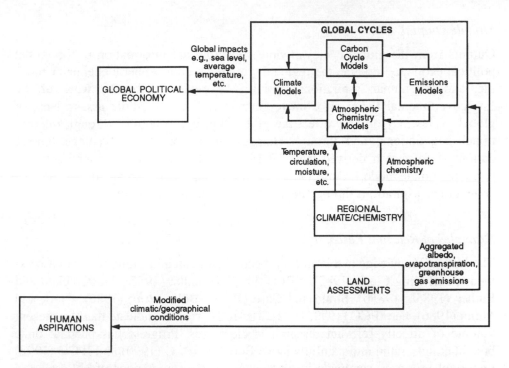

Figure 9. Global cycles module.

mately 1.2 °C. However, the actual consequences of doubling atmospheric CO_2 are complicated by the presence of important feedback effects, including water vapor, snow and ice albedo, and clouds. When these effects are included, the change in long-term equilibrium temperature as a result of doubled CO_2 is generally given as 1.5 °C to 4.5 °C. Other greenhouse gases absorb even more solar radiation than does CO_2. On the other hand, their emissions rates are far lower, so that their overall effect is generally considered to be less than that of CO_2. Aerosols operate on climate by changing the planet's albedo rather than absorbing energy. As a consequence, increasing concentrations of these atmospheric constituents tend to cool rather than warm the earth. Climatic characteristics such as precipitation, evaporation, cloudiness, and windiness are more difficult to estimate than temperature.

Module Inputs

The principal inputs to the global cycles module come from the land assessments module, which provides information on trace gas fluxes and albedo changes. Other important information comes from the regional climate and atmospheric chemistry module, which provides information regarding local and regional atmospheric processes, which in turn affect global-scale processes such as the earth's radiation balance and albedo.

Module Outputs

Outputs from the global cycles module are important information to the global political economy module, the regional climate and atmospheric chemistry module, and the human aspirations module. As evidenced by the work of the Intergovernmental Panel on Climate Change (IPCC), scientific assessments of global cycles may directly affect the global political economy, through international negotiations, and through changes in comparative advantage among nations. Information derived from GCMs provides an essential framework for regional climate models. Expectations of future changes in climate may well moderate human aspirations through the geography and climate submodule.

Examples of Relevant Research

The state of knowledge in this area has been documented in a long series of assessments, including Clark (1982), NRC (1983), Trabalka (1985), MacCracken and Luther (1985a, 1985b), Strain and Cure (1985), Bolin *et al.* (1986), Clark and Munn (1986), and IPCC (1990, 1992). The IPCC reports indicate that the present volume of directly relevant research is enormous. Progress has been at once breathtakingly rapid and painfully slow. Between IPCC (1990) and IPCC (1992) important progress was made in understanding the interactions of chlorine and stratospheric ozone depletion, and the role of sulfur aerosols on radiative forcing. Findings in both of these areas led to a significant retreat in the application of global warming potential (GWP) coefficients. Although the concept of these coefficients initially looked like a relatively simple construct, these recent results have proven that it is extremely complex in practice.

Potential Contributions

Significant areas of understanding continue to elude researchers. The carbon cycle remains as unbalanced in 1992 as it was in 1982. The use of the 'neutral biosphere' assumption for forecasting remains a problem in that it implicitly assumes that every hectare deforested creates a net sink of equal magnitude, and every hectare saved from deforestation destroys a sink of equal magnitude. One of the most important potential contributions remains a complete scientific explanation of the disposition of atmospheric carbon emissions.

Atmospheric chemistry has continued to provide surprises at every turn. While it is impossible to foresee another discovery of the magnitude of stratospheric ozone depletion, the area remains a potential source of unexpected results.

General circulation models remain the heart of the global climate change issue. Their analysis of global mean temperature changes resulting from potential buildup of atmospheric CO_2 remains one of the most important reference points in the field of global cycles. But feedback effects have remained sufficiently complex and resistant to simulation that the range of quoted uncertainty about this variable remains the same as a decade ago. GCMs continue to have difficulties replicating the present subcontinental-scale climate. Despite the problem that every increase in knowledge has led to new questions, hope remains for progress in narrowing the range of uncertainty.

Forecasting emissions remains as difficult today as it was a decade ago. Much of the problem stems not from the scientific methods employed, but from the potential for humanity to alter its behavior. Improved understandings of the costs and benefits have emerged in the past few years and remain a potentially fruitful area for further efforts. Reconstructions of historical records of emissions, including those of human and natural systems, also show promise of yielding useful knowledge.

Data Issues

The study of global cycles requires data sets from global to local in scale. Much of the work in this field relies on remote sensing data, yet 'ground truth' observations remain essential to validating results. Despite the enormous amount of work that has been undertaken to date, we are just beginning to gather reliable global data on deforestation rates. We still lack sufficient data to reconcile the carbon cycle accounts, and scientists are only now beginning to gather a consistent data set capable of fully describing activities within a column of atmosphere from the earth's surface to space. Balancing the methane budget between sources and sinks across processes and regions remains an elusive goal.

Conclusions

We have now completed a single pass through the entire land-use/cover diagram. Our diagram puts human decision-making at the center of processes to be understood. Further, it recognizes that decision-making operates at a hierarchy of levels, from the family farm to the World Bank and the United Nations. It points out that there are corresponding natural systems at each level, emphasizing that an important policy observation, 'Think globally: Act locally,' has an equally valid converse, 'Think locally: Act globally.'

It is worth pointing out that our wiring diagram attends solely to changes in land use/cover, not to the plethora of other environmental changes that humans make locally, regionally, and globally. It is therefore not an analog to the U.S. Committee on Earth and Environmental Sciences (CEES) wiring diagram.

In our excursion through the diagram we have identified disciplines that may have important contributions to make in understanding the processes to be described in each module and submodule. However, we do not mean this to imply that the present research of those disciplines will eventually provide neatly interlocking parts of the overall picture of land-use/cover change. Rather, each module contains opportunities for traditional disciplinary research, multidisciplinary research (where disciplines collaborate but disciplinary paradigms remain identifi-

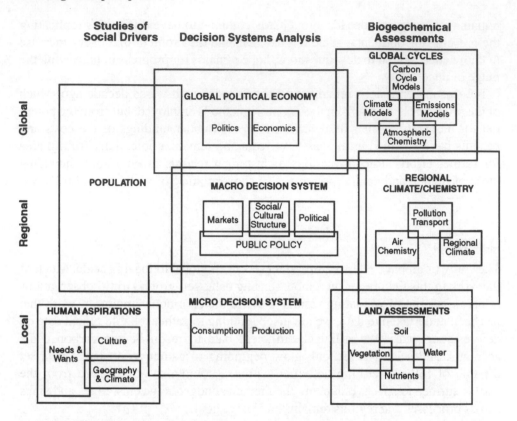

Figure 10. Convergence on meaning in study of land-use/cover change.

able), and interdisciplinary research (where new paradigms are created that cannot be identified with disciplinary specialties).

If the need exists for new interdisciplinary research processes within modules, the need to develop the interfaces between adjacent modules is even more acute. The graphic representation of these interfaces is convenient for labeling certain information flows but may mislead us about the nature of communication among the modules. Such communication cannot simply be the transfer of information along a conduit from one body of researchers to another. Communication among researchers in adjacent modules must surely be focused on the creation of shared meaning as much as on the transmission of information. In this respect the relationships among submodules and modules may be better represented as a series of overlapping fields (Figure 10).

We recommend that the CEES or other influential body initiate a series of interdisciplinary workshops to expand understanding and application of each module of the wiring diagram and to further specify the information flows between modules. The goal of these workshops will be to develop an international cooperative research program on land use and global environmental change involving both natural and social scientists, evolving from the scientific linkages specified in the wiring diagram. Participants would be experts in the fields specified by only a single module or (where appropriate) a pair of modules. Any broader group would be inclined to rethink the entire wiring diagram, and several such groups would produce massive confusion. To avoid ethnocentric assumptions about land-use behaviors, each workshop should fully represent both North and South perspectives. To keep the discussions focused on a common goal, the workshop moderator(s) should have a comprehensive understanding of the overall framework.

The initial task of these experts would be to explore the subject matter of each module in depth with the goal of establishing priorities for elaborating theoretical, methodological, and data issues while reducing irrelevant detail. Ongoing research should be identified and extant data sources brought into an international data network. A second task would be to identify potential intervention points in natural and human systems for conscious modification of land use or its consequences. Policy recommendations of the nature of 'produce a more efficient automobile' have not proved helpful; identifying intervention points will prove far more useful.

Based on these expanded discussions of the modules, it would be possible to design an authentically interdisciplinary research program that would combine social science insights about information and commodity flows and natural science measurements of materials flows among the land, air, water, and products for human use. A major goal of such a research program would be to evaluate the feasibility, consequences, and sustainability (political and ecological) of the potential interventions identified in the work groups. We estimate that such a program would be a multiyear effort that would yield significant results.

Acknowledgments

I circulated drafts of this report to all working group members for comments and received particularly helpful additional contributions from Stan Buol, Mike Fosberg, Evert van Imhoff, and Richard Rockwell. As moderator of the working group, Richard Rockwell played a major role in shaping this paper. To plug some of the gaps in our knowledge base I had to seek help from outside of the group. This help was generously provided by my colleagues Jae Edmonds and Norman Rosenberg. Leisha Davis helped to structure the report in a logical and consistent fashion and provided invaluable editorial and bibliographic input. Scott Miller provided information on land-use data. Jenniffer Leyson created the graphics. Finally, I am grateful to Gerry Stokes and the Global Studies Program at Pacific Northwest Laboratory for supporting my work on this project.

References

Benedick, R. E. 1991. *Ozone Diplomacy: New Directions in Safeguarding Environmental Cooperation*. Harvard University Press, Cambridge, Massachusetts.

Blaikie, P. M. 1985. *The Political Economy of Soil Erosion in Developing Countries*. Longman, London.

Bohannan, P. 1970. *Tiv Farm and Settlement*. Johnson Reprint Corporation, New York.

Bolin, B., B. R. Doos, J. Jæger, and R. A. Warrick, eds. 1986. *The Greenhouse Effect, Climatic Change, and Ecosystems.* SCOPE 29, John Wiley and Sons, New York.

Cantor, R., S. Henry, and S. Rayner. 1992. *Making Markets: An Interdisciplinary Perspective on Economic Exchange.* Greenwood Press, Westport, Connecticut.

Carson, C. 1984. The underground economy. *Survey of Current Business 64,* 21–37.

Clark, W. C., ed. 1982. *Carbon Dioxide Review: 1982.* Oxford University Press, New York.

Clark, W. C., and R. E. Munn, eds. 1986. *Sustainable Development of the Biosphere.* Cambridge University Press, New York.

Clarke, J. J., and D. W. Rhind. 1992. *Population Data and Global Environmental Change.* ISSC-UNESCO, Paris, France.

Cummings, R. G., L. A. Cox, and A. M. Freeman. 1986. General methods for benefits assessment. In *Benefits Assessment: The State of the Art* (J. Bentkover, V. T. Covello, and J. Mumpower, eds.), D. Reidel Publishing Company, Dordrecht, The Netherlands.

Douglas, M. 1962. Lele economy compared with the Bushong: A study of economic backwardness. In *Markets in Africa* (P. Bohannan and G. Dalton, eds.), Northwestern University Press, Evanston, Illinois.

Douglas, M. 1976. Relative poverty—Relative communication. In *Traditions of Social Policy, Essays in Honour of Violet Butler* (A.H. Halsey, ed.), Basil Blackwell, Oxford, U.K.

Douglas, M. 1978. *Cultural Bias.* Royal Anthropological Institute Occasional Paper No. 35, London.

Douglas, M., and B. Isherwood. 1979. *The World of Goods.* Basic Books, New York.

Etzioni, A. 1988. *The Moral Dimension: Toward a New Economics.* Free Press, New York.

Evenson, R., and E. Y. Kislev. 1975. *Agricultural Research and Productivity.* Yale University Press, New Haven, Connecticut.

FAO (U.N. Food and Agriculture Organization). 1990. *Soil Map of the World.* UNESCO, Paris, France.

Feingold, B. H. 1979. *The Wilderness Experience: The Interaction of Person and Environment.* Ph.D. dissertation, University Microfilms No. 80000914, California School of Professional Psychology, University of Michigan, Ann Arbor, Michigan.

Ferman, P. R., and L. A. Ferman. 1973. The structural underpinning of the irregular economy. *Poverty and Human Resources Abstracts 8,* 3–17.

Frank, A. G. 1967. *Capitalism and Underdevelopment in Latin America.* Monthly Review Press, New York.

Geertz, C. 1963. *Agricultural Involution: The Process of Ecological Change in Indonesia.* Association of Asian Studies, University of California Press, Berkeley, California.

Gritzner, J. 1988. *The West African Sahel: Human Agency and Environmental Change.* University of Chicago Committee on Geographic Studies, Chicago, Illinois.

Gross, J. L., and S. Rayner. 1985. *Measuring Culture: A Paradigm for the Analysis of Social Organization.* Columbia University Press, New York.

Haas, G. E., B. L. Driver, and P. J. Brown. 1980. Measuring wilderness recreation experiences. In *Proceedings of the Wilderness Psychology Group Annual Conference,* University of New Hampshire, Durham, New Hampshire.

Haas, P. M. 1990. *Saving the Mediterranean: The Politics of International Environmental Cooperation.* Columbia University Press, New York.

Hayami, Y., and V. R. Ruttan. 1985. *Agricultural Development: An International Perspective*. Johns Hopkins Press, Baltimore, Maryland.

Hecht, S. B., and A. Cockburn. 1989. *The Fate of the Forest:* Johns Hopkins Press, Baltimore, Maryland.

Holland, P. W., and S. Leinhardt, eds. 1979. *Perspectives on Social Network Research*. Academic Press, New York.

HRAF (Human Relations Area Files). 1992. *Human Relations Area Files Publications 1991–1992*. New Haven, Connecticut.

IGBP (International Geosphere–Biosphere Program, Committee on Global Change). 1988. *Toward an Understanding of Global Change*. National Academy Press, Washington, D.C.

Inglehart, R. 1990. *Culture Shift in Advanced Industrial Society*. Princeton University Press, Princeton, New Jersey.

IPCC (Intergovernmental Panel on Climate Change). 1990. *Scientific Assessment of Climate Change*. Island Press, Washington, D.C.

IPCC (Intergovernmental Panel on Climate Change). 1992. *U.S. Climate Change 1992: The Supplementary Report to the IPCC Scientific Assessment* (J.T. Houghton, B.A. Callander, and S.K. Varney, eds.). Cambridge University Press, Cambridge, U.K.

ISSC (International Social Science Council). 1990. *A Framework for Research on the Human Dimensions of Global Environmental Change*. ISSC-UNESCO, Paris, France.

Jasanoff, S. 1986. *Risk Management and Political Culture: A Comparative Study of Science in the Policy Context*. Russell Sage Foundation, New York.

Jeffries, V. H., and E. Ransford. 1980. *Social Stratification: A Multiple Hierarchy Approach*. Allyn and Bacon, Boston, Massachusetts.

Keilman, N. W. 1990. *Uncertainty in National Population Forecasting: Issues, Backgrounds, Analyses, Recommendations*. Swets and Zeitlinger, Amsterdam, The Netherlands.

Levinthal, D. 1988. A survey of agency models of organization. *Journal of Economic Behavior and Organizations 9*, 153–185.

LTC (Land Tenure Center). 1992. *Land Tenure Center Available Publications List 1992*. Land Tenure Center, Milwaukee, Wisconsin.

MacCracken, M. C., and F. M. Luther. 1985a. *Detecting the Climatic Effects of Increasing Carbon Dioxide*. DOE/ER-0235, National Technical Information Service, U.S. Department of Commerce, Springfield, Virginia.

MacCracken, M. C., and F. M. Luther. 1985b. *Projecting the Climate Effects of Increasing Carbon Dioxide*. DOE/ER-0237, National Technical Information Service, U.S. Department of Commerce, Springfield, Virginia.

Maslow, A. H. 1970. *Motivation and Personality*. Harper and Row, New York.

Matthews, E. 1983. Global vegetation and land use: New high-resolution data bases for climate studies. *Journal of Climate and Applied Meteorology 22*, 474–487.

McHale, J., and M. C. McHale. 1975. *Human Requirements, Supply Levels, and Outer Bounds: A Framework for Thinking About the Planetary Bargain*. Aspen Institute for Humanistic Studies, New York.

McKenney, M. S., and N. J. Rosenberg. 1991. *Climate Data Needs from GCM Experiments for Use in Assessing the Potential Impacts of Climate Change on Natural Resource Systems*. Report to the U.S. Department of the Interior, Office of Policy Analysis, Resources for the Future, Washington, D.C.

Morgenthau, H. J. 1974. *Politics Among Nations: The Struggle for Power and Peace*. Knopf, New York.

Mueller, D. C. 1979. *Public Choice*. Cambridge University Press, New York.

NAPAP (National Acid Precipitation Assessment Program). 1991. *Acidic Deposition: State of Science and Technology, Vol. 1, Emissions, Atmospheric Processes, and Deposition*. Government Printing Office, Washington, D.C.

NRC (National Research Council). 1983. *Changing Climate*. National Academy Press, Washington, D.C.

NRC 1990. *Research Strategies for the U.S. Global Change Research Program*. National Academy Press, Washington, D.C.

NSB (National Science Board). 1985. *Science Indicators: 1985*. Government Printing Office, Washington, D.C.

Olson, J. S., J. A. Watts., and L. J. Allison. 1985. *Major World Ecosystem Complexes Ranked by Carbon in Live Vegetation: A Database*. Oak Ridge National Laboratory, Oak Ridge, Tennessee.

Olson, M. 1965. *The Logic of Collective Action: Public Goods and the Theory of Groups*. Harvard University Press, Cambridge, Massachusetts.

Pahl, R.E. 1984. *Divisions of Labour*. Basil Blackwell, Oxford, U.K.

Pingali, P., Y. Bigot, and N. P. Binswanger. 1987. *Agricultural Mechanization and the Evolution of Farming Systems in Sub-Saharan Africa*. Johns Hopkins University Press, Baltimore, Maryland.

Pirages, Dennis. 1989. *Global Technopolitics: The International Politics of Technology and Resources*. Brooks/Cole, Pacific Grove, California.

Plott, C. R. 1986. Laboratory experiments in economics: The implications of posted-price institutions. *Science 232*, 732–738.

Rappaport, R. A. 1967. *Pigs for the Ancestors: Ritual in the Ecology of a New Guinea People*. Yale University Press, New Haven, Connecticut.

Rayner, S. 1991. A cultural perspective on the structure and implementation of global environmental agreements. *Evaluation Review 15(1)*, 55–102.

Repetto, R., W. Magrath, M. Wells, C. Beer, and F. Rossini. 1989. *Wasting Assets: Natural Resources in the National Income Accounts*. World Resources Institute, Washington, D.C.

Schultz, T. W. 1990. *Restoring Economic Equilibrium: Human Capital in the Modernizing Economy*. Blackwell, Cambridge, Massachusetts.

Schwarz, M., and M. Thompson. 1990. *Divided We Stand: Redefining Politics, Technology and Social Science*. Harvester-Wheatsheaf, Brighton, U.K.

Smith, V. 1986. Experimental methods in the political economy of exchange. *Science 234*, 167–73.

Stankey, G. H. 1972. A strategy for the definition and management of wilderness quality. In *Natural Environments: Studies in Theoretical and Applied Analysis* (J.V. Krutilla, ed.), Johns Hopkins University Press, Baltimore, Maryland.

Stockwell, W. R., P. Middleton, J. S. Chang, and X. Tang. 1990. The second generation acid deposition model chemical mechanism for regional air quality modeling. *Journal of Geophysical Research 95*, 16343–16367.

Strain, B. R., and J. D. Cure. 1985. *Direct Effects of Increasing Carbon Dioxide on Vegetation*. DOE/ER-0238, National Technical Information Service, U.S. Department of Commerce, Springfield, Virginia.

Taylor, C. L., and M. C. Hudson. 1973. *World Handbook of Political and Social Indicators,* 2nd edn. Inter-university Consortium for Political and Social Research, Ann Arbor, Michigan.

Trabalka, J., ed. 1985. *Atmospheric Carbon Dioxide and the Global Carbon Cycle.* DOE/ER-0239, National Technical Information Service, U.S. Department of Commerce, Springfield, Virginia.

Trauger, D. B., *et al.* 1986. *Nuclear Power Options Viability Study*, Vol. III. ORNL/TM-978013, National Technical Information Service, Springfield, Virginia.

Turner, B. L., II, and S. B. Brush, eds. 1987. *Comparative Farming Systems.* Guilford Press, New York.

USSCS (United States Soil Conservation Service). 1975. *Soil Taxonomy: A Basic System of Soil Classification for Making and Interpreting Soil Surveys.* John Wiley and Sons, New York.

van Dop, H. 1986. The CCMS air pollution model intercomparison study. *Atmospheric Environment* 20, 1261–1271.

Watkins, S. C. 1992. Fertility determinants. In *Encyclopedia of Sociology,* Vol. 2 (E.F. Borgatta and M.L. Borgatta, eds.), Macmillan, New York.

Whitten, G. Z., H. Hogo, and J. P. Killus. 1980. The carbon bond mechanism: A condensed kinetic mechanism for photochemical smog. *Environmental Science and Technology 14*, 690–700.

Wilson, M. F., and A. Henderson-Sellers. 1985. A global archive of land cover and soils data for use in general circulation models. *Journal of Climatology 5*, 119–143.

WMO (World Meteorological Organization). 1985. *Atmospheric Ozone: Assessment of Our Understanding of the Processes Controlling Its Present Distribution and Change.* WMO Global Ozone Research and Monitoring Project Report No. 16, WMO/UNEP, Geneva, Switzerland.

Worcester, R. M., and S. Barnes. 1991. *Dynamics of Societal Learning About Global Environmental Change.* ISSC–UNESCO, Paris, France.

Young, O. R. 1982. *Resource Regimes: Natural Resources and Social Institutions.* University of California Press, Berkeley, California.

3

Toward a Typology and Regionalization of Land-Cover and Land-Use Change: Report of Working Group B

John McNeill (rapporteur), Diogenes Alves, Lourdes Arizpe, Olga Bykova, Kathleen Galvin, John Kelmelis, Shem Migot-Adholla, Peter Morrisette, Richard Moss, John Richards, William Riebsame, Franklin Sadowski, Steven Sanderson, David Skole, Joel Tarr, Michael Williams, Satya Yadav, Stephen Young

Making sense of changes in global patterns of land use or land cover requires drastic simplification. The complexity of changes on the world scale easily defies the most acute and informed observer. But simplification is no simple process. To be useful it must bring out into sharp relief what is most important and relegate to obscurity what can most safely be ignored. Hard decisions and expert judgments are necessarily involved. In making these decisions and judgments we are acutely aware that others might well have been made in their stead, and perhaps with equal justification. We regard our work as a first step that we hope will help others to take long strides toward a more complete understanding of global patterns of land-use and land-cover change.

The Objective: A Comprehensive Schema

Our aim is to provide a comprehensive analytic framework that will lead to a systematic typology, and ideally a scheme of regionalization, of land-use/land-cover change. We call this framework a schema. Regrettably but inevitably, the framework cannot include every instance of change; we have instead tried to capture the major and most important varieties of land-use and land-cover change. The major changes are those of great magnitude, defined by the area involved or the numbers of people affected. The most important changes (a set that obviously intersects with the set of major changes) are distinguished by their criticality from the points of view of human and scientific concern. For instance, certain land-cover changes, even ones small in magnitude, have extreme social costs, while others have few or none. The costly ones we regard as more important from the human point of view. Other changes, not necessarily large in magnitude, have vast consequences from the point of view of biogeochemical flows or of biodiversity. In concentrating on the largest-scale and most important types of land-use and land-cover change we intend also to isolate and identify the most interesting types from the point of view of research

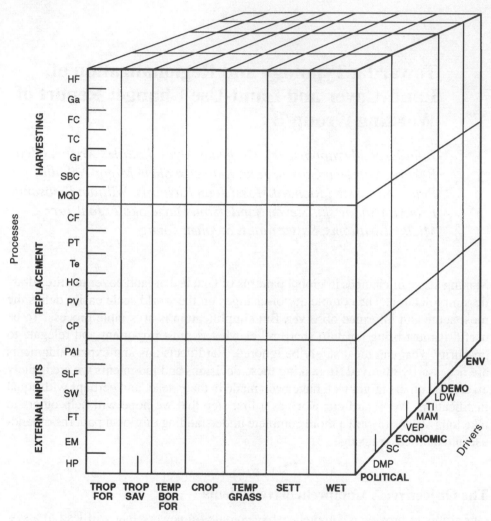

Figure 1. The cube

Land cover types		Drivers	
TROP FOR	Tropical forest	Political	
TROP SAV	Tropical savanna and grassland	DMP	Decision-making process
TEMP BOR FOR	Temperate and boreal forest	SC	State capacity
CROP	Cropland	Economic	
TEMP GRASS	Temperate grassland	VEP	Vulnerability to external
SETT	Settled and built-up area		pressure
WET	Wetland	MAM	Market allocation mechanism
		TI	Technological intensity
		LDW	Level and division of wealth
		DEMO	Demographic
		ENV	Environmental

needs. We hope our schema not only will lead to a deeper understanding of global change, but also will point to questions and cases that deserve high priority as research subjects. In the new and necessarily confused world of global change research, helping to sharpen the definition of research agendas is no small ambition.

Towards a Typology: The Axes and the Cube

Specific instances of change in land use or land cover may be defined by any set of characteristics. Possible examples include the duration, intensity, purpose(s), and consequence(s) of change. But confining ourselves to the most distinguishing characteristics, and keeping matters simple, we have selected only three, which taken together give a reasonably precise idea of any given land-use or land-cover change. These are:

- The *land-cover type* in which any given change takes place
- The *driving forces* producing change
- The specific *processes of conversion* of land cover.

Any single transition in land use or land cover can be analyzed from these three perspectives. Imagined in three-dimensional space, these three characteristics form axes, which taken together define a block or cube, in which every example of land-cover/land-use change occupies certain spaces (see Figure 1). After some elaboration of each of these axes, we will discuss how this schema works. In the appendix to this chapter, the schema is applied to actual cases of land-use or land-cover change.

Land-Cover Types

Scholars have created numerous classification systems of land cover. Each has its merits and its faults, and is appropriate for certain uses but not others. Ours has its

Caption for Fig. 1. (*cont.*)

Processes			Replacement	
Harvesting			CF	Clearing/firing
HF	Hunting/fishing		PT	Plowing/tilling
Ga	Gathering		Te	Terracing
FC	Fuelwood cutting (industrial and domestic		HC	Hydrological control (irrigation, drainage)
TC	Timber cutting		PV	Planting or vegetation change
Gr	Grazing		CP	Construction, paving, earth shaping
SBC	Slash-and-burn cultivation		External Inputs	
MQD	Mining/quarrying/ drilling		PAI	Plant or animal introductions
			SLF	Supplementary livestock feed
			SW	Supplementary water
			FT	Fertilizer/trace elements
			EM	Energy/machinery
			HP	Herbicides/pesticides

faults (it is not all-inclusive, for example), but has the merit of reflecting the concerns of scientists interested in global land-use and land-cover change. We think it is the most appropriate system of classification for our purposes. It includes seven major types of land cover (see Table 1). Some of these are zones of natural vegetation; others represent land uses. But in every case, they describe what actually occupies the land.

The hard decisions and expert judgments behind this classification system of land-cover types rest upon criteria that ought to be made clear. First, we have included cover types that are important to global change research from the points of view of (1) the physical climate system on the global scale, (2) regional and continental energy and water balances, (3) global biogeochemistry, (4) atmospheric chemistry, and (5) biodiversity. Second, we have emphasized cover types important to the human dimensions of global change research as revealed in (1) land quality, soil fertility, and biodiversity (again); (2) sustainable development, sustainable agriculture, and resource development issues; and (3) land tenure, land access, and land-use issues. Beyond these two general guidelines governing our choices, we have also given weight to considerations of the current rate and magnitude of changes, to the simple areal extent of cover types, and, bowing to necessity, to gaps and uncertainties in available data. How these criteria for selection led to the seven cover types represented in our classification scheme is explained briefly in the second column of Table 1. The land-cover types in the table are arranged from most to least important according to the criteria listed above.

The underlying choices were hard, but the resulting classification scheme is simple. While this is a virtue not to be discarded rashly, the system may easily be expanded and/or subdivided to provide more complete coverage of the globe or to distinguish more finely among land-cover types. Indeed, one could easily create a classification system that recognized the gradients between desert and grassland and forest. For the moment, however, we prefer the simplest formulation that captures the most important varieties of land cover.

Driving Forces

The motors of land-cover/land-use change are countless. Some act slowly (and often obscurely) over centuries, while others trigger events quickly and visibly. In every case, several forces are at work, sometimes operating independently but simultaneously, sometimes operating synergistically. No aspect of global change is more complicated than the driving forces.

Our system of classification of driving forces, like our land-cover classification system, is one among many possible such schemes. It represents a theoretically informed set of choices and distinctions, designed specifically for the purposes of capturing the most consequential human and environmental causes of land-cover/land-use change. The major categories of driving forces are political, economic, demographic, and environmental (see Table 2). Within these broad categories, we identify specific attributes that strongly influence land-use and

Table 1. Important land-cover types

Cover Type	Rationale	Location
1. Tropical forest	• Large conversion extent • High rate of change • Wet regimes, high trace gas flux • Climate/water influence • Biodiversity • Difficult soil management • Sustainable development • Largest frontier • Developing countries • Typology for agricultural and commercial enterprise, geopolitical/	Amazon, West Africa, Southeast Asia, Central Africa, Central America
2. Tropical savanna and grassland	• Large extent of occupation • Large conversion extent • High rate of change • Frequent burning, non-CO_2 trace gases • Sustainable development • Dwindling frontier	Brazil, Sahel-Sudan, South Africa
3. Temperate and boreal forest	• Commercial timber harvest is dynamic • Agriculture unknown • Potential sink • Land intensive • Indirect effects	U.S., USSR, Europe, Canada, Scandinavia
4. Cropland	• Land intensive = trace gas • Land extensive = CO_2 • Rapidly increasing area • Tenure conflict • Sustainable development issues • Impact on frontier	Global
5. Temperate grasslands	• Unknowns in key regions: central Asia • High soil carbon • Potential degradation	
6. Settled and built up	• Extent unimportant • Sphere of influence could be important: core/periphery	
7. Wetlands	• High loss rate, large area uncertainty	

land-cover patterns. Each of these attributes can be expressed as a variable or set of variables. For example, almost every land-use or land-cover change takes place within a political context. At some level, international, state, or local, a decision-making process—or perhaps several of them—is involved. This process might be one with high public participation in an open system in which power is decentralized; or it might be just the opposite, a process in which few people are involved, exclusion rather than openness prevails, and power is centralized. Each polarity—

Table 2: Driving forces behind land-use change

Attribute	Variable	Kinds of Indicators
1. Political		
Decision-making process	Degree of public participation (open/closed, centralized/ decentralized	• Unitary or federal structure • Number of special-interest groups
State capacity	Public sector pressure/ influence	• Public sector expenditure/GDP • Public/total land area
2. Economic		
Vulnerability to external pressure (economic/ political)	Open vs. closed economy	• Exports/GDP • Partner concentration
	Primary sector dependence	• Primary sector exports/ total exports
	Type of exchange rate management	• PEA (Population economically active) agriculture/PEA total • Real exchange rate • Debt service ratio
Market allocation mechanism	State controlled, market driven	• Agricultural subsidies • Public sector expenditure/GDP
Technological intensity	High-low	• Energy intensity of GDP
Level and division of wealth (asset inequality)	Wealth/poverty-induce consumption	• Primary sector/GDP • Energy consumption/ capita • PEA agriculture/PEA total • Percent absolute poor in total population
3. Demographic		
Population pressure on the land	High-low	• Cultivated/arable land • Change in population density • PEA agriculture/PEA total
4. Environment		
Natural resource quality	Scarcity	• Stock, yield, flow

centralized vs. decentralized, e.g.—can be visualized as a continuum. Every polity falls somewhere along these continua. To know where, one must use indicators that provide an approximation of the degree of centralization or the degree of public participation in a given polity. The third column of Table 2 provides a list of appropriate indicators for each of the variables we find useful in characterizing driving forces. Most of these indicators are quantitative measures, although some are cruder instruments permitting only yes/no statements. Inevitably, some indicators are more amenable than others to precise measurement. This is part of the devilish complexity of dealing with the driving forces of global change.

Our system of classification differs from others in several respects. First, it gives great weight to the economic and political differences among societies. This is essential, because the same climatic or even demographic pressures can produce sharply different results in different political and economic circumstances. Recognizing the distinctions among political and economic systems, even on a fairly rudimentary basis, is a step away from crude demographic or climatic deter-minisms. Second, we have incorporated technology into the broader category of economic attributes rather than designating it as a major driving force in its own right. We do not consider technology to be a major determinant of land-cover or land-use change on the global scale. Technologies appear to us as social responses to needs and opportunities, devised, diffused, and used wherever they do well what people want done. It is the needs and opportunities that motivate people to adopt a given technology that seem to us fundamental.

This is a view that will leave many scholars uneasy, including some members of our working group. Certainly there are many historical examples that seem to run counter to it. To choose only one, the adoption of the heavy plow in the Russian steppes after the 16th century permitted the extensive conversion of grasslands to croplands in southern Russia and the Ukraine for several centuries. Without the right technology, that land-cover change could not have happened. But, equally, without the demographic expansion of Russian population, the new market oppor-tunities that large-scale grain producers enjoyed, and the increasing military domi-nance of the Russian state over the pastoral peoples of the steppe, the change would not have happened either. In this example, all elements of the mix were necessary to produce the land-use change. In the modern world we find this to be true more rarely. More commonly it seems to be the case that social, political, economic, and demographic circumstances summon applicable technologies, and that the tech-nologies themselves are not major variables. The chief exception is in transport technology. The presence or absence of transport infrastructure is so important that it commands special attention. The conspicuous difference in land-cover change between Zaire's tropical forests and Brazil's is in large part a direct result of the absence or presence of roads. However, even in this case, the construction of Brazil's Amazonian roads was in large part the result of political decisions.

A second noteworthy feature of our classification system for driving forces is the absence of any overt consideration of culture. All scholars working on land-use/land-cover change grant culture some importance, but most despair of forming any useful generalizations about it. Great discrepancies arise when one studies what people say, what they believe, and how they behave. The most accessible of these— what people say—is probably least relevant to land-use/land-cover change, and the most relevant is unhappily also the most difficult to ascertain and measure.[1] Furthermore, culture is so localized, so fragmented, and so mutable in most societies

---

[1] L. Arizpe, the Working Group B leader, prefers the view that what people say is especially important to under-standing cognitive and political negotiations that have to take place to change people's land-use practices. (See also Rockwell, this volume.)

that it defies scholars' attempts to give it due weight. Only in local studies, where variations are minimized, is culture easily factored in to explanations of land-use or land-cover change. We have no simple solution to these problems. To some extent, cultural differences are present behind the political and economic variables that figure prominently in our classification scheme. Indeed, catching the manifestations of cultural traits indirectly, through political, economic, and even demographic variables, provides a more manageable approach to culture's impact on land use than the direct one, which so confounds measurement and generalization.

This axis, like the land-cover type axis, could easily be amended to be more inclusive or more refined in its distinctions. Our choices and judgments represent a compromise between simplicity and full accuracy.

Land Conversion Processes

This third axis of the cube in Figure 1 represents purposeful human activities aimed at increasing the productivity of land (whatever their real consequences). We have arrived at 19 discrete processes that have direct impacts on land cover and land use, and also have clear purposes, if often unclear side effects. These 19 processes are grouped for analytic convenience into three main categories: (1) harvesting, which is the appropriation of natural products resulting in modification but not conversion of land cover; (2) replacement, i.e., the conversion from one land cover to another; and (3) transfer, meaning processes that import from an external source additional resources or energy in an effort to improve or intensify production. The specific processes in each of these three broad categories are listed in Figure 1.

To our knowledge this system of classification is unique in trying to lend coherence to the welter of specific human actions affecting land use and land cover. Like the other axes, it is potentially subject to expansion or contraction. The present set of choices aims not to be all-inclusive, but merely to represent the principal mechanisms of land-use and land-cover change abroad in the world today.

From the Three Axes to a Typology

Every case of land-use or land-cover change takes place within a specific land-cover type or types, has certain specific driving forces that cause it, and consists of certain specific processes of conversion from one land use or cover to another. Hence every case, if analyzed from the perspectives represented by the three axes of Figure 1, can be plotted within three-dimensional space. Every case occupies a set of minicubes, each of which represents the intersection of a given land-cover type, a given driving force, and a given land conversion process. This set of minicubes we call a constellation, to emphasize that in practice the set is not likely to be contiguous. Obviously, several driving forces and often several conversion processes will be at work in any given case. As a hypothetical example, let us say that in a tropical forest zone, population pressure, security concerns of a military government, and foreign capital investment are driving conversion from forest to settled land and to cultivated land. This case would occupy 12 minicubes to account for all intersections of tropical forest land cover with the demographic, political (state capacity), and economic (vulnerability to external pressure) drivers and with the following processes: clearing/firing, plowing/tilling, slash-and-burn cultivation, and planting or vegetation change.

Translating the characteristics of a given case into the appropriate minicubes is not a mere mechanical business. Careful judgments are required, especially in identifiying the key driving forces. Is an internal colonization scheme sponsored by the military best captured as 'high state capacity' or as some other political attribute? In many instances modest amounts of research would be needed to generate the data and the judgments necessary to translate a case reliably into a constellation of minicubes. But if this can be accomplished, this method of analysis summarizes precisely what is going on, where, and why in any given case.

The next step is to amass and analyze cases. Imagine 20 or, better yet, 100 cases plotted in three-dimensional space and translated into 100 constellations. One is now in a position to ask, and answer systematically, a series of questions that will lead to the distillation of large numbers of cases into a few prototypical 'situations.' With 100 constellations to compare, one could easily examine how these relate to one another in space. Do the constellations cluster together? Which constellations overlap and where? Are there minicubes, or sets of minicubes, that show up again and again in the constellations? Do similar sets of driving forces and conversion processes operate in different cover types, or are the same cover types playing host to similar driving forces and/or conversion processes? There is probably no end to the variety of questions researchers might wish to ask of the data. This is perhaps the most appealing promise of our construct: the possibility of reducing the daunting irregularity of existing ecological, economic, political, social, and demographic data to a common template so that systematic and precise comparisons can be made and a clearer idea of what is afoot in global land use and land cover can be obtained. Naturally something is lost in reducing the irregular data that most closely correspond to the real world to a common template. But without this step, however brutal, comparison, typing, and aggregation among individual cases rest on impressions and intuitions: more an art than a science. We will return to this question of data and their handling below, when we take up the subject of further research.

With 100 cases plotted as constellations, and then compared and analyzed, it would be possible to map cases or groups of typical cases ('situations') on any scale desired, local, regional, or global. The visual display of the information yielded by our methods will be a simple and, we hope, extremely useful and revealing step.

The Time Dimension

Translating land-use or land-cover change data into easily comparable constellations of minicubes provides only a snapshot of what is happening. A more complete picture must be a moving picture, one that incorporates time. Of course time is implicit in the three axes described above: the driving forces operate gradually, at great or lesser speeds; the conversion processes themselves take time; and land-cover types, even if left undisturbed by human agency, change over time—usually very slowly. But this implicit recognition of time does not sufficiently capture the importance of historical change. To do that one needs a series of cubes to yield a series of snapshots, permitting one to analyze the gradual evolution of land uses and land covers. Many apparent changes in land use or cover are but brief fluctuations, soon to be changed again, and sometimes changed back. Over the long haul forests may be replaced by fields and return again to forest, as many landscapes in the Yucatan and New England attest. To know the difference between fluctuations and trends, between temporary conditions and new realities, one must systematically incorporate time into one's analysis.

Theoretically this is a simple procedure. Imagine a series of cubes, representing 5- or 50- or 100-year intervals. With sufficient data (on which question see below), one could plot cases over time, generating a series of constellations. If the selected intervals were short enough, one could visualize constellations evolving, as it were, before one's eyes, just as a series of photographs taken at very brief intervals and then seen in rapid succession becomes a movie. With the help of our schema—and sophisticated computer graphics—the trajectories of these constellations could be plotted in space and then compared and analyzed, just as described above for individual constellations. Comparing and analyzing quantities of trajectories would permit the same sort of generalization and aggregation as with the constellations. If one could detect similar trajectories through time among certain constellations, this would hold out the promise of some predictive capacity for our analytic schema.

All of this hinges on the availability of sufficient and suitable data. Historical data may be recoverable for many parts of the world for the last century, although certainly there will be many gaps. For some unusual places, it will be possible to uncover adequately precise historical data for several prior centuries. With sufficiently intense and directed efforts at data collection, it should be possible to do this with great precision for the present and future. The utility of analyzing trajectories is so much greater than that of analyzing snapshots that we fervently hope data will be recovered where possible for the past and collected widely in the future, whatever the time and labor required.

Research Implications of the Schema

Before any but the most rudimentary of the projected benefits of this schema may be realized, three major paths of research must be probed and ideally followed to their logical conclusions. These are the further refinement of the three axes, the development of reliable data sets, and the development of procedures and technology for integrating the three dimensions represented (or four, including time). We will take these up briefly in order.

The systems of classification portrayed on each of our three axes represent decisions made in the course of five days' discussion. We do not pretend that they are the product of carefully considered and prolonged investigation. Consequently, each one merits further review and is surrounded by questions in need of close scrutiny. The land-cover type axis, for instance, ought perhaps to be articulated further so as to incorporate soil types and climatic variables, or ought perhaps to be arranged in gradients. The driving forces axis confronts more difficult problems in classifying human behavior as it affects land use and cover, and our efforts here presumably require significant modification. The proper status of technology and culture need more attention than we could give them. How could one best quantify those attributes that are not easily susceptible to objective measurement? More generally, have we captured the important ways in which land-use and land-cover change is caused? The land conversion processes axis is beset with uncertainties as well. Have we compromised suitably between inclusiveness and simplicity? Does the general classification system of harvesting, replacement, and transfer hold up under scrutiny? How can one best determine whether a given process is worthy of attention? Further work on all three axes is called for. It will require teams of scholars from several disciplines.

Equally necessary is the collection and organization of sufficient data. To make the best use of our schema one would need as many cases as possible with full ecological, demographic, political, economic, and social data sets, for the past as well as the present. And those data must be converted into forms suitable for translation into the systems of classification of the three axes. That hope is unrealistic; but the closest approximation to it will yield the most reliable and useful results. Fundamental questions remain about the character and scale of data needed. Where are the gaps in existing data? Which gaps are most crucial? Which can be filled at lowest cost? At what scale will this schema operate best—for example, Amazonia in its entirety, a square 200 km by 200 km, or a single mountain valley in the Andes?

Finally, it also seems clear that further work is necessary on the integration of the three (or four) dimensions. There are technical problems in the processing and analysis of spatially arrayed data, and complex computations and visualization graphics will be required.

Conclusion

Our working group sought to create a typology of world situations in land-use/land-cover change that took proper account of the myriad causes, processes, and circumstances involved. Deviating from this goal somewhat, we created what we regard as a systematic method, called here a schema, by which this could be done, and by which a far deeper understanding of land-cover and land-use change in the past, present, and the future is obtainable. But while we have sketched the underlying concepts and the governing procedures, we have not yet made operational our schema. That requires far more work in refining procedures and collection of data than five days, however feverish, permitted.

Appendix A: Case Studies

Amazonia

Diogenes Alves

Two 'snapshots' of the tropical forest in the Brazilian Amazon at two different times illustrate two different situations of land-cover change. The first is the early 1900s; the second, the intensive colonization cycle during the 1960s, 1970s, and 1980s.

Early 1900s in the Brazilian Amazon

In the early years of this century, the process of human settlement in the Brazilian Amazon was limited to areas along the major rivers. There were no roads, and most transportation was by river. Agriculture was concentrated in the eastern part of the region. Rubber tapping was the most important source of revenue, but development of rubber plantations in South Asia ended the 'rubber cycle' in the Amazon.

Land-cover type:
• Tropical forest

Drivers:
• Economic (vulnerability to external pressure in the form of competition from Asian rubber)

Processes:
• Harvesting: timber cutting, grazing, gathering
• Replacement: clearing/firing, planting
• External inputs: plant or animal introductions

Amazon Colonization after 1960

An intensive process of colonization took place in the Brazilian Amazon during the 1960s, 1970s, and 1980s. A combination of several factors stimulated immigration, agriculture, and several other economic activities that led to land-cover changes in the region.

Beginning in the 1960s, and continuing throughout the 1970s, the development of infrastructure, especially roads, improved access to the area. Federal and state governments offered incentives for farmers and industries to move to the Amazon, where land was usually cheaper than in other parts of the country.

Both intensive and extensive agriculture can be found in the region. In several areas of significant settlement, soils are poor. Incentives and subsidies used to be proportional to area cleared, leading to unnecessary clearing and later abandonment. Traditional gathering activities such as rubber tapping are still popular in the Amazon. Much of the clearing has been related to the conversion of forest into pasture.

The development of infrastructure has been stopped for more than a decade, incentives have been cut, and the development of agriculture is not always competitive due to the quality of the land and the distance to major consumer centers. These factors are causing new changes in land use, such as the abandonment of some areas and urban expansion due to migration from rural areas, that are still insufficiently studied.

Land-cover type:
• Tropical forest

Drivers:
• Political (decision-making processes and state capacity through the incentives program)
• Economic (market allocation mechanism)

Processes:
• Harvesting: timber cutting, gathering, grazing, mining
• Replacement: clearing/firing, planting, dam construction
• External inputs: plant/animal introductions, fertilizers, machinery.

The Burma Delta under British Rule

John F. Richards

In 1852 the East India Company, the ruling colonial government in India, defeated the Konbaung regime in Burma and annexed the lower half of the kingdom. Immediately thereafter British rule opened the Irrawaddy Delta lands to a new form of intensive exploitation under world capitalism. Between Burma's initial conquest and the world depression in 1930, the Burma delta tropical wetland forest was converted to a domesticated wetland. By 1930 there were 4 million hectares of intensively cultivated wet rice fields in the delta. How would we model this massive transformation in land use by using the three-axis cube?

Land-cover type:
• Wetlands

The lower reaches of the delta were subject to daily tidal inundation from the sea. Along the nine branches of the Irrawaddy and their many creeks, species of mangrove (*Rhizopora*) and palms grew luxuriantly. Slightly inland were heavy formations of the kanazo tree (*Heritiera fomes*), which often grew to heights of 45 m. Annual flooding in the rainy seasons meant that these areas were completely covered for several months of the year. Abundant wildlife, including tigers and elephants, roamed these tracts. Kanazo faded into drier, mixed scrub delta forests. At conquest the entire delta was very thinly populated with subsistence rice farmers, salt-makers, fishermen, and bandits.

Processes:
• Replacement: cutting, firing, tilling, hydrological control, planting

Increasing numbers of pioneer settlers engaged in intensified replacement activities. They cleared land by cutting and firing. They controlled water by digging drainage channels and erecting *bunds* (embankments) around rice fields. They tilled the land and planted rice as the main cash crop. The colonial state repaired and extended embankments along the Irrawaddy to protect thousands of hectares from annual flooding.

Driving Forces:
• Political (state capacity)
• Economic

The decision to open the delta to settlement and cultivation was taken by a foreign, colonial government that permitted virtually no public participation in this decision. Certain special interest groups were considered. For example, Indian immigrant brokers, merchants, and money-lenders and British merchant houses that were prepared to handle the wholesale trade in rice were favored. This was an open export economy with no protection against the vagaries of the world market. Burma rice soon became one of the two major exports in bulk, along with timber taken from the hilly teak forests.

The state asserted control over all 'unoccupied' delta lands. This control was codified in the Burma Land and Revenue Act of 1876. Land was free to any Burmese who would clear and cultivate the jungle. Pioneer settlers could obtain full rights of ownership if they grew rice and paid taxes to the state for 12 years. They could also obtain written permission for occupancy, hold the land tax-free for 7 years, and then obtain ownership. Ownership conveyed permanent, heritable, alienable rights to the Burmese small holder.

Previously unknown technologies were deployed by the new colonial government. Public works officials designed and built new port facilities at Rangoon and subsidized the rapid development of steamship service up the Irrawaddy River.

British engineers using immigrant Indian laborers repaired existing river embankments and built miles of new embankments to protect delta fields from annual flooding. Newly built canals connected various tributaries of the great river. Permitted to enter freely, immigrant Indians supplied labor for these projects. Other Indian groups, especially the Chettiars from Madras, brought in capital to act as money-lenders for peasant settlers. They also financed the processing of paddy and its transport to the warehouses and docks of Rangoon. British agency houses took over the business of overseas shipment and sale of Burmese rice.

Burma was certainly not overpopulated in the mid-19th century. Land and other natural resources were ample for the population. New state policies made vacant land in the delta an attractive resource for Burmese farmers from upper Burma. Income differentials were such that individual families responded with alacrity to new monetary incentives. No new technologies were employed in clearing land, managing water, or growing wet rice. The extant bundle of agrarian techniques was more than sufficient for the task at hand.

The Aral Region

Olga Bykova

The Aral region includes the Aral Sea basin; the basins of the Syrdarya, Amudarya, Tedjen, and Murgab rivers; the Karakumsky Canal; small rivers running from the West Tien Shan and Kopet-Dag; closed basins; and areas between the rivers and around the Aral Sea. Its area in the USSR is 1.4 million km^2 and its total area 2 million km^2. Its population of around 31–32 million is growing rapidly at a rate of over 2% annually.

Land-cover type:
- Sand and stone deserts in the temperate and subtropical climatic zones

Drivers:
- Political (high state capacity)
- Economic (high technological intensity)
- Demographic (rapid population growth)

Processes:
- Harvesting: grazing, fishing
- Replacement: irrigation, drainage
- Transfer: planting, application of herbicides and pesticides, fertilizing, use of machinery

In the early 1960s, the government decided to begin large-scale expansion of irrigation within Central Asia to meet several goals:

- To increase cotton production to meet internal demands for cloth and increase exports
- To increase production of fruits and vegetables
- To provide the population with meat and rice
- To create new jobs for the growing population.

In the following decades, the total area of irrigated land increased in the Uzbek and Tajik Soviet Republics by 1.5 times, in the Kazakh Republic by 1.7 times, and in the Turkmen Republic by 2.4 times. During the same period, agricultural subsidies grew several-fold; energy consumption increased by 6 times; application of fertilizers increased by 3.5–6 times; and the number of tractors increased by 3.2 times.

The new large-scale irrigation, and the application of chemical inputs and other human impacts, have caused intensive transformation of the natural environment, economy, and population of the Aral region:

- River runoff in the Aral Basin has decreased from 56 km^3 in 1960 to 7–11 km^3 in the mid-1970s and practically nothing in the 1980s, the runoff regime being transformed by water consumption.
- The hydrographic network has spread with the creation of irrigation and drainage canals.
- The total area of the Aral Sea has decreased from 67,000 to 41,000 km^2 while salinity has increased from 10 to 28–30%.
- Salt transfer from the seabed has increased greatly, according to recent estimates reaching 40–150 million tons/year.
- Irrigation has led to a significant rise of the groundwater level, leading in turn to intensive secondary salinization of the soil. Moderately and highly salinized soils now occupy 50% of the irrigated area.
- The lowered water level in the rivers and in the Aral Sea has caused intensive desertification of the coastal and delta areas.
- Drainage runoff into the rivers has changed their chemical composition. Mineralization of the Amudarya's waters increased from 0.8 g/l in 1960 to 2.8 g/l in 1985.
- Soil and water pollution from fertilizers, pesticides, and herbicides is very high; the content in most of the rivers exceeds sanitary norms several-fold.
- The climate in the Aral region has become more continental, and dust storms have become more frequent and affect larger areas.
- The diversity of mammals has decreased from 70 to 30 species and of birds from 173 to 38, while 54 plant species are endangered, including relic and endemic ones.

The social consequences and indicators have included a growth in child mortality rate, which in some areas exceeds 110 per 1000. Further, disease and mortality rates have grown in the adult population. There also has been a significant decrease in the size and quality of the cotton crop. The Aral Sea's fishing industry is defunct. Pasture lands have decreased, with a consequent decrease in the total number of livestock and the production of wool and karakul (astrakhan fur).

It is estimated that remediation of the area will cost approximately 37 billion rubles.

This report is based on: Glazovsky, N. F. 1990. *The Aral Crisis: The Origin and Possible Way Out.* Nauka, Moscow.

California's Central Valley

Peter M. Morrisette

The Central Valley of California is one of the most intensively irrigated agricultural regions in the world. Over 7 million acres of land are under irrigation, supporting a nearly $8 billion annual industry. Irrigation water in the Central Valley is provided through two publicly supported water systems: the federally operated Central Valley Project (CVP) and the state-operated State Water Project (SWP). Irrigation water provided by the Bureau of Reclamation through the CVP is highly subsidized, reflecting neither the full operational cost of providing the water nor environmental or opportunity costs. State water is not subsidized, although its price also does not reflect the environmental or opportunity costs of using that water for irrigation. The development of irrigated agriculture in California's Central Valley represents a fundamental change in land cover from a semiarid grassland to intensively used cropland. This conversion has not been without economic and environmental costs.

Land-cover type:
• Cropland

Processes:
• Replacement (plowing, tilling, hydrological control)
• External inputs (plant introduction, supplementary water, fertilizer/trace elements, energy and machinery, herbicides and pesticides)

Hydrological control includes the valley's extensive network of canals, aqueducts, and reservoirs used for irrigation and drainage.

Driving forces:
• Political (decision-making process)
• Economic (technological intensity, market allocation mechanism)

The Central Valley, with its massive irrigation network and its use of external inputs such as fertilizer, energy, and machinery, represents perhaps the zenith of modern technological agriculture. While it would be a mistake to discount the

71

importance of technology, however, the evolution of the agricultural landscape of the Central Valley is probably best understood in the context of the role of special interests and their manipulation of the water allocation system.

For example, because water provided to farmers from the CVP is so heavily subsidized (the price reflects only about one-fourth of the delivery costs and does not take environmental costs into account), there is little incentive to use this water efficiently. The National Research Council, in a 1989 report entitled *Irrigation-Induced Water Quality Problems: What Can Be Learned from the San Joaquin Valley Experience*, argues that the low cost of water is 'the most pervasive economic issue contributing to irrigation-related water quality problems and affecting the choice and success of solutions' (p. 5). The report further notes that 'the subsidized low cost of water results in more water being used, encourages farmers to cultivate less desirable lands, and leads to increased agricultural runoff' (p. 5). This system of water allocation, which undervalues water and does not internalize external costs, is defended by agricultural interests in the Central Valley and a bureaucracy with a vested interest in maintaining the status quo.

Nevertheless, change is occurring. For example, because of concern over serious environmental problems, the 222,750 hectare Westlands Water District in the San Joaquin Valley has been forced to internalize the environmental costs of disposing of toxic drainage water. The Westlands district now has an incentive to develop less costly means of dealing with its drainage problem. One possible alternative would be for farmers in the district to sell water that would otherwise be used on lands with drainage problems to the SWP or to Southern California's Metropolitan Water District, where it would be used for higher value municipal uses. Despite the fact that water would be used more efficiently and environmental damage would be mitigated, there are major institutional barriers to this alternative. While progress in breaking down these barriers has so far been slow, it is likely that the economics and politics of water in California (increasing demand relative to supply) and growing concern over environmental problems will necessitate such a change.

Returning to the cubic typology of world situations, the minicubes that would best represent the case of irrigated agriculture are those identified by the intersection of croplands (cover type) with external inputs (processes) and economic/market mechanism and political/decision-making process (drivers). By itself, the exercise of using the cube in this case does not tell us anything that was not already obvious; what will be interesting, however, is to see what other cases might be similar. In other words, what areas share these minicubes with the Central Valley?

4

Land-Use and Land-Cover Projections:

Report of Working Group C

Jennifer Robinson (rapporteur), Stephen Brush, Ian Douglas, T.E. Graedel, Dean Graetz, Winifred Hodge, Diana Liverman, Jerry Melillo, Richard Moss, Alexei Naumov, George Njiru, Joyce Penner, Peter Rogers, Vernon Ruttan, James Sturdevant

The Need for Projections of Land-Use and Land-Cover Change

How will land use and land cover change under rising pressure from human demands for subsistence and increasing standards of living over the coming 10 to 50 years? What will be the consequence of the land-cover changes for the status and functioning of the biosphere and the earth system? How might such changes interact with natural or anthropogenic climate change?

Answers to these questions are critical if humankind is to appreciate and manage its impact on the structure and function of the biosphere and earth system and to assess options for the near future. In recognition of these needs, we seek predictive global understanding of the effects of social change on land use and land cover. This understanding is to be based on contemporary assessment and historical review of the relationships among driving forces of social change as affecting land use and land cover.

Specific Needs Related to Earth System Science and Global Change

Atmospheric general circulation models (GCMs) play a pivotal role in global change modeling. As listed in Table 1, over the last decade, GCMs have been complemented and extended by terrestrial biophysical, chemical, and biogeochemical models (collectively, global X models, or GXMs, where X specifies subject domain), and a comprehensive framework of linked models for studying the physical and biochemical consequences of human alteration of the earth is emerging.

As shown in Table 1, land use and land cover, through their control on albedo, transfer of momentum, and cycles of water and nutrients, establish the boundary conditions for many of these models.[1] Changes in land use and land cover alter these boundary conditions and change the fundamental behavior of the system. Modelers can study such effects by doing sensitivity studies and altering boundary conditions,

---

[1]Ocean circulation models are a significant exception.

Table 1: Global physical and biological system models that address the land surface

Model Type	Subject Matter	Examples	Institutions	Parameters Forced By Land Use/ Cover
				by Land Use/ Cover
General Circulation Models (GCM)	Atmospheric general circulation, climate change	~20; see Cess, 1990 for listing	Diverse	Surface albedo, latent heat flux, transfer of momentum
Global Chemical Transport Models (GCTM)	Reactions and transport of chemically and radiatively important gas and aerosol species	Spivakovsky *et al.*, 1990; Penner *et al.*, 1991; Taylor *et al.*, 1991; Lilieveld & Crutzen, 1990	GISS, LLNL, Harvard, NCAR, SUNY	Sources and sinks of trace gases and aerosols
Global Biogeochemical Models (GBM)	Carbon storage in the terrestrial biosphere, primary productivity, N and P links to C cycle	Skole, this volume; Houghton *et al.*, 1983; Esser, 1987; Emanuel *et al.*, 1984; Goudriaan & Ketner, 1984; Olson, 1981	UNH, WHOI, WHR, ORNL	Balances of C, N, P, etc., as affected by harvesting, herbivory, fertilization, water balance, microclimate
Global Vegetation Models (GVM)	Biophysical relations of plant canopies	Dickinson *et al.*, 1986; Xue *et al.*, 1991; Pollack & Thompson, in preparation	NCAR, UNM, GSFC, UKMO, Macquarie	Vegetation type as affecting transpiration, interception of precipitation, root uptake, soil and vegetation temperatures
Global Natural Vegetation Models (GNVM)	Natural vegetation as function of climate	Prentice, 1990; Guetter and Kutzbach, 1990; Emanuel *et al.*, 1985	GISS	Vegetation type

GISS: Goddard Institute of Space Studies, LLNL: Lawrence Livermore National Laboratory, SUNY: State University of New York, UNH: University of New Hampshire, WHOI: Woods Hole Oceanographic Instutuion, WHR: Woods Hole Research, ORNL: Oak Ridge National Laboratory, NCAR: National Center for Atmospheric Research, UNM: University of New Mexico, GSFC: NASA Goddard Space Flight Center, UKMO: U.K. Meteorological Office.

but they lack a vehicle for generating comprehensive, self-consistent land-use projections that are compatible with model structure. The absence of a framework for understanding land-use/cover trends furthermore means that the causes of change are understood only vaguely, on a level too general to provide guidance, either for policy or for formulating scientific priorities for research and data acquisition.

The present GXM framework does not encompass all aspects of global change. It largely fails to capture the fact that global change is caused by humans and will occur on landscapes that are increasingly altered by human activity. Analysis of the effects of changing climate and atmospheric chemistry has not systematically reckoned with the probable doubling of human numbers and commensurate increase in land-use intensity over the next ~50 years envisioned in scenarios for a doubling of atmospheric CO_2. Climate impact analysis has concentrated on biological links to crop productivity and changes in the potential vegetation of undisturbed ecosystems. Where the direct effects of human modification of the land surface have been studied, it has mostly been through extreme, unrealistic scenarios, such as replacement of the entire Amazon forest by pasture or complete desertification of the Sahel.

Furthermore, the present framework fails almost completely to address *cumulative change*[2]—change that is manifested through the geographically dispersed, progressive result of human activity, such as soil loss, habitat loss, and loss of biological diversity. This deficiency is serious, not only because cumulative change can be significant, but also because it is likely to exacerbate, and may overwhelm, the effects of climate change.

Basic Recommendation

We recommend the development of a new model genre, the global land-use/cover model (GLM). GLMs should:

- Organize available data on land use/cover and the driving forces that affect it into a global framework
- Comprehensively assess data adequacy and define priorities for research
- Produce projections that can be verified (or falsified), subjected to sensitivity testing, and improved on the basis of verification exercises
- Provide comprehensive land-use/cover forcings for other GXMs
- Generate insight into the mechanisms that cause land-cover change and the direction and magnitude of measures required to curtail undesired change
- Establish a framework for systematic study of cumulative global change, and for realistic study of feedback from climate change in a biosphere that is already strongly disturbed by human activity
- Evolve with the GXM framework
- Permit exploration of the effects of direct regulation of land use/land cover and

---

[2]In the terminology of Turner *et al.* (1990) as contrasted to systemic change, which is manifested through the great integrating mechanisms of the atmosphere and fluid envelope, and which is relatively well modeled.

of measures to alter the driving forces behind land-use/land-cover change.

GLMs should be responsive to new opportunities and needs arising from both modeling advances (e.g., better climate projections) and data improvements (e.g., improved land-cover information from remote sensing). They must be undertaken with the sort of scientific rigor that permits iterative improvement, through testing, criticism, and refinement.

Model Form

GLM development could either build on past global social system models or start afresh. Either way, it should develop a realistic range of projections and acknowledge the errors introduced and propagated through (1) use of dubious data, (2) compounding of error in calculations, and (3) errors inevitably introduced by imperfectly specified models—problems that can jointly degrade precision and introduce inaccuracies to the point where models have little credibility.

Use of Past Social System Models

Various global social and agricultural systems models—e.g., the World Integrated Model (Mesarovic and Pestel, 1974), the Model of International Relations in Agriculture (MOIRA, Linneman *et al.*, 1979), the Latin American World Model (Herrera and Scholnik, 1976), and the International Institute for Applied Systems Analysis Food and Agriculture Project (IIASA/FAP) model (Fischer *et al.*, 1988)—have projected changes in agricultural land and its relation to driving forces such as population, technology, and prices. None of these models can be used as is to drive GXMs because none is grid based. The units of analysis are nations in MOIRA and IIASA/FAP and regional blocks of nations in most other models. Further modifications might be required because the models grossly aggregate or omit structural factors and driving forces that appear to be critical in land-use/cover dynamics (e.g., land tenure, climate change, accessibility) and do not explicitly include some major land uses and covers (e.g., forests, grazing lands).

Is it worth adapting existing models? The fact that past global models were controversial is insufficient cause for dismissal. Disciplinary parochialism and entrenched thinking often cause people to attack bold efforts that cross disciplinary lines. Trivial criticism is all the more likely because in business-as-usual scenarios, all longer-term global social system models predict regional collapse, induced by resource limitations, before the mid-21st century (Meadows *et al.*, 1982)—a result that people tend either to accept uncritically or to reject out of hand, depending on their predispositions.

More worrisome than controversy is the apparently general breakdown of scientific rigor in global social system modeling. Sympathetic critics (Liverman, 1986, 1989; Fox and Ruttan, 1983; Robinson, 1985) have found that various global social systems models are over- (or under-) sensitive to driving forces and do poorly in reproducing observed behavior. The amount of time and effort that modelers have devoted to sensitivity testing and validation is unimpressive, and the direction of model development has not been open to improvement through learning from experimental results.

The breakdown of scientific rigor appears to be related to the huge number of free parameters and simplifying assumptions that must be made to construct a global social system model. Structural uncertainty is a worse problem than data uncertainty. Modeler judgment is ultimately responsible for the formulations for a very large number of structural features, including trade, resource feedbacks, technological change, ecological stability under sustained exploitation, policies, markets, and resource substitution. The structural assumptions made are of paramount importance in determining model results (Barney, 1980; Cole, 1973; Maddox, 1972; Meadows and Robinson, 1985; Nordhaus, 1973). Reviewers do not agree on what simplifications are appropriate. Some (Robinson, 1980) have demonstrated that omission of resource feedbacks produces overly optimistic results. Others (e.g., Cole, 1973) have shown that even the most pessimistic of outcomes can be turned to sustained growth if one assumes sufficient increases in productivity (at no cost) via technological change. The sheer number of arbitrary assumptions made and the ideological implications of structural representation make balanced testing extremely difficult.

Past global models should be used where they are useful. Parts of past modeling efforts may be useful for GLM development. For example, data bases on land-support capacity developed for MOIRA and the IIASA/FAP model may be very useful for specifying possibilities and constraints for land use, and production or demand functions and elasticities from various models may be worth study. Structural understanding of more complex global models may also provide important insights on how to build simpler models. Review of more recent models, e.g., global demographic and forest sector models, is also worthwhile.

We think it is likely that new attempts at global modeling will suffer from the science problems of previous modeling attempts. To avoid some of the pitfalls, we recommend, at least initially, a simpler, more tractable, and more feasible approach.

Empirical Investigations

We conclude that relatively simple empirical studies are preferable to more complex global modeling. Many of the basic forces that drive land use/cover—including population size and geographical distribution, technological options, income and income distribution, and transport and other infrastructure—change on time scales of decades. Therefore a structure that registers the momentum of present trends and captures simple, strong relationships and the impact of developmental sequences is likely to succeed reasonably well as a structure for projections for periods of decades.

For example, as shown in Figure 1, Hayami and Ruttan (1985) have found highly structured relationships in worldwide patterns of agricultural intensification

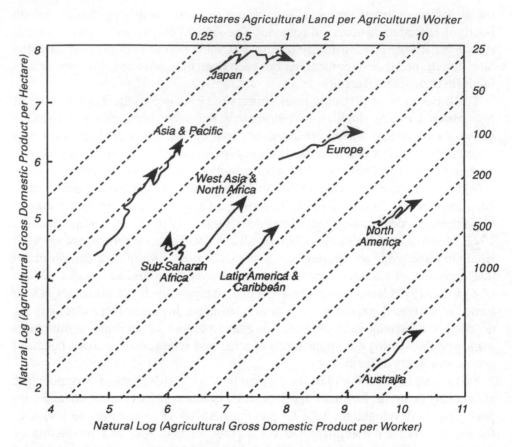

Figure 1. Land and labor productivity changes for different regions over 1961–85. Agricultural gross domestic product deflated from local currencies to 1980 values and then converted to U.S. dollars (redrawn from Craig *et al.*, 1991).

by studying the relationship between output/ha and output/worker. Likewise, many of the technological relationships elucidated by Grübler (this volume) appear to offer a basis for reasoned quantitative projections for a decade or two into the future, and demographic projections are widely used in social planning to understand probable longer-term changes in social needs.

The credibility of projections will decline past 10 years. Nonetheless, we recommend, for two reasons, that all projections be made for 50 years. First, extended projections may reveal something about system structure and behavior. Absurd results (e.g., 200% of the land in a given grid cell converted to agriculture accompanied by expansion of negative forests) may suggest feedbacks that will set and redirect the system as it moves far outside the range of observed behavior. Second, extended projections are required by GXMs, and projections produced by an explicit, geographically comprehensive, and consistent method, despite their deficiencies, are preferable to *ad hoc* predictions based on expert judgment.

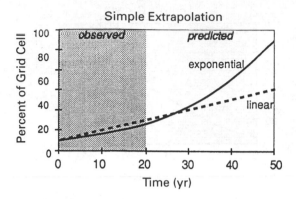

Figure 2. Persistence forecasting in its simplest form.

Projection Methods

The simplest projection technique, as illustrated in Figure 2, is to assume persistence of present rates of change, i.e., to extrapolate observed rates of land conversion into the future. This implicitly assumes that the forces that dominated the calibration period will persist throughout the projection period. It can be done assuming either constant increments or constant proportions (linear growth and exponential growth, respectively). Because the total land area must remain constant, growth of one land use must be balanced by decline of another. In most cases, we expect to see growth of agricultural land at the expense of forest or grassland.

Simple extrapolation has the problem that it is unconstrained by physical reality; it can yield predictions of more than 100%, or less than zero, of a grid cell being devoted to a given land use or cover. Especially for 50-year forecasts, it will be useful to modify extrapolations by constraining the amount of land conversion based on soils, topography, and other information, to the fraction of the 200-km grid cell that can potentially be devoted to that use. For any geographically bounded area, be it defined politically, economically, institutionally, or in some other way, the task becomes one of forecasting the proportions of area that are or could be occupied by the various land covers.

As shown in Figures 3 and 4, various functional forms can be used to define the way a system approaches a limit, including logarithmic, logistic (sigmoidal) growth, and Michaelis-Menten type relationships (see Figure 4). As shown in Figure 5, the land-constrained representation might be further elaborated to take account of the potential to expand the area of cultivable land through investment (e.g., in irrigation) or to reduce the amount of cultivable land by nonsustainable use (e.g., salinization, desertification).

Empirical Models

Models based on the assumption of persistence provide no information about the underlying driving forces. As a move toward causal explanation and perhaps bet-

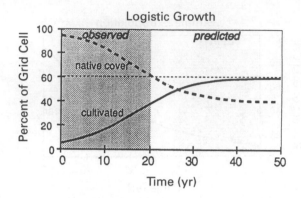

Figure 3. Forecast with constrained growth assuming logistic growth form (biological analog).

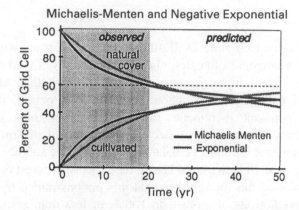

Figure 4. Forecast with constrained growth assuming Michaelis-Menten relationships (analogy to catalytic reaction with linear increase of substrate).

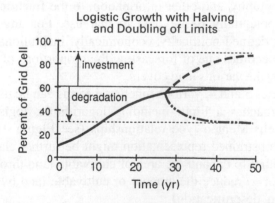

Figure 5. Extension of (solid line) logistic growth model to show results of (dashed line) augmentation of cultivable land area through investment and (dot-dashed line) reduction of cultivable land area through land degradation. The former is assumed to increase the cultivation limit from 60% to 90% of the grid cell; the latter, to decrease it from 60% to 30%.

ter prediction, we recommend the further development of empirical models, which relate postulated driving variables to trends in land use and land cover. Specifically, we recommend the following sequence of activities:

1. Seek empirical relationships in historical data between land-use change and variables that are postulated to drive land use, such as population, income, level of technology employed, and prices for agricultural inputs and products.
2. Project these driving variables.
3. Predict land use based on predicted trends for drivers and the empirically fitted relationships to land use/cover.

Initial driving forces to be considered are listed below. Technology is difficult to quantify, and we recommend that productivity and/or energy use be used as a surrogate.

- Population: births, deaths, migration
- Income: growth, distribution
- Technology: augmentation, diffusion
- Prices: land, energy, labor, land substitutes.

As with projections based on the assumption of persistence, empirical relationships can be modified to take account of resource constraints. For example, instead of relating absolute changes in land use to postulated driving variables, one can relate relative changes, scaled to the potential land in a given use (e.g., change in fraction of arable land in cultivation), to driving variables. The resulting projections might take a form like that shown in Figure 6.

Incorporating Additional Structure

The above does not exhaust the possibilities for simple structural algorithms that may help explain patterns of changing land use and land cover. Additional factors and structural relationships that may make empirical models more realistic and increase their predictive power should be considered, including:

- Limits to expansion posed by large landholders operating at low use intensity (e.g., parks and reserves, timber companies, and *latifundia*), which prevent the expansion of small holders
- Exclusion or limitation of settlement by disease
- Overshooting of limits and expansion of cultivation into land unsuited for cultivation
- Migration between cells when available land is exhausted, and the impacts of road construction on migration
- Complementarity and substitutability of capital and labor as inputs to production (e.g., the Cobb-Douglas relationship)
- Balance between the cost of clearing, as affected by vegetation characteristics, and the value of cleared land, as affected by soils, access to markets, and other factors.

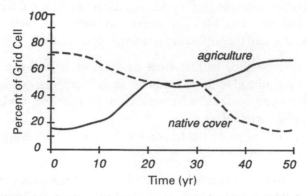

Figure 6. Hypothetical projection of fraction cover as predicted by empirical projection. Cover predicted as a function of population, income, productivity, and prices over time.

Data Requirements

Land Use and Land Cover

Both persistence-based and empirically formulated models require data on land-use and land-cover change. Ideally, this would include data on all the following processes:

- Deforestation/forest regrowth
- Pasture expansion/pasture contraction
- Cropland expansion/cropland contraction
- Growth of areas devoted to settlement
- Land degradation
- Wetlands conversion/expansion of irrigation.

In practice, we cannot expect consistently to find the level or precision or quantity of data required for fitting relationships to all the above trends. Concentration on the most important processes, such as conversion of land to agriculture, should be an effective way of dealing with data limitations. Regional considerations may call for consideration of additional land uses. For example, pasture formation has been an important land-use/land-cover change in much of Latin America over the last two decades. Logging has been important in initiating agricultural colonization in South and Southeast Asia.

We recommend opportunistic use of any data on land use/cover available including national census data, remote sensing observations, U.N. Food and Agriculture Organization (FAO) data, and historical reconstructions. In most cases, these data will use not grid-based units but the irregular shapes of political boundaries. Data should be forced into a grid-based format through extrapolation and interpolation within the context of a geographical information system.

Nations may be smaller or larger than the usual grid cell dimension and may be contained in one cell or spill over to several. For many of the larger nations, regional and local data are available, e.g., census records. Thus grid cells will not all have data of equal precision.

Constraints to Land Expansion

The primary physical constraints to the expansion of agricultural and other land-use practices are from soil resources, topography, and climate. Social and biological factors may also be limiting, but physical limits are better mapped and easier to specify. Thus they should be emphasized initially. For each cell, data on soils, topography, climate, and water resources should be used to predict the potential area available for agricultural expansion. The FAO Agro-Ecological Zones Project (FAO, 1978–81) and FAO/UNFPA (1980) provide a good basis for delineating agricultural potential for Africa, Latin America, and Southeast and Southwest Asia.

Social Drivers

We anticipate that data are available for most of the postulated driving variables for at least the last 20 years for most nations, and for 50 years or more for some. Some data sources are listed in Table 2 (see also Working Group A, this volume).

Data Management

GLM development is one of many badly needed research activities that link social systems to physical and biological systems by projecting subnational social system data onto a physical map base. The development of linkages is impeded, for both GLM development and other research, by a general scarcity of appropriate global data sets. Among the data needs are:

- Social system data sets that extend below the national level and extend backwards in time for two or more decades
- Political boundary files for social system data below the national level, adjusted to account for changing boundaries and reconciled with social system data sets
- Global data bases on natural resources, e.g., soils, climate, topography, and water resources
- Biological data, as required to convert physical resource data into agricultural production potential
- Global land-use and land-cover data.

Development and maintenance of the required data sets is a massive job. It is strongly desirable that the task be done so as to yield an integrated data base that is publicly available and organized in a fashion that is useful for purposes beyond GLM development. Specifically, we recommend that data base development be done through an institution, such as the Consortium for International Earth Science Information Network (CIESIN), that is charged with integrating and facil-

Table 2: Data sources for driving variables (in approximate order of declining priority)

Data	Sources (in approximate order of priority)
Birth, death, urban migration	U.N. Population Division, U.S. Department of Agriculture, national censuses
Income distribution and growth	World Bank Indicators
Productivity growth	FAO, U.S. Department of Agriculture, national agricultural returns
Land prices	National data
Energy prices	U.N. energy statistics
Labor prices	U.N. International Labor Organization
Fertilizer prices	FAO, U.S. Department of Agriculture, International Fertilizer Institute

itating the use of information from government-wide earth monitoring systems and enabling the wide distribution of objective information appropriate to the needs of diverse audiences.

Testing

Validity Testing

Ideally, GLM structure should be applicable globally across both space and time. Validity testing reveals the extent to which a postulated model departs from this ideal. For example, the various functions fitted for a grid cell (or group of grid cells) A can be used to predict the historical pattern of grid cell(s) B, where A and B are similar and data are available for both. The result can be tested against actual historical observations for B. To test the temporal stability of fitted relations, for example, functions fitted for the last decade can be used to predict the trend for the preceding decade or vice versa and the result compared to historical records.

Analysis of Statistical Residuals

Failure to predict may be as informative as prediction itself. If institutional, cultural, or other factors that were omitted from model formulation have a strong impact on land-use evolution, they should, for specific regions, show up as statistical structure in the unexplained variance of validation tests (statistical residuals). It is vital to both the science and the utility of the modeling effort that residuals be examined in detail, taking systematic account of expert opinion. How this serves the modeling effort depends on the nature of the omitted variable. If it is predictable, the exercise suggests an extension of model structure that will increase predictive power. If it is something such as war or abrupt policy change that is inherently difficult to predict, the exercise increases confidence by attributing the model's failures to known causes, and at the same time provides better understanding of the model's inherent limitations.

Sensitivity Testing

In order to understand the behavior of complex, nonlinear models, modelers often treat the model as a black box and feed parameter variations into it, first singly, and then in combinations. The functional forms proposed above are sufficiently simple and well known that their sensitivities can be predicted using formal mathematics. No black box is needed, and sensitivity testing in the usual sense of the word is unnecessary. Thoughtful translation of mathematics into applied terms, however, is necessary for understanding and explaining model results.

Dimensional Matching: GXMs and GLMs

To be useful, global land models must be scaled compatibly with GXMs and formulated in a way that can be supported by the available data. Satellite remote sensing is ultimately preferred for *monitoring* land use/cover. We conclude, however, that GLM development should not wait for comprehensive land-use/cover histories derived from remotely sensed data, but should commence working with the available social-system-based land-use/cover data, supplemented, where possible, by remotely sensed data. There are two reasons for this decision. First, it may be many years before satellite-based land-use/cover estimates are available and adequately calibrated, and could be at least a decade before useful historical records have been constructed from remotely sensed data. Second, and more importantly, social and economic (e.g., census) data provide information on the cause as well as the effect of long-term land-use patterns. Thus while such data may be less precise than remotely sensed data, they provide a better basis for understanding causal relationships and developing predictive models.

Temporal Scale

Except locally, land use and land cover seldom change at rates of more than about 5% per year. Changes at less than decadal scale will be difficult to discern within the margin of error in the present GXM context. Thus while better understanding of the baseline condition is important, projections should focus, at minimum, on the decadal time scale.

Climate change studies require on the order of a 30-year time horizon before the climate signal stands out over the noise of natural system variability. Fifty-year time horizons have commonly been adopted in studies of doubled CO_2 or its equivalent in mixed greenhouse gases. Thus time scales of 0 to 10 and 0 to 50 years into the future are appropriate for initial work on projection of future land use/cover. The historical record should be studied as far back in time as data availability permits.

Spatial Scale

We recommend an initial grid scale on the order of 200×200 km for land-use projections. Although global biogeochemical models could use much higher resolution data, it is unlikely that GCMs, global chemical transport models (GCTMs), or global vegetation models will achieve higher resolution within the time required to construct a pilot GLM. Because higher (horizontal) spatial resolution must inevitably be accompanied by a shorter time step, even massively parallel computer systems will be hard pressed to push GCMs to beyond 100- or 50-km resolution over the next decades (Verstraete, 1989). Social systems data, in most cases, seem to be available in sufficient detail to support 200-km resolution, but it is not clear that present global data sets will support much higher resolution.

We conclude that proportional representation of land use and land cover within the grid cell—in other words, study of subgrid-scale variability—is achievable and important in the present context and should be actively pursued. GLM development should look toward more highly resolved surface resolution, and experimentation with questions of resolution and scale is appropriate even at early stages of development. Higher resolution can now be coerced from GCMs through nested models and/or statistical inference. Within a few decades, GXM grid cell resolution on the order of 100 to 50 km is anticipated.

Tuning GLM Output to Specific GXM Needs

A generic GLM cannot directly provide the detailed input needs of all GXMs, and in many cases, GLM projections will need to be transformed to produce appropriate model inputs. This we illustrate with reference to GCTMs, which are especially demanding in their data inputs.

Table 3 itemizes the data needs and priorities for GCTMs. As shown, GCTMs require information on land-use-driven sinks for multiple gas and aerosol species, as affected by a variety of land-use and land-cover processes. Moreover, the temporal and spatial resolution at which data are required varies depending on the species. Data on short-lived species (aerosols, carbon monoxide, oxides of nitrogen) are required on a weekly to monthly basis. Seasonal or annual flux estimates are adequate for longer-lived species (e.g., methane, nitrous oxide). Because sources for methane (CH_4) are strongly concentrated in intense source regions, such as rice paddies and anoxic wetlands, 200-km resolution is inadequate for evaluating the CH_4 source.

Present understanding of source strength is often insufficient to support efficient deduction of anthropogenic sources from areas of land in given land-use activities. It is, for example, likely that a very large fraction of nitric oxide, nitrous oxide, or CH_4

Table 3: Data needs (200 × 200 km) associated with land use/land cover for a GCTM on a scale of 1 (least) to 4 (greatest)

Variable	Impact	Impact Severity	Data Adequacy	Data Need[1]
Broad land-cover type	nonmethane hydrocarbons, albedo, roughness, water movement, temperature	4	2	8
Rice	CH_4, N_2O, NO_x	3	2	6
Wetlands	CH_4	3	2	6
Termites	CH_4	1	3	3
Ruminants	CH_4	3	2	6
Agricultural systems	CH_4, N_2O, NO_x	3	3	9
Forest harvest/regrowth	CO_2, nutrients	3	3	9
Fuelwood extraction	CO_2, nutrients	2	3	6
Harvest fodder	CO_2, nutrients	1	3	3
Biomass burning	CO_2, CH_4, NO_x, nutrients	3	3	9

[1]Computed as the product of impact severity and data adequacy.
Needs determined by a survey of meeting participants.

emissions from agriculture is associated with particular cultivation practices, and thus that land-use information is required, not merely on the crop, but on the type of cultivation. Unfortunately, the basic processes involved are complex and are only beginning to be understood (cf. Slemr and Seiler, 1991), and it may be quite some time before an appropriate typology is developed. Further field measurements are needed even to specify which types of cultivation are critical, much less what weighting factors are required to account for between-type variation in source strengths.

Tuning GLMs to specific GXM needs is complex and needs further study. We recommend both a systemic analysis of the problem by an interdisciplinary group, and continuous, structured dialog between land-use modelers and GXM users of land-use projections.

Climate Feedback in GLMs

Climate affects land use and land cover more strongly than land use/cover affects climate. Conceptually, climate feedbacks can be accommodated in the GLM framework, as shown in Figure 7. Can this be done in practice?

Climate can greatly affect land use. Climate impacts are manifested both through the effects of mean precipitation, temperature, etc., and by the frequency of extreme events such as frosts, droughts, etc. Mean climate determines what can generally be grown; extreme events shape the risks associated with a particular land-use technology and the resulting chances of failure. The two are linked: small changes in cli-

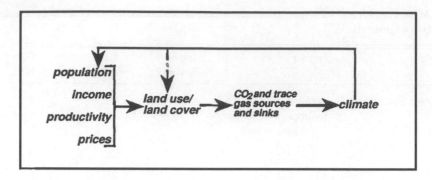

Figure 7. Climate feedbacks to a GLM.

matic means result in significant changes in the frequency of extreme events. Climate impact analysis can hardly function on temperature alone, and at present the field is seriously impeded by the limited skill of GCMs in reproducing present mean patterns of precipitation (Grotch, 1988; Kalkstein, 1991) and observed frequencies of extreme events (EPA, 1989). This situation is expected to improve over the coming decades. With good climate projections, climate feedbacks on land use/cover could begin to be addressed by quantifying the effects of climate change on:

- Land suitability for various productive uses, including changed possibilities for extension of land use into new geographical regions and changes that make land unsuitable for its historical use
- Land productivity and stability of land-use systems (including impacts manifested through prices of products and inputs), as affecting both intensification and extensification of land use.

Assessment must also consider the speed and effectiveness with which land users and social institutions adjust to climate change.

In part, these questions can be addressed within the proposed GLM framework by altering the constraints to land expansion. Satisfactory resolution of social adaptation and market ramifications requires a framework of analysis. In the future, that deficiency could be resolved, either by extending the GLM framework or by using GLM projections as an input to climate impact analyses.

Closing Considerations

Figure 8 summarizes the flow of work suggested in the above sections. In closing we bring up some overriding considerations about the manner in which work is done.

The empirically based suggestions outlined above represent a crude, brute-force approach, designed to initiate urgently needed research. They are unsatisfactory in many ways. They cannot take account of the rich variety and detail of institutional, cultural, and political factors that affect land use and land cover (see Sanderson and Rockwell, this volume). They are very likely to miss 'surprises' and turning

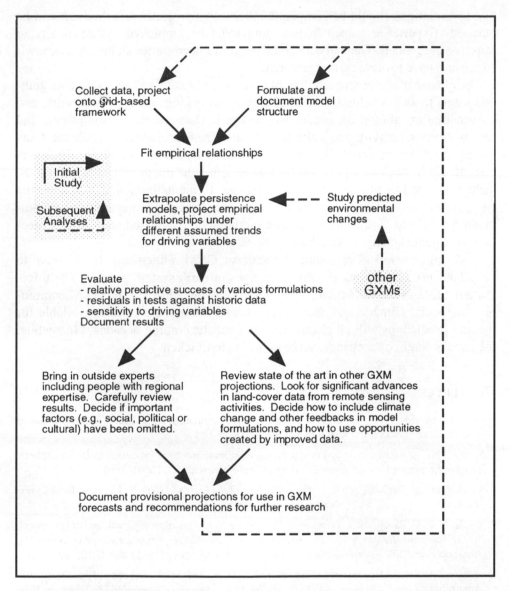

Figure 8. Work flow for GLM development.

points. If projected growth rates are sufficient to push cultivation to the limits, results will be dominated by the limits that are specified. Nonetheless, we anticipate that the construction of a GLM will move the science forward. The simplicity of the proposed approach should yield a product whose flaws are transparent and easy to interpret and begin correcting.

For simplicity to be an advantage rather than a liability, it is important that a projection exercise be done self-critically and experimentally, with as much emphasis on evaluation and interpretation as on projection *per se*. Sensitivity and validity testing should be undertaken along with projections; the results of alternative formulations should be compared. Alternative projection methods should be pursued. Experience gained in modeling must be employed systematically to improve the models. Gaps in data and structural uncertainties should be acknowledged and used to drive further research.

The classic modeler's pitfall, that of lavishing time and resources on data gathering and model development and skimping on testing, analysis of results, and documentation, should be assiduously avoided. Data sources, assumptions, and results from sensitivity and validity tests and from projections should be thoroughly documented and made available to GXM modelers and others. The exercise should be seen as a pilot project whose aim is as much to develop methods, techniques, and institutions that can be applied to maintaining and managing the earth system in the 21st century as it is to produce land-use projections. We recommend that global land-cover projection methods be reassessed and GLM projections be updated on a five-year basis.

Updating, as well as providing for iterative GLM refinement, should serve to keep land-use projections abreast of changes in earth system science. The information needs of the GXM modeling community, the ability of the GXM community to predict climatic and other forcings, and the amount of data available for making predictions will all change rapidly over the coming decades. Meanwhile, the pace of land-cover change can be expected to quicken.

References

Barney, G. O. 1980. *The Global 2000 Report to the President,* Vol. 2, Sec. IV, *Comparison of Results*. Seven Locks Press, Cabin John, Maryland, 659–681.

Cess, R. 1990. Intercomparison and interpretation of climate feedback processes in 19 atmospheric general circulation models. *Journal of Geophysical Research 95D,* 16601–16615.

Cole, H. S. D., ed. 1973. *Models of Doom: A Critique of the Limits to Growth*. Universe Books, New York.

Craig, B. J., P. G. Pardy, and J. Roseboom. 1991. Patterns of agricultural growth and development. In *Agricultural Research Policy: International Quantitative Perspectives* (P.G. Pardy, J. Roseboom, and J.R. Anderson, eds.), Cambridge University Press, Cambridge, U.K., 183.

Dickinson, R. E., A. Henderson-Sellers, P. J. Kennedy, and M. F. Wilson. 1986. *Biosphere-Atmosphere Transfer Scheme (BATS) for the NCAR Community Climate Model*. Technical Note TN-275, National Center for Atmospheric Research, Boulder, Colorado.

Emanuel, W. R., G. C. Killough, W. M. Post, and H. H. Shugart. 1984. Modeling terrestrial ecosystems in the global carbon cycle with shifts in carbon storage capacity by land-use change. *Ecology 65,* 970–983.

Emanuel, W. R., H. H. Shugart, and M. P. Stevenson. 1985. Climate change and the broad-scale distribution of terrestrial ecosystem complexes. *Climatic Change 7,* 29–43.

EPA (U.S. Environmental Protection Agency). 1989. Climate variability. In *The Potential Effects of Global Climate Change on the United States,* Office of Policy Planning and Evaluation, Climate Change Division, U.S. Environmental Protection Agency, Washington, D.C., 29–55.

Esser, G. 1987. Sensitivity of global carbon pools and fluxes to human and potential climatic impacts. *Tellus 39B,* 245–260.

FAO (U.N. Food and Agriculture Organization). 1978–1981. *Report on the Agro-Ecological Zones Project*, Vol. 1 (1978), *Methodology and Results for Africa*, Vol. 2 (1978), *Results for Southwest Asia*, Vol. 3 (1981), *Methodology and Results for South and Central America*, Vol. 4 (1980), *Results for Southeast Asia*. World Soil Resources Reports 48(1–4), FAO, Rome, Italy.

FAO/UNFPA (U.N. Food and Agriculture Organization and U.N. Fund for Population Activities). 1980. *Land Resources for Populations of the Future*. Report on the Second FAO/UNFPA Consultation, FAO, Rome, Italy.

Fischer, G., *et al.* 1988. *Linked National Models: A Tool for International Food Policy Analysis*. Kluwer Academic Publishers, Dordrecht, The Netherlands.

Fox, G., and V. W. Ruttan. 1983. A guide to LDC food balance projections. *European Review of Agricultural Economics 10*, 325–356.

Goudriaan, J., and P. Kctncr. 1984. A simulation study for the global carbon cycle including man's impact on the biosphere. *Climatic Change 6*, 167–192.

Grotch, S. 1988. *Regional Intercomparison of GCM Predictions and Historical Climate Data*. U.S. Department of Energy, Washington, D.C.

Guetter, P. J., and J. E. Kutzbach. 1990. A modified Köppen classification applied to model simulations. *Climatic Change 16*, 193–215.

Hayami, Y., and V. W. Ruttan. 1985. *Agricultural Development: An International Perspective*. Johns Hopkins University Press, Baltimore, Maryland, 121, 131.

Herrera, A. O., and H. D. Scholnik. 1976. *Catastrophe or New Society? A Latin American World Model*. International Development Research Center, Ottawa, Canada.

Houghton, R. A., J. E. Hobbie, J. M. Melillo, B. Moore, B. J. Peterson, G. R. Shaver, and G. M. Woodwell. 1983. Changes in the carbon content of the terrestrial biota and soils between 1860 and 1980: A net release of CO_2 to the atmosphere. *Ecological Monographs 53*, 235–262.

Kalkstein, L., ed. 1991. *Global Comparisons of Selected GCM Control Runs and Observed Climate Data*. Office of Policy Planning and Evaluation of Climate Change Division, U.S. Environmental Protection Agency, Washington, D.C.

Lilieveld, J., and P.J. Crutzen. 1990. Influence of cloud photochemical processes on tropospheric ozone. *Nature 343*, 227–233.

Linneman, Y., J. de Hoogh, M.A. Keyser, and H.D.J. van Heemst. 1979. *MOIRA: Model of International Relations in Agriculture*. North Holland, Amsterdam, The Netherlands.

Liverman, D. M. 1986. The sensitivity of global food systems to climatic change. *Journal of Climatology 6*, 355–373.

Liverman, D. M. 1989. Evaluating global models. *Journal of Environmental Management 29*, 215–235.

Maddox, J. 1972. *The Doomsday Syndrome*. McGraw-Hill, New York.

Meadows, D. H., and J. M. Robinson. 1985. *The Electronic Oracle: Computer Models and Social Decisions*. John Wiley and Sons, Chichester, U.K.

Meadows, D. H., J. Richardson, and G. Bruckmann. 1982. *Groping in the Dark: The First Decade of Global Modeling*. John Wiley and Sons, New York.

Mesarovic, M., and E. Pestel. 1974. *Mankind at the Turning Point*. E.P. Dutton, New York.

Nordhaus, W. D. 1973. World dynamics: Measurement without data. *Economic Journal 83*, 1156–1183.

Olson, J. S. 1981. Carbon balance in relation to fire regimes. In *Fire Regimes and Ecosystem Properties* (H.A. Mooney, N. L. Christensen, J. E. Lotan, and W. A. Reiners, eds.), General

Technical Report WP-26, U.S. Department of Agriculture, Alexandria, Virginia, 327–378.

Penner, J. E., C. S. Atherton, J. Dignon, S. J. Ghan, and J. J. Walton. 1991. Tropospheric nitrogen: A three-dimensional study of sources, distributions, and deposition. *Journal of Geophysical Research 96D*, 959–990.

Pollard, D., and S. L. Thompson. *Description of a Land-Surface Transfer Model (LSX) as Part of a Global Climate Model*. National Center for Atmospheric Research, Boulder, Colorado, in preparation.

Prentice, K. C. 1990. Bioclimatic distribution of vegetation for general circulation models. *Journal of Geophysical Research 95D*, 11811–11830.

Robinson, J. M. 1980. The comparisons. Chapter 31 in *The Global 2000 Report to the President*, Vol. 2, *The Technical Report* (G. O. Barney, ed.), U.S. Government Printing Office, Washington, D.C., 661–681.

Robinson, J. M. 1985. Global modeling and simulations. In *Climate Impact Assessment* (R. W. Kates, J. H. Ausubel, and M. Berberian, eds.), SCOPE 32, John Wiley and Sons, New York, 469–492.

Slemr, F., and W. Seiler. 1991. Field study of environmental variables controlling the NO emissions from soil and the NO compensation point. *Journal of Geophysical Research 96D*, 13017–13031.

Spivakovsky, R., R. Yevich, J. A. Logan, S. C. Wofsy, and M. B. McElroy. 1990. Tropospheric OH in a three dimensional chemical tracer model: An assessment based on observations of CH_3CCl_3. *Journal of Geophysical Research 95D*, 18441–18471.

Taylor, J. A., G. P. Brasseur, P. R. Zimmerman, and R. J. Cicerone. 1991. A study of the sources and sinks of methane and methyl chloroform using a global three-dimensional Lagrangian tropospheric tracer transport model. *Journal of Geophysical Research 96D*, 3013–3044.

Turner, B. L., II, R. E. Kasperson, W. B. Meyer, K. M. Dow, D. Golding, R. C. Mitchell, and S.J. Ratick. 1990. Two types of global environmental change. *Global Environmental Change*, 14–22.

Verstraete, M. 1989. Land surface processes in climate models: Status and prospects. In *Climate and Geo-Sciences* (A. Berger, S. Schneider, and J.C. Duplessy, eds.), Kluwer Academic Publishers, Dordrecht, The Netherlands, 321–340.

Xue, Y., P. J. Sellers, J. L. Kinter, and J. Shukla. 1991. A simplified model for global climate studies. *Journal of Climate 4*, 345–364.

III

CHANGES IN LAND USE AND LAND COVER

Introduction

These chapters examine global and world-regional changes in three broad land types: forest/woodland, grassland, and settlement. The first two are clearly of global importance by any measure. The third, settlement, represents the most intensive form of land use and one that, though still quite small in the area it occupies, is (1) expanding rapidly; (2) possibly expanding at the expense of prime lands valuable for other uses, such as cultivation and wetlands; (3) in the extreme form of megacities, the home for an ever-increasing share of the world's population; and (4) the source of large impacts on land cover locally, throughout the urban hinterlands, and at considerable distances. Each chapter defines the land type in question and reviews what is known about its modern trajectories of change, globally and world-regionally.

Taken literally, two of these terms—'forest/woodland' and 'grassland'—denote forms of land cover, whereas 'settlement' refers to land use. Use and cover, of course, are not equivalent, and there is room for some overlap. For example, urban settlements, as Douglas points out, include not only such land covers as buildings and pavement, but large areas of grass and tree cover. In practice, though, these terms tend at present to function as hybrids that cluster certain kinds of land cover with certain human uses. Such is necessarily the case because key sources of global data do not clearly distinguish between cover and use. An author studying one must often settle for data that refer, strictly speaking, to the other. It is an unsatisfactory situation, and separate data collections on use and cover are needed. Also needed are improved classifications and definitions of use and cover types to ensure the comparability of data collected in different parts of the world.

Indeed, the poor quality of much of the data on land uses and land covers is one of the lessons that emerges most clearly from these papers. The data are often not collected according to categories and definitions consistent enough to make different sets comparable, and even concerning what they ostensibly record they are often of dubious reliability. The national-level figures published annually in the U.N. Food and Agriculture Organization's *Production Yearbooks* are indispensable for many global-level studies because of the lack of other sources, but many questions have been raised about their accuracy. The bases and meaningfulness of U.N. Environment Program estimates on the global extent of desertification, to cite another example, have been widely challenged in recent years. Another problem increasingly apparent is that many of the standard bodies of data afford information only on conversion between categories and not on important forms of modification. Such problems cast some doubt on even broad global and regional generalizations and severely hamper efforts at more fine-grained analyses—of the land transformations themselves, of their further environmental consequences, and of their human driving forces.

Some things can nonetheless be said with confidence, and these chapters detail them as best they can with existing data. At the long-term global scale, certain trends are evident: the advance of more intensive land uses at the expense of less intensive ones, the rapid expansion of cultivated land and settlement, and the shrinkage of the forest. All of the categories are also being significantly modified by more intensive use. Spatial variation in the rates and even the directions of change, however, remains a marked feature. Trajectories of conversion and modification of forest and grasslands differ considerably between the developed and developing worlds and among various areas within each. Global aggregations of change conceal much regional diversity.

5

Forests and Tree Cover

Michael Williams

Felling trees for the combined objectives of obtaining wood for construction, shelter, and toolmaking; of providing fuel to keep warm, to cook food, and to smelt metals; and above all, of creating land for growing food, has culminated in one of the main processes whereby humankind has modified the world's surface cover of vegetation.

Despite the importance and magnitude of this process, the distribution, quantitative extent, and rate of change in the area of the forest, through both deforestation and reforestation, have been and remain subjects of great debate and uncertainty.

The point is illustrated well by the debate on the basic issue of what constitutes 'forest.' Most writers make a distinction between closed and open forest that corresponds to our intuitive experience of forest environments. The U.N. Food and Agriculture Organization (FAO) defines closed forest as 'land where trees cover a high proportion of the ground where grass does not form a continuous layer on the floor' and open forest (sometimes called woodland) as 'forests in which trees are interspersed with grazing lands.' Alternatively, from time to time, various percentages of crown cover have been suggested as diagnostic of forest. Thus, in a recent publication by FAO (1990), *forest* is an ecological system with a minimum crown coverage of the land surface of 10%, and *wooded land* is a part of *nonforested* land. Neither definition is the same as a legal definition, for example, where an area is proclaimed to be forest under a national forest act or ordinance.

These and similar issues are reviewed in this paper, and judgments are made about data quality and reliability. Evidence and method are constantly being refined, and there is support for the view that estimates are firming up. Nonetheless, there is still room for greater accuracy and refinement than now exists, based on temporal and regional analysis.

Distribution and Extent

The Distribution of Forests

There have been many attempts to map the distribution of the world's forests and woodlands. Of all of these, perhaps two should command attention. First, there is the massive *Weltforstatlas* (Heske, 1971), which shows the distribution of forests by country and continent, as well as their breakdown into botanical species. Detailed as it is, however, it does not serve current purposes well as it cannot be used to get any overall idea of different types of forest nor accurate *statistical* measures of extent.

Perhaps more useful, because of its simplicity, is the world map published by FAO in 1976 (Lanly, 1976). It is based on the climatic characteristics of six forest types, and it has been widely adopted as a basis for description and calculations. A major deficiency of this map (and indeed of many others) is the distortion at the polar extreme of the Northern Hemisphere, which gives an exaggerated prominence to the coniferous forest belt. In actual fact it occupies about one-third of the total of about 4.1×10^6 km^2 of the world's forests, the remaining two-thirds being broadleaf forests, divided fairly evenly between temperate mixed and tropical varieties.

Forests and World Ecosystems

Forests, of course, are only one of many ecosystems that cover the world, all of which play a role in the radiation balance of the earth and in various biogeochemical cycles related to climatic change. Consequently, climate modelers have been in the forefront of reconstructing past and present ecosystems. A range of sources has been analyzed digitally. This is an advance on the previous practice of aggregating small-scale vegetation map data, which have inherent problems of diverse aims, subjective classifications, and boundary delimitation, all of which are compounded when a variety of maps are aggregated in order to create the world picture. In any case the resultant qualitative map cannot be satisfactorily incorporated into quantitative modeling studies using digitized data.

Thus, during the last decade, there have been at least four attempts to overcome problems inherent in qualitative maps, and these are of interest to those in search of measures of deforestation. The first of these were two attempts designed to assist in surface albedo studies (CLIMAP, 1981; Hummel and Reck, 1979). Their content reflects their purpose. They are general descriptions of land cover (e.g., tropical woodland/grassland, deciduous forest, arable land) and do not conform to generally accepted classifications of vegetation such as those of Ellenberg and Mueller-Dombois (1967), Fosberg (1961), and the U.N. Educational, Scientific, and Cultural Organization (UNESCO, 1973). Neither has a scale as both are basically files, so that grid size acts as the best indication of spatial precision (e.g., CLIMAP has $2° \times 2°$ cells). In order to extend the utility of these maps and to help in the elucidation of other global issues (e.g., carbon density or biomass for climatic modeling studies), the Oak Ridge National Laboratory modified Hummel and Reck (Olson and Watts, 1983). They produced a map on the scale of 1:30 million by reorganizing data into a $0.5° \times 0.5°$ grid and designating 12 general vegetation types, these being further annotated by climatic (e.g., tropical, subtropical, boreal) and elevational (e.g., lowland, montane) characteristics, resulting in 43 vegetation types.

Subsequently, Matthews (1983) attempted to widen the utility of vegetation mapping by making it applicable to a variety of climatically related research, such as primary productivity, surface roughness, and ground hydrology, in addition to albedo and biomass. She constructed two separate data bases, one of natural vegetation, the other of current land use. In the vegetation data base an attempt was made to reconstruct the preagricultural vegetation, and the land use data base was used to calculate the amount of vegetation remaining. Both data bases were constructed on $1° \times 1°$ cells so that they could be quantified. The vegetation classification used was the UNESCO hierarchical system based first on life form, and then subdivided down into density, seasonality (evergreen or deciduous), altitude, climate, and vegetation structure. It was compiled from over 40 atlases, with all the problems inherent in that. Of the potential 225 vegetation types, 178 were used, together with 119 land-use types.

More recently a more accurate view of the distribution of the forest has come from the map of world land cover compiled by the Institute of Geography of the USSR Academy of Sciences and Moscow State University from sources independent of the suspect FAO data (Vanvariova, 1986). The categories include arable land, subdivided into constant and periodic cultivation; plantations, subdivided into irrigated and dryland; pasture and rangeland, subdivided into improved and unimproved; and forest, all categories being subdivided further as upland or lowland. At a scale of 1:15 million there is a fair degree of resolution, certainly sufficient for digitization.

The utility of these digitized data files will depend on the purpose to which they are put. The methodology is in place; agreement on the identification and classification of different types of land cover and changed land use still needs resolution.

The Extent of the Forest

The *extent* of the forest is largely the outcome of the accurate delineation of its *distribution*. There has been little agreement about the amount of the 'contemporary' forest (Mather, 1987). Between 1923 and 1985 there have been at least 26 calculations of closed forest land, and these are arranged chronologically in Figure 1. They range from 60.5×10^6 km^2 to 23.9×10^6 km^2. They are randomly distributed around a mean of 41.27×10^6 km^2, and show no discernible trend over time. The estimates are compiled from different sources, utilizing different definitions, and they certainly cannot, therefore, be used as an indicator of current deforestation rates (Allen and Barnes, 1985; Sedjo and Clawson, 1984). However, there is a general firming-up of the estimates during the 1980s between about 47 and 37 $\times$ 10^6 km^2, still a wide variation.

The only long time series data for forest area come from the FAO returns of land cover (not land use) for the 35 years between 1950 and 1985. However, the FAO figures from the Third and erstwhile Second Worlds must be used with caution.

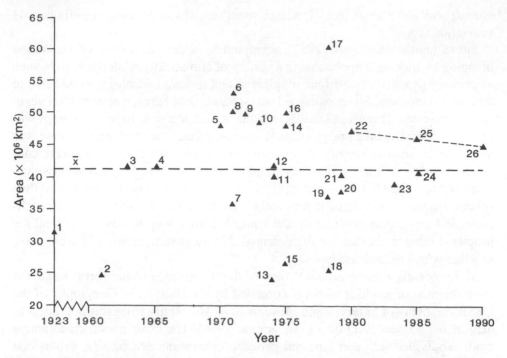

Figure 1. Estimates of forest extent, 1923–90. The figures in the diagram refer to the following sources: (1) Zon and Sparhawk, 1923; (2) Weck and Wiebecke, 1961; (3) FAO, 1963; (4) FAO, 1966; (5) Olson, 1970; (6) Bazilevich *et al.*, 1971; (7) Brüning, 1971; (8) Whittaker and Woodwell, 1971; (9) Leith, 1972; (10) Whittaker and Likens, 1973; (11) Persson, 1974; (12) Brüning, 1975; (13) Windhorst, 1974; (14) Olson, 1975; (15) Eckholm, 1975; (16) Leith, 1975; (17) Eyre, 1978; (18) Ross-Sheriff, 1980; (19) Openshaw, 1978; (20) Steele, 1979; (21) FAO *Production Yearbook*, 1980a; (22) WRI, 1987; (23) Matthews, 1983; (24) FAO *Production Yearbook*, 1985; (25) WRI, 1987; (26) WRI, 1988–89.

First, forest and woodland in the Yearbooks is defined as 'land under natural or planted stands of trees, whether productive or not, and includes land from which forests have been cleared but will be reforested in the foreseeable future.' But when is the 'foreseeable future,' and will there be replanting? Secondly, Houghton *et al.* (1991) have shown conclusively that the Yearbooks of 1976, 1981, and 1986 contain retrospective ten-year estimates that give different totals from each other and do not tally with the totals for individual Yearbooks for the years before 1975. Until the aggregates are analyzed for each country or group of countries, the FAO data must be regarded as another set of imperfect estimates. On face value the data suggest that temperate forests (North America, Europe, mainland China, the USSR, and Oceania) are in a steady state, with slight increase (Figure 2), and tropical forests are declining slightly. However, the aggregate total masks many regional fluctuations.

The problems of definition that bedeviled global estimates, particularly the distinction between closed and open forest, become more critical as the scale of analysis shrinks. For example, Table 1 shows calculations by Barney (1980),

100

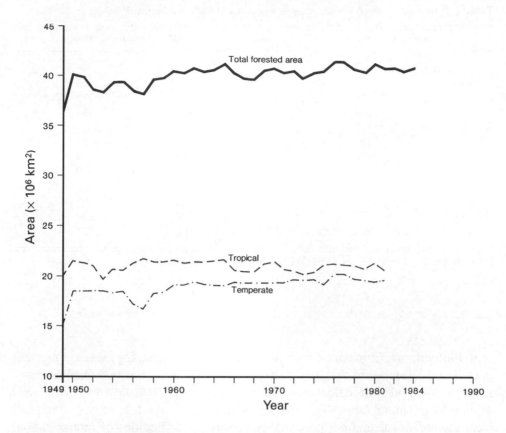

Figure 2. The area of temperate and tropical forests, 1950–90 (from FAO *Production Yearbooks*, 1950–90).

Postel (1984), the World Resources Institute report (1987) (hereafter WRI), and Postel and Heise (1988) for three major continental areas: North America, Africa, and Latin America. There is little agreement except in a most general way.

Perhaps the most consistent view of forest extent at the present time is that of the World Resources Institute, which is based on a number of recent sources from FAO and the U.N. Economic Commission for Europe (Table 2). The table shows that in 1980 closed forests occupy 2792×10^6 ha (27.92×10^6 km^2) and occupy about 21.3% of the earth's surface. The other woodlands of 1707×10^6 ha (17.07×10^6 km^2) include open forest and forest left 'fallow' after shifting cultivation, and they constitute another 13% of the earth's surface. The report strikes a note of caution about the latter as 'estimates of open woodland are highly uncertain and the actual density of trees in such areas is often very low' (WRI, 1988: 70). The extent of closed forest is greater in temperate areas of the world than in the tropical.

Any attempt to compare this calculation with two previous WRI calculations for 1980 and 1985 (and indeed the time series in Figure 1, above) is fraught with difficulties. First, the earlier calculations (see WRI, 1987) divided the data into 'Developed' and 'Developing' rather than 'Temperate' and 'Tropical,' as at pre-

Table 1: Estimates of closed forest and open woodland in three continents, 1980–88 ($\times 10^6$ ha)

	1980	1984	1987	1988
North America				
Closed	470	470	419	459
Open	176	176	215	275
Total	646	646	670	734
Africa				
Closed	190	217	218	221
Open	570	486	500	499
Total	760	703	718	720
Latin America				
Closed	679	679	692	693
Open	217	217	250	240
Total	836	836	942	933

Data from Barney, 1980; Postel, 1984; WRI, 1987; Postel and Heise, 1988.

sent. Furthermore, China and South Africa were moved from the Developing category into the Temperate category. (They are the only countries that changed category.) Second, previous calculations included between 620 (1980) and 675 (1985) million ha of natural 'shrubland and degraded forests' which have been excluded from the recent calculation, presumably because of difficulties of interpretation. Bearing these changes in mind, the 1985 data are reassigned to the new categories, and they are shown in brackets in Table 2. The significance we should place on regional changes is open to question; however, the world total of closed and open forests as now defined has apparently fallen from just over 4727×10^6 ha (47.27×10^6 km^2) in 1980 to just over 4553×10^6 ha (45.53×10^6 km^2) in 1985 to just under 4499×10^6 ha (44.99×10^6 km^2) at present. These three totals are joined by a dashed line in Figure 1.

Perhaps the main hope for consistent baseline data lies in regarding the FAO Yearbook data as a comprehensive global measure against which other sources such as satellite data (e.g., NOAA's Global Vegetation Index) and other compilations (e.g., USDA, 1990) can be compared. Given the current uncertainty about the extent of the forest and the similar uncertainty about the rate of tropical deforestation (and the rate of temperate reforestation), clearly the last has not been heard on the topic of the extent of the forest.

Change

It is the incidence and rate of change in the forest cover that is the focus of this paper. This can be conveniently divided into past change and current change.

Table 2: Distribution of world's forest and woodlands, ca. 1990 and 1985 (parentheses) ($\times 10^6$ ha)

Region	Land Area	Closed Forest	Other Woodland			Total Forest and Woodland	Shrubland (1985)
			Total	Open	Forest Fallow		
Temperate							
North America	1835	459 (469)	275 (215)	x (215)	N.A.	734 (684)	N.A.
Europe	472	145 (153)	35 (21)	x (21)	N.A.	181 (174)	N.A.
USSR	2227	792 (792)	138 (128)	x (128)	N.A.	930 (920)	N.A.
Other countries	1883	194 (194)	115 (85)	x (85)	N.A.	309 (279)	(30)
Total temperate	6417	1590 (1608)	563 (449)	x (449)	N.A.	2153 (2057)	(30)
Tropical							
Africa	2190	217 (218)	652 (660)	486 (500)	166 (160)	869 (878)	(450)
Asia & Pacific	945	306 (347)	104 (159)	31 (83)	73 (76)	410 (506)	(45)
Latin America	1680	679 (692)	388 (420)	217 (250)	170 (170)	1067 (1112)	(150)
Total tropical	4815	1202 (1257)	1144 (1239)	734 (833)	410 (406)	2346 (2496)	(645)
Total world, 1990	13,077	2792	1707	734	410	4499	
Total world, 1985	13,077	– (2865)	(1688)	(1282)	(406)	(4553)	(675)
Total world, 1980	13,077	2948	1779	1372	407		
	4727	624					

x = Not Available
N.A. = Not Applicable
Based on WRI, 1987: 59 and 1988–9: 70.

Table 3: Estimate of preagricultural and present area of major ecosystems ($\times 10^6$ km^2)

Ecosystem	Preagric.	Present	Change
Forests			
Tropical rainforest	12.8	12.3	–0.5
Other forest	34.0	27.0	–7.0
Total	46.8	39.3	–7.5
Woodland	9.7	7.9	–1.8
Shrubland	16.2	14.8	–1.4
Grassland	34.0	27.4	–6.6
Tundra	7.4	7.4	0.0
Desert	15.9	15.6	–0.3
Cultivation	0.0	17.6	+17.6

Revised from Matthews, 1983 (Matthews, personal communication).

Past Change

It is not the aim here to discuss in detail the evidence for past changes in forest cover. That has been done elsewhere (Williams, 1989a, 1989b, 1990, in press). Nevertheless, it is clear that change implies modification from some stock at some known datum point, but what stock?

If the Holocene Period is taken as a starting point, one is struck immediately by the immense and far-reaching climatic fluctuations. Radiocarbon-dated sequences of the late Quaternary palynological record have been synthesized for wide areas of the Northern (e.g., Huntley and Birks, 1983; Velichko *et al.*, 1984; Wright, 1983) and Southern (Kutzbach *et al.*, 1988) hemispheres. They show that forest dynamics, particularly in the middle latitudes, were bewildering as retreating or advancing ice associated with warming or cooling caused shifts in forest location, while in the upper latitudes new species invaded the deglaciated areas. The maps in McDowell et al. (1990) of the distribution of forbs, sedge, spruce, birch, and oak at ten time intervals in eastern North America between 18,000 and 500 years before the present (B.P.) make the point abundantly clear. We simply cannot be dogmatic and say that a particular forest of a particular composition extended over a particular area for all time.

Particularly useful in calculating global change of forest cover has been the work of climate modelers who have attempted to reconstruct the amount of past forest, as noted above. The detailed study by Matthews (1983), described earlier, can be used to attempt to calculate the decrease in major vegetation types (Table 3). A total of 56.5×10^6 km^2 of forest and woodland has been reduced by 8.8×10^6 km^2. Of this, 7.5×10^6 km^2 has come from the forest, of which only 0.5×10^6 km^2 came from the tropical rainforest, with the bulk (7.0×10^6 km^2) coming from the temperate forests. Most of the decline in the temperate forest has come from the clearing in Europe and eastern America of the cold deciduous (2.57×10^6 km^2) and cold deciduous with coniferous forests (1.53×10^6 km^2) to the north, with intensive small-scale

Table 4: Estimated area cleared (×10³ km²)

Region or country		Pre-1650	1650–1749	1750–1849	1850–1978	Total High Estimate	Total Low Estimate
North America	H	6	80	380	641	1107	1107
Central America	H	18	30	40	200	288	
	L	12					282
Latin America	H	18	100	170	637	925	
	L	12					919
Oceania	H	6	6	6	362	380	
	L	2	4				374
USSR	H	70	180	270	575	1095	
	L	42	130	250			997
Europe	H	204	66	146	81	497	
	L	176	54	186			497
Asia	H	974	216	596	1220	3006	
	L	640	176	606			2642
Africa	H	226	80	−16	469	759	
	L	96	24	42			631
Total highest	H	1522	758	1592	4185	8057	
Total lowest	L	986	598	1680	4185		7449

From Williams, 1990.

peasant farming over millennia followed by commercial farming more recently. The temperate evergreen forests and subtropical drought-resistant forests of Asia have been reduced by 0.82×10^6 km^2 and 1.0×10^6 km^2, respectively, because of intensive subsistence farming. Woodlands have declined by 2.13×10^6 km^2, much of that in the dry African miombo, where widespread subsistence farming has been practiced for centuries; around the Mediterranean basin from both widespread sub-sistence farming and small-scale commercial agriculture; and in the dry eucalypt and mallee lands of Australia, where vast areas have been cleared for large-scale commercial agriculture. Thus, the total area of forest in the world has possibly decreased by 16.0% and the woodlands by 18.6%, a massive amount, but not, per-haps, the worldwide devastation that is commonly supposed.

The numbers could be left at that, except that it is possible to tackle the problem of past change from another angle. Historical evidence can be assembled to esti-mate the possible extent of clearing and modification through time as in Table 4. The data for five of the major world regions (mainly the more recently settled parts of the world) can be assembled with some confidence, but the three oldest-settled continents (Europe, Asia, and Africa) present many difficulties. The best that can be done is to indicate the magnitude of likely change that could be expected to occur based on the possible size of the population at the time (itself an estimate) and assigning one-fifth of a hectare of permanently cleared land to each person, on the assumption that new cultivated land would be most likely to be taken out of the productive forest. These figures are shown as high (H) and low (L) estimates.

Much in Table 4 is subjective and therefore must be treated with caution. Nevertheless, the overall total of 7.4×10^6 km^2 based on the low estimate, or 8.1×10^6 km^2 based on the high estimate, goes some considerable way to accounting for the estimated decrease of 9.14×10^6 km^2 calculated by Matthews from other evi-dence. We cannot be more accurate than that, particularly as we know that the land affected by conversion is not necessarily changed permanently; it can revert to for-est, as so often happened in Europe during times of the Black Death, population decrease, and economic recession (Jager, 1951), or in the Yucatan Peninsula in Mayan times (Whitmore et al., 1990). The whole calculation is currently being reexamined in the light of other and conflicting evidence (Richards, 1990).

In all of this discussion it is important to emphasize that the reduction of forest and woodland by possibly 7.5 and 1.8×10^6 km^2, respectively, during the postglacial era (Matthews, personal communication) has been accomplished by *human forces*. The slow and almost imperceptible increase in population before modern times and then its rapid rise from roughly 1600 onward has led to a steady decrease of the world's forests as humankind has needed more land for growing food, timber for construction and shelter, and fuelwood to keep warm, cook food, and smelt metals.

The transformation has probably been greatest with the development of seden-tary agriculture, but has not been confined to agriculturalists alone. Fire, 'the first great force employed by man' (Stewart, 1956), was used by pastoralists to extend pastures, revitalize herbage, and to herd game for hunting. My current work reveals, for example:

- How much further back in time the process can be pushed.
- How much more widespread it was. For example, the Neolithic in Europe is well documented (and getting better every year with more penetrating anthropological studies, e.g., Bogucki, 1988; Gregg, 1988), but so too is the Amerind impact in the Americas (Chapman et al., 1982; Delcourt, 1987).
- A growing number of studies of the early modern era (ca. 1600–1900) are adding greater precision to transformations in specific places at specific times, although what was probably the biggest deforestation of all in the past, that of China, remains an intractable gap in knowledge.

Finally, two major points can be made about past deforestation. First, far more forest was cleared in the past than is being cleared at present. Second, the continuity of this process of terrestrial transformation over time does help to illuminate our understanding of what is happening. In a reflexive way, therefore, the causes and nature of past deforestation throw light on present processes, and the present situation throws light on past processes.

Contemporary Change: Viewpoint and Definition

Since World War II, the upsurge of world population from 2.5 billion in 1950 to a projected 5.29 billion in 1990, together with the widespread availability and use of trucks, tractors, and chain saws, has put an unprecedented strain on the world's forest resources. No wood is too inferior to be harvested and used, and no location is too remote to be exploited. The regional impact has varied in large measure according to the development status of the countries concerned. It has largely (though by no means exclusively) been concentrated in the developing world, which is mainly coterminous with the tropical world. Hence attention has focused primarily on tropical deforestation.

Our understanding of tropical deforestation is bedeviled by a lack of objective data, and consequently by claim and counterclaim. Given the seemingly contradictory conclusions drawn from comparing global totals, trends of forest areas, and on-the-spot evidence of clearing, it is not surprising that there is a vigorous debate about the magnitude of change (Allen and Barnes, 1985; Fearnside, 1982; Lugo and Brown, 1982; Melillo et al., 1985; Molofsky et al., 1986; Myers, 1982; Sedjo and Clawson, 1984). Some paint a bleak picture of future trends, which are multifaceted and exponential (Barney, 1980; Caulfield, 1985; Eckholm, 1976, 1982; Myers, 1980, 1982, 1984). Others have a more upbeat view and regard deforestation as but a continuation of past trends of economic development and attempts to raise well-being, believing that reports are greatly exaggerated and that all will be well if the forests are managed carefully (Clawson, 1981; Sedjo and Clawson, 1984; Simon, 1981).

But disagreement is not only a product of differing philosophies about change; more importantly, as before, it is the result of differing definitions of what constitutes forest, and also of what 'deforestation' means. For example, in the

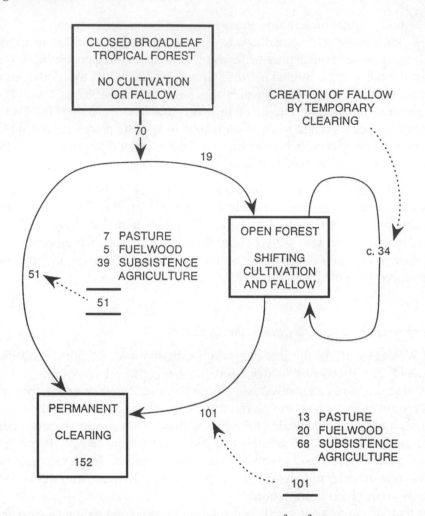

Figure 3. Annual tropical forest pathways of conversion ($\times 10^3$ km^2) (based on Houghton *et al.*, 1986; reprinted with permission from Nature, © 1986, Macmillan Magazines Ltd.).

FAO/UNEP report (Lanly, 1982), 'deforestation' is defined as the transformation of forested land either into a shifting cultivation cycle or into permanently cleared land. Curiously, the definition does not include logging, all the more strange as one of the primary concerns of FAO is stocks of timber. Somewhat different is the definition in the report by Myers (1980) for the National Academy of Sciences. 'Conversion' includes any modification of the forest, from permanent clearing through logging to selective harvesting, largely because the concern of this report is the totality of the environmental benefits performed by the forest ecosystem, from gene pools to wildlife and biomass stocks and their effect on climate. In Myers's 1989 report for Friends of the Earth (Myers, 1989) the term 'deforestation' refers

generally to complete destruction of forest cover through clearing for agriculture

... [so] that not a tree remains, and the land is given over to non-forest purposes ... [and to] very heavy and unduly negligent logging ... [that results in a] decline of biomass and depletion of ecosystems services ... so severe that the residual forest can no longer qualify as forest in any practical sense of the word (Myers, 1989: 5).

Pathways of Change

As contentious as the definitions, and indeed part of the reason for the confusion, is the role of shifting agriculture in causing either permanent change or distur-bance leading ultimately to ecosystem change and degradation (i.e., deforesta-tion). The number of shifting cultivators is not negligible; there may be as many as 250–300 million shifting and 'shifted' (displaced) agriculturalists (Denevan, 1980). An illustration of the problem of disentangling their role is given in Figure 3, which is based on an interpretation by Houghton et al. (1985) of pathways of change implicit in the National Academy of Sciences report (Myers, 1980) with later amendments by Myers (1984). From a stock of closed broadleaf, unculti-vated, and nonfallow forest, an annual amount of 70,000 km² is 'deforested,' of which 51,000 km² is cleared totally and passes into the category of permanently cleared land, and the remaining 19,000 km² goes into the category of fallow (shift-ing) closed forest. From the fallowed forest a total of 101,000 km² is converted annually to permanently cleared land as pressures build up and fallows are short-ened. In fact, the net reduction of the fallow forest category is only 82,000 km² because of the annual addition of 19,000 km² of newly disturbed nonfallow forest; nonetheless, the reduction is still larger than the annual reduction of nonfallow (untouched) forests. If these figures are correct, a total of 152,000 km² is being cleared annually in the tropics with 51,000 coming from the nonfallow forests and 101,000 coming from the fallow forests that were once a part of the shifting culti-vation cycle. Similar calculations of pathways can be hypothesized for the FAO/UNEP data (Woodwell et al., 1986), but the information in Myers (1989) does not allow such interpretations.

On a larger scale, Houghton et al. (1991) have suggested that the area of forest lost does not automatically go into increased agricultural land. Using the raw FAO data, they argue that in the case of tropical Latin America at least one-third goes into 'other' categories, and in Africa the fraction appearing in increased agricul-tural area is only 10%. Clearly, then, much forest must either become totally degraded or, more likely, become some form of open woodland. The internal dynamics of the pathways of change are crucial to understanding rates of defor-estation, and obviously have a regional component.

Rates of Change

Pathways of conversion and total amounts of conversion are inseparable ques-tions. Estimates since 1978 vary from 110,000 km²/yr to 204,000 km²/yr. The lower figure is generally acknowledged to represent the complete removal of trees,

Table 5: Deforestation estimates for closed tropical forests, selected countries ($\times 10^3$ ha)

Country	FAO Estimate	Annual Rate of Loss (%)	Recent Estimate	Annual Rate of Loss (%)	Period of Recent Estimate
Brazilian Amazon	1480	0.4	8000	2.2	1987
Cameroon	80	0.4	100	0.6	1976–86
Costa Rica	65	4.0	124	7.6	1977–83
India	147	0.3	1500	4.1	1975–82
Indonesia	600	0.5	900	0.8	1979–84
Myanmar	105	0.3	677	2.1	1975–81
Philippines	92	1.0	143	1.5	1981–88
Thailand	379	2.4	397	2.5	1978–85
Vietnam	65	0.7	173	2.0	1976–81
	3013		12,014		

Based on WRI, 1990: 102.

whereas the upper figure depends upon a wider definition of deforestation which could include modification to some degree. Thus, an area nearly equal to that destroyed could be severely disturbed or degraded. In this calculation definitions are all-important.

The latest calculation of annual change by Myers (1989) is 142,200 km^2. Another recent estimate based on satellite imagery over Amazonia has put the rate as high as 270,000 km^2/yr (*Christian Science Monitor*, 1988), but this high figure has not been confirmed from other sources. Possibly more realistic is the analysis by the World Resources Institute of recent studies of several key countries, including Brazil, Cameroon, India, Indonesia, Myanmar (Burma), the Philippines, Thailand, and Vietnam, which are compared to the 'official' estimates of FAO (Table 5). The new studies examine the loss of closed forest only and exclude open forest. The discrepancies are enormous, and if they are accurate then up to 204,000 km^2 of tropical forest are being lost annually, an amount that is 79% greater than the FAO estimates.

There is a prevailing tendency to establish a particular annual global total and to think of it as one point in a linear trend of increase. However, there is evidence that there are marked spatial and temporal variations. At the moment, for example, the rate of clearing in the legal Amazon (i.e., the area defined as the Amazon by Brazil, which is slightly different from and larger than the geographical natural region) is slowing down in response to internal economic forces and external pressures. The National Institute of Space Research (Instituto Pesquisas Espaciais, or INPE) calculates that the average rate of clearing (including land flooded by dams) has fallen from 21,500 km^2/yr during 1978–89 to 18,842 km^2/yr in 1987/88/89 to 13,818 km^2/yr during 1989/90 (D.S. Alves, personal communication). In all this uncertainty we can be sure of one thing; the debate on the rate of deforestation is not over (see also Skole, this volume).

This concentration on the developing world should not blind us to the fact that there have also been great fluctuations either way in the forested area of individual countries in the developed world.

Proximate Sources of Change

There are many explanations but little agreement as to why change is occurring in the contemporary forest. The arguments can be divided, very crudely, into two schools of thought: (1) the population increase and (2) socioeconomic inequality, as exemplified by the attempted overarching explanations contained in the work of Grainger (1989) and Guppy (1984). As ever, megatheories need to be tempered by regional case studies. Moreover some sources of change appear to operate more strongly (or at least have been more written about) in some parts of the world than in others (Williams, 1989b). Superficially, at least, this gives the treatment of sources of change a regional emphasis. Thus:

- Agricultural development associated with population increase/resettlement results in planned or spontaneous colonization schemes in Amazonia, Indonesia, and Malaysia.
- Ranching and pasture development are important in Central and Latin America.
- Fuelwood gathering is most important in Africa and, to a lesser extent, in India.
- Logging is important in South and Southeast Asia and of declining importance in West Africa.
- Forest death or *Waldsterben* is most common in western and central Europe and the northeastern United States, but is not confined to these two regions.
- Careful forest management and afforestation seem most common in the developed world.

These regional manifestations of predominant/proximate sources of change are examined in more detail below.

Agricultural Development and Population Resettlement

Everywhere in the developing world, expanding population is having a twofold impact on the forests. On the edges, sedentary cultivators are nibbling away in order to create more land to grow food, while in the forest itself the expanding numbers of shifting cultivators are forced to shorten rotations, leading to permanent change. From an analysis of Landsat imagery between 1973 and 1976 in an area of roughly 45,000 km^2 in south central Thailand, Morain and Klankamsorn (1978) show clearly that it is the edge of the forest that succumbs first in a piecemeal fashion, but the thinning from the interior of the forest is more difficult to detect. The same is true for other parts of the world (Lanly, 1969; Salati and Vose, 1983).

In contrast to this spontaneous and diffuse deforestation, which is difficult to calibrate and is the primary cause of the debate on rates (Melillo et al., 1985), there are the planned and deliberate schemes of governments to promote agricultural development and resettlement in order to alleviate population pressures elsewhere, although the schemes themselves may also prove to be the catalyst for larger, spontaneous migrations. In Indonesia the government has attempted to shift over 2 million peasant farmers from overcrowded Java to forested lands in Sumatra, Irian Jaya, and Kalimantan—the transmigration project (Kartawinata et al., 1981; Ranjitsinh, 1979; Rich, 1986; Secrett, 1986). Attracting much less attention, but equally devastating, are the activities of the Federal Land Development Agency (FELDA) and associated government agencies in peninsular Malaysia, which have embarked on a deliberate policy of expanding primary production of food and cash crops in order to increase national wealth for a rapidly growing population (Bahrin and Perera, 1977; Brookfield et al., 1990). As a result the rainforest has been reduced from 84,832 km^2 in 1966 to 67,351 km^2 in 1982, a decrease from 64% of the land surface of the peninsula to 51%. At least another 10% will be cleared before government aims are fulfilled, but the ability of the government here, or anywhere, to put a brake on associated spontaneous clearing is doubtful.

Perhaps the largest and most notorious colonization and resettlement projects are those in Latin America, particularly in Brazil and in adjacent countries that share portions of the same rainforest ecosystem in the Amazon Basin. In Brazil, government-backed schemes to move settlers from the overpopulated Northeast to the Amazon Basin (Barbira-Scazzocchia, 1980; Denevan, 1973; Fearnside, 1979, 1985; Goodland and Irwin, 1975), together with an extensive program of highway construction and paving, have opened up the forest to planned and spontaneous migration alike (Goodland and Bookman, 1977; Hemming, 1985; Mahar, 1989; Moran, 1983; Schmink and Woods, 1985; Smith, 1981). In particular, the spontaneous migration is swelled by the movement north of peasants displaced by mechanization and modernization of farming in the more productive agricultural regions of temperate Brazil in the states south of São Paulo.

One region where attention has been focused particularly has been the state of Rondônia, virtually unsettled and undisturbed in 1960 except for about 10,000 Indians and a few rubber gatherers. By 1988, about 58,000 km^2 were cleared and more than 1 million people were living there, and the settlement process continues (Mahar, 1989). The herringbone pattern of main and branching minor roads cut into the forest shows up clearly on Landsat and advanced very high resolution radiometer (AVHRR) imagery, and it is a favorite symbolic image of deforestation (Malingreau and Tucker, 1987; Mueller, 1980; Tucker et al., 1984, 1986; Woodwell et al., 1986). But Rondônia is not the only place undergoing rapid change. Mahar (1989) calculates that 570,326 km^2 have been cleared throughout the legal Amazon between 1975 and 1988, and a report by INPE, analyzing images drawn from an 80-day period during 1987, 'conservatively estimated' that over 200,000 km^2 of the forest, or 4% of the Amazon region, was burning between July and October (Margolis, 1988). Burning, however, does not necessarily translate into deforestation, as the section 'Rates of Change' suggests.

To the south in the Zona Central of Paraguay, 40,000 km^2 of tropical rainforest in 1945 has been reduced to about 13,000 km² today (Kleinpenning and Zoomers, 1987; Nickson, 1981). On the western edge of the Amazon Basin in the eastern portions of Venezuela, Bolivia, Columbia, Ecuador, and Peru, a mixture of planned highway development and spontaneous migration has resulted in perhaps as much as 720,000 km² of forest disappearing between 1940 and 1987 (e.g., Bromley, 1972; Crist and Nissly, 1973; Eidt, 1962; Hegen, 1966; Hiraoka and Yamamoto, 1980). Much of this clearing happened during the 1960s and 1970s before the current upsurge of interest in tropical forest conversion, and most of it has been forgotten in the face of current concerns.

Ranching and Pasture Development

While the stated aim of many of the Latin American projects is the resettlement of peasant proprietors on agricultural small holdings, a whole amalgam of economic, social, and fiscal reasons work toward the new clearings' eventually being converted to pasture. Pasture is the easiest means of keeping the land from reverting to secondary forest, cleared land has a speculative value far in excess of crop production in high-inflation economies, pastures are encouraged by tax laws, and cleared land is the surest title to ownership in a situation of chaotic land title registration and not a little corruption (Fearnside, 1983; Furley and Leite, 1985; Shane, 1986).

In Central America, however, pasture development is probably the major initial causal factor in forest clearance, which is carried out by a minority of hacienda proprietors who own disproportionately large areas of land. Here the socioeconomic equity arguments hold undisputed sway. Landholders wish to supply cheap beef for domestic use, but particularly for export, mainly to the United States for pet and fast foods (DeWalt, 1983; Guess, 1979; Myers, 1981; Myers and Tucker, 1987; Nations and Komer, 1982). Land clearing by individual peasants is not absent, of course, in a region with one of the fastest growing populations in the world (e.g., Lewis and Coffey, 1985). Perhaps as much as 25,000 km² is cleared annually for ranching.

Fuelwood and Charcoal

Clearing for fuelwood and charcoal is of major global importance. Approximately 1.5 to 2 billion people (30% to 40% of the world's population) rely on wood, not only for warmth but for the daily preparation of the very food that they eat. In many parts of the developing world, fuel is scarcer and more expensive than the food that is eaten, and sometimes it can consume one-fifth to one-half of the monetary budget of urban households and up to four-fifths of the annual working year as the countryside is scoured for the last remnant of woody fiber to burn (Arnold and Jogma, 1978; Eckholm, 1975; FAO, 1983). This is particularly true around urban and industrial areas. For example, the closed forests within 100 km of nine major Indian cities have been reduced from 96,625 km² to 72,278 km² between 1972 and 1982 (Bowonder et al., 1987). Rural dwellers rarely cause the same sort of

Table 6: Export and import of forest products, 1988 (× million $ U.S.)

Country	Exports	Imports	Net Trade
Canada	17,440	1893	15,547
Sweden	7405	1114	6291
Finland	8184	5496	2688
USSR	3041	660	2381
Indonesia	2873	286	2587
Malaysia	2572	318	2254
Brazil	1760	299	1461
Austria	2492	1134	1358
Portugal	933	–	933
Norway	1387	765	622
Chile	635	–	635
Australia	352	958	–606
Switzerland	738	1450	–692
Spain	736	1475	–739
Egypt	–	754	–754
Belgium-Luxemburg	1649	2409	–760
Hong Kong	381	1219	–838
Denmark	288	1357	–1069
Korean Republic	396	1809	–1413
Netherlands	2609	3710	–1701
France	3217	5496	–2279
China	772	3554	–2782
Italy	1184	4688	–3504
United States	10,723	14,305	–3582
Germany	6043	9928	–3885
United Kingdom	1501	10,484	–8983
Japan	1031	10,988	–9957

Countries with a net trade of less than ±$500 million are excluded.
From FAO, 1988.

absolute deforestation; rather they collect deadwood or cause a 'thinning' of the forest. In the extreme, however, that too can become complete deforestation, especially in the drier, more open forests.

Deficiencies are particularly acute in Andean Latin America, the Caribbean Islands, most of the Indian subcontinent, and particularly Nepal (Alam et al., 1985; Cecelski et al., 1979), but most of all in Africa, which depends on wood for up to 90% of all energy requirements (Anderson and Fishwick, 1984; Eckholm et al., 1984, Munslow et al., 1988) and where depletion far exceeds the rate of growth in many savanna areas.

Currently just over one-half (55%) of all wood known to be extracted from the forests of the world (3.4×10^9 m^3) is fuelwood, and just as demand has doubled during the last 20 years, so the eminently predictable increase in world population makes it unlikely that the demand will slacken in the future. In addition, the rise of oil prices in the 1970s added pressure to fuelwood resources. Nearly 85% of the demand is in the developing world. Energy is essential in a developing economy (Earl, 1975), as the history of fuelwood use in the 19th-century United States shows clearly (Williams, 1982, 1989a)—a history that has been repeated with variations in 20th-century Brazil (Dean, 1987). A switch to alternative fuels such as petroleum and kerosene by the 2 billion fuelwood burners is feasible in terms of the extra amount of these fuels consumed—a mere 3.5% of the current world petroleum production. However, the income and the hard currency needed to pay for this are usually not forthcoming (Foley, 1985). In all, it is thought that as much as 20,000 to 25,000 km^2 of woodland and forest is cleared annually by fuelwood gathering.

Timber Extraction

In the face of the ever-increasing demands for fuelwood, industrial roundwood accounts for an increasingly smaller proportion of the total drain on the forest. It is, however, still sizeable at about 1.5×10^9 m^3 per annum, and it is increasing by about 25×10^6 m^3 every year. In terms of value, the world trade in wood products was worth over $84 billion in 1988. Next to petroleum and natural gas, wood is the third most valuable primary commodity in world trade.

In the industrial economies of the developed world, extraction and regeneration are roughly in equilibrium, regrowth exceeding extraction in Canada, New Zealand, the USSR, and Scandinavia, but probably marginally so in the United States, Western Europe, and Japan (Table 6). But this internal conservation is often achieved at the expense of producers in the tropical world, who are ready to supply hardwood for hard currency. If the big softwood exporters such as Canada, Sweden, Finland, the USSR, and the United States are excluded, the next largest are the hardwood tropical exporters, Malaysia, Nigeria, Indonesia, and the Philippines.

The decline of production in Indonesia and the Philippines through sheer over-exploitation is now being played out again in Malaysia. Sabah and Sarawak are the main areas of cutting, and the rate of extraction is roughly four times the natural regrowth. Although the valuable hardwoods account for only 2–10% of any unit area of the forest, careless and indiscriminate logging destroys up to 60%, and the soil is compacted or eroded (Myers, 1984). For example, a study of selective logging in the Paragominas region of Para State, Brazil (Uhl and Vieira, 1989), found that while only 1–2% of the trees were purposefully harvested, 26% of the trees were killed or damaged (12% lost their crowns, 11% were uprooted by bulldozers, 3% suffered substantial bark scarring), and the forest canopy was reduced by almost one-half. Roads made by the loggers also scar the forest and become pathways of exploitation by spontaneous migrations of slash-and-burn cultivators. The degradation of the ecosystem can be completed if fire sweeps through the logged area, as happened with a combination of logging, shifting cultivation, and drought in East Kalimantan in 1983 when about 3500 km^2 were destroyed or heavily damaged (Malingreau et al., 1985). In the Ivory Coast, a substantial exporter in Africa, current exploitation rates (combined with a rapid population growth and clearing) have reduced exports from 4–5 million m^3 to below 3 million, and will most likely eliminate existing forest resources by the end of this decade (Bertrand, 1983; Lanly, 1982; Postel and Heise, 1988). *In toto*, about 44,000 km^2 of the tropical forest is logged over annually and largely destroyed or degraded, in addition to the ca. 110,000 km^2 cleared for agriculture.

Various models have been constructed to predict the consumption and extraction of wood products worldwide under a number of different assumptions or scenarios. The results are contradictory; all predict rising demand, but some predict constant prices while others expect a doubling of price. Some think that global warming will result in more rapid tree growth in higher latitudes and hence an expansion of coniferous forests, while others again think that acid rain will cause destruction, an increase in cutting, and a short-term rise in production (Dykstra and Kallio, 1987). Generally the assumption is that the world demand for timber will exceed the maximum level available from the forests on a sustainable basis (FAO, 1981). Whatever credence one puts on these global models, regional models have a seemingly greater validity. For example, Grainger's TROPFORM model of trends in tropical hardwood supply, demand, trade, and natural growth over the 40-year period 1980–2020 suggests that supplies will be more limited than assumed formerly, peaking in the first decade of the next century and then falling. With a fall in supply the sources of timber would shift: from South and Southeast Asia, as supplies are cut out and local demand rises, to untapped but expensive sources in Africa and, very significantly, in Latin America.

Forest Death

In all of the above instances the focus has been on the tropical, developing world, but it would be remiss not to point out that the forests of the developed temperate world are not immune to destruction. Felling for more cropland in response to changing agricultural prices during the 1970s has been mentioned already in relation to the United States, although current 'set-aside' schemes under way in the United States and western Europe might eventually rectify that. Ultimately more important, but less easy to predict or prevent, are the potential consequences on forest extent of acid rain pollution caused by gaseous emissions and heavy metals.

The many possible processes and biological pathways of *Waldsterben*, or forest death, are not fully understood (White, 1988), and there are many hypotheses. Equally difficult is to separate the rhetoric from the reality of what is happening (Park, 1987). Nevertheless, it is incontestable that possibly between 70,000 and 100,000 km^2 of coniferous forests in central and eastern Europe are affected, and it is highly possible that triple that figure are at risk, perhaps as much as one-fifth of all forests there (Mandelbaum, 1985; Nilsson, 1986). Climatic stress after a recent spell of three dry years and two mild winters, accompanied by increasing pollution from eastern Europe, has also put broadleafed forests at risk throughout western Europe.

The problem is also becoming evident in the northeastern United States, Canada, the Appalachians, and the coastal ranges of California (Park, 1991; Postel, 1984; WRI, 1986) and has even been detected in China, Malaysia, and Brazil (McCormick, 1985).

Reforestation

Finally, the purely negative aspect of decline of the world's forest stocks should not be emphasized as being the only change. There is a countertrend, albeit small, of afforestation. But there are problems in ascertaining its extent. Whereas deforestation is often concealed or underestimated in many countries, afforestation is publicized and exaggerated, and its success is optimistically assessed as a positive and desirable part of public works programs. Moreover, it is known that many trees that are planted do not survive. For example, 11.7×10^6 ha were planted in the 13 southern states of the United States between 1925 and 1979, but spot checks over large areas showed that between 13 and 18% of the replantings had failed (Williston, 1979). Therefore, some claims must be treated with caution and even skepticism. Bearing that in mind, it is calculated that the average annual amount of recorded afforestation during the 1980s was ca. 150,000 km^2, almost exactly the same as the amount of deforestation (WRI, 1990). However, of that total, 128,420 km^2 or 85.6% was in the cool coniferous, temperate mixed, and warm temperate moist forests of the Northern Hemisphere, including China, Japan, and Korea, and in the forests of South Africa, Australia, and New Zealand. In the tropical world Brazil had an annual rate of 5610 km^2, but the only other countries rated above 1000 km^2 were India (1730) and Indonesia (1640).

The very clear implication of these figures is that depletion is being matched (or nearly so) by replanting and careful forest management, particularly in the cool coniferous forests of the big exporting countries of Canada, the United States, the USSR, and Scandinavia. Depletion in the tropical forests, where trees are more difficult to propagate anyhow, is not being compensated for by replanting.

The whole topic of afforestation and the part it plays in the balance of forest resources is underresearched.

Conclusion

The process of change in the forest will not end in the future. Nothing is likely to stop the world's population from being over 6 billion at the turn of the century, and possibly 9 billion by 2100. For those living in or near the tropical forest it will continue to be 'the mantle of the poor,' providing land, fuel, and shelter for as long as it lasts; others will wish to restrict its use and preserve it. Deforestation is fast becoming a matter of humanitarian concern mixed with long-term environmental ethics.

In all of this, there is a need for much more accurate data, mainly through the medium of more rapid interpretation and monitoring of remote sensing (principally Landsat) evidence. These data, however, are expensive to collect, need expert staff to interpret, and are probably accumulating faster than they can be assessed. Side-looking airborne radar (SLAR) will assist in overcoming cloud cover in tropical areas. It is probable that the development of geographical information systems will lead to precise inventories of what is known of past and current forest conversions with a view to calibrating the rate of change.

Of all national territories, only Brazil has a national remote sensing program, though there is now hope of wider world cooperation between half a dozen of the major tropical countries affected (WRI, 1988). The need is to add hard knowledge of land-use/cover change where at the moment there is much polemic and the scantiest of objective insight into the realities of forest and tree cover.

References

Alam, M., J. Dunkerley, K. N. Gopi, W. Ramsey, and E. Davis. 1985. *Fuelwood in Urban Markets: A Case Study of Hyderabad.* Concept Publishing Company, New Delhi, India.

Allen, J. C., and D. F. Barnes. 1985. The causes of deforestation in developing countries. *Annals of the Association of American Geographers 75*, 163–184.

Anderson, D., and R. Fishwick. 1984. *Fuelwood Consumption and Deforestation in African Countries.* Staff Working Paper No. 704, World Bank, Washington, D.C.

Arnold, J. E. M., and J. Jogma. 1978. Fuel and charcoal in developing countries. *Unasylva 29*, 2–9.

Barbira-Scazzocchio, F., ed. 1980. *Amazonia.* Occasional Paper No. 3, Centre for Latin American Studies, Cambridge University Press, Cambridge, U.K.

Bahrin, T. S., and P. D. A. Perera. 1977. *FELDA: 21 Years of Land Development.* Federal Land Development Authority, Kuala Lumpur, Malaysia.

Barney, G. O., ed. 1980. *Global 2000 Report to the President: Entering the Twenty-First Century.* Vol. 2, *Technical Report.* Council of Environmental Quality and U.S. Department of State, Government Printing Office, Washington, D.C.

Bazilevich, N. E., L. F. Rodin, and N. N. Rozov. 1971. Geographical aspects of biological productivity. *Soviet Geography, Review and Translation 12*, 293–317.

Bertrand, A. 1983. La deforestation en zone de fôret en Côte d'Ivoire. *Bois et Fôrets des Tropiques 202*, 3–17.

Bogucki, P. I. 1988. *Forest Farmers and Stockherders: Early Agriculture and Its Consequences in North-Central Europe.* Cambridge University Press, Cambridge, U.K.

Bowonder, B., S. S. R. Prasad, and N. V. M. Unni. 1987. Deforestation around urban centres in India. *Environmental Conservation 14*, 23–28.

Bromley, R. J. 1972. Agricultural colonization in the upper Amazon Basin. *Tijdschrift voor Economisch en Sociale Geografie 63*, 278–294.

Brookfield, H., F. J. Lian, L. Kwai-Sim, and L. Potter. 1990. Borneo and Malay Peninsula. In *The Earth as Transformed by Human Action: Global and Regional Changes in the Biosphere Over the Past 300 Years* (B.L. Turner II, W.C. Clark, R.W. Kates, J.F. Richards, J.T. Mathews, and W.B. Meyer, eds.), Cambridge University Press, New York, 495–512.

Brüning, E. F. 1971. Förstliche Produktionslehre. *Europäische Hochschulschriften 25(1)*.

Brüning, E. F. 1975. Ökosysteme in den Tropen. *Umshau 47*, 405–410.

Caulfield, C. 1985. *In the Rainforest*. Heinemann, London.

Cecelski, E., J. Dunkerley, and W. Ramsey. 1979. *Household Energy and the Poor in the Third World*. Resources for the Future, Washington, D.C.

Chapman, J., P. A. Delcourt, P. A. Cridlebaugh, A. B. Shea, and H. R. Delcourt. 1982. Man-land interaction: 10,000 years of American Indian impact on native ecosystems in the lower Little Tennessee Valley. *Southeastern Archaeology 1*, 115–121.

Christian Science Monitor, 10 October 1988.

Clawson., M. 1981. Entering the 21st century: The Global 2000 report to the President. *Resources 66*, 19–21.

CLIMAP Project Members. 1981. *Seasonal Reconstructions of the Earth's Surface at the Last Glacial Maximum*. GSA Map and Chart Series MC-36, Geological Society of America, Boulder, Colorado, 172–229.

Crist, R. E., and M. C. Nissly. 1973. *East from the Andes*. University of Florida Social Sciences Monograph No. 50, University of Florida Press, Gainesville, Florida.

Dean, W. 1987. *Firewood in Paulista Industrialization and Urbanization, 1900–1980*. Paper given to conference of American Society for Environmental History, Duke University, Durham, North Carolina, April 1987.

Delcourt, H. R. 1987. The impact of prehistoric agriculture and land occupation on natural vegetation. *Ecology 34*, 341–346.

Denevan, W. M. 1973. Development and imminent demise of the Amazon rainforest. *Professional Geographer 25*, 130–135.

Denevan, W. M. 1980. Latin America. In *World Systems of Traditional Resource Management* (G.A. Klee, ed.), Halstead Press, New York, 217–244.

DeWalt, B. R. 1983. The cattle are eating the forest. *Bulletin of Atomic Scientists 39*, 18–23.

Dykstra, D. P., and M. Kallio. 1987. Scenario variations. In *The Global Forest Sector: An Analytical Perspective* (M. Kallio, D.P. Dykstra, and C.S. Binkley, eds.), John Wiley and Sons, New York.

Earl, D. E. 1975. *Forest Energy and Economic Development*. Clarendon Press, Oxford, U.K.

Eckholm, E. 1975. *The Other Energy Crisis: Firewood*. Paper No. 1, Worldwatch Institute, Washington, D.C.

Eckholm, E. 1976. *Losing Ground: Environmental Stress and World Food Prospects*. Pergamon Press, Oxford, U.K.

Eckholm, E. 1982. *Down to Earth: Environment and Human Needs*. W. W. Norton for International Institute for Environment and Development, New York.

Eckholm, D., G. Foley, and G. Bernard. 1984. *Fuelwood: The Energy Crisis That Won't Go Away*. Earthscan, London.

Eidt, R. C. 1962. Pioneer settlement in eastern Peru. *Annals of the Association of American Geographers 52*, 255–278.

Ellenberg, H., and D. Mueller-Dombois. 1967. Tentative physiognomic-ecological classification of plant formations of the earth. *Berichte–Geobotanischen Institutes der Eidgenossischen Technischen Hochschule Stiftung Rubel 37*, 21–55.

119

Eyre, S. R. 1978. *The Real Wealth of Nations*. St. Martin's Press, New York.

Fearnside, P. M. 1979. The development of the Amazon rainforest: Priority problems for the formulation of guidelines. *Interciencia 4*, 338–342.

Fearnside, P. M. 1982. Deforestation in the Brazilian Amazon: How fast is it occurring? *Ambio 7*, 82–88.

Fearnside, P. M. 1983. Land-use trends in the Brazilian Amazon as factors in accelerating deforestation. *Environmental Conservation 10*, 141–148.

Fearnside, P. M. 1985. Agriculture in Amazonia. In *Amazonia* (G.T. Prance and T.E. Lovejoy, eds.), Pergamon Press, Oxford, U.K.

Foley, G. 1985. Wood fuel and conventional fuel demands in the developing world. *Ambio 14*, 253–258.

FAO (U.N. Food and Agriculture Organization). 1949 and annually. *Production Yearbook*. FAO, Rome, Italy.

FAO. 1963. *World Forest Inventory*. FAO, Rome, Italy.

FAO. 1981. *Forest Product Prices, 1961–1980*. FAO, Rome, Italy.

FAO. 1983. *Wood for Energy*. Forest Topics Report No. 1, FAO, Rome, Italy.

FAO. 1988. *Yearbook of Forest Products, 1977–1988*. FAO, Rome, Italy.

FAO. 1990. *Forest Resources Assessment 1990: Guidelines for Assessment*. FAO, Rome, Italy.

Fosberg, R. F. 1961. A classification of vegetation for general purposes. *Tropical Ecology 2*, 1–28.

Furley, P. A., and L. L. Leite. 1985. Land development in the Brazilian Amazon with particular reference to Rondônia and the Ouro Preto colonization project. In *Change in the Amazon Basin: The Frontier After a Decade of Colonization* (J. Hemming, ed.), Manchester University Press, Manchester, U.K.

Goodland, R. J. A., and J. Bookman. 1977. Can Amazonia survive its highways? *Ecologist 7*, 376–380.

Goodland, R. J. A., and H. S. Irwin. 1975. *Amazon Jungle: Green Hell to Red Desert?* Elsevier, Amsterdam, The Netherlands.

Grainger, A. 1989. *Modeling Deforestation in the Humid Tropics*. Paper given to seminar on Deforestation or Development in the Third World, Saariselka, Finland, June 26–30, 1989, Finnish Forest Research Institute.

Gregg, S. A. 1988. *Foragers and Farmers: Population Interaction and Agricultural Expansion in Prehistoric Europe*. University of Chicago Press, Chicago, Illinois.

Guess, G. 1979. Pasture expansion, forestry and development contradictions: The case of Costa Rica. *Studies in Comparative International Development 14*, 42–55.

Guppy, N. 1984. Tropical deforestation: A global view. *Foreign Affairs 62*, 928–965.

Hegen, E. E. 1966. *Highways in the Upper Amazon Basin: Pioneer Lands in Southern Colombia, Ecuador and Northern Peru*. University of Florida Press, Gainesville, Florida.

Hemming, J. (ed.). 1985. *Change in the Amazon*, Vol. 1, *Man's Impact on Forests and Rivers*, Vol. 2, *The frontier after a Decade of Colonization*. Manchester University Press, Manchester, U.K.

Heske, F. (ed.). 1971. *Weltforstatlas—World Forestry Atlas—Atlas des forêts du mond—Atlas forestal del mundo*. Herausgegeben von der Bundesforschungsanstalt fur Forst- und Holzwirtschaft, Paul Parey Verlag, Reinbek bei Hamburg, Germany.

Hiraoka, M., and S. Yamamoto. 1980. Agricultural development in the upper Amazon of Ecuador. *Geographical Review 52*, 423–445.

Houghton, R. A., R. D. Boone, J. M. Melillo, C. A. Palm, G. M. Woodwell, N. Myers, B. Moore III, and D.L. Skole. 1985. Net flux of carbon dioxide from topical forests in 1980. *Nature 316(6029)*, 617–620.

Houghton, R. A., D. S. Lefkowitz, and D. L. Skole. 1991. Changes in the landscape of Latin America between 1850 and 1985. I. Progressive loss of forests. *Forest Ecology and Management 38*, 143–172.

Hummel, J., and R. Reck. 1979. A global surface albedo model. *Journal of Applied Meteorology 18*, 239–253.

Huntley, B., and H. J. B. Birks. 1983. *An Atlas of Past and Present Pollen Maps of Europe, 0–13,000 Years Ago*. Cambridge University Press, Cambridge, U.K.

Jager, H. 1951. *Die Entwickslung der Kulturlandschaft im Kreise Hofgeismar*. Gottinger geographische Abhandlungen No. 8, Gottingen, Germany.

Kartawinata, K., S. Adisoemarto, S. Riswan, and A. D. Vayda. 1981. The impact of man on a tropical forest of Indonesia. *Ambio 10*, 115–119.

Kleinpenning, J. M. G., and E. B. Zoomers. 1987. Environmental degradation in Latin America: The example of Paraguay. *Tijdschrift voor Economische en Sociale Geografie 78*, 242–250.

Kutzbach, J. E., W. F. Ruddiman, F.A. Street-Perrott, T. Webb III, and H.E. Wright, Jr., eds. 1988. *Global Climates 9000 and 6000 Years Ago*. University of Minnesota Press, Minneapolis, Minnesota.

Lanly, J.-P. 1969. Regression de la fôret dense en Côte d'Ivoire. *Bois et Fôrets de Tropiques 127*, 45–49.

Lanly, J.-P. 1976. Tropical moist forest inventories. *Unasylva 28(112–113)*, 42–51.

Lanly, J.-P. 1982. *Tropical Forest Resources*. Forest Paper No. 30, FAO, Rome, Italy.

Leith, H. 1972. Über die Brimerproduktion der Pflanzendecke der Erde. *Angewandte Botanik 46*, 1–37.

Leith, H. 1975. Primary production of the major vegetation units of the world. In *Primary Production of the Biosphere* (H. Leith and R.H. Whittaker, eds.), Ecological Studies No. 14, Springer-Verlag, Berlin, 203–205.

Lewis, L. A., and W. J. Coffey. 1985. The continuing deforestation of Haiti. *Ambio 14*, 158–160.

Lugo, A. E., and S. Brown. 1982. Conversion of tropical rainforests: A critique. *Interciencia 7*, 89–93.

Mahar, D. 1989. *Government Policies and Deforestation in Brazil's Amazon Region*. World Bank, Washington, D.C.

Malingreau, J. P., and C. J. Tucker. 1987. The contribution of AVHRR data for measuring and understanding global processes. In *Large-Scale Deforestation in the Amazon Basin*, Proceedings of the International Geoscience and Remote Sensing Symposium (IGARSS '87), Ann Arbor, Michigan, May 18–21, 443–448.

Malingreau, J. P., G. Stephens, and L. Fellows. 1985. Remote sensing and forest fires: Kalimantan and North Borneo in 1982–83. *Ambio 14*, 314–321.

Mandelbaum, P. 1985. *Acid Rain—An Economic Assessment*. Plenum Press, New York.

Margolis, M. 1988. Threat from Amazon burn-off. *The Times* [London], 6 September, 9.

Mather, A. S. 1987. Global trends in forest resources. *Geography 77*, 1–15.

Matthews, E. 1983. Global vegetation and land use: New high-resolution data bases for climatic studies. *Journal of Climate and Applied Meteorology 22*, 474–487.

McCormick, J. 1985. *Acid Earth: The Global Threat of Acid Pollution*. Earthscan, London.

McDowell, P. F., T. Webb III, and P. J. Bartlein. 1990. Long-term environmental change. In *The Earth as Transformed by Human Action: Global and Regional Changes in the Biosphere Over the Past 300 Years* (B. L. Turner II, W. C. Clark, R. W. Kates, J. F. Richards, J. T. Mathews, and W. B. Meyer, eds.), Cambridge University Press, New York, 143–162.

Melillo, J. M., C. A. Palm, R. A. Houghton, G. M. Woodwell, and N. Myers. 1985. A comparison of recent estimates of disturbance in tropical forests. *Environmental Conservation 12*, 37–40.

Molofsky, J., C. A. S. Hall, and N. Myers. 1986. *A Comparison of Tropical Forest Surveys*. U.S. Department of Energy, Washington, D.C.

Morain, S. A., and B. Klankamsorn. 1978. Forest mapping and inventory techniques through visual analysis of Landsat imagery: Examples from Thailand. In *Proceedings, 12th Symposium on Remote Sensing of the Environment*, Manila, Philippines. American Meteorological Society, Boston, Massachusetts, 417–426.

Moran, E. F. 1983. Government-directed settlement in the 1970s: An assessment of Trans-Amazon Highway colonization. In *The Dilemma of Amazonian Development* (E.F. Moran, ed.), Westview Press, Boulder, Colorado.

Mueller, C. 1980. Frontier-based agricultural expansion: The case of Rondônia. In *Land, People and Planning in Contemporary Amazon* (F. Barbira-Scazzocchio, ed.), Occasional Paper No. 3, Center for Latin American Studies, Cambridge University Press, Cambridge, U.K.

Munslow, B., Y. Katerere, A. Ferf, and P. O'Keefe. 1988. *The Fuelwood Trap: A Study of the SADCC Region*. Earthscan, London.

Myers, N. 1980. *Conversion of Tropical Moist Forests*. Report prepared for the Committee on Research Priorities on Tropical Biology of the National Research Council, National Academy of Sciences, Washington, D.C.

Myers, N. 1981. The hamburger connection: How central America's forests become North America's hamburgers. *Ambio 10*, 3–8.

Myers, N. 1982. Response to the Lugo-Brown critique of 'Conversion of tropical moist forests.' *Interciencia 7*, 358–360.

Myers, N. 1984. *The Primary Source: Tropical Forests and Our Future*. W.W. Norton, New York.

Myers, N. 1989. *Deforestation Rates in Tropical Forests and Their Climatic Implications*. Friends of the Earth, London.

Myers, N., and R. Tucker. 1987. Deforestation in central America: Spanish legacy and North American consumers. *Environmental Review 11*, 55–71.

Nations, J. D., and D. I. Komer. 1982. Indians, immigrants and beef exports: Deforestation in central America. *Cultural Survival Quarterly 6*, 8–12.

Nickson, R. A. 1981. Brazilian colonization of the eastern border region with Paraguay. *Journal of Latin American Studies 13*, 111–131.

Nilsson, S. 1986. *Development and Consequences of Forest Damage Attributed to Air Pollutants and Changes in Climate*. International Institute of Applied Systems Analysis, Laxenburg, Austria.

Olson, J. S. 1970. Carbon cycles and temperate woodlands. In *Ecological Studies* (D.E. Reichle, ed.), Chapman and Hall, London, 226–241.

Olson, J. S. 1975. *World Ecosystems*. Seattle Symposium, Washington, D.C.

Olson, J., and J. A. Watts. 1983. *Major World Ecosystem Complexes Ranked by Carbon in Live Vegetation* (Map, scale = 1:30 million). Oak Ridge National Laboratory, Oak Ridge, Tennessee.

Openshaw, K. 1978. Woodfuel—A time for reassessment. *Natural Resources Forum 3(1)*, 35–71.

Park, C. C. 1987. *Acid Rain: Rhetoric and Reality*. Routledge, London.

Park, C. C. 1991. Trans-frontier air pollution: Some geographical issues. *Geography 76*, 21–35.

Persson, R. 1974. *World Forest Resources: Review of the World's Forest Resources in the Early 1970s*. Research Notes 17, Department of Forest Survey, Royal College of Forestry, Stockholm, Sweden.

Postel, S. 1984. *Air Pollution, Acid Rain and the Future of Forests*. Worldwatch Paper No. 58, Worldwatch Institute, Washington, D.C.

Postel, S. 1985. Protecting forests. In *The State of the World, 1984*. Worldwatch Institute/W.W. Norton, New York.

Postel, S., and L. Heise. 1988. *Reforesting the Earth*. Worldwatch Paper No. 83, Worldwatch Institute, Washington, D.C.

Ranjitsinh, M. K. 1979. Forest destruction in Asia and the South Pacific. *Ambio 8*, 192–201.

Rich, B. M. 1986. The World Bank's Indonesia Transmigration Project: Potential for disaster. *Indonesia Reports 15*, 2–5.

Richards, J. F. 1990. Land transformation. In *The Earth as Transformed by Human Action: Global and Regional Changes in the Biosphere Over the Past 300 Years* (B.L. Turner II, W.C. Clark, R.W. Kates, J.F. Richards, J.T. Mathews, and W.B. Meyer, eds.), Cambridge University Press, New York, 163–178.

Ross-Sheriff, B. 1980. Forest projections. In *Global 2000 Report to the President,* Vol. 2, *Technical Report* (G.O. Barney, ed.), Council of Environmental Quality and U.S. Department of State, Government Printing Office, Washington, D.C., 118–135.

Salati, E., and P. B. Vose. 1983. The depletion of tropical rainforests. *Ambio 12*, 67–71.

Schmink, M., and C. H. Woods (eds.). 1985. *Frontier Expansion in Amazonia*. University of Florida Press, Gainesville, Florida.

Secrett, C. 1986. The environmental impact of transmigration. *The Ecologist 16*, 77–88.

Sedjo, R. A., and M. Clawson. 1984. Global forests. In *The Resourceful Earth: A Response to Global 2000* (J.L. Simon and H. Kahn, eds.), Basil Blackwell, Oxford, U.K., 128–171.

Shane, D. R. 1986. *Hoofprints in the Forest: Cattle Ranching and the Latin America's Tropical Forests*. Institute for the Study of Human Relations, Philadelphia, Pennsylvania.

Simon, J. L. 1981. *The Ultimate Resource*. Princeton University Press, Princeton, New Jersey.

Smith, N. J. H. 1981. *Rainfall Corridors: The Trans-Amazon Colonization Scheme*. University of California Press, Berkeley, California.

Steele, R. C. 1979. Some social and economic constraints to the use of forests for energy and organics in Great Britain. In *Biological and Sociological Basis for a Rational Use of Forest Resources for Energy and Organics* (S. G. Boyce, ed.), Proceedings of Man and the Biosphere workshop, Michigan State University, May 6–11. FAO, Rome, Italy.

Stewart, O. C. 1956. Fire as the first great force employed by man. In *Man's Role in Changing the Face of the Earth* (W. L. Thomas, ed.), University of Chicago Press, Chicago, Illinois, 115–133.

Tucker, C. J. B., B. N. Holben, and T. E. Goff. 1984. Intensive forest clearing in Rondônia, Brazil, as detected by satellite remote sensing. *Remote Sensing of the Environment 15*, 255–261.

Uhl, C., and I. C. G. Vieira. 1989. Ecological impacts on selective logging in the Brazilian Amazon: A case study from the Paragominas region of the State of Para. *Biotropica 21*, 98–106.

UNESCO (U.N. Educational, Scientific and Cultural Organization). 1973. *International*

Classification and Mapping of Vegetation. UNESCO, Paris, France.

USDA (U.S. Department of Agriculture). 1990. *World Agriculture: Trends and Indicators, 1970–89*. Economic Research Service, Statistical Bulletin No. 815, U.S. Department of Agriculture, Washington, D.C.

Vanvariova, L. F. 1986. *Map of World Land Cover* (Map, scale =1:15 million). Institute of Geography, USSR Academy of Sciences/Moscow State University, USSR.

Velichko, A. A., H. E. Wright, Jr., and C. W. Barnosky, eds. 1984. *Late Quaternary Environments of the Soviet Union*. University of Minnesota Press, Minneapolis, Minnesota.

Weck, J., and C. Wiebecke. 1961. *Weltforstwirtschaft und Deutschlands Forst- und Holzwirtschaft*. Munich, Germany.

White, J. C. 1988. *Acid Rain: The Relationship between Sources and Receptors*. Elsevier, New York.

Whitmore, T. M., B. L. Turner II, D. I. Johnson, R. W. Kates, and T. R. Gottschang. 1990. Long-term population change. In *The Earth as Transformed by Human Action: Global and Regional Changes in the Biosphere Over the Past 300 Years* (B. L. Turner II, W. C. Clark, R. W. Kates, J. F. Richards, J. T. Mathews, and W. B. Meyer, eds.), Cambridge University Press, New York, 25–39.

Whittaker, R. H., and G. E. Likens. 1973. Primary production: The biosphere and man. *Human Ecology 1*, 357–369.

Whittaker, R. H., and G. M. Woodwell. 1971. Measurement of net primary productivity of forests. In *Productivity of Forest Ecosystems* (P. Duvigneaud, ed.), UNESCO, Paris, France, 159–175.

Williams, M. 1982. The clearing of the United States forests: The pivotal years, 1810–1860. *Journal of Historical Geography 8*, 12–28.

Williams, M. 1989a. *The Americans and Their Forest*. Cambridge University Press, New York.

Williams, M. 1989b. Deforestation: Past and present. *Progress in Human Geography 13*, 176–208.

Williams, M. 1990. Forests. In *The Earth as Transformed by Human Action: Global and Regional Changes in the Biosphere Over the Past 300 Years* (B.L. Turner II, W.C. Clark, R.W. Kates, J.F. Richards, J.T. Mathews, and W.B. Meyer, eds.), Cambridge University Press, New York, 179–201.

Williams, M. History of deforestation. In *Deforestation: Environmental and Social Impacts* (J. Thornes, ed.), Chapman Hall, London, in press.

Williston, H. L. 1979. *A Statistical History of Tree Planting in the South, 1925–79*. U.S. Department of Agriculture Forest Service, Atlanta, Georgia.

Windhorst, H. W. 1974. Das Ertagspotential des Wälder der Erde. *Geographische Zeitschrift 39*.

Woodwell, G. M., R. A. Houghton, T. A. Stone, and A. B. Park. 1986. Changes in the area of forests in Rondônia, in the Amazon Basin, measured by satellite imagery. In *The Changing Carbon Cycle: A Global Analysis* (J.R. Trabalka and D.E. Reichle, eds.), Springer-Verlag, New York, 242–257.

WRI (World Resources Institute). 1986, 1987, 1988, and 1990. *World Resources: An Assessment of the Resource Base that Supports the Global Economy*. World Resources Institute/International Institute for Environment and Development, Basic Books, New York.

Wright, H. E., Jr., ed. 1983. *Late-Quaternary Environments of the United States*, Vol. 2, *The Holocene*. University of Minnesota Press, Minneapolis, Minnesota.

Zon, R., and W. N. Sparhawk. 1923. *Forest Resources of the World*. McGraw Book Co., New York.

6

Grasslands

Dean Graetz

In ecological terms, grasslands are strictly defined by structure and composition. In line with the nature of available global data sets, grasslands can also, less rigorously, be defined by land use to include all land covers used for livestock production. This functional definition adds to the naturally occurring grasslands, savannas, arid grassy shrublands, and the human-made pastures that are intimately connected with the croplands.

Over the last 15 years, the rate of conversion of grasslands to croplands has slowed in the developed world. In the developing world the rates are high and increasing, but quite variable from one country to another. The rates of conversion are highest in grassland-rich nations and are small or negative in tropical nations where forests are being converted to croplands and grasslands. Of global significance is not just the area converted but also the area modified, or changed in ecological condition. Grasslands are being degraded through overuse, with such consequences as soil erosion, changed floristic composition, and diminished productivity.

Of the major physical processes of global change—climatic change, alteration of biogeochemical cycles, and land use—the most influential agent of future change in grasslands will be land use. The most probable outcome is an extensive further transformation of the grasslands, either by conversion to croplands or by degradation through unsustainable pastoral use. These changes will be principally, but not exclusively, driven by the subsistence requirements of a burgeoning human population. These land-use changes will have far more extensive and degrading impacts on the grasslands than the forecast consequences of climate change. After all, the brunt of the modern global human population explosion is yet to come.

This chapter develops four themes. The first section examines the nature of grasslands, including savannas and pastures, detailing their global distribution in response to the influential factors of climate, soils, and disturbance. The second examines the nature and extent of direct anthropogenic change, with the third theme being a case study illustrating such change, in the Sahelian zone of Africa. The last section is a summary and a prognosis for change in grasslands and pastures until the year 2040, a 50-year horizon.

Grasslands as a Land-Cover Type

A Definition

Grasslands are ecologically defined by the vegetation attributes of structure and floristic composition. They are far more commonly defined by the criterion of land use as any open land that is used for livestock production.

In ecological terms, grasslands are landscapes that have a ground story in which grasses are the dominant vegetation life form, whether this dominance is defined by cover or by biomass. This definition focuses attention on the structurally distinct, treeless grasslands of common experience, such as the prairies of North America or the steppes of central Asia. To meet the objectives of this volume, however, the sparse woodlands with grassy understories, the savannas, and the very sparsely shrubbed grasslands of arid regions are also included within this category (see Plate 10).

The range of floristic composition possible under this generic grassland category is enormous because there are some 10,000 species of grasses. The extensive grasslands of the various continents have been characterized by name: the pampas, campo, and llanos of South America; the prairies of North America; the steppes of Central Asia; the veldts and maras of Africa; and the spinifex of Australia. The productivity and biotic diversity of the grasslands are apparent in the higher trophic levels. In particular, the grasslands have, for the last 5 million years or so, supported diverse and abundant populations of herbivores upon which humans have become dependent within the last 100,000 years or so (De Wet, 1981; Reader, 1990; Stebbins, 1981).

The grasslands as so defined may be called native or natural to distinguish them from those that result from human activity. For at least the last 10,000 years, humans have cleared forests to create grasslands, croplands, or, most commonly, a mixture of the two (Buringh and Dudal, 1987). These grasslands may be composed of (locally) native species or exotic species introduced to the area. In the last century, such grasslands have usually been subsidized directly with fertilizers or indirectly by the inclusion of nitrogen-fixing legumes into the grass community. These subsidized, human-made grasslands are called (improved) 'pastures' to indicate their exotic nature, biogeochemical subsidy, and use for livestock production.

Most, but not all, of the natural grasslands are also used by humans for livestock production, but the spatial and temporal patterns of use vary greatly across the globe. Some extensive grassland types, such as the hummock grasslands (spinifex) of the arid sand plains of interior Australia, are not and never will be used for livestock. Conversely, global classification according to the use of land for livestock production will undoubtedly be too broad to be useful in the analysis of land-cover change because livestock production is too closely integrated with cropping, whether for cash or subsistence, as well as with the use of native grasslands.

126

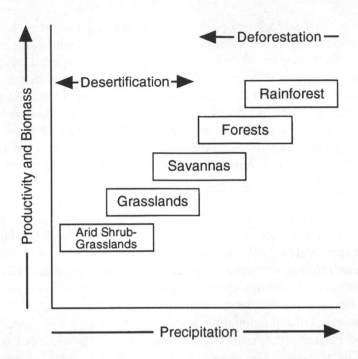

Figure 1. A generalized abstracted relationship between the ecological variables relating to structure (biomass density) or functioning (primary productivity) of the five major vegetation types (biomes) and precipitation. As precipitation increases so does productivity, and therefore biomass, with the two extremes being the low sparsely shrubbed grasslands of the deserts and the tall closed forests, be they tropical, temperate, or boreal.

It is critical to any analysis focused on the interaction of the grasslands with the physical processes of global change to retain the definition of grassland by ecological (vegetation structural and composition) attributes rather than by its principal use, livestock production. To make this choice results in a major data complication because many global data sets are aggregated by land use rather than by a clearly defined vegetation type. Nonetheless, the most important reason for preferring the ecological definition of grasslands to that based on land use, is that it is not possible directly to relate land use as such to the major physical processes of global environmental change. Land use cannot be directly related to these forms of global change because it is only a qualitative descriptor. Land-use categories are abstract typologies that, although useful, cannot be meaningfully included in process models seeking to forecast the time and space patterns of global change. It is land cover, rather than land use, that has mechanistic meaning in the processes of global environmental change. Therefore land cover, rather than land use, needs for such purposes to be described in terms of quantitative attributes rather than qualified using typologies.

An insightful description of the relationship between grasslands and the other major biomes can be had by locating them within the two-dimensional space of vegetation structure and precipitation. Figure 1 is a generalized, abstract depiction of the relationship between the ecological variables relating to the structure (biomass density) or function (primary productivity) of the major vegetation types (biomes) and precipitation. As precipitation increases, so does productivity, and therefore biomass, with the two extremes being the low, sparsely shrubbed grasslands of the deserts and the tall closed forests, be they tropical, temperate, or boreal.

The grasslands most commonly occur toward the drier end of this simple ordination of vegetation types. As precipitation declines, long-lived, woody shrubs tend to replace grass. As precipitation increases, the biomass of trees tends to increase and suppress the grasses of the ground story by competition for water and, at the wetter end of the continuum, light. At the closed forest extreme, this competitive exclusion of grasses by trees is almost complete.

The domains of two land transformation processes are indicated on Figure 1. Grasslands are being altered, often irreversibly, at the dry end of the spectrum under a variety of land uses. The consequences of these many land uses are collectively (and inaccurately) called desertification. For a variety of reasons, the domain of desertification is extending in both directions along the rainfall axis, that is, into the more arid areas, such as the Sahel of Africa, as well as into the temperate, relatively more productive grasslands, like Mongolia.

At the wet end of the precipitation axis, grasslands have been, and are still being, created by deforestation. For millennia, humans have cleared forests to replace them with grasses. The grasslands may be used to produce annual food plants (e.g., wheat, corn, barley, millet, sorghum) or pasture species for livestock forage. Most of the world's croplands and improved pastures, particularly in Europe, are the result of clearing temperate woodlands and forests (see selected case studies in Wolman and Fournier, 1987). Contemporary concern is focused on the rapidity of this process in the tropical lowland rainforests of South America, Africa, and Asia (Williams, this volume).

Why Grasslands?

Much of what is collectively called land degradation (desertification) is more easily understood if the interaction of the many factors that determine the distribution and persistence of grasslands is appreciated.

The location of grasslands along the precipitation gradient in Figure 1 is globally correct. At any finer scale, however, it is necessary to invoke additional factors to explain the occurrence, structure, and functioning of a grassland as well as its response to land use.

The first, and most consequential, point is that as a structural type, grasslands are the exception rather than the rule. Given identical conditions, long-lived woody vegetation (trees and shrubs) will ultimately outcompete the relatively short-lived grasses. The competitive advantages of woody species are significant, and grasslands only exist where the dominance by the woody species is suppressed. The three principal factors that prevent dominance by woody plants in grasslands are disturbance (fire), soil (physical and chemical properties), and (freezing) temperatures (Axelrod, 1985).

Fire is by far the most influential factor in the grasslands of the tropics and sub-tropics, but it is less so in temperate climatic zones. Grasses can withstand repeated burning, whereas most woody plants cannot (Collins and Wallace, 1990). Indeed, this characteristic alone seems to account for their evolutionary success over the last 25 million years. In grasslands where fire at some frequency is needed to suppress tree and shrub species, any reduction of this fire frequency, for example by the reduction of fuel loads (grass biomass) through grazing, ultimately releases the suppression of the woody species. Woody plants then outcompete the grasses and create a positive feedback loop: the greater the biomass of woody species, the lower the grass biomass (fuel load); the lower the fuel load, the lower the fire frequency, leading to increased woody biomass.

The observations supporting the above widely held view are found throughout in the scientific literature. The interaction of fire frequency and grass tree competition is most comprehensively documented for the savannas (Huntley and Walker, 1982; Tothill and Mott, 1985). The role of fire in the formation and maintenance of grasslands from woodlands and forests in temperate regions is best documented for the prairies of North America and has recently been summarized by Delcourt and Delcourt (1991).

Some soil factors can, in certain circumstances, perpetuate grasslands by providing conditions that reduce the competitive advantage of woody plants. Soil chemical conditions, such as extremely low nutrient status, favor grasses rather than trees or shrubs; a familiar example is the hummock grassland of Australia that occupies deep, infertile sand plains. Similarly, some soil types also suppress the growth and persistence of woody plants; for example, the deep, sandy soils that provide little or no soil water storage; or those that swell and shrink with wetting, thus hindering the woody roots of trees and shrubs; or those that are subject to regular flooding or waterlogging, for example, the llanos of Brazil.

The last factor is the frequency and duration of below-freezing temperatures. This single environmental factor can satisfactorily explain the nature and distribution of woody vegetation at global scales. If temperature and water balance are combined, the global distribution of grasslands can be predicted (Woodward and Williams, 1987; Woodward and McKee, 1991). The synergistic interaction of these two climatic factors is thought to inhibit the establishment and persistence of woody plants in otherwise temperate zones, such as the vast steppes of Central Asia. In addition, the influence of fire, no matter how infrequent, will contribute to the suppression of woody plants and the increase in grasses.

There are three important points to be drawn from this discussion. The first is that since disturbance or soil or climatic factors acting in combination with global-scaled gradients of temperature and moisture availability (Figure 1) determine the existence and persistence of grasslands, then the consequences of human use for the extent of these grasslands can be relatively insignificant. The exception is the conversion of grasslands to croplands, as has happened over much of the Great Plains of North America.

Where the factors of soil or freezing temperatures are principally responsible for suppressing competition by woody species, then pastoral overuse by humans will serve only to degrade the grasslands, i.e., alter their condition or ecological status, rather than to eliminate them. Much of the Mongolian steppe is apparently in this condition, having undergone significant and extensive changes in species composition over the last 50 years or so. Nevertheless, a depauperate grassland, no matter how species-poor, eroded, or lowly productive, is still structurally and floristically a grassland. Grasslands that owe their existence to unusual soil or climatic factors are therefore somewhat buffered in this sense from the direct consequences of human use.

This dictum does not hold where the persistence of grasslands is determined by the frequency of fire. Fire is the predominant determinant of the structure and floristic composition of grasslands and savannas in the tropics, subtropics, and arid regions (e.g., de Van Booysen and Tainton, 1984). If human use of the grasslands changes the fire frequency, and thus the competitive advantage of woody species, then grasslands can 'expand' at the expense of savannas as fire frequency is increased, or 'contract' as they become invaded by woody species when fire frequency falls. Humans and climate synergistically determine the fire frequency, and both agents have effectively shaped the extent and composition of grasslands over geologic, prehistoric, and historic time scales.

The second point that follows from the role of fire is the recognition that boundaries between grasslands and savannas are diffuse and dynamic, and for the purposes of this chapter irrelevant. Savannas must be included with grasslands.

The last point is that because the largest areas of grasslands are determined structurally and floristically by fire frequencies and humans can dramatically alter fire frequencies, then these grasslands can be significantly altered in both extent and condition by land use. We can conclude that most grasslands and savannas are but poorly buffered from the direct consequences of human use.

Where Are the Grasslands?

I think there exist no satisfactory high-resolution maps of global vegetation types based on structural attributes. The appropriate grassland volumes in the authoritative *Ecosystems of the World* series (numbers 8A, 8B) are not yet published. Instead, the global data set assembled by Matthews (1983) is here taken as sufficiently illustrative of the points made above even though it has a spatial resolution of 1° latitude × 1° longitude. Plate 11 depicts the distribution of three major vegetation types (biomes) that are collectively called grasslands in this chapter. In the first category are the extensive, treeless grasslands, which may be temperate (the steppes of Europe and Asia) or tropical (Africa, South America, Australia). In the second category are the savannas, the grasslands with a woody overstory, which often occur next to the treeless grasslands. The last category is the arid shrubby grasslands. The continuum of these three vegetation categories along the gradient of precipitation is most continuous, and thus obvious, within the continents of Africa, South America, and Australia.

Table 1: Comparison of the FAO and ISRIC estimates of world grassland area ($\times 10^6$ km^2)

FAO[1] (as of 1986)	
Permanent pasture	32.128
ISRIC[2] (as of 1988)	
Tropical grassland	2.115
Temperate grassland	10.467
Savannas	10.695
Arid shrublands	12.001
Total	35.278

[1]FAO, 1990.
[2]Table 12.1 in Bouwman, 1990.

The Extent of Grasslands

It is not a simple task to produce an acceptable summary of the areal distribution of any global land cover. The difficulty arises from the diverse and inconsistent mapping criteria used by various authors, the primary data sources and methods used, and the dynamic status of some components of land cover. An illustration of the variability in the data is provided by a comparison of the estimates of the area of the major biomes made by five authors over the time span 1973–83 (Bouwman, 1990, Table 3.7).

For this chapter, two sources are selected and compared. The first is that provided by the U.N. Food and Agriculture Organization (FAO), which is compared with that from the International Soil Reference and Information Center (ISRIC). The former, though widely used, is based on nationally supplied figures and is therefore suspect for several reasons (Buringh and Dudal, 1987). In contrast, the latter appears to be meticulously based on independently acquired and assimilated data sets, not the least important being soils (Table 1). Given the difficulties of comparing an inventory based on land use (FAO includes grasslands under permanent pastures) with one offering defined vegetation types, the agreement is surprisingly good. The ISRIC figure is the preferred value, for reasons elaborated above; its value of 35×10^6 km^2 for the area of grasslands represents just under 25% of the global land area.

The Significance of Grasslands

In global terms, the dependence of humans on grass and grassland products is striking when the grass crop plants (the grains) and the herbivore products of the grasslands (meat and fiber) are aggregated. The oceans are deserts relative to the land in general, and to the grasslands in particular. The croplands, the lands that were once forests or more recently were once grasslands, are the most significant land cover for humankind, but the spatial integration of grasslands with croplands is almost always intimate.

The grasslands have been the theater in which the hominid evolutionary play has been staged over the last 1–2 million years. It has been proposed that early humans lived in close association, as scavengers, with the great herds of herbivores that migrated ceaselessly across the grasslands of East Africa. Changes in the distribution and abundance of humankind driven by population growth have been helped by the domestication of wild herbivores in nomadic pastoralism, or of grasses in sedentary agriculture. The dependence of modern human society on grasses and grasslands is diverse but it remains essential; 'Should its harvest fail for a single year, famine would depopulate the world' (J.J. Ingalls, 'In Praise of Grass' [poem], 1872).

The Nature and Extent of Anthropogenic Change

The most severe human impact on the grasslands is desertification. Most commonly the consequence of overgrazing and cropping, it is essentially irreversible on human time scales. The cumulative outcomes are obvious in the Sahel, where they represent 30 years or less of overuse, or in the moonscapes of the Middle East, where they represent 2000–3000 years of human occupancy. Other savannas and grasslands are apparently under growing land-use pressure through grazing animals, cropping, and, above all, the extensive use of fire. These land-cover changes will have global consequences through trace gas emissions as well as local aggregated to regional-scale diminution or destruction of life-supporting ecosystems.

The Pattern in Time and Space

Since the evolution of bipedal hominids, human use of grasslands has followed a sequence that has been driven by multiplicative interaction of population densities, affluence, and available technology. For perhaps the last 1 million years, the human users of the grasslands have been societies of hunter-gathers that have harvested the available resources in an opportunistic, nomadic way. The early users were not without impact, for they undoubtedly increased the fire frequency to enhance their survival and consequently shaped the structure, composition, and functioning of the grasslands. This impact can only be generally inferred from paleoecological evidence.

The hunter-gatherers were and still are being displaced from the most productive grasslands to the arid extremes by subsistence pastoralists herding domesticated herbivores. These were also opportunistic nomads, harvesting the best forage and casual water supplies wherever the seasons produced them. This human group had more significant and more complex impacts than the one it displaced. Only in subtropical areas did the use of fire remain widespread; in the less productive regions, fire was avoided because it meant a loss of forage. The physical and grazing impact of herbivores was probably slight overall but concentrated (and cumulative) around the permanent water supplies.

The significant change in the nature and extent of the degrading impact of this society on the grasslands comes about when the affluence level is changed from a subsistence to a cash or market system. This in turn is usually followed by a great jump in technological capability. The change in affluence shifts the focus of animal husbandry from surviving or persisting in the face of variability to producing a regular or increasing income despite seasonal variability.

The effect of technological change for pastoral societies has principally been to override the natural controller of herd levels and incentive for nomadism, the supply of permanent waters, through the construction of artificial waters. Nomadism gives way to a sedentary existence. The synergistic contribution of these two factors usually results in an increase in the human density, at least in the short term, because sedentary family units are usually greater than those of nomads.

The nomadic pastoralists were, and still are, being displaced by the sedentary pastoralists to the less productive grasslands, the desert margins. Here they compress the ambit of the hunter-gathers exiled earlier; a case in point is the plight of the present-day Bushmen in Botswana (e.g., Reader, 1990). In the last stage of the sequence, the sedentary pastoralists either become, or are absorbed by, invading waves of agriculturalists who convert the most productive grasslands to croplands and may subsidize this conversion with fertilizer. The scale and extent of this transformation of the land cover again are determined by the levels of affluence and technological capability, particularly the input of fossil fuels. The extremes of this spectrum are the Corn Belt of North America as compared with the subsistence-level farming of the Sahel.

The model of the evolutionary sequence of grassland use outlined above is widely held to be generally true. Certainly the processes and consequences of the displacement of pastoral land use by agriculture, particularly in Africa, are frequently reported; see, for example, the case studies in Little and Horowitz (1987). It is difficult to find all of the stages in the sequence operating in any one grassland area now. The difficulty derives from the strong influence that the cultural and economic context on the adoption and implementation of new technologies (Grübler, this volume). Bearing this in mind, however, the various (temporal) stages in this proposed sequence can be identified (spatially) across the grasslands of the world. These changing spatial patterns are a response to the largest form of land transformation—the expansion of agriculture. Agriculture, as a land use, involves the most dramatic and (mostly) irreversible transformations of land cover: from forests or grasslands to croplands. By simplifying the biotic diversity of natural landscape to monocultures, it focuses the innate productivity of that landscape almost exclusively to service human needs. There is debate whether the development of agriculture permitted a dramatic expansion in size of sedentary human populations or sedentary agriculture arose in response to the burgeoning needs of growing human populations (Buringh and Dudal, 1987; Simmons, 1987). Nevertheless, whether cause or consequence, expanding human populations have been, and are still, associated with the expansion of, and latterly with the intensification of, agriculture (e.g., Alayev *et al.*, 1990; Berry *et al.*, 1990; Richards, 1990). Agriculture, with its transformation of land cover and its subsidy with

Table 2: Estimated changes in the areas of the major land-cover types between preagricultural times and the present ($\times 10^6$ km^2)

Land-Cover Type	Preagricultural Area	Present Area	% Change
Total forest	46.8	39.3	−16.0
Tropical forest	12.8	12.3	−3.9
Other forest	34.0	27.0	−20.6
Woodland	9.7	7.9	−18.6
Shrubland	16.2	14.8	−8.6
Grassland	34.0	27.4	−19.4
Tundra	7.4	7.4	0.0
Desert	15.9	15.6	−1.9
Cultivation	0.0	17.6	+1760.0

From J.T. Matthews (personal communication).

nutrients, energy, and water, is by far the most effective land use to support growing human populations. But it is not without cost.

A Summary of Change

J.T. Matthews has compiled global estimates of the changes in land cover from the preagricultural period to the present (Table 2). These figures have not gone unchallenged, but they enjoy sufficient support to illustrate the general global trends of the past. They indicate the forcing role of agricultural land use on land cover with the conversion of forests and grasslands to croplands. The estimates of the proportion of grasslands converted to croplands is approximately 20% globally; this proportion varies by continent, being highest in North America, approximately 40%, and least in South America (Singh *et al.*, 1983).

With a dramatic change in time scales, it is possible to analyze land-cover changes over a 15-year period, 1971–86, using the systematic assessments provided by FAO (1987). As with the previous data set, these figures are not without problems of accuracy and precision as well as category. The FAO includes grasslands in a land-use definition of pastures and collates information only for area without any assessment of ecological condition. Even with these limitations, however, the FAO data set is the only one with which to examine modern trends in the land cover globally (Table 3). The 'World' row of figures in Table 3 provides the best reference point for assessing anthropogenic change, status, and trend in the grasslands.

Under the FAO classification, some 24.6% of the ice-free land surface was classified as pastures (here equated to grasslands) as of 1986. The trend has been a loss of <1% in grassland area globally during 1971–86. In contrast, the greatest expansion of a land-cover type was croplands. The largest loss was in forest cover.

Table 3: Comparison of the relative proportions (%) of various geographic units occupied by the four FAO land-cover categories and the proportional changes (%) in relative area, 1971–86

	Crops	Pasture	Forest	Other	Δ Crop	Δ Pasture	Δ Forest	Δ Other
World	11.27	24.57	31.18	32.99	4.64	−0.28	−2.92	1.56
Africa	6.24	26.58	23.25	43.93	8.58	−0.54	−6.22	2.79
Kenya	4.16	6.57	6.46	82.8	12.32	−2.35	−10.9	0.6
Niger	2.96	7.26	2.01	87.77	37.96	−10.29	−25.84	0.82
Nigeria	34.4	23.02	16.03	26.54	4.8	5.68	−23.56	8.72
S. Africa	10.79	66.65	3.7	18.87	−0.6	−1.62	8.8	4.8
S. America	8.06	26.75	51.85	13.34	24.98	6.22	−5.7	−0.31
Argentina	13.17	52.11	21.78	12.94	6.5	−1.18	−1.65	1.43
Brazil	9.08	19.75	66.27	4.9	38.94	14.07	−5.42	−19.42
Venezuela	4.32	19.84	35.53	40.32	8.67	5.93	−12.19	9.44
Asia	16.84	25.33	20.17	37.66	2.67	−2.63	−1.88	1.71
India	56.76	4.06	22.63	16.55	2.63	−8.91	5.24	−11.65
China	10.48	34.21	12.53	42.77	−4.28	0	4.14	−0.07
Mongolia	0.84	78.74	9.7	10.73	68.65	−11.92	1.19	1935.2
Europe	29.52	17.74	32.92	19.82	−2.27	−5.67	3.23	3.78
N. America	12.83	14.85	33.57	38.76	1.58	2.31	−0.02	−1.34
E. Europe and USSR	11.95	16.75	41.78	29.52	−0.46	−0.02	3.13	−3.94
Developed	12.3	23.05	33.86	30.79	0.99	−1.32	0.14	0.45
Developing	10.52	25.66	29.24	34.58	7.93	0.41	−5.35	2.29

Data from FAO, 1987.

This change in area of pastures most probably represents only conversion to croplands by cultivation and does not include the transformation by severe degradation of grassland to wastelands, which may fall into the category of 'Other.' I hold the phenomenon of degradation or desertification of grasslands to be far more serious and widespread than that of conversion to croplands. What few figures are available will be presented later (Schlesinger *et al.*, 1990).

Continuing with the global perspective, the FAO figures for nations aggregated into just two groups, developed and developing, illustrate the basic duality in the patterns of land-cover change. In the developed countries, with an aggregate status of 12.3% croplands, 23.05% pastures, and 33.86% forests, the trend was for a small (~1%) gain in cropland area, an equivalently small (~1%) loss of pastures, and a small increase in forest area. In contrast, the developing countries exhibited a dramatic increase in croplands, a much smaller increase in pastures, and a large decrease in forest cover.

The land-use/land-cover model advanced earlier in the chapter was for grasslands to be converted to croplands and for the forests to be converted to either croplands or pastures (grasslands). At the global aggregate level (the 'World' row of Table 3) this model is not supported, but it is at continental and national scales. For example, the South American continent recorded a dramatic increase in the area of croplands during 1971–86 with a moderate increase in the area of pastures, both associated with a 5% loss of (rain-) forested area. The forested nations of Brazil and Venezuela reinforce the trend of increasing croplands and pastures and decreasing forests in contrast to the grassland-rich nation of Argentina, in which both pastures and forests are being converted into croplands.

The conversion model is repeated in Asia as illustrated by the examples of India and Mongolia, the former being in particularly perilous state (Plate 12). My own experience is that the absence of change recorded for China is not representative.

Africa, like South America, illustrates the two scenarios of the conversion of both grasslands and forests to croplands. The most dramatic changes are illustrated by Niger, a Sahelian nation, with substantial conversions of both land covers. In Kenya the rates of conversion are less, but both semiarid countries are in a parlous position; for the latter see Berry *et al.* (1990). Nigeria, on the other hand, presents the same conversion profile as the tropical nations of South America, with a dramatic conversion of forests into both croplands and pastures. In complete contrast is the land-cover change profile of South Africa, which can be taken to represent the expression of a more rigid governmental planning and regulation of land use.

The Sahel of Africa: A Case Study

In the section on anthropogenic change above, an evolutionary model of land use was outlined that described the changing patterns of dependence on, and use of, the grasslands driven by population growth and technological change. I believe that this model, which is not original, is generally applicable because the separate stages can be identified and dated for human societies using the grasslands of Europe, the Americas, etc., where the great changes have occurred in the last few centuries. The transformation of grasslands is still in progress in the tropical and subtropical zones, particularly of Africa. A case study of modern times below illustrates all of the stages in a compressed period using a simple ecological systems view of the interactions of humans, land use, and land cover (Graetz, 1991).

The Sahel of Africa, its name derived from an Arabic word meaning 'the shore' (of the desert), is the southern boundary of the Sahara desert. It is a fringing belt of semiarid landscape, 5000 km long, stretching from the Atlantic Ocean to the Red Sea. The extensive land degradation of the Sahel has been reported to the world in the 1960s and again in the early 1980s. Each time the plight of the land and its users was highlighted because of drought. Although some regard drought as the cause of desertification (e.g., Sandford, 1983), this linkage has been convincingly disputed by Sinclair and Fryxell (1985), who consider drought to be a catalyst. Droughts serve to collapse systems that were already stressed through overuse between the drought years. In this view, the cause of the two epochs of desertification observed since 1960 is not drought but overgrazing (see the review by Mainguet, 1991).

In ecological terms, the Sahel is a transition zone between the hyperarid Sahara desert in the north and the humid savannas in the south. Precipitation varies between 100–200 mm in the more arid northern grasslands and 400–600 mm in the southern savannas. The Sahel is approximately 500 km wide, with a rainfall gradient of approximately 1 mm/km (Le Houérou, 1980). Its rainfall results from a continental-scale weather pattern, the Inter-Tropical Convergence Zone (ITCZ), which produces a rainfall belt that moves from south to north. This moving rainfall swathe is followed by migratory insects, birds, and the large mammalian herbivores. The most spectacular of these migrations are those of the wildebeest in the Serengeti of Tanzania and Kenya and the kob in Sudan. In both cases more than 1 million animals move from the higher-rainfall (600 mm/yr) savannas, tall perennial grasslands of relatively low quality in the dry season (January–March), to the lower-rainfall (<400 mm/yr) but higher-quality grasslands in the wet season. These grasslands of the more arid northern edge of the Sahel support a very high animal biomass (density) for a short time while the grass remains green. Other smaller herbivores that do not migrate en masse have adjusted to regular seasonally limited supplies of quality food and water by a range of adaptations that include dietary changes (grazing-browsing) and small-scale migration from patch to patch of temporary water and food generated by the spatially patterned rainfalls.

The natural nomadic movement of the wildebeest and kob provides for periods in both wet and dry seasons when there is no, or very little, grazing. During these periods, the perennial grasses reproduce and build up reserves to maintain vigor, essential for the persistence of these grasslands under grazing. A second conclusion is that this nomadic strategy allows a larger population of herbivores to exist than would be possible under a sedentary strategy. Under a sedentary system, herbivore populations would be forced to exist on abundant but low-quality grasses for most of the year, and this would reduce growth and reproductive rates. The nomadic strategy gives an opportunity for the harvesting of high-quality forage for a sufficiently long period to improve reproductive success and build reserves to survive on the poor resources of the savannas during the long dry season.

Pastoral societies across the Sahel have until very recently exercised a nomadic strategy very similar to that of the large native ungulates. Cattle herds moved north into the more arid areas following the rain to graze annual grasses of very high abundance and quality, returning to graze the lower grasslands and agricultural stubble of the sedentary farmers. Here, as with the large ungulates, the native grasslands and the agricultural lands experienced periods of no grazing. Sinclair and Fryxell (1985) assemble evidence that suggests that this nomadic system has been in operation for many centuries, possibly 2000–4000 years. The exact behavior and reaction of this nomadic system under drought is not well known, but it is recognized that at least three major droughts have occurred in the last 100 years as severe as that which, in 1969–75, drew attention to Sahelian desertification.

The 1969–75 drought in the Sahel was significant not because of its severity but because it followed upon substantial changes in the nature, size, and behavior of the nomadic pastoral societies. It was the impact of these social changes that began

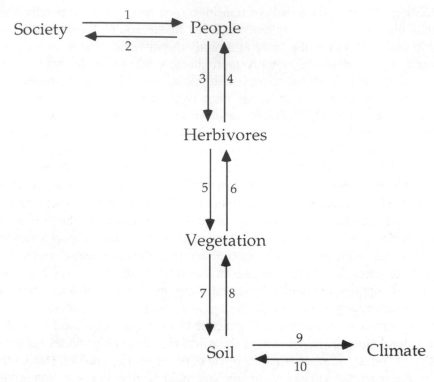

Figure 2. A generalized model of the structure of a Sahelian nomadic pastoral system. The ecosystem is represented by just four components, people, herbivores, vegetation, and soils, and two influences, climate and society. The ten reciprocal interactions individually comprise several ecosystem processes, such as herbivory, infiltration, and predation (adapted from Graetz, 1991).

the desertification process. The drought merely exacerbated and expedited a degradation process that was already in train. The crucial changes in the pastoral society were precipitated by post–World War II aid that, by its nature and direction, encouraged the expansion of cultivated agriculture in the southern fringes of the Sahel at the expense of the traditional dry season pastures of the pastoralists. The second aid intervention was the settling of the nomads. In a process encouraged by the provision of permanent waters, where previously the natural waters had been ephemeral, or enforced by new nationalist governments, nomadism declined dramatically in the early 1960s.

Cultivation decreased the resources available to the pastoralist, and settlement enforced sedentary, year-long grazing of grasslands by a population of humans and cattle that grew rapidly. This rapid growth in the populations of both pastoralists and cattle, ≈3% per year, resulted from the removal of a previously limiting population control, the high mortality rates of the newborn. This control on both human and livestock populations was suppressed by access to medical and veterinary services following settlement. Lamprey (1983) has summarized the intricate social and ecological interaction that develops, and how all increased human activity is transmitted to, and focused on, the land.

The net result of these initial, catalytic social changes has been the overgrazing of the landscape. The cumulative impact of the unrelenting overgrazing, in concert with the additional stress of drought, has been desertification over extensive areas. Ironically, where bores were sunk by international aid projects to supply permanent water, one major result has been that now large numbers of cattle die of hunger whereas before only a few had died of thirst.

It is possible to represent the Sahelian system as a simple model with four components: human society, herbivores, herbage and soils, and climate (Figure 2). In this system there are ten interactive links between components that vary in strength. Each of these links may involve one or more ecological processes, such as herbivory, predation, etc., and depending upon context, the paired links operate as either positive or negative feedback loops. The proposition is that, within this simple system model, desertification results from changes in the strength and sign of the interactive links. No new patterns of linkage are established. The transformation is effected by a change in the nature of two feedbacks, from weak to strong and from negative to positive. That is, the beginning of the desertification process will, by positive feedback, increase the probability that further desertification will occur.

The Sahelian nomadic pastoral system has persisted for a considerable time. Presumably it has often experienced droughts as severe as that of 1969–75. Therefore we may conclude that this system evolved as a stable, resilient strategy. This conclusion is supported by the similarity in grazing strategy between the nomads and the large migratory herbivores, the wildebeest and the kob. In its undisturbed state the system was driven by climate (rainfall), interaction 9 in Figure 2, and controlled by negative feedbacks between vegetation and herbivores, interaction 6, and to a lesser extent between herbivores and people, interaction 4. In the former case, the negative nature of the feedback was determined by the nutritional limits to herbivore health, growth, and reproduction offered by the low-quality but reliable dry season grasslands. The grasslands acted as a 'passively harsh' environment for herbivores. In the latter case the needed but hazardous dependence of the pastoralist on the herbivores must also be seen as a limit to human populations through similar controls of disease, growth, and reproduction.

The patterns of nomadic movements of the livestock can be interpreted as a strategy to harvest the higher-quality but less predictable arid grasslands. This nomadic strategy incorporated an additional control in the form of availability of water for livestock. Its availability determined the distribution and density of herbivores sympathetically with the productivity of the grasslands. Consistent with this model, the modern epoch of desertification can be explained by one small but catalytic change. It was that the isolation, and therefore the integrity, of the nomadic pastoral society was breached by the interaction with alien economic systems starting in 1950. This produced two responses. First, a pulse of population growth of pastoralists and then livestock passed from the higher levels to the lower. Second, a controlling negative feedback between vegetation and herbi-

Table 4: Distribution of rangelands among major geographic areas and percentage of desertification in each

Rangelands	Area ($\times 10^6$ km^2)	Desertified (%)
Sudano-Sahelian Africa	3.80	90
Southern Africa	2.50	80
Mediterranean Africa	0.80	85
Western Asia	1.16	85
Southern Asia	1.50	85
USSR in Asia	2.50	60
China and Mongolia	3.00	70
Australia	4.50	22
Mediterranean Europe	0.30	30
South America and Mexico	2.50	72
North America	3.00	42
Total	25.56	62

Data from UNEP.

vores, lack of drinking water, was overridden by the supply of permanent water. This allowed the pulse of disturbance to pass down the system and be transmitted to the vegetation and soils where before it would have been suppressed.

Under the additional stress of the first drought (1969–75), the interaction exceeded some threshold by which interaction 8 (Figure 2), previously weak and positive, was transformed into strong negative feedback on plant growth and establishment by soil erosion by wind and water. This continued in a self-reinforcing way until a new domain of resilience had been reached. Most interactions had been spatially uncoupled, productivity at all levels had declined and become episodic, and the landscape looked like a desert.

The collapse of the primary productive system, in the form of diminished production, etc., was transmitted back up the system (1975–?), and interaction 2 became a strong positive feedback on further aid, intervention 1. The cycle repeated itself by 1980–? Given the continual input of ecologically inappropriate intervention, and the unchecked population growth, the cycle will continue at an ever-increasing frequency.

There is no other alternative behavior of the Sahelian pastoral system given that, since the time of the first drought, the human population has almost doubled (Le Houérou, 1992).

Desertification, here the destructive interaction of human land use and grasslands, is 'a tale of two feedbacks' (Graetz, 1991). A nomadic pastoral system was stabilized by the negative feedbacks between the grazed, the grazer, and the grazier. The nomadic system of the Sahel was destabilized by the effective removal of two negative controlling feedbacks. The resultant pulse of disturbance and the drought precipitated a negative feedback of soil erosion that desertified the productive base of the system. This pulse of response, a new and severe negative

Table 5: Summary of the human-induced soil degradation data assembled by UNEP and ISRIC

	Used Area ($\times 10^6$ ha)	Desertified Area[1] ($\times 10^6$ ha)	Desertified Proportion (%)	Desertified by Overgrazing[2] (%)
World	8056	1943	24	35
Africa	1799	494	27	49
North and Central America	1044	158	15	24
South America	1386	244	18	28
Asia	2440	745	31	26
Europe	843	199	24	23
Oceania	544	103	19	80

[1]Includes degradation by vegetation removal, overexploitation, overgrazing, and agricultural activities.
[2]Proportion of total desertified area attributed to overgrazing.
Adapted from WRI, 1992.

feedback, passed back up the system only to solicit a further positive feedback of destabilizing aid to renew the cycle.

Grasslands: A Summary and a Prognosis

The prognosis for the grasslands depends upon the validity of the diagnosis. The validity of the diagnosis is determined by its predictive capability under changing circumstances, be they changes in atmosphere, climate, or land use.

The first point to be made is that a focus on the grasslands as a distinct category is misplaced because the grasslands are part of a continuum of global land cover along which human land use is rapidly changing. Instead the focus should span the grasslands and include the vegetation types that put it in ecological, as well as in land-use, context.

Inventories of the areal extent of grasslands are of little value unless they are based on measurement rather than cartographic extrapolation as they are now. In general, though, it can be said that grasslands (and croplands) are being generated at the expense of tropical forest cover, and grasslands are being transformed into croplands in the tropical savannas and temperate grasslands (Werger, 1983). Globally these changes, according to FAO data, are small: <1% over the last 15 years. This rate varies nationally, however, being very high in Africa, notably in Niger, Kenya, and Nigeria, where it is driven by rapidly expanding populations, and in South America, especially Argentina, where it is presumably driven by market forces.

The critical issue of grassland change globally is not just the area of conversion but also the transformation or changes in ecological 'condition.' Grasslands are being degraded through overuse. The expressions of this are soil erosion, changed floristic composition, and diminished productivity. Collectively these are known as desertification.

The extent and severity of desertification globally is but poorly quantified. Two global assessments have been made in the last decade: by Mabbutt (1984) and, more recently, by the ISRIC (see WRI, 1992). Both of these studies were in collaboration with the U.N. Environment Program (UNEP). If the figures can be believed, and there are reasons that they should be treated with caution, the extent of desertification is staggering. Both sets of data are presented for comparison and to make two substantial points.

The assessment of Mabbutt (1984) has been summarized as Table 4 and catalogs the relative areas of desertification for the rangelands, the native (grassland) pastures that are used for livestock production. The estimated global aggregate for desertification of these land-cover types is 62%. This is an arresting figure, as are the assessments for individual regions of livestock production. The forceful conclusion from Table 4 is that desertification of grazing lands is a very substantial phenomenon.

The most recent assessment of UNEP and ISRIC (WRI, 1992) has a different focus (Table 5). It assesses the human-induced soil degradation for all land uses but only during the period 1945–90. This soil degradation is attributed to five agents: vegetation removal, overexploitation, overgrazing, agricultural activities, and industrial pollution. I have ignored the last agent and pooled the other four as desertification.

The two assessments, Tables 4 and 5, are not strictly comparable by geographical units, levels of severity, or categories of interest. The first appraisal of Table 5 is that it presents desertification as a less severe problem than does Table 4. Nonetheless a global aggregate of 25% is very significant, as is the proportion of this degraded area that was generated by overgrazing, 35%. This latter statistic is very high for Oceania, mostly due to the contribution of Australia, and for Africa, and it is appreciable for all other regions.

The first point I wish to make is that the degradation or desertification of grasslands represents a far more significant and consequential change than current conversions of grasslands to croplands (Schlesinger *et al.*, 1990).

The second point is that both of the global appraisals presented here are unsatisfactory. The UNEP-ISRIC assessment is the best available because it systematically collates and generalizes the available data sets. But it is not based on systematic measurement. It is not built on a systematically acquired data base. It, and all the other global assessments, cannot serve to monitor changes in the state of the earth's land cover. In 20 years' time, these data sets will be of no use in determining either the trend or the direction of land-cover change.

This present lack of objective and systematically acquired data about global land cover is an appalling state of affairs. Yet, it does not need to be so. The data do exist, although they have not yet been used on global scales. The surface of the earth has been observed by satellites on a systematic basis since 1972. Today there are several operational satellites continuously observing the earth with sensors that provide data that are more than adequate to measure the extent, severity, and dynamics of landscape degradation in the grasslands and savannas. The absurdity of this situation is best emphasized by noting that, as the result of our past priorities for scientific research, there are far more detailed and accurate maps of the surface of the moon than there are of the earth.

The consequences of desertification are cumulative, and there is no obvious way that the ecological destruction can be repaired or rehabilitated or even stabilized. In terms of ecosystem transformation, much of which is destructive, the global prognosis is a one-way street; more of the same at an accelerating rate as human populations grow and accessible technologies change. The impacts of humans on ecosystems are much amplified by their available technology (Le Houérou, 1992).

Can we be more precise? Based on our diagnosis of the present situation of the global grasslands, we can put forward a prognosis for the next 50 years—to the year 2040. The prognosis will be developed by considering each of the three relevant processes of global change separately and then the probability, strength, and significance of the possible interactions between each of the three.

The increasing concentration of CO_2 in the global atmosphere is the driver of global change about which there is the least uncertainty. The atmospheric CO_2 concentration has increased by 25% over preindustrial levels and by approximately 10% since World War II. Setting aside the contribution of this and other trace gases to global warming, CO_2 is considered here only for its 'fertilizer' effect on global vegetation in general and the grasslands in particular.

There is no scientific consensus about the magnitude or significance of the CO_2 fertilization effect at global or even landscape levels. At the level of single plants under laboratory conditions, the physiological mechanics of response to increased CO_2 are well understood and predictable; but they fail when translated to the field (Agren *et al.*, 1991; Long and Hutchin, 1991). The 10% increase of the last 50 years has not had a noticeable effect on vegetation. The most obvious and parsimonious conclusion is that the influence of the climatic factors of precipitation, temperature, etc., have a more pronounced influence on plant growth than do enhanced CO_2 levels.

I propose that the CO_2 fertilizer effect, as one driver of global change, can be safely discounted as a significant influence on global vegetation generally, and the grasslands in particular, over the next 50 years. The CO_2 response of vegetated landscapes appears to be conservative under the dynamics of daily weather patterns, and any enhancement of plant growth through increased water-use efficiency is rapidly limited by nutrient levels. The adjustment of ecosystems to increasing CO_2 alone is predicted to be a very gradual process. The CO_2 concentration is not forecast to double until at least the year 2100, or even later as the biosphere and the oceans appear to be slowing the rate of atmospheric increase.

It is widely accepted that the increasing atmospheric concentrations of CO_2, methane, the chlorofluorocarbons, nitrous oxide, and other trace gases are producing a global warming of the atmosphere—a cumulative, indirect impact of human activities. The climatic consequences of this global warming are much less predictable. Nevertheless, two summary points can be made. The first is that ecosystems will be affected by changes in the frequency of extreme events, such as

Table 6: Continental distribution of improved pastures

Area	Pasture ($\times 10^6$ km^2)
Africa	7.8
North and Central America	3.6
South America	4.6
Asia	6.4
Western Europe	0.6
Eastern Europe	0.2
USSR	3.7
Oceania	4.5
World	31.5

From Bouwman, 1990; reprinted by permission of John Wiley & Sons, Ltd., copyright © 1990, John Wiley & Sons, Ltd.

droughts and floods, and in sequences of these events far more than by changes in mean values. Second, global mean values are meaningless; nothing responds at a global level. Global climate changes predicted by general circulation models must be translated to regional, biome-level scenarios that are applicable to the grass-lands and the savannas. This has not yet been satisfactorily achieved. We are left with the general prediction that climate change will significantly and directly affect the grasslands of the globe, probably in very different ways, and that the most noticeable impact will be changes in the frequencies of extreme events—droughts, floods, fires—all of which have the potential to reshape dramatically the ecological nature of the grasslands. Climate change, however poorly defined, must be considered as a very significant determinant of the future of grasslands.

The last driver of global change is the cumulative, direct impact of human activities on the biosphere in general, and the grasslands in particular. As outlined above, the area of grasslands (and croplands) is being 'expanded' by tropical deforestation. Deforestation has serious global consequences when measured in terms of biogeochemical and biospheric changes as argued earlier. It is, however, best treated under forests rather than grasslands (Williams, this volume).

The human-made grasslands, the 'improved pastures,' are an integral component of the agricultural production systems of the developed world, and consequently they are very difficult to quantify and characterize separately. In global inventories, the various categories of 'pasture' sometimes include native grass-lands with the improved pastures. For example, the geographic distribution of 'pastures' provided by Bouwman (1990) aggregates to a global value comparable to that of natural grasslands as defined in this chapter (Table 6).

Land-use change affects the improved pastures, but there are no adequate global data sets to determine precisely the rates of change of aggregate areas, or the location of changes over the globe. Therefore qualitative projections will have to suffice. These are for an expansion of these human-made grasslands following deforestation, particularly in South America, and an intensification of their management and use in the developed countries. Conversely it has also been forecast that with the economic 'rationalization' of agriculture in Europe, appreciable areas of cropland, and especially pastures, will be afforested. The (nitrogenous) fertilization of these improved pastures is substantial and apparently increasing with that for the adjacent croplands. The levels of fertilization will be solely determined by economic factors, principally the balance between the costs and returns. Changes in the area and intensity of use of improved pastures will have global significance because of the relative contribution of nitrous oxide, a consequence of fertilization, to the atmosphere (see Penner, this volume).

The absence of global data sets spanning at least 50 years restricts the projections of land-use/land-cover changes in the savannas and grasslands over the next 50 years to a qualitative, but highly plausible, status. The two changes that are of principal concern are the conversion of grassland to cropland in the savannas and grasslands, and the degradation of arid grassland. The loss of grassland to agricultural land is being driven in South America, the USSR, and Central Asia by economic forces and in Africa by the expansion of subsistence requirements. As has been argued elsewhere in this chapter, the consequent land-cover changes have global significance.

The figures for the extent of desertification (Tables 4 and 5), however rubbery, are the only quantitative evidence I can produce to support the very widespread concern among ecologists that the greatest threat posed to grassland ecosystems, as well as to the rainforests and savannas, is their diminution and, probably, effective destruction because of human land use, principally, but not exclusively, driven by the subsistence requirements of a burgeoning human population.

It is unfortunately obvious that very few people are able to contemplate realistically the biospheric consequences of a doubled human population. Some cannot; most will not. Most of the population growth will be in the tropical belt. This is the geographic location of the land-cover types that, for independent climatological, biogeochemical, and biological reasons, have been identified in this chapter as the most critical areas of land-cover change. During the next few decades, the grasslands, savannas, and rainforests will continue to be transformed, but at an ever-increasing rate. In terms of absolute population increases, the full impact of the global population is yet to come (Demeny, 1990).

Direct, human-induced change is forecast to have far more extensive and degrading impacts on the grasslands than climate change as now understood. Further, climate change is expected to exacerbate the human-induced change rather than ameliorate it (Verstraete and Schwartz, 1991; Le Houérou, 1992; Mainguet, 1991) (Plate 13).

References

Agren, G. I., R. E. McMurtrie, W. J. Parton, J. Pastor, and H. H. Shugart. 1991. State-of-the-art models of production-decomposition linkages in conifer and grassland ecosystems. *Ecological*

Applications 1, 118–138.

Alayev, E. B., Y. P. Badenkov, and N. A. Karavaeva. 1990. The Russian Plain. In *The Earth as Transformed by Human Action* (B. L. Turner II, W. C. Clark, R. W. Kates, J. F. Richards, J. T. Mathews, and W. B. Meyer, eds.) Cambridge University Press, Cambridge, U.K., 543–560.

Axelrod, D. I. 1985. Rise of the grassland biome, central North America. *Botanical Review 51*, 163–202.

Berry, L., L. A. Lewis, and C. Williams. 1990. East African Highlands. In *The Earth as Transformed by Human Action* (B. L. Turner II, W. C. Clark, R. W. Kates, J. F. Richards, J. T. Mathews, and W. B. Meyer, eds.), Cambridge University Press, Cambridge, U.K., 533–541.

Bouwman, A. F. 1990. *Soils and the Greenhouse Effect*. John Wiley and Sons, Chichester, U.K., 574 pp.

Buringh, P., and R. Dudal. 1987. Agricultural land use in space and time. In *Land Transformation in Agriculture* (M. G. Wolman, and F. G. A. Fournier, eds.), SCOPE 32, John Wiley and Sons, Chichester, U.K., 9–43.

Collins, S. L., and L. L. Wallace, eds. 1990. *Fire in the North American Tallgrass Prairies*. University of Oklahoma Press, Norman, Oklahoma, 175 pp.

Delcourt, H. R., and P. A. Delcourt. 1991. *Quaternary Ecology: A Palaeoecological Perspective*. Chapman and Hall, London, 241 pp.

Demeny, P. 1990. Population. In *The Earth as Transformed by Human Action* (B.L. Turner II, W. C. Clark, R. W. Kates, J. F. Richards, J. T. Mathews, and W. B. Meyer, eds.), Cambridge University Press, Cambridge, U.K., 41–54.

de Van Booysen, P., and N. M. Tainton, eds. 1984. *Ecological Effects of Fire in South African Ecosystems*. Springer-Verlag, New York.

De Wet, J.M. 1981. Grasses and the cultural history of man. *Annals of the Missouri Botanical Garden 68*, 87–104.

FAO (U.N. Food and Agriculture Organization). 1987. *Production Yearbook*. FAO, Rome, Italy.

FAO. 1990. *Production Yearbook*. FAO, Rome, Italy.

Graetz, R. D. 1991. Desertification: A tale of two feedbacks. In *Ecosystem Experiments* (H. Mooney, E. Medina, D. W. Schindler, E.-D. Schulze, and B. H. Walker, eds.), SCOPE 45, John Wiley and Sons, Chichester, U.K., 59–88.

Huntley, B. J., and B. H. Walker, eds. 1982. *Ecology of Tropical Savannas*. Springer-Verlag, Berlin, Germany, 669 pp.

Lamprey, H. F. 1983. Pastoralism yesterday and today: The overgrazing problem. In *Tropical Savannas* (F. Bourliere, ed.), Elsevier, Amsterdam, The Netherlands, 643–666.

Le Houérou, H. N. 1980. The rangelands of the Sahel. *Journal of Rangeland Management 33*, 41–46.

Le Houérou, H. N. 1992. Climatic change and desertization. *Impact of Science on Society 42*, 183–201.

Little, P. D., and M. M. Horowitz, eds. 1987. *Lands at Risk in the Third World: Local-Level Perspectives*. Westview Press, Boulder, Colorado, 285 pp.

Long, S. P., and P. R. Hutchin. 1991. Primary production in grasslands and coniferous forests with climate change: An overview. *Ecological Applications 1*, 139–156.

Mabbutt, J. A. 1984. A new global assessment of the status and trends of desertification. *Environmental Conservation 11*, 103–113.

Mainguet, M. M. 1991. *Desertification: Natural Background and Human Mismanagement.* Springer-Verlag, Berlin, Germany, 306 pp.

Matthews, E. 1983. Global vegetation and land use: New high-resolution data bases for climate studies. *Journal of Climate and Applied Meteorology 22*, 474–487.

Reader, J. 1990. *Man on Earth.* Penguin, Harmondsworth, U.K., 256 pp.

Richards, J. 1990. Land transformation. In *The Earth as Transformed by Human Action* (B. L. Turner II, W. C. Clark, R. W. Kates, J. F. Richards, J. T. Mathews, and W. B. Meyer, eds.), Cambridge University Press, Cambridge, U.K., 163–178.

Sandford, S. 1983. *Management of Pastoral Development in the Third World.* John Wiley and Sons, New York, 225 pp.

Schlesinger, W. H., J. F. Reynolds, G. L. Cunningham, L. F. Huenneke, W. M. Jarrell, R. A. Virginia, and W. G. Whetford. 1990. Biological feedbacks in global desertification. *Science 247*, 1043–1048.

Simmons, I. G. 1987. Transformation of the land in pre-industrial time. In *Land Transformation in Agriculture* (M. G. Wolman and F. G. A. Fournier, eds.), SCOPE 32, John Wiley and Sons, Chichester, U.K., 45–77.

Sinclair, A. R. E., and J. M. Fryxell. 1985. The Sahel of Africa: Ecology of a disaster. *Canadian Journal of Zoology 63*, 987–994.

Singh, J. S., W. K. Lauenroth, and D. G. Milchunas. 1983. Geography of grassland ecosystems. *Progress in Physical Geography 7*, 46–79.

Stebbins, G. L. 1981. Co-evolution of grasses and herbivores. *Annals of the Missouri Botanical Garden 68*, 75–86.

Tothill, J. C., and J. J. Mott, eds. 1985. *Ecology and Management of the World's Savannas.* Australian Academy of Science, Canberra, Australia, 384 pp.

Verstraete, M. M., and S. A. Schwartz. 1991. Desertification and global change. *Vegetatio 91*, 3–13.

Werger, M. J. A. 1983. Tropical grasslands, savannas, woodlands: Natural and manmade. In *Man's Impact on Vegetation* (W. Holzner, M.J.A. Werger, and I. Ikusima, eds.), W. Junk, The Hague, The Netherlands, 107–137.

Wolman, M. G., and F. G. A. Fournier, eds. 1987. *Land Transformation in Agriculture.* SCOPE 32, John Wiley and Sons, Chichester, U.K., 521 pp.

Woodward, F. I., and I. F. McKee. 1991. Vegetation and climate. *Environment International 17*, 535–546.

Woodward, F. I., and B. G. Williams. 1987. Climate and plant distribution at global and local scales. *Vegetatio 69*, 189–197.

WRI (World Resources Institute). 1992. *World Resources: 1992–93.* Oxford University Press, New York, 385 pp.

7

Human Settlements

Ian Douglas

Settlement refers to the occupation of land for human living space (CRGEC, 1991). Settlements are places where people live and to which they bring materials for consumption and for transformation into other objects. Activities in settlements place demands on other lands and produce wastes out of proportion to their size. Settlements are the location of concentrated emissions to the atmosphere and of discharges of contaminants to surface and groundwater. 'In the places where man's activities are most densely concentrated—his settlements—the environmental impact is greatest and the risks of environmental damage are most acute' (United Nations, 1974: 12).

Settlement as Land Cover

As land cover, settlement represents the most profound alteration of the natural environment by people, through the imposition of structures, buildings, paved surfaces, and compacted bare soils on the ground surface. Settlements also create demands that lead to other land-cover changes, such as the storage of water in reservoirs (Main, 1990); the removal of vegetation and soil to extract sand, gravel, brick clays, and rock; the replacement of vegetation by planted cover in gardens, parks, sports grounds, and golf courses; the alienation of ground for landfill and waste treatment; and the use of land for transportation routes.

The area of land actually covered by the structures of settlements is small, but it nevertheless marks the most massive change in the flows of energy, water, and materials on the earth's surface. The ability to bring water into settlements is fundamental, and from the beginnings of the first cities, water supply was a key factor, often involving long-distance transfers of water to urban settlements. Many rural settlements also depend on an ability to manage the flow of water, particularly through irrigation schemes. Any settlement, whatever its size, has some human-made structure, if only the simplest tent, that alters the flows of energy, water, and materials. In more complex settlements, an infrastructure links individual buildings or shelters. It may take the form of simple pathways where the ground surface is bare and compacted or of the paved roads and underground pipe and cable networks characteristic of the modern city. There is a constant process of

149

Table 1: Urban impacts

Action	Impacts			
	Within City	Around City	Downwind and Downstream	Rural and long-distance (including transfrontier)
Airborne pollutant emissions	Smog particulates, heavy metal fallout (lead, etc.)	Contamination of agroecosystem and natural habitats	Pollutant plumes downwind, acid rain	Acid rain, change to global CO_2 budget
Waterborne pollutant emission	Loss of aquatic life in streams, contamination of aquifers and groundwater supplies	Contamination of local water supplies, productive and recreational fish stocks, and irrigation water	Pollution of once good surface water supplies for downstream villages and towns	Deterioration of potential irrigation water, cumulative impact on major rivers (Ganges)
Solid wastes	Littering, illegal dumping, pollution of groundwater and streams, attraction of vermin, flies, and disease vectors	Dumping of night soil, alienation of terrain for landfill operations, illegal disposal of rubbish	Siltation of streams, loss of channel capacity, in combination with increased urban storm runoff, greater frequency of flooding	Possible contamination of base flow to rivers, dumping of sewage sludge in oceans
Noxious chemicals in gaseous and liquid forms	Danger of Bhopal-type disasters	Illegal disposal of chemical wastes, high danger of Minimata-type incidents	Contamination of ecosystems, especially from mine tailings and industrial wastes	Damage to aquatic systems and food chains, especially through heavy metals
Growth, delivery, marketing, and consumption of food	Traffic congestion around market areas, problems of food wastes and vermin	Intense competition for available market garden land, high use of pesticides, and possible contamination of groundwater	Wastes from food processing plant, attraction of birds and vermin to wastes	Impact of fertilizer residues on major rivers, pressure to grow crops on marginal land with consequent risks of soil erosion, siltation of reservoirs and streams

Building materials and construction activity	Severe soil erosion and ground disturbance during construction, urban flooding and siltation	Extraction of sand and gravel, leaving derelict areas unsuited for agriculture	Dust from urban area carried downwind, gravel workings affect stream channels and bank stability	Removal of forest trees to meet demands for wood, tropical forest removal affects genetic resources, water balance, and global CO_2 budget
Fuel supplies	Air pollution from coal and wood burning, fire risk from paraffin stoves	Removal of growing timber, including windbreaks to meet fuelwood demands	Drift of pollutants downward, illegal dumping of ashes into streams	Reduction of forest cover to produce charcoal, competition with rural dwellers for fuelwood, impact of mining
Water supplies	Lowered aquifer and subsidence from pumping of groundwater	Alienation of land for water storages, modification of runoff by land	Reduced base flow in some streams, increased effluent disturbances	Competition between irrigation and urban uses of water

interaction within and between these groups of structures and infrastructures and between them and the natural elements. The mere existence of built features alters the flows of the natural environment. The energy coming from the sun, the rainwater from the clouds, and the wind are all forced into different pathways. Human activities within these settlements produce noise, waste, traffic, and smoke. Air and water become polluted, and biological systems are modified. As settlements grow, the degree and complexity of environmental change caused by settlements increase, with the great cities of a million inhabitants or more being the places where people have most modified nature. Urban impacts occur within cities, around cities, downwind and downvalley, and even long distances away (Table 1). Despite the enormous alterations of natural systems, the people of great cities are not immune to the effects of extreme events in nature. Indeed, they are quite vulnerable to environmental change, at all levels from local to global.

Settlements as Land Use

Cities and, to a lesser extent, all settlements depend on the rural areas around them. The traditional models of land use around urban settlements, such as that of von Thünen (1966), illustrate the way in which settlements affect the surrounding countryside. Nearby land is used, depending on culture and level of economic activity, for fuelwood scavenging, for market gardening and weekend vegetable growing, for recreational activities such as horse riding and golf, for semirural housing on large blocks of land (the 2.5-ha lot phenomenon), and for amenity, landscape conservation values. All are urban land uses, but they involve little urban land cover.

People in settlements need food, raw materials, and relaxation, the gaining of all of which involves exploiting the natural resources of the vicinity. The supply chains of modern cities are extensive and complex, so that events in one part of the world have implications for survival or well-being in cities thousands of kilometers away. Equally, urban demands affect land cover far beyond their immediate neighborhood. Settlement expansion is thus both a compact direct change in land cover and a widespread force affecting land cover and land use in other areas.

The Changing Pattern of Human Settlements

Until quite recently, nearly all people lived in rural areas. In 1800, only 3% of the world's population lived in towns of 5000 or more (United Nations, 1974), and many of the small towns were then, as in parts of India today, effectively large villages in their socioeconomic activity. By 1900, 14% of the world's population was living in towns. The urban numbers increased rapidly after 1950, so that 40% of the world's population was urban by 1974 and almost 50% by 1991.

While the percentage of people in urban areas has been growing, the size of large settlements has been increasing enormously (Table 2), their nature has been changing, and to some extent urban and rural areas have become less distinct, par-

Table 2: Low-latitude city population growth (in millions of people)

	1950	1960	1970	1980	2000
Mexico City, Mexico	2.9	5.1	8.9	15.0	31.0
São Paulo, Brazil	2.4	4.4	8.0	13.5	25.8
Shanghai, China	5.8	7.4	10.0	13.4	22.7
Beijing, China	2.2	4.4	7.0	10.7	19.9
Rio de Janeiro, Brazil	2.9	4.4	7.1	10.6	19.0
Greater Bombay, India	2.9	4.0	5.8	8.3	17.1
Calcutta, India	4.4	5.5	6.9	8.8	16.7
Jakarta, Indonesia	1.7	2.7	4.5	7.3	16.5
Cairo/Giza/Imbaba, Egypt	2.5	3.7	5.5	7.4	13.1
Madras, India	1.4	1.7	3.0	5.4	12.9
Manila, Philippines	1.6	2.3	3.6	5.7	12.3
Bangkok, Thailand	1.4	2.2	3.2	4.9	11.9

Source: United Nations, 1985.

ticularly in the more affluent countries. Until about 200 years ago, towns were often fortified outposts of the military power controlling an area, set apart from the countryside by walls and gates. They provided safety and protection for commerce and religion and a base for government, but they visibly and symbolically divided the *civis* (the civilized) from the *rures* (the rustic). The Industrial Revolution and improvements in transportation saw the transformation of rural areas into dormitory zones and the establishment of mines, factories, and transportation nodes on greenfield sites between cities. The freeway age led to the further mixture of urban and rural with the building of major retail and leisure facilities at intersections on freeways and turnpikes bypassing cities. The growth of semiurbanized corridors between major cities is now a worldwide phenomenon, from the San Diego–Santa Barbara coastal zone of California to the Kelang Valley axis of Malaysia or the Guangzhou–Hong Kong axis in China. In the last-named country, however, the rural-urban contrast of the past remains. Most of rural Asia and Africa show little imprint of 20th-century urban growth, especially away from the major paved roads. It is important to distinguish those rural areas, largely concentrated in Europe and North America, that are provided with the full range of urban infrastructure (electricity, telecommunications, paved roads, piped treated water, education, and health services) from those, mainly in Asia, Africa, and Latin America, that are devoid of all services but the most rudimentary schooling and health care.

Rural Settlements

A fundamental difference exists between the rural settlements of the developed world, which are almost all fully integrated into the modern economy and are virtually urban in lifestyle and facilities, and the vast majority of traditional villages

153

of poorer nations, which lack modern amenities and differ vastly in lifestyle from the urban settlements of their countries. A provisional typology of rural settlements might recognize the following categories:

- Types found mainly in the developed world:
  Modified villages: urban workers' commuter villages, tourist resorts
  Resource exploitation settlements: mining or logging camps, fishing villages
  Tourist complexes: resort areas, ski villages, beach settlements
  Dispersed habitat: individual farmsteads (e.g., in the United States), isolated ranches and sheep stations
- Types found mainly in low-latitude areas:
  Traditional villages: Asian and African villages
  Planned settlements: land development schemes, Ujamaa villages
  Dispersed habitat: Amazonian colonization
  Transitory settlements: shifting cultivators, hunters and gatherers
  Plantation/ranching complexes: tea, coffee, rubber, oil palm estates

Rural Settlements in Developed Countries

Rural populations of the developed world are decreasing. Around the great conurbations, particularly in densely settled areas of Europe, rural areas are great playgrounds for city dwellers, who surge into the countryside at weekends and holidays, putting great stress on key nodes and routes. The impacts on land cover are substantial. Country lanes become choked with vehicles, and major walking and trekking routes, such as the Pennine Way in England, become overused and severely eroded through excessive trampling. This accessibility of rural areas to a highly mobile urban population puts high demands on the rural infrastructure. Parking places have to be provided for cars and buses. Catering establishments must serve more meals and dispose of the resulting waste. Recreational facilities, such as golf courses, boating lakes, and theme parks from France's Eurodisney to Florida's Disneyworld and England's Camelot create virtual urban impositions in the countryside. The surface area that is roofed or paved grows as recreational provision increases. The demand for all types of waste disposal increases. The local demand for water and electricity is raised. In this sense, the rural settlements become an adjunct to the city, an integral part of the urban system. The environmental impact is registered on the countryside, however, and often puts a strain on the existing facilities. It may, for example, affect local groundwater tables and streamflows beyond their natural capability to adjust and self-clean.

In addition to the direct impact of the urban population, rural settlements have been transformed by changes in agriculture, especially as agrobusiness has become an integral part of the global economy. Farms in the developed countries vary greatly in size and investment. Some in Europe are still small-scale family enterprises with little investment in infrastructure. Others are virtually factory installations in the countryside with large sheds and barns, machinery yards, storage tanks, and complex waste disposal problems from intensive animal production units. Livestock numbers of beef and dairy cattle and pigs (hogs) in the United States have remained constant since 1969, while those of chickens and sheep have declined and turkeys have increased (Logan, 1990). Despite this stability in numbers, confined animal operations (intensive feedlot rearing) have increased. The small, integrated, animal/grain farm has given way to large specialized grain and livestock operations. This concentration of livestock means that the grain nutrients normally returned to the same land as manure are now concentrated in localized areas.

The modern large-scale farm is environmentally akin to a small town, with a considerable impermeable area, large inputs of energy and chemicals, and problems of surface waste and water disposal. Even remote pastoral stations in inland Australia have similar characteristics, often being better equipped than small outback towns, with a full range of basic services, such as extended storage of food, machinery parts, and other supplies; vehicle repairs; communications; and some medical services (Holmes, 1988). Essentially the modern agricultural trend is to have physically large agricultural settlements with relatively few people, much capital investment in machinery, much transformation of the earth's surface, and, often, high concentrations of animals, which, in terms of waste, have as much impact on the environment as small urban settlements. All of these trends mean that although rural populations are declining in most of Europe and North America, the area occupied by such settlements is not decreasing as second homes, recreational buildings, and farm infrastructure occupy more and more land.

Rural Settlements in Developing Countries

Although some plantations and large land development schemes in tropical countries show the characteristics of the modern farms or pastoral stations described in the previous section, the majority of developing-country settlements are villages where people have a close relationship to the land and are largely dependent on their own resources and activity for their livelihoods. These settlements usually have buildings made of local materials, perhaps wood or mud bricks, with a thatched roof. Their energy resources may be locally gathered fuelwood or animal dung. They have access to the outside world by footpath, riverboat, or rough cart track. They are often unsanitary and unhealthy places.

Rural villages lack infrastructure and have to take their supplies of water and food as they find them. Dysentery from poor-quality water supplies is a major rural handicap, and, as the 1991 epidemic in Latin America shows, cholera is readily spread in such areas. Other diseases are directly related to rural poverty and poor housing. In Latin America, between 15 and 18 million people are affected by Chagas' disease; it is transmitted by blood-sucking bugs of the *Triatomidae* family, which have adapted to the specific environment of poor-quality housing (Bos, 1988). The people-environment relationships in such situations involve both the effects of the environment on people and the effects of the people on their own environment. Poor water, often contaminated by the villagers themselves, or by people in other villages upriver, weakens individuals and their ability to improve housing and living conditions generally. This combination of poor health and poor housing leads to persistent disease risks and instability. People in such conditions are not well equipped to withstand extreme natural events, such as the serious flooding caused by tropical cyclones in Bangladesh or volcanic eruptions and associated mudflows in the Philippines.

Even though tropical rural peoples are much more vulnerable to perturbations of the natural environment than are their urban counterparts, they have large collective impacts on the natural world. In the great areas of paddy rice cultivation and irrigation agriculture of Asia, the whole landscape is a pattern of fields, water channels, and settlements. Water flows are tightly controlled, and the waste from the settlements is returned to the fields. Settlements in these areas are often compact, to maximize the amount of land available for cultivation. In areas cultivated or grazed less intensively, rural settlements may sprawl and have crop production intermingled with the houses, as in the typical Malay-Indonesian *kampong* (village). A tightly knit series of dwellings with only compacted ground between them will modify the radiation balance and hydrological cycle much more than will a dispersed village with gardens around all the houses and trees shading much of the roofed area.

The areas occupied by rural settlements in the richer and poorer counties are difficult to estimate. As modern commercial farms in rich countries are like small industrial plants, they may have a large impermeable area in relation to population numbers, as for example length of road per inhabitant. Rural settlements in the developed world may therefore occupy a large area per person. Taking 3 people per ha as the density in rural settlements, the area occupied in the more developed world would be about 103 million ha. With the much more restricted development of rural settlement in the less developed world, a density of 25 per ha may be more appropriate, giving 106 million ha occupied by rural settlements or 1.5% of the world land area. The nature of most rural settlements is changing rapidly and they are likely to consume more land, especially all of the nontraditional types of settlement. Addition of such basic amenities as schools and health clinics will add to the impermeable land cover in rural villages.

Urban Settlements

Urban settlements cover a relatively small part of the earth's surface. Recent estimates suggest that urban areas containing 43% of the world's population occupy only 1% of the total land surface (Miller, 1988; see also Grübler, this volume). Globally it has been forecast that 24 million ha of cropland will be transformed to urban-industrial uses by the year 2000. This is only about 2% of the world total, but it is equivalent to the present-day food supply of some 84 million people (Simmonds, 1989). The loss of agricultural land to urbanization is most severe in the developing countries, which are expected to have 17 of the 24 million ha of urban land by the year 2000. More than 476,000 ha of land a year will become built up in developing countries in the remaining years of the 20th century (WRI, 1988).

Problems of Defining Urban Land

The urban land surface itself is far from uniform. Only the central districts of large urban areas have almost continually paved and roofed surfaces. Many suburban areas are dominated by green vegetation, especially in gardens and open spaces. Detailed investigations of four cities in the eastern United States showed average tree canopy cover to be 24–37%, with 46% of the urban areas being devoted to residential suburbs and 14% taken up by vacant land (Rowntree, 1984). In Dayton, Ohio, 58% of the city's land is not covered with artificial surfaces, and 37% of the nonsurfaced land is covered with tree crowns (Sanders and Stevens, 1984). Indeed, urban areas in the midst of intensively cropped farming country may provide far more variety of plant life and greater protection for wildlife than the adjacent rural zones. In Britain, outside London, open space for private or public use, excluding private gardens, occupies an average of 16% of the urban area (Patmore, 1970). Urban land uses are constantly changing, tending to intensify as land values increase. In the suburbs, older houses with large gardens are pulled down and replaced with four or five smaller houses or apartment buildings. The result is an increase in the impermeable, paved surface and thus in the modification of air, water, materials, and energy flows. The nature of the urban surface is particularly important in the disposal of precipitation falling on a city. The design of drainage systems has to take account of the percentage of the surface that is impermeable and will yield water directly to drains and artificial drainage channels. In North American and Australian cities the percentage of the urban area that is impermeable is on the order of 33% in industrial and commercial areas and 19% in residential areas (Nouh, 1986). Such proportions change as urban land uses are altered, the impermeable area usually tending to increase.

The Paris agglomeration illustrates the problem of defining urban settlements and of characterizing urbanized land. The agglomeration now covers some 2300 km^2 of the Ile de France region and has 8.7 million inhabitants at a density of 38 persons per ha. Within the agglomeration, however, 42% of the land remains green space, including some agricultural land between suburban towns. Only 34% of the agglomeration is covered with residential buildings, 17% with urban services and transportation facilities, and as little as 7% with commercial and industrial installations (Bastié, 1984). The whole agglomeration covers only 0.4% of France's national territory.

Within the Paris agglomeration, Bastié (1984) recognizes three different interpretations of the idea of Paris, with diverse land uses and population densities. Parts of the core of Paris are extremely crowded, the 2ème Arrondissement having 740 people per ha (Chabot, 1948). 'Paris the beautiful,' the classic central area from Notre Dame to the Arc de Triomphe and from Montmartre to Montparnasse, is 20 km^2 in area with 315 residents per ha. Beyond it is the 'Ville de Paris' proper, the area within the *boulevard periphérique*, totaling 90 km^2 if the Bois de Boulogne and de Vincennes are excluded, with 244 residents per ha. The whole agglomeration is the third Paris, with an average density of 38 people per ha. The inner suburbs have a density of about 150 per ha, while the outer suburban zone has only about 19 per ha. This great diversity in population density affects the degree to which human-made structures and resource use affect the environment. Most of the gardens and open spaces are in the outer zone of greater Paris.

Mexico City demonstrates a similar diversity of urban population density. The four central parts of the city occupy 137 km^2 with a density of 215 people per ha. The whole area of the inner zone of the Federal Territory has 580 km^2 with 70 people per ha, with some wards, such as Ixtacalco, having as many as 227 per ha. The municipalities and districts of the outer zone cover 1397 km^2 with a density of 14 people per ha. The whole metropolitan area occupies 2114 km^2 and has an average density of 42 people per ha. Again, much variation in the intensity of residential land occupancy occurs across the city. Other urban uses occur irregularly over the city, with large paved airport areas on the periphery and railway marshaling yards, transportation depots, and large roofed areas of factories closer to the city center. The global figure of 2114 km^2 for the urban area of Mexico City is inevitably an inadequate indicator of the area of urban settlement that will exert an influence on the environment. Much of the land will still be vegetated and will not alter the flows of energy, water, and materials.

Grübler (this volume) uses a figure of 130 million ha for world built-up land, or about one-half the figure of 247 million ha given here (Table 3). The important point from his work, however, is that only about 10% of this built-up area is actually covered with structures. The direct climate-modifying impact on the radiation balance is thus inevitably relatively restricted, only about 0.1 or 0.2% of the world's land surface being so completely modified by artificial, impermeable cover.

Official statistics for the land areas of urban complexes are thus of relatively little value as indicators of the environmental effects of urban settlements. A much better approach would be to use remotely sensed data to detect areas that have spectral signatures significantly different from those for the surrounding countryside. Landsat data can be used to detect outer suburban, inner suburban, and densely built-up urban core areas. Such analyses are few, however, and they are not usually comparable one with another. In one such analysis for hydrological purposes, a Landsat image of northwestern England was analyzed to determine the impervious area of the 147 km^2 Croal Catchment in Greater Manchester. Although topographic maps showed 24% of the catchment to be urbanized, only one-half that area was actually impervious. Of the fully built-up urban core, 0.9 km^2 was devoid of green space and virtually entirely impervious, and 6.5 km^2 of the inner urban area was 82% impervious, but the bulk of the urban area, covering 28 km^2, was only 45% impervious (Adi, 1990).

The U.S. Geological Survey has mapped urban land uses from satellite and

Table 3: Generalized approximation of urban land areas for the continents

	Urban Population (Millions)	Urban area ($\times 10^6$ ha)	Urban Growth Rates	
			Population per year	Area per year
World	2286	247	2.6	
More developed	898	138	1.0	
Less developed	1388	108	3.4	
Africa	229	17	4.8	
East Africa	42	3	6.2	
Central Africa	32	2	4.6	
North Africa	74	5	3.8	
South Africa	23	3	3.6	
West Africa	58	4	5.5	
Americas	532	46	2.0	
Latin America	326	25	2.6	
Caribbean	20	2	2.5	
Central America	80	6	2.9	2.5
Temperate South America	42	4	1.5	3.0
Tropical South America	184	13	2.6	3.3
North America	206	21	1.2	
United States	186	19	1.2	1.9
Asia	928	66	3.1	
East Asia	398	24	2.3	
China	250	14	2.9	3.4
South Asia	529	35	3.7	
Eastern South Asia	127	11	3.8	
Southern Asia	325	20	3.6	
India	232	14	3.5	5.7
West Asia	77	7	3.5	
Europe	376	67	0.7	
Eastern Europe	76	15	1.2	
Hungary	7	1.3 (1.2)[1]	1.1	
Northern Europe	73	15	0.3	
Finland	3	0.7 (0.5)	0.8	
United Kingdom	52	1.9	0.2	0.2
Southern Europe	99	18	1.1	
Western Europe	128	29	0.4	
Oceania	19	0.8	1.5	
Australia	15	0.6	1.3	
Papua New Guinea	0.7	0.1	4.7	
USSR	201	50	1.5	

[1]Figures in parentheses are actual figures for urban land areas in United Nations, 1985.
Areas are estimated from figures for the urban population from United Nations (1985) unless otherwise indicated. Densities of urban population are taken for characteristic cities in the continents or regions indicated and include people living in the rural parts of urban administrative areas. The areas stated are thus likely to be much larger than the intensely occupied, modified, paved, and roofed areas likely to have significant impacts on the environment.

high-altitude aerial photograph data, gaining an effective interpretation of land uses occupying areas of at least 4 ha. A special land-use and land-cover classification distinguishes various forms of urban land use, which can be mapped and expressed in areal statistics (Reed and Lewis, 1978). Such land-use information can then be used to predict the spatial distribution of air pollution and changes in urban climate (Pease *et al.*, 1980).

Landsat thematic mapper data, including thermal data, are frequently used in urban area analysis (Leak and Venugopal, 1990), although many studies cover only small parts of major metropolitan areas, and some do not adequately subdivide land-use densities within urban areas (Haack, 1983; Haack *et al.*, 1987). Other applications, especially the use of the finer-resolution data from the System probatoire d'observation de la terre (SPOT) satellite, enable land-use change to be detected, but satellite-based remote sensing technology cannot yet be used to monitor land use at the level of accuracy required by developers, engineers, planners, and real estate interests. Aerial photographs are required (Martin and Howarth, 1989). The satellite data do, however, offer great potential for defining rates of rural-to-urban land conversion for whole metropolitan regions. The data on urban areas collected here (Table 3) are from official sources and probably refer more to the sizes of administrative areas designated as urban by national authorities than to actual built-up areas.

Urban Land Area Estimates

In the United States, urban areas containing 76% of the population occupy only 2% of the land area (183 million people in 18.34 million ha of urban land, or a density of 10 per ha; Miller, 1988). In 1980, Los Angeles had 11.33 million people living on 1.23 million ha of urban land, or a density of 9.4 persons per ha (Light, 1988). Sydney, Australia, covers 0.15 million ha with 2.9 million people at a density of almost 20 per ha (Horvath and Tait, 1986). The urban population of England and Wales in 1978 occupied 1.75 million ha at a density of 22.3 per ha (Anderson, 1984). Singapore, in contrast, has a total area of 61,000 ha with a population of 2.7 million at a density of 50 per ha. Small areas within cities may have very high densities. For example, the squatter settlement of the 137-ha Tondo district of Manila, almost entirely covered with single-story structures, had 1000 people per ha in 1970 (Watt *et al.*, 1977). The variations in density influence the percentage of the urban area that remains vegetated; the intensity of energy, materials, and food consumption; and the rates per unit area at which gaseous emissions, wastewater, and solid wastes accumulate.

Rates of Addition of Urban Land

Some authors express considerable concern about the loss of agricultural land to urban development, while others complain that planning controls restrict the supply of land for new urban developments. The situation varies greatly from one country to another, and the poorer countries have the additional problem of unofficial squatter settlements becoming established on the edges of cities. For the

Table 4: Expansion of urbanized areas in the 20th century ($\times 10^6$ ha)

	Asia		Europe	North America		Latin America	
	Colombo, Sri Lanka	Pearl River Delta, China	Urban Area of England & Wales	Whole of USA	Northeastern U.S. Urban Complex	Mexico City	Santiago, Chile
1900	0.002		0.81				
1910							
1920			0.89				
1930			1.05				
1940			1.29			0.011	
1950			1.46	3.3^1	0.84^1		
1960	0.004		1.62	6.5^2	1.39^2		0.021
1970			1.78	9.1	1.79		0.031
1980		0.014	1.98			0.100	0.044
1990		0.022^3	2.18				0.057
Forecast							
2000			2.42			0.127	

[1]Value considered low by Clawson and Hall, 1973.
[2]Value considered high by Clawson and Hall, 1973.
[3]Source: Yeh et al., 1989.

United States, Hart (1976) found a total urban area in 1967 of 246,765 km², which was increasing at a rate of 4645 km² per year. The continuation of this rate of increase would result in only 4% of the U.S. land being under urban use by the year 2000, an amount that in absolute terms Hart considered relatively trivial, but deserving careful management.

Even in Britain, where 92% of the population is urban, the urban area grew by only 3% (from 8 to 11%) from 1939 to 1987 (Nicholson-Lord, 1987), and from 1986 to 1988 some 7645 ha of rural land were converted to urban uses (Elkin and McLaren, 1991). This is considerably less than the estimated annual rate of some 24,000 ha in the 1930s and 16,000 ha in the period 1962 to 1967 (Patmore, 1970).

In Canada, the decline in Ontario farmland has caused concern, but the 477 km² of farmland converted to urban uses between 1966 and 1976 by the expansion of major Ontario urban centers represent just 5% of the 9520 km² withdrawn from farmland in those ten years (Johnston and Smit, 1985). The growth of Mexico City took 53,000 ha of agricultural land from 1960 to 1980 (FAO, 1985) and was expected to require another 271,432 ha of land from 1980 to 2000 (Table 4), of which 152,002 ha would be used for housing (U.N. Center for Human Settlements, 1987). Around Lima, Peru, from 1966 to 1971, 14,000 ha of irrigated land were lost to urban growth (Blitzer *et al.*, 1981). This loss of productive farming land introduced indirect external costs of urban expansion. Not only does extra farmland have to be found elsewhere, but the food has to be carried greater distances and be stored longer at higher energy costs. In Canada, replacing the food production from one hectare of good Ontario farmland lost to urban growth requires about three hectares in the prairie, thus increasing the rate of environmental change overall. In addition, the urban expansion creates demands for clays to make bricks and for sand, gravel, and crushed rock to make concrete, cement, and roadstone, which lead to large land-use changes in rural areas that are not included in the statistics of urban land use.

India lost about 1.5 million ha of land to urban growth from 1955 to 1985, and a further 800,000 ha were expected to be so used by the year 2000 (Chhabra, 1985). Between 1941 and 1971, the growth of New Delhi consumed more than 14,000 ha of agricultural land (WRI, 1988). The giant cities of China, which account for 28% of the annual gross national industrial output, contain only 8.2% of the population and occupy only 1.5% of the land area (Chen, 1988). Even that percentage of the land area may be exaggerated, as Chinese municipalities often include large areas of the surrounding countryside, and the true urban populations and built-up areas are much less (Tang and Jenkins, 1990). Nevertheless, where Chinese government policies have deliberately encouraged urban and industrial growth, especially in the Guangzhou–Hong Kong corridor of the Pearl River Delta, rates of urban land conversion are high. From 121.23 km^2 in 1978, the area expanded by over 6% per year to 170.13 km^2 in 1984, and it is growing at least as rapidly into the 1990s (Yeh *et al.*, 1989). Elsewhere the land for new urban development may come in part from the reuse of former industrial and transportation land, such as the redevelopment of many waterfront and dockland areas. British data (Table 5) show that nearly as much urban land as rural land was converted to new urban uses in the three years 1986 to 1988.

In some areas, urban expansion changes especially sensitive environments, particularly wetlands. Most cities have expanded onto their floodplains, narrowing the channels available for the passage of stormwater. Coastal and lakeside cities also encroach on marshes and mangrove swamps. Toronto, Canada, expanded its industrial, utility, and residential activities onto marshes from 1931 to 1976, reclaiming 271 of the 482 ha of marshes. Since then, marshland loss along the waterfront has accelerated (Lemay and Mulamootil, 1984). A culturally significant loss of this type has occurred around Mexico City, where the traditional garden plots, or *chinampas*, on Lake Xochimilco have been affected by pollution and tourism to such an extent that the area is now regarded as potential residential and commercial development land (Outerbridge, 1987). Most marked has been the virtual disappearance of the lakes in the Valley of Mexico. Lake Texcoco, originally 14,500 ha in extent, now covers only 1500 ha (Ibarra *et al.*, 1987). By contrast, some older cities are clearing now-derelict industrial sites along their river valleys and converting once impermeable areas back to grass and tree cover. The work of the Groundwork Trust in the United Kingdom is a good example of such restoration. Nevertheless, the overriding trend is for the areas occupied by cities to grow faster than the urban population, which in turn grows faster than that of countries as a whole. Countries of rapid population growth are therefore more likely to have high rates of conversion to settlement-related land covers and uses.

Table 5: British land-use changes to urban areas (annual average hectarage, 1986–88)

	New Use					
	Residential	Transport & Utilities	Industry & Commerce	Community Services	Vacant	Total
Rural	4155	1660	905	455	470	7645
Agriculture	3508	1345	670	285	105	5920
Forestry & open land	310	215	100	50	80	755
Minerals & landfill	60	40	55	10	210	380
Outdoor recreation	240	50	45	95	30	450
Defense	35	10	35	15	50	140
Urban	3445	890	1740	390	1065	7530
Residential	1230	60	90	50	190	1625
Transport & utilities	75	470	185	15	270	1020
Industry & commerce	335	55	825	25	515	1750
Community services	190	10	30	185	55	470
Vacant	1610	300	605	115	35	2660
All uses	7600	2550	2645	840	1535	15,175
Net change in land in particular use						
1986	5700	2415	865	320	−1480	6410
1987	5770	2465	640	340	−1560	6620
1988	6255	2775	850	400	−1675	7190

Note: Figures may not sum to totals owing to rounding.
From Elkin and McLaren, 1991.

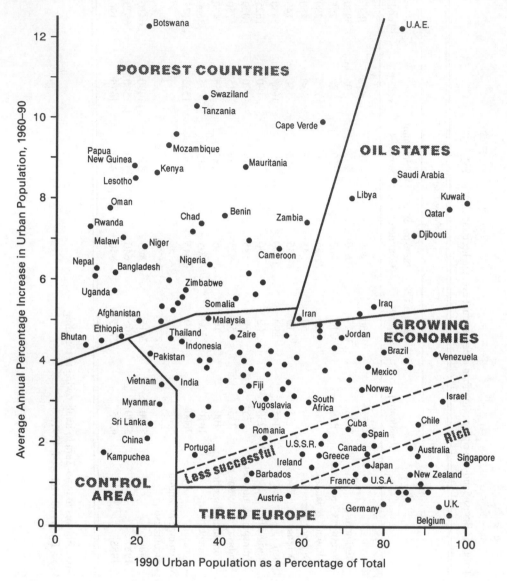

Figure 1. Typology of urban growth.

Identifying the Areas of Rapid Settlement Expansion

Expansion of Urban Settlements

Urban population growth rates by country provide a surrogate measure for the expansion of urban areas. By plotting the annual percentage increase in the urban population against the percentage of the total population of the country living in urban areas, a reasonably meaningful typology of urban growth can be obtained

(Figure 1). The upper left-hand segment of the graph contains many of the world's poorest countries, mainly in Africa, but also including Bangladesh, Nepal, Bhutan, and Afghanistan. In all of these countries, urban growth is rapid and is likely to cause the conversion of much agricultural or grazing land to urban uses. Urban infrastructure development is unlikely to keep pace with urban population growth. Both local environmental impacts, such as deterioration of water quality in streams and an increased potential for harboring disease vectors, and offsite land-cover changes, such as the loss of woodland and forest to meet urban fuelwood demands, are likely to occur.

The top right-hand part of the graph has a wide scattering of points representing the oil states. The small countries of the Persian Gulf show high urban growth with already a large percentage of the urban population living in cities. Such growth is fed by the arrival of migrant workers from neighboring countries such as Pakistan. Iran and Iraq have large rural populations, but their cities are growing more rapidly than those of the newly industrializing nations such as Brazil, Mexico, or Malaysia. Expansion of cities in these arid lands has enormous implications for the water resources of those countries. The costly experiments with center-pivot irrigation in Libya and Saudi Arabia are an example of the land-cover adjustments induced by rapid dryland urbanization.

The growing economies sector comprises a vast range of countries, from Norway, whose economic situation was transformed by North Sea oil, to the big, complex economies of India and Brazil. A few European countries (Portugal, Yugoslavia, and Romania) that lie within this sector may more properly be seen as part of the group below, the less successful 'Westernized' or industrial economies, including Cuba, Chile, and the USSR as well as Ireland, Spain, and Greece. Some countries of Western Europe such as France are associated with the urban growth rates of Canada, Australia, the United States, and New Zealand, labeled 'rich.' At the bottom, the 'tired Europe' category groups the old industrial nations, including the United Kingdom, Belgium, and Germany. This last group represents the area where conversion of land to urban uses is likely to be least marked and, as these countries have tight planning controls, least prone to introduce major unwanted environmental impacts. Another form of control may be affecting the countries in the lower left of the graph. These Asian regimes all seem to have strict controls on the movement of their peoples and may also have been affected by recent internal conflicts.

Overall Expansion of Settlements

An indication of the overall differences in rates of expansion of settlements may be gained by examining the countries that have the fastest rates of increase of both urban and rural populations. In the group of most rapid growth (Table 6) are Kenya, Oman, Qatar, and the United Arab Emirates. Kenya is by far the largest of these countries and an area where land conversion issues will be of critical importance. The countries listed in this table are those toward which attention should be

Table 6: Countries with rapidly expanding human settlements

Extreme urban and rural growth:[1]

Kenya	Qatar
Oman	United Arab Emirates

High urban and rural growth:[2]

Bangladesh	Niger
Botswana	Nigeria
Comoros	Papua New Guinea
Cote d'Ivoire	Rwanda
Djibouti	Somalia
Gambia	Sudan
Honduras	Tanzania
Iran	Togo
Madagascar	Uganda
Malawi	Zimbabwe
Nepal	

Moderate urban but high rural growth:[3]

Algeria	Mongolia
Burkina Faso	Pakistan
Costa Rica	Paraguay
Ethiopia	Syria
Mali	Thailand

High urban but moderate rural growth:[4]

Angola	Liberia
Benin	Mozambique
Burundi	Sierra Leone
Chad	Solomon Islands
Gabon	Swaziland
Laos	Yemen
Lesotho	Zambia

---

[1] Urban population growing more than 7.5% per year;
rural population growing more than 3.0% per year.
[2] Urban population growing more than 5.0% per year;
rural population growing more than 2.0% per year.
[3] Urban population growing more than 4.0% per year;
rural population growing more than 2.0% per year.
[4] Urban population growing more than 5.0% per year;
rural population growing more than 1.0% per year.

directed in terms of land-cover and land-use change. They represent only 19.5% of the world's total land area, but include nearly all of tropical Africa, where hydrologic and water resource issues are already raising extreme concerns for the sustainability of agriculture and livestock production. As Africa is the one continent that has not shared in the world's general expansion of food production per capita, the environmental management problems and threats to ecosystem stability implied by this listing are clear. While at present Africa has only 1/10th of the world's urban population in about 1/14th of the global urban area, these proportions are bound to grow rapidly. For that continent, therefore, land-cover conversion issues are particularly pressing, especially in terms of urban expansion.

Global Change Issues Related to Human Settlements

The importance of human settlements in global environmental change is related less to the area covered with buildings and other structures than to the role of settlements and settlement-related activities in promoting land-cover change generally, and the varying rates of conversion of land to urban and rural settlement uses. In much of the world, urban populations are rising by over 4% per year, which means a virtual doubling within 20 years. As industrialization and provision of infrastructure facilities will increase with such growth, the area occupied by human settlements may double in 20 to 25 years. Pushed further, such extrapolation would show human settlements as a major land-use type by the year 2100 and the actual area covered by structures reaching 1 or 2% of the global land area. Institutional constraints may emerge long before then to slow population growth and intensify rather than extend the use of land by settlements. There are, however, few signs that conversion of land to settlement uses is reaching a stage of no growth, even in the wealthiest and most stable countries. Further attention needs to be paid to the likely expansion of changes in hydrology, local radiation balances, and the emissions of atmospheric pollutants arising from the increasing concentration of the world's expanding population into ever more crowded great conurbations. In particular, low-latitude urbanization will be a key environmental issue for the 21st century.

References

Adi, S. 1990. *The Influence of Urbanization on Flood Magnitude and Storm Runoff in the Bolton Area, Greater Manchester.* Thesis, University of Manchester, U.K. 165 pp.

Anderson, M. A. 1984. Complete urban containment—A reasonable proposition? *Area 16*, 25–31.

Bastié, J. 1984. *Géographie du grand Paris.* Masson, Paris, France, 208 pp.

Blitzer, S., J. E. Hardoy, and D. Satterthwaite. 1981. Shelter: People's needs and governments' response. *Ekistics 48*, 4–13.

Bos, R. 1988. Chagas' disease vector control: The role of improved housing. *WHO Rural and Urban Development Network Newsletter 5*, 2–4.

Chabot, G. 1948. *Les Villes.* Colin, Paris, France, 224 pp.

Chen, X. 1988. Giant cities and the urban hierarchy in China. In *The Metropolis Era,* Vol. 1, *A World of Giant Cities* (M. Dogan and J.D. Kasarda, eds.), Sage Publications, Newbury Park, California, 225–251.

Chhabra, R. 1985. India: Environmental degradation, urban slums, political tension. *Draper Fund Report 14*, 1–6.

Clawson, M. and Hall, P. 1973. *Planning and Urban Growth: An Anglo-American Comparison.* The John Hopkins University Press, Baltimore, Maryland.

CRGEC (Committee for Research on Global Environmental Change). 1991. *Report of the Working Group on Land-Use Change.* Social Science Research Council, Washington, D.C., 20 pp.

Elkin, T., and D. McLaren. 1991. *Reviving the City.* Friends of the Earth, London, 278 pp.

FAO (U.N. Food and Agriculture Organization). 1985. Urbanization: A growing challenge to agriculture and food systems in developing countries. In *The State of Food and Agriculture 1984,* Agriculture Series No. 18, FAO, Rome, Italy, 79–124.

Haack, B. N. 1983. An analysis of thematic mapper simulator data for urban environments. *Remote Sensing of the Environment 13,* 265–275.

Haack, B., N. Bryant, and S. Adams. 1987. An assessment of Landsat MSS and TM data for urban and near-urban land-cover digital classification. *Remote Sensing of the Environment 21,* 201–213.

Hart, J. F. 1976. Urban encroachment on rural areas. *Geographical Review 66,* 1–17.

Holmes, J. H. 1988. Remote settlements. In *The Australian Experience* (R.L. Heathcote, ed.), Longman Cheshire, Melbourne, Australia, 68–84.

Horvath, R. J., and D. Tait. 1986. A social geography of Sydney: The 1981 Social Atlas. In *The Changing Face of Sydney,* Conference Papers 6, National Geographical Society of New South Wales, Sydney, Australia, 61–69.

Ibarra, V., S. Puente, F. Saavedra, and M. Schteingart. 1987. *Urban Environmental Quality: The Case of Mexico City.* Project Ecoville Working Paper No. 39, Institute of Environmental Studies, University of Toronto, Toronto, Canada, 68 pp.

Johnston, T., and B. Smit. 1985. An evaluation of the rationale for farmland preservation policy in Ontario. *Land Use Policy 2,* 225–237.

Leak, S. M., and G. Venugopal. 1990. Thematic mapper thermal infra-red data in discriminating selected urban features. *International Journal of Remote Sensing 11,* 841–857.

Lemay, M., and G. Mulamootil. 1984. A study of changing land uses in and around Toronto waterfront marshes. *Urban Ecology 8,* 313–328.

Light, I. 1988. Los Angeles. In *The Metropolis Era,* Vol. 2, *Mega-Cities* (M. Dogan and J.D. Kasarda, eds.), Sage Publications, Newbury Park, California, 56–96.

Logan, T. J. 1990. Sustainable agriculture and water quality. In *Sustainable Agricultural Systems* (R. Edwards, R. Lal, P. Madden, R.H. Miller, and G. House, eds.), Soil and Water Conservation Society, Ankeny, Iowa, 582–613.

Main, H. A. C. 1990. Dams and urbanization, Nigeria. *Singapore Journal of Tropical Geography 11,* 87–99.

Martin, L. R. G., and P.J. Howarth. 1989. Change-direction accuracy assessment using SPOT multispectral imagery of the rural-urban fringe. *Remote Sensing of the Environment 30,* 55–66.

Miller, G. T., Jr. 1988. *Living in the Environment,* sixth ed. Wadsworth, Belmont, California.

Nicholson-Lord, D. 1987. *The Greening of the Cities.* Routledge and Kegan Paul, London, 270 pp

Nouh, M. 1986. Effect of model calibration in the least-cost design of stormwater drainage systems. In *Urban Drainage Modelling* (C. Maksimovic and M. Radjokovic, eds.), Pergamon, Oxford, U.K., 61–71.

Outerbridge, T. 1987. The disappearing chinampas of Xochimilco. *The Ecologist 17,* 76–83.

Patmore, J. A. 1970. *Land and Leisure.* Penguin Books, Harmondsworth, U.K., 338 pp.

Pease, R. W., C. B. Jenner, and J. E. Lewis. 1980. *The Influences of Land Use and Land Cover on Climate: An Analysis of the Washington-Baltimore Area That Couples Remote Sensing with Numerical Simulation.* USGS Professional Paper 1099-A, U.S. Geological Survey, Washington,

D.C., 39 pp.

Reed, W. E., and J. E. Lewis. 1978. *Land Use and Land Cover Information and Air-Quality Planning*. USGS Professional Paper 1099-B, U.S. Geological Survey, Washington, D.C., 43 pp.

Rowntree, P. A. 1984. Forest canopy cover and land use in four eastern United States cities. *Urban Ecology 8*, 55–67.

Sanders, R. A., and J. C. Stevens. 1984. Urban forest of Dayton, Ohio: A preliminary assessment. *Urban Ecology 8,* 91–98.

Simmonds, I. G. 1989. *Changing the Face of the Earth: Culture, Environment, History*. Blackwell, Oxford, U.K., 487 pp.

Tang, W.-S., and A. Jenkins. 1990. Urbanization: Processes, policies and patterns. In *The Geography of Contemporary China* (T. Cannon and A. Jenkins, eds.), Routledge, London, 203–223.

U.N. Center for Housing, Building and Planning. 1974. *Human Settlements: The Environmental Challenge*. Macmillan, London, 209 pp.

U.N. Center for Human Settlements. 1987. *National Human Settlements: Institutional Arrangements, Selected Case Studies*. United Nations, Nairobi, Kenya, 373 pp.

U.N. Population Division. 1985. *Compendium of Human Settlement Statistics 1983*, 4th ed. ST/ESA/STAT/SER.N/4, United Nations, New York, 541 pp.

von Thünen, J.H. 1966. *The Isolated State* (P. G. Hall, ed.), English ed. of *Der Isolierte Staat*, trans. C.M. Wartenberg. Pergamon Press, Oxford, U.K.

Watt, K. E. F., L. F. Molloy, C. K. Varshney, D. Weeks, and S. Wirosardjono. 1977. *The Unsteady State: Environmental Problems, Growth and Culture*. University Press of Hawaii, Honolulu, Hawaii, 287 pp.

WRI (World Resources Institute). 1988. *World Resources 1988–89*. Basic Books, New York, 372 pp.

Yeh, A.G.-O., K.C. Lam, S.M. Li, and K.Y. Wong. 1989. Spatial development of the Pearl River Delta development issues and research agenda. *Asian Geographer 8*, 1–9.

IV

ENVIRONMENTAL
CONSEQUENCES

Introduction

Many land-cover alterations set in motion further processes of environmental change. These often inadvertent changes in other systems account for much of the concern over, and scientific interest in, land transformations. Many land-cover changes that do not appear to constitute environmental degradation in their own right begin to do so when the full sum of their consequences is added up.

The chapters in this section deal with three areas of secondary impacts of land-cover change: on atmospheric chemistry/air quality, on soils, and on hydrology and water quality. An author was sought but not found to write an overview of land-cover impacts on biodiversity, another of their most notable areas of impact. (The possible direct connections of land-cover change with climate are covered in some detail by Melillo in the final section of this volume as part of a case study in natural science modeling of environmental interactions.)

Taken together, the three chapters illustrate the full range of scales over which those impacts operate, from the very local to the globally systemic. The impacts on soils examined by Buol—erosion and chemical and physical deterioration—are largely registered in the same locale as the land-cover changes; by some definitions, indeed, they would be classed as land-cover changes themselves. The impacts of land transformation on hydrology and water quality discussed by Rogers may be felt some distance away within the same hydrologic unit, in the form of pollution or changes in flows downstream. The emissions of a number of key trace gases, as Penner shows, operate over a range of spatial scales depending on the reactivity of the species involved, but many are global in reach. Time scales of impact vary as well; many changes in soil, for example, as well as some in the atmosphere, are reversible only over very long periods. Other things being equal, the greater the distance between source and impact, the more difficult may be its management. Yet even in the case of soils, adequate management is not simple. Buol emphasizes the depletion of soil nutrients as a key impact through its consequences for the vegetation cover, and though fertilizer inputs for avoiding such depletion exist, they may not be applied because of socioeconomic constraints on the soil user. Not all changes, it should be added, require management. As Rogers indicates in detail in the realm of hydrology, the costs of responding may well exceed the benefits gained thereby. An accounting of the rough balance of the two is an essential step in analysis.

Here, as in the area of land-use/cover changes, the global data available are currently less than satisfactory. Penner's review shows that land transformations are a key element in the anthropogenic releases of many trace gases, but also that large uncertainties still exist regarding their magnitude, especially where biological processes are involved. Buol outlines the rough global distribution of major impacts on soils, but points out the scale problems in generalization involved as well as in the incomplete nature of the data base—issues equally apparent in Rogers's review of water flows and quality.

8

Atmospheric Chemistry and Air Quality

Joyce E. Penner

Introduction

Our ability to impact the local environment has been evident for some time. Since the pioneering studies of Haagen-Smit in the 1950s to explain the photochemical formation of smog, we have been aware that fossil fuel emissions can impact local air quality. During the 1960s and 1970s, we became increasingly aware of the impacts of the long-range transport of fossil fuel emissions. These impacts manifest themselves on a more continental scale, the regional scale of acid rain and acid deposition, for example.

During the last two decades as well, it has become increasingly clear that human activities can also result in changes to much larger areas, even the entire globe. On these scales, not only may the chemistry of the atmosphere change, but the climate of our planet may change as well. The most obvious changes to climate are driven by increases in CO_2 (see Table 1 for chemical names and formulas), a species whose lifetime is long and whose impact on temperatures affects reaction rate constants and the amount of water vapor and clouds in the atmosphere. Much of the increase in CO_2 is driven by increases in fossil fuel use, but land-use change through removal of forested regions was historically of more importance than fossil fuel burning and remains important today. More recently, trends in more minor species, for example, CH_4, N_2O, and the halocarbons, have been documented. These species also interact radiatively in the atmosphere to alter temperatures. They interact chemically in the atmosphere to impact other species whose lifetimes are much shorter (e.g., tropospheric OH and stratospheric O_3). Identifying the sources responsible for the trends in these species is an important problem in atmospheric chemistry. As we shall see, many of the sources of these species are related to specific land-use (or land-cover) categories. Additionally, all of these gases have important sources that are related to other human activities (e.g., fossil fuel burning, etc.).

In addition to the well-documented changes in CO_2, CH_4, N_2O, and the halocarbons, a number of other components of the earth's atmosphere may be changing. The abundance of CO, for example, with its substantial increased concentration in the Northern Hemisphere, is known to have been perturbed by fossil fuel burning, but biomass burning, which is mainly in tropical regions, is also thought to be important. Increases in tropospheric O_3 concentrations have been documented for

Table 1: Chemicals named in this chapter

CH_3CCl_3	methyl chloroform
CH_3Br	methyl bromide
CH_3Cl	methyl chloride
CH_3O_2	methyl peroxy free radical
CH_4	methane
Cl	chlorine free radical
CO	carbon monoxide
CO_2	carbon dioxide
COS	carbonyl sulfide
CS_2	carbon disulfide
DMS	dimethylsulfide (CH_3SCH_3)
HNO_2	nitrous acid
HNO_3	nitric acid
HNO_4	pernitric acid
HO_2	hydroperoxyl free radical
HO_x	odd hydrogen
H_2O	water vapor
H_2O_2	hydrogen peroxide
H_2S	hydrogen sulfide
H_2SO_4	sulfuric acid
M	N_2 or O_2; acts as a third body in reaction
NH_3	ammonia
NH_4^+	ammonium ion
$(NH_4)_2SO_4$	ammonium sulfate
NO	nitric oxide
NO_2	nitrogen dioxide
NO_3	nitrogen trioxide
NO_3^-	nitrate ion
NO_x	$NO + NO_2$
NO_y	reactive nitrogen
N_2	diatomic nitrogen
N_2O	nitrous oxide
N_2O_5	dinitrogen pentoxide
$O(^1D)$	electronically excited oxygen atom
O_3	ozone
OH	hydroxyl free radical
PAN	peroxyacetyl nitrate
RO_2	generic organic peroxy free radical
SO_2	sulfur dioxide
SO_4^{2-}	sulfate ion
SO_x	reactive sulfur

a number of specific sites. These documented site-specific increases imply increases that are at least regional in character; because of spatial heterogeneity, it is difficult to quantify the global increase in O_3. Increases in NO_y, NMHCs, SO_x, and aerosols have surely occurred as well, but these increases have not been documented. The short lifetimes and heterogeneous spatial distributions exhibited by these shorter-lived atmospheric components have made it impossible to detect or quantify a trend with any confidence. Quantification of the sources of these species, however, and of the fractions of sources due to human activity implies that the abundances of these shorter-lived species must have changed relative to those present during preindustrial times. A major challenge for atmospheric chemists is to quantify the magnitude of the changes in the anthropogenic sources relative to the natural sources. Identifying the sources of these species that are associated with particular land-use and land-cover categories is an important part of this task. A second major challenge is to quantify the global or regional extent of the species abundance change resulting from the change in sources.

The sources of many of these gases and of atmospheric aerosols are varied, but each includes components that are related to specific land-use and land-cover categories. Thus, in addition to changes from increased industrialization and fossil fuel burning, changes in land use or land cover are inexorably tied to changes in the concentrations and abundance patterns for these gases and for atmospheric aerosols. Estimates of the role of land use in species budgets are uncertain because data bases have been poorly documented with respect to both the land cover or vegetation type and with respect to the total carbon abundance and trace species emissions from various land-cover types. Nevertheless, current data show that land use or land cover can be significant in the budgets for many species. Knowing this, we may infer that changes in land use or land cover will lead to changes in chemical fluxes that are also significant. We know, for example, that approximately 20% of the CO_2 source is thought to be due to land-clearing practices today, primarily in the tropics. In preindustrial times, changes in land use were responsible for most of the increase in the abundance of atmospheric CO_2. The role of changing land-use practices in the budgets of other species and the changes in these budgets over time are not known, but may also have been important.

Trace species in the troposphere may be divided into two categories: those with atmospheric lifetimes longer than a few months to a year, and those with lifetimes significantly less than one year (see Table 2a and 2b). Increases in the sources of the former will lead to significant global accumulations. Their long lifetimes assure that changes in their tropospheric abundances will penetrate into the stratosphere, causing changes in the chemical cycles that determine stratospheric O_3. The species in this category include CO_2, CH_4, N_2O, and COS. Carbon monoxide has a lifetime of several months and is treated in this category as well. Short-lived atmospheric trace species have distributions that exhibit significant regional and temporal variability. Increases in the sources for these species can contribute to regional increases leading to degradation of air quality and acid rain, for example. In some cases, observational evidence exists for more widespread, global changes. Also, model calculations have shown that larger-scale changes in the concentrations of some of these species may be taking place (Penner, 1990; Penner *et al.*, 1991a; Langner and Rodhe, 1990; Erickson *et al.*, 1991). Photochemical interactions of species in this category can therefore lead to larger

Table 2a: Long-lived species with sources related to land use categories

Species	Typical abundance	Recent trend (%/yr)	Lifetime (yr)	Importance
CO_2	350 ppm	0.5	50–200	Greenhouse gas; Affects stratospheric O_3
CH_4	1.7 ppm	1.0	10	Greenhouse gas; Affects tropospheric and stratospheric O_3; Sink for tropospheric OH; Source of stratospheric H_2O
N_2O	310 ppb	0.2–0.3	130–170	Greenhouse gas; Source of stratospheric NO_x; Affects stratospheric O_3
COS	500 ppt	?	2	Forms aerosol in stratosphere which acts to cool climate and reacts heterogeneously to affect O_3
CO	120 ppb (NH)	~1 (NH)	0.15–0.3	Produces O_3 in high NO_x areas; Sink for tropospheric OH
	60 ppb (SH)	0 (SH)	—	

NH, Northern Hemisphere; SH, Southern Hemisphere.

Table 2b: Short-lived species with sources related to land use categories

Species	Typical surface abundance		
	Ocean	Rural	Urban[1]
NMHCs	20–1000 ppt C	15–100 ppb C	50–1500 ppb C
NO_x	10–50 ppt	1–5 ppb	50–500 ppb
SO_2	25–75 ppt	1–5 ppb	50–2000 ppb
DMS	50–150 ppt	–	–
Tropospheric Aerosol	100–500 cm^{-3}	10^3–10^4 cm^{-3}	10^4–10^6 cm^{-3}

[1]The upper ranges refer to concentrations in severe air pollution episodes as summarized by Finlayson-Pitts and Pitts (1986).

and even global-scale changes in tropospheric trace gas and aerosol concentrations. Specifically, important changes to tropospheric OH and O_3 concentrations may be occurring, with the former affecting trace species lifetimes and the latter affecting climate. Also, increases in the tropospheric aerosol abundance may have had a significant influence on climate. The short-lived species of concern in this category include the NMHCs, NO_y, SO_x, and tropospheric aerosols.

Table 2 gives a summary of the species treated in this chapter. Table 2a summarizes data concerning the concentrations and estimated recent trends for each of the longer-lived species, as well as their estimated lifetimes. The table also summarizes the primary importance of each component for atmospheric chemistry. Table 2b summarizes data for the concentrations of the shorter-lived species. Since these are so variable, I have separately listed the typical concentrations observed in clean ocean areas, rural environments, and urban environments. There are no observationally confirmed trends for these species.

In the following sections, I review the importance of each of the species listed in Table 2 for atmospheric chemistry, air quality, and climate. I then review current estimates of the sources for each species, deriving the fraction of each source that is due to specific land-use practices or land-cover categories. For species whose trends are known, it is possible to project increases into the future if the estimated sources from human activity and land-use change can be projected and if the known atmospheric sinks and the interactions in atmospheric chemistry and climate change are taken into account. Regional trends in the short-lived species can be projected as well, assuming the estimated sources and sinks are correct. However, significant uncertainties continue to surround the estimated budgets for most of these species. These uncertainties are also discussed below.

Long-Lived Atmospheric Trace Species

Carbon Dioxide

The effects of CO_2 on atmospheric chemistry are mainly through its temperature effects. It is the most abundant gas in the atmosphere after N_2, O_2, H_2O, and Ar, and it has been increasing recently at a rate of 0.5%/yr. It is the most important of the greenhouse gases, explaining roughly 60% of the increased radiative forcing in today's atmosphere relative to the preindustrial atmosphere (Shine *et al.*, 1990). Besides increasing tropospheric temperatures, increases in CO_2 also lead to decreases in stratospheric temperatures. This causes changes to atmospheric reaction rates that ameliorate, to some extent, the projected decreases in stratospheric O_3 due to increasing chlorofluorocarbon concentrations (Penner, 1980). Although CO_2 concentrations are observed to be increasing, projections are difficult because the sources and sinks of CO_2 are not well known.

Atmospheric CO_2 is exchanged with the biosphere on short time scales so that its observed seasonal changes (which are close to 100 Gt/yr; Gt $= 10^{15}$ g) are primarily due to seasonal changes in the amount of CO_2 sequestered by the terrestrial and oceanic biosphere. On longer time scales, CO_2 is taken up by the ocean and mixed into the deep ocean waters. This represents a net sink for atmospheric CO_2 because the time scales for mixing into the deep ocean are so long (several hun-

Environmental Consequences

Environmental Consequences

Table 3: Estimated sources and sinks of CO_2 (Gt C/yr)

Budget Factor	Release	Accumulation
Fossil fuel source	5.7 ± 0.5	
Land clearing of forested areas	1.6 ± 1.0	
Accumulation in atmosphere		3.4 ± 0.2
Uptake by ocean		2.0 ± 0.8
Missing sink or net imbalance	1.6 ± 1.4	

dred to a thousand years). The current uptake of CO_2 by the oceans is estimated to be about 2 Gt C/yr.

The abundance of CO_2 has increased in the atmosphere by about 25% over that present during preindustrial time. The estimated role of land-use change in contributing to this increase is uncertain because the change in land use over time is poorly documented. Also, the change in carbon stocks within forested areas is poorly known. Estimates of the degradation of forest carbon stocks due to illicit logging, fuelwood gathering, and burning remain poor even today (Houghton, 1991). Estimates of the amount of carbon present in tropical closed forests vary by a factor of two (see estimates by Brown and Lugo, 1984, and Houghton et al., 1985). Taking the estimates for fossil fuel use from Marland (1989) and the estimates for release of CO_2 due to land-use change from Houghton and Skole (1990), I estimate that approximately 35% of the increase in atmospheric CO_2 between 1850 and 1985 can be attributed to land-use change, primarily deforestation. Deforestation causes an immediate increase in atmospheric CO_2 if the biomass at the deforestation site is burned. A slower release occurs as roots, stumps, slash, and twigs decay from a site that has been logged or burned and if wood products that have been removed from the site oxidize. Organic carbon associated with the soil at the site can also be oxidized and released as CO_2. As plants and trees become reestablished, of course, CO_2 is again removed from the atmosphere.

The current rate of release of CO_2 to the atmosphere from fossil fuel burning and from other industrial processes is about 5.7 ± 0.5 Gt C/yr (Marland, 1989). This may be compared to estimates for the release of CO_2 to the atmosphere from land clearing practices, which range from 0.6 to 2.6 Gt C/yr (Watson et al., 1990). Thus, current estimates of CO_2 emissions from land-use change range from 10% to almost 50% of those from fossil fuel and industry. At present, most of the land clearing takes place in wooded areas in the tropics.

The emissions of CO_2 from fossil fuel burning and from land clearing, while uncertain to some extent, do not fully balance the budget of CO_2 in the atmosphere. It is estimated that an additional 1.6 ± 1.4 Gt C/yr is being released into the atmosphere that is not accounted for by the known atmospheric accumulation rate and the estimated uptake rate in the ocean (Watson et al., 1990). This global imbalance is displayed in Table 3.

Table 4: Species which are removed or oxidized by reaction with OH

Species	Importance
CH_4	Greenhouse gas; Affects tropospheric O_3 and OH; Affects stratospheric O_3
CO	Urban pollutant; Affects tropospheric O_3 and OH cycles
CH_3CCl_3 CH_3Br CH_3Cl	Greenhouse gases; Release chlorine or bromine in stratosphere which destroys O_3
NMHCs	Urban pollutants; Enhance tropospheric O_3 in high NO_x areas; Sink for tropospheric OH
DMS H_2S SO_2	Form sulfate aerosols which contribute to acid rain and reflect solar radiation; Form cloud condensation nuclei which alter cloud properties and cool the climate

Methane

After CO_2, CH_4 is the greenhouse gas with the most impact on increases in the radiative forcing of climate (Shine *et al.*, 1990). It also plays a vital role in atmospheric chemistry through its effect on tropospheric O_3 and OH (Logan *et al.*, 1981). These two species, O_3 and OH, are important because they determine the oxidizing (and cleansing) capacity of the atmosphere. Much of the oxidizing capacity of the troposphere is determined by its HO_x content (defined as the sum of OH, HO_2, HNO_2, HNO_4, H_2O_2, CH_3O_2, and other organic radicals) and by the balance of species within the HO_x pool, particularly the OH concentration. Thus, for example, reaction with OH is the single most important scavenger for a variety of species in the troposphere (see Table 4). (NO_3 is a second important scavenger whose reactions are particularly critical at night.) The oxidation products of CH_4 act as a source for O_3 and HO_x. But since CH_4 also reacts directly with OH it mainly decreases the tropospheric OH concentration. In the stratosphere, the oxidation of CH_4 leads to the production of H_2O and odd hydrogen radicals. Odd hydrogen radicals destroy O_3 in the stratosphere through several different catalytic cycles. CH_4 reaction with Cl radicals can also interfere with the chlorine catalytic cycle which destroys O_3 in the stratosphere. This effect can lead to a net increase of stratospheric O_3 concentrations when CH_4 increases.

Methane has a current atmospheric abundance of about 1.7 ppm with about 0.1 ppm higher concentrations in the Northern Hemisphere (Watson *et al.*, 1990). Two hundred years ago the concentration of CH_4 was only about 0.8 ppm. Most of the doubling in the atmospheric concentration of CH_4 has taken place since 1900. The concentration of CH_4 in the atmosphere represents the balance achieved between a variety of sources and its removal by chemical reaction with OH. In addition, the reactions of CH_4 with Cl and $O(^1D)$ in the stratosphere each represent secondary,

Environmental Consequences

Table 5: Sources of methane (Tg CH_4/yr)

Source	Total	Range
Anthropogenic: fossil or industrial		
Coal mining, gas drilling, etc.	80	45–100
Anthropogenic: land use by man		
Landfills	40	20–70
Enteric fermentation	80	65–100
Rice paddies	110	25–170
Biomass burning	40	20–80
Land cover related:		
Natural wetlands	115	40–200
CH_4 hydrate destabilization	5	0–100
Termites	20	2–100
Fresh waters	5	1–25
Oceans	10	5–20
Total source strength	505	222–965

minor sinks as does removal by soils. We have fairly good knowledge of the total source strength for CH_4 because the sum of the sources of CH_4 must equal the sum of the removal by chemical reaction and soils plus its estimated growth rate. The reaction rate coefficient for the reaction of CH_4 with OH was recently measured by Vaghjianai and Ravishankara (1991). They found that the rate coefficient was approximately 25% slower than the rate used in previous analyses. With the new rate, approximately 430 Tg CH_4/yr is removed by reaction with OH (Fung *et al.*, 1991; Crutzen, 1991). The removal of CH_4 by soils is estimated to be 30 ± 15 Tg/yr (Born *et al.*, 1990; Crutzen, 1991), and the estimated growth rate for CH_4 in the atmosphere is approximately 45 ± 5 Tg/yr. The total source must therefore equal approximately 505 Tg CH_4/yr (Crutzen, 1991), with an uncertainty range of 400–610 Tg CH_4/yr.

Table 5 presents a summary of source estimates by category, based largely on Cicerone and Oremland (1988) but updated to account for more recent estimates of the source from termites (Lassey *et al.*, 1992). Among these sources, land use is directly related to the sources from rice paddies (agriculture), landfills (urbanization), biomass burning (forests, savannas, wood fuel burning, and agricultural wastes), and enteric fermentation (pasture or grassland).

The largest single source for CH_4 is its production through biological processes under anaerobic conditions in natural wetlands. As shown in Table 5, this source is estimated as 115 Tg CH_4/yr, based on analyses by Matthews and Fung (1987) as updated by Fung *et al.* (1991) (see also Aselmann and Crutzen, 1989). This estimate may need to be revised because new data indicate that the flux of methane from the Amazon basin is almost a factor of two higher than previously thought (Bartlett *et al.*, 1990). In addition, analyses of data from the Arctic Boundary

Layer Experiment (ABLE-3) missions indicate that the flux of CH_4 from northern wetlands is smaller than previously thought (Bartlett *et al.*, 1992).

The anaerobic microbial processes that lead to methane emissions in natural wetlands also operate in rice paddies. The estimated magnitude of the source from rice paddies is similar to estimates from wetlands, namely, 110 Tg CH_4/yr. This estimate is tentative because few data have been gathered from Asia, where over 90% of the world's area of rice paddies exist. In addition, estimates for both rice paddies and natural wetlands are uncertain and subject to change, because the biological processes responsible for the production of CH_4 are sensitive to the physical, chemical, and biological characteristics of the soils. Therefore, for example, changes in agricultural practices (fertilization, density of rice plants, etc.), nutrient availability, and extent of rice paddies will affect this source. Of course, changes in the areal extent of either rice paddies or wetlands will also affect these estimates.

Enteric fermentation, which is associated with the use of land for pasture and grazing, is the third largest source of methane. It is estimated to be approximately 80 Tg CH_4/yr with a range of 65–100 Tg CH_4/yr (Cicerone and Oremland, 1988). This source arises from ruminant animals. Emissions have been estimated by accounting for the world's population of cattle, sheep, and wild animals (Lerner *et al.*, 1988; Crutzen *et al.*, 1986).

The sources associated with fossil fuels (gas drilling, coal mining) and with methane hydrate destabilization may be estimated from studies of the ^{14}C content of atmospheric CH_4 (Wahlen *et al.*, 1989). These studies imply a combined source of approximately 100 ± 20 Tg CH_4/yr. The remaining sources (biomass burning, termites, landfills, oceans, and freshwater) vary in size from 5 to 40 Tg CH_4/yr, but together add up to a large fraction of the major sources. Estimation of the sources of CH_4 is thus made difficult by the large number of small but significant sources. According to Table 5, nearly all of the methane sources are associated with either anthropogenic activity (fossil fuel use or land use) or land cover. Land use (associated with rice paddies, landfills, biomass burning, and enteric fermentation) accounts for roughly one-half of the total CH_4 source strength. Current uncertainties are consistent with land-use practices, being associated with a range of from 25% to 80% of all sources of CH_4.

Nitrous Oxide

Nitrous oxide is important because it acts as a greenhouse gas in the atmosphere. It is also a precursor to stratospheric NO_x. It has increased from a preindustrial value of about 285 ppb to its current level of 310 ppb. The increase of N_2O in the atmosphere over the last 200 years explains roughly 4% of the increased radiative forcing experienced due to increasing greenhouse gases over this period (Shine *et al.*, 1990). Increases in N_2O can also lead to decreases in stratospheric O_3, because N_2O serves as the major stratospheric source of NO through reaction with $O(^1D)$.

Stratospheric NO_x ($NO + NO_2$) destroys O_3 through the well-known catalytic cycle:

$$NO + O_3 \rightarrow NO_2 + O_2$$
$$NO_2 + O \rightarrow NO + O_2$$

$$\text{Net: } O_3 + O \rightarrow O_2 + O_2$$

Continuous records at the Atmospheric Lifetime Experiment/Global Atmospheric Gases Experiment (ALE/GAGE) network of sites since 1978 show a global trend of increase of about 0.2 to 0.3%/yr and a 0.75 ppb deficit in N_2O in the Southern Hemisphere relative to the Northern (Watson *et al.*, 1990; Prinn *et al.*, 1990). Analyses in the 1980s indicated that Northern Hemisphere station trends were larger than those for the Southern Hemisphere stations, suggesting an increasing interhemispheric disparity (Khalil and Rasmussen, 1983), but recent analysis suggests a persistent difference (Prinn *et al.*, 1990). The trend in N_2O may be explained by a growing tropical source (perhaps attributable to tropical land disturbance) and by a growing northern midlatitude source (attributable to a combination of anthropogenic sources) (Prinn *et al.*, 1990).

The sources of N_2O are very poorly known. Because N_2O has such a long lifetime (about 150 years), even small sources can be important. Stratospheric photolysis was recently estimated to remove close to 13 Tg N/yr (Ko *et al.*, 1991), though earlier work tended to support a removal rate of close to 10 Tg N/yr (Watson *et al.*, 1990). The calculated sink from photolysis and the observed increase of N_2O (3–4.5 Tg N/yr) must be balanced by the sources of N_2O. In addition to the suspected anthropogenic sources mentioned above, N_2O is produced by a wide variety of biological processes both in the ocean and in soils. The influence of human activities on N_2O fluxes becomes particularly difficult to estimate when factors such as atmospheric deposition of nutrients (supplied by increases in deposition of nitrate from pollution sources, for example) come into play. Additional factors, such as estimation of the importance of nitrogen fertilizer on the flux of N_2O, add further complication.

The main anthropogenic source of N_2O was once thought to be combustion (e.g., Hao *et al.*, 1987), but this source was vastly overestimated in these earlier studies due to an error in sampling techniques (Muzio and Kramlich, 1988). Recently, a new study has summarized a variety of anthropogenic sources (Khalil and Rasmussen, 1992), including automobiles and coal-fired power plants (fossil fuel), nylon production, sewage disposal, biomass burning, cattle and feedlots, fertilizer application, and contaminated aquifers. Of these, the last four may be identified with land-use practices (see Table 6).

Microbial activity in soils through either denitrification (in anaerobic soils) or nitrification (in aerobic soils) is also a source of N_2O. Normally associated with the 'natural' source of N_2O, this source may be perturbed if the character of the soils is perturbed due to a change in land use or practice. For example, one measurement showed that the flux of N_2O from pasture land in Brazil was five times

Table 6: Sources of nitrous oxide (Tg N/yr)

Source	Total	Range
Anthropogenic: fossil fuel or industrial		
Combustion of fossil fuel	0.5	0.1–1.3
Nylon production	0.4	
Sewage disposal	1.0	0.2–2.0
Anthropogenic: land use by man		
Biomass burning	0.5	0.1–2.0
Cattle and feedlots	0.3	0.2–0.6
Fertilizer application	1.1	0.01–2.2
Contaminated aquifers	0.5	0.5–1.3
Land cover related:		
Tropical forest soils	3.0	2.2–3.7
Temperate forest soils	1.1	0.7–1.5
Temperate grassland	sink?	
Tropical grassland	source?	
Oceans	2.0	1.4–5.0
Total source strength	10.4	5.8–20.0

larger than the fluxes from undisturbed tropical forest soils, floodplain soils, and soils in cleared and burned areas (Matson *et al.*, 1990). Other studies, however, have shown that the fluxes from tropical pasture lands were no larger than those from forests (Davidson *et al.*, 1991; Sanhueza *et al.*, 1990). Thus, the magnitude of the soil source of N_2O is still largely unknown because of the wide variety of factors influencing the rate of emissions and the measured variability in emissions (Watson *et al.*, 1990; Bowden *et al.*, 1990). The factors that influence the variability in the flux include vegetation types, soil and climatic conditions, soil moisture, and differences in the processes that produce N_2O (nitrification, denitrification). Fertilizer addition increases the production of N_2O, but the amount of stimulation appears to depend on the availability of nitrogen in the soil, the type of fertilizer, farming methods, etc. (Keller *et al.*, 1988). Additionally, soils that have experienced high rates of N deposition over several decades may emit higher fluxes of N_2O when fertilizer is added than soils that have low concentrations of NH_4^+ (Bowden *et al.*, 1991). These complications, and others, limit our knowledge of the flux of N_2O from soils, though substantial progress has been made (Matson and Vitousek, 1990).

The quantification of the ocean source of N_2O is also poor because the partial pressure of N_2O in surface waters varies spatially and temporally. However, data suggest that N_2O is supersaturated in upwelling regions and undersaturated in areas in Antarctica and within gyres or oligotrophic areas.

Table 6 presents estimates for the contribution of each source category to the total N_2O source. For this table, I used the estimates for each category summarized

Table 7: Sources of carbonyl sulfide (Tg S/yr)[1]

Source	Total	Range
Anthropogenic: fossil fuel or industrial		
Fossil fuel burning and industry	0.07	0.02–0.2
Industrial production of CS_2	0.2	<0.37
Anthropogenic: land use by man		
Biomass burning (wood fuel, agricultural wastes, savannas, and forests)	0.1	0.04–0.2
Land cover related:		
Terrestrial sources (soils, vegetation, wetlands)	0.2	0.–0.4
Marshes	0.01	<0.03
Release of CS_2 from marshes	0.04	0.02–0.08
Oceans (from COS + CS_2)	0.5	0.1–1.2
Volcanoes	0.01	<0.025
Release of CS_2 from volcanoes	0.01	<0.05
Total source strength	1.14	0.2–2.6

[1]Sources of CS_2 are assumed to form one COS molecule after photochemical oxidation.

by Watson *et al.* (1990) and/or Khalil and Rasmussen (1992). It is clear from the table that land-use and land-cover change may play a large role in the budget for N_2O. Sources related to land use and land cover—soils, biomass burning, fertilizer application, cattle and feedlots, and contaminated aquifers—constitute about 80% of the total N_2O source listed in Table 6.

Carbonyl Sulfide

Carbonyl sulfide is present in the atmosphere at concentrations of about 500 ppt and has a lifetime against reaction with OH in the troposphere, of approximately ten years, although uptake by plants represents a more important sink and leads to an estimated lifetime of ~2 years. In the stratosphere it may either react with OH or be photolyzed to form SO_2. SO_2 reacts with OH and, after a sequence of reactions, forms SO_4^{2-}, the major component of background stratospheric aerosols. Besides direct volcanic input of sulfur compounds (which is episodic in nature), COS is the most important source of the stratospheric SO_4^{2-} aerosol. Stratospheric aerosol scatters solar radiation and can thereby cool the planet, if changes in the sources of COS occur over time. Stratospheric aerosols also act as a site for heterogeneous chemical reactions that may affect the stratospheric ozone layer. Table 7 summarizes the known sources of carbonyl sulfide.

The largest single source of atmospheric COS is emission from the oceans, both directly and through photochemical conversion of CS_2 to COS by a rapid reaction with OH. Sources of COS are poorly estimated, and sources and sinks of CS_2 are largely unknown, but their combined flux was estimated at 1–2% of the ocean's

DMS flux (Bates *et al.*, 1992). This leads to an estimate for the total ocean COS source ranging from 0.4 to 1.2 Tg S/yr. A more recent estimate of the ocean source of COS surveyed measured concentrations of COS and CS_2 in sea water and estimated the flux directly as only 0.25 Tg/yr (Chin and Davis, 1993). The estimate in Table 7 reflects an average of the values based on DMS flux and the estimate by Chin and Davis (1993).

The known anthropogenic sources of COS include the burning of fossil fuels, especially coal, and industrial processes. Industrial processes may also provide a source of CS_2. Fossil fuel burning and industrial sources of COS are thought to provide a source of about 0.07 Tg S/yr (Khalil and Rasmussen, 1984), while industrial sources of CS_2 provide a source of COS of 0.2 Tg S/yr. The only other identified anthropogenic source of COS is biomass burning. This source is associated with the burning of savannas and forests, the burning of agricultural wastes, and the use of forests for wood fuel. Biomass burning may provide between 0.04 to 0.2 Tg S/yr (Crutzen and Andreae, 1990).

COS is also emitted by and taken up by plants. The estimated source of COS from terrestrial emissions may be identified with land-use or land-cover types through its dependence on soils, vegetation, and wetlands. The magnitude of this source is still unknown. The value estimated in Table 7 is from Bates *et al.* (1992) but may be an overestimate (Castro and Galloway, 1991). The sink of COS through its uptake by growing vegetation also appears to be significant (Brown and Bell, 1986; Goldan *et al.*, 1988). Plants act as a sink for COS by absorbing it through their open leaf stomata. The magnitude of this estimated loss is between 0.1 and 0.3 Tg S/yr (Goldan *et al.*, 1988).

Marshes and volcanoes also provide small sources of COS and CS_2 to the atmosphere. Estimates of these sources were provided by Khalil and Rasmussen (1984). These estimates are small, but also uncertain (see Chin and Davis, 1993).

As Table 7 shows, if combined, the anthropogenic sources from fossil fuels and biomass burning may provide ~30% of the estimated total source of COS. If the terrestrial sources are assumed to be also subject to change (through changes in land use), the total COS source strength subject to change is roughly 50% of the total estimated source. Measurements of COS in the troposphere are too sparse to indicate the presence of any persistent trend. However, evidence is beginning to accumulate that suggests an interhemispheric gradient in COS. The observed gradient implies larger sources in the Northern Hemisphere that are perhaps of anthropogenic origin (see Johnson *et al.*, 1990; Bingemer *et al.*, 1990, Chin and Davis, 1993). Increases in anthropogenic sources might be associated with increasing concentrations of COS leading to an increase in the background stratospheric aerosol concentration, although this is not supported by spectroscopic measurements of COS over the last decade (Rinsland *et al.*, 1992). Although increased aerosol concentrations of ~5%/yr over the last ten years were recently reported by Hofmann (1990) the explanation for the increased aerosol concentrations may be aircraft emissions or solid rocket exhaust rather than increases in COS (Bekki and Pyle, 1992). An increase of this magnitude would be expected to have only a small impact on tropospheric and stratospheric temperatures, although larger changes may have important consequences (cf. Pollack *et al.*, 1981; Turco *et al.*, 1980). Changes to the stratospheric aerosol layer may, however, be important because these aerosols act as a site for heterogeneous chemical reactions that lead to stratospheric O_3 depletion (Brasseur *et al.*, 1990; Rodriguez *et al.*, 1991). A better understanding of the budget for COS is needed in order to evaluate whether changes to COS might contribute to changes to stratospheric O_3 through these mechanisms.

Reactive Species in the Troposphere

In this section I first describe the relevant sources and sinks of CO, NMHCs, and NO_x. These species, together with CH_4, are all involved in the photochemical interactions that determine the concentrations of O_3 and OH in the troposphere. Tropospheric O_3 is of concern because it is a greenhouse gas. It also acts as a respiratory irritant and can damage plants. Through its photolysis to form $O(^1D)$ and through the subsequent reaction of $O(^1D)$ with H_2O, it is also the most important source of tropospheric OH.

Following the discussion of CO, NMHCs, and NO_x, I discuss the sources of SO_x in the troposphere. The nitrogen oxide, NMHC, and sulfur emissions contribute to acid rain as well as to the formation of aerosols. The sources of tropospheric aerosols, which are related to the sources of NMHCs and NO_x as well as to the sources of SO_x, are discussed last.

Carbon Monoxide

The data base that defines the present levels of CO in the atmosphere is not large (see Watson *et al.*, 1990; Cicerone, 1988), but it does demonstrate that the Northern Hemisphere abundance, which is around 120–150 ppb, is roughly twice the average Southern Hemisphere abundance (50–60 ppb). Data have also shown that CO exhibits a seasonal cycle with a winter maximum. The seasonal cycle has increasing amplitude at higher latitudes. Recent measurements indicate an increase of about 1%/yr in the Northern Hemisphere, but trend data for the Southern Hemisphere are ambiguous (Cicerone, 1988).

The dominant sink process for CO is its reaction with OH. This reaction also serves as a major conversion pathway of OH to other forms of HO_x. This latter process is important in controlling the concentration of OH, and, therefore, it has been postulated that increases in CO contribute to decreases in global tropospheric OH concentrations (see, for example, Penner *et al.*, 1977). The situation is far more complex, however, because of the projected simultaneous increases in CO, CH_4, NO_x, and the NMHCs, all of which contribute to determining OH abundances. Because of regional diversity in the sources of these gases and because of uncertainties in their budgets, the relationship between OH and increasing trace species trends is regional in character and demands a three-dimensional treatment

(Penner, 1990; Penner *et al.*, 1989). As noted above, changes in atmospheric OH can lead to a suite of changes in other species (see Table 4).

The reaction of CO with OH also initiates a photochemical sequence that, in the presence of sufficient NO_x, produces tropospheric O_3:

$$CO + OH \rightarrow CO_2 + H$$
$$H + O_2 + M \rightarrow HO_2 + M$$
$$HO_2 + NO \rightarrow OH + NO_2$$
$$NO_2 + h\nu \rightarrow NO + O$$
$$O + O_2 + M \rightarrow O_3 + M$$

Net: $CO + 2O_2 + h\nu \rightarrow CO_2 + O_3$

(The term $h\nu$ indicates that the reaction involves the absorption of a photon; M is O_2 or N_2, a third body whose collision facilitates the reaction.) This reaction sequence is similar to the sequences of photochemical reactions that produce urban smog. In regions of low NO_x, however, the reaction sequence initiated by the reaction of CO with OH destroys tropospheric O_3:

$$CO + OH \rightarrow CO_2 + H$$
$$H + O_2 + M \rightarrow HO_2 + M$$
$$HO_2 + O_3 \rightarrow OH + 2O_2$$

Net: $CO + O_3 \rightarrow CO_2 + O_2$

An important problem for tropospheric chemistry is to define those regions with sufficient NO_x to produce ozone.

The source estimates for CO appear in Table 8. Production of CO during the oxidation of CH_4 is the largest single source. Roughly 50% of this source is related to land-use classes through the dependence of CH_4 emissions on rice paddies, landfills, biomass burning, and enteric fermentation. The remainder depends on CH_4 from land cover and anthropogenic sources. A second major source is the oxidation of natural NMHCs. This is related to land cover through its dependence on the type of vegetation (see below); Logan *et al.* (1981) estimate it to be 560 Tg CO/yr. A third major source is the burning of fossil fuels. According to Logan *et al.* (1981), this source contributes about 450 Tg CO/yr to the atmosphere. A related source, the oxidation of anthropogenic NMHCs, was estimated to add 90 Tg CO/yr. The source of CO from the burning of fuelwood, biomass (in forests and savannas), and agricultural waste is currently estimated to range from 280 to 950 Tg CO/yr in tropical areas (Hao *et al.*, 1990). Adding to this 25 Tg CO/yr from forest wildfires (Logan *et al.*, 1981) gives an average of 640 Tg CO/yr from biomass and wood burning. Direct emissions from vegetation are another minor source, as is the ocean (Logan *et al.*, 1981). Finally, soils can act as both a minor source and a minor sink. The soil sink is presumed to dominate in the budgets of both Logan *et al.* (1981) and Seiler and Conrad (1987).

By combining the source estimates for CO that derive from CH_4 emissions

Table 8: Sources of carbon monoxide (Tg CO/yr)

Source	Total	Range
Anthropogenic: fossil fuel or industrial		
Fossil Fuel	450	400–1000
Anthropogenic NMHCs	90	0–180
CH_4 oxidation (from fossil/industrial sources)	90	45–110
Anthropogenic: land use by man		
Biomass burning	640	280–950
CH_4 oxidation (from land use by man)	310	135–390
Land cover related:		
Natural NMHCs	560	280–1200
Plants	130	50–200
CH_4 oxidation (from land cover)	190	115–395
Oceans	40	20–80
CH_4 oxidation (from oceans)	10	5–20
Total source strength	2510	1330–4525

associated with land cover with the source estimates for CO that derive from NMHC oxidation and biomass burning, I estimate that almost 60% of the total current sources are related to land use or land cover.

Nonmethane Hydrocarbons

A very large variety of NMHCs is found throughout the troposphere. The abundances of these diverse species range from the tens of ppt level to the ppb level and higher. The NMHCs are often conveniently lumped into the categories of alkanes, alkenes, aromatics, aldehydes and other oxygenated hydrocarbons, and the biogenically produced compounds isoprene and terpenes. As pointed out above, the NMHCs participate in the OH and O_3 photochemical cycle. They generally react with OH, so that an increase in their concentrations will decrease OH concentrations. However, as these species are oxidized, they also produce odd hydrogen species and organic radicals, so that an increase in the NMHC concentrations contributes to an increase in the concentrations of peroxy radicals and odd hydrogen radicals as a whole. When HO_2 and RO_2 concentrations are increased, in the presence of sufficient NO, the formation of NO_2 is increased. As noted above in the discussion of CO, this leads to the production of ozone after photolysis of NO_2.

The alkenes, isoprene, and terpenes can also react with O_3 directly, leading to its removal. The reaction of alkenes, isoprene, and terpenes with NO_3 can also be significant.

The lifetimes of the NMHCs vary from a few minutes (for isoprene) to several months for some of the lower-molecular-weight alkanes. These latter species can

Table 9: Sources of NMHCs (Tg C/yr)

Source	Total	Range
Anthropogenic: fossil fuel or industrial		
Fossil fuel, industrial and evaporative emissions	58	
Anthropogenic: land use by man		
Biomass burning (includes fuel wood agricultural wastes and forest and savanna burning)	34	12–50
Land cover related:		
Isoprene	400	350–450
Terpenes and other biogenics	400	350–450
Oceans	60	36–82
Total source strength	950	840–1100

be transported long distances from their sources, while the former can be significant in the formation of ozone and in reaction with the hydroxyl radical on local scales. The estimated biogenic emissions of isoprene and the terpenes are so large and so widespread, however, that their impact on global photochemistry may also be of importance. In particular, the oxidation of isoprene and the terpenes can lead to the formation of PAN. This species can be long-lived at lower temperatures and is thought to play a role in the transport of reactive nitrogen from its source regions to the remote troposphere (Singh and Hanst, 1981). As noted above, the NMHCs are also significant sources of atmospheric CO.

Table 9 presents estimates for the emissions of NMHCs. The sources from biomass burning and emission by plants are all associated with land-use or land-cover categories. Anthropogenic sources (which includes biomass burning and industrial and fossil fuel sources) are close to 100 Tg/yr (Singh and Zimmerman, 1992). Of this total, Singh and Zimmerman estimate that 58 Tg/yr are from industrial and fossil fuel sources; this is the value quoted in Table 9, while the estimates of Hao *et al.* (1990) are used for biomass burning. On a global basis, the sources of isoprene range from 350 to 450 Tg C/yr (Singh and Zimmerman, 1992; J. Dignon, personal communication). Terpenes and the emissions of other nonspecified hydrocarbons are about as large as those of isoprene (Singh and Zimmerman, 1992). Finally, the ocean is a source of propene and ethene (as well as a small source of higher alkanes) that is estimated at between 36 and 82 Tg C/yr.

As noted from Table 9, because most of the emissions of NMHCs are related to vegetative sources, more than 80% of the total emissions of NMHCs are related to land use or land cover, implying a large potential for change if there are changes in land cover.

Reactive Nitrogen

Reactive nitrogen (NO_y) consists of a suite of nitrogen-containing species, the most important of which are NO, NO_2, HNO_3 and aerosol NO_3^-, PAN and other organic nitrates, NO_3, N_2O_5, and HNO_4. The fraction of reactive nitrogen that is present as NO_x (the sum of NO and NO_2) is particularly important in both the global troposphere and local urban areas. This is because NO_x plays a major role in the formation of tropospheric O_3 and smog As discussed above, in regions of high NO_x, photochemical sequences initiated by the reaction of CO with OH or the reaction of NMHCs or CH_4 with OH lead to O_3 formation. In regions of low NO_x (such as remote ocean areas), the photochemical sequences lead to O_3 destruction. NO_x concentrations also partly control the concentration of OH. Increases in NO_x lead to increases in OH up to NO_x concentrations of a few tenths of a ppb. Increases in NO_x above these levels lead to decreases in OH (Logan *et al.*, 1981). At high levels of NO_x, such as experienced in some urban areas, the effect of NO_x on OH concentrations can lead to decreases in O_3 as well. This means that a decrease in these emissions would actually increase some local, urban ozone levels. This effect has led to the adoption in the Clean Air Act Amendments of 1990 of hydrocarbon-only controls for some urban areas that experience high ozone concentrations, even though the emissions of NO_x in these areas, when not controlled, must contribute to ozone formation in downwind areas and regionally.

A large fraction of NO_y in the troposphere is present as HNO_3 and, in the marine boundary layer, as aerosol NO_3^-. These two components are important because they are major contributors to acid deposition. Also, their deposition to nitrogen-poor ecosystems and ocean areas can provide an important nutrient for these systems. The long-range transport of reactive nitrogen may be inferred from measurements of HNO_3 and particulate nitrate at Mauna Loa Observatory (Galasyn *et al.*, 1987). These measurements found high levels of HNO_3 and other pollutants in summer, when transport paths could be shown to originate from North America (Moxim, 1990). Also, nitrate concentrations in remote regions of the Northern Hemisphere are higher than those measured in the Southern Hemisphere (Savoie *et al.*, 1989). The inference that these larger concentrations are due to the higher fossil fuel use in the Northern Hemisphere may be derived from model studies of the tropospheric nitrogen budget (Levy and Moxim, 1989; Penner *et al.*, 1991a). This conclusion is also supported by the observation of increasing concentrations of nitrate in ice cores in Greenland (Neftel *et al.*, 1985). These increasing concentrations parallel the increase in fossil fuel use between 1930 and the mid-1970s (Hameed and Dignon, 1988).

A second major component of reactive nitrogen in the troposphere is PAN, other higher homologs of PAN, and other organic nitrates (Calvert and Madronich, 1987; Atherton and Penner, 1988). These species transport reactive nitrogen that is emitted in regions with high sources (e.g., urban areas) to remote areas (Singh *et al.*, 1986). PAN and the other organic nitrates may then decompose and contribute to local NO_x, HNO_3, and aerosol nitrate concentrations and the other, minor reactive nitrogen components.

There are six major sources of reactive nitrogen in the troposphere: fossil fuel emissions, biomass burning (which includes savannas, forests, agricultural wastes,

Table 10: Sources of reactive nitrogen (Tg N/yr)

Source	Total	Range
Anthropogenic: fossil fuel or industrial		
Fossil fuel burning	22	15–25
Aircraft emissions	0.4	0.1–0.7
Anthropogenic: land use by man		
Wood burning	0.2	0.1–0.8
Forest and savanna burning	9	2–20
Agricultural waste burning	1.5	1–2.5
Land cover related:		
Soil microbial emissions	10	1–20
Lightning discharge	10	2–100
Production in the stratosphere	1.0	0.5–1.5
Total source strength	54.1	21–170

and wood fuel burning), aircraft emissions, lightning discharges, soil microbial activity, and production in the stratosphere from the reaction of N_2O with $O(^1D)$ followed by transport into the troposphere. The biomass sources and the source from microbial activity in soils are both associated with land use or land cover. Estimates for the spatial distribution of the global emissions from several of the most important of these sources were developed by Penner *et al.* (1991a) for a three-dimensional model study of reactive nitrogen. Table 10 presents an update of the global source estimates from that study. The largest and the best-determined single source of reactive nitrogen appears to be fossil fuel burning at about 22 Tg N/yr for 1980 (Hameed and Dignon, 1988; Dignon, 1992).

Biomass burning is another major source of reactive nitrogen. Recent estimates of the mean contribution from this source vary from 2 to 19 Tg N/yr (Penner *et al.*, 1991; Dignon and Penner, 1991; Crutzen and Andreae, 1990; Hegg *et al.*, 1990). The estimate of 9 Tg N/yr for forest and savanna burning in Table 10 was obtained by using the relationship between fuel nitrogen and NO_x emissions (Clements and McMahon, 1980) as in Dignon and Penner (1991) but correcting the emission factor of NO_x from biomass burning in savannas (Lobert *et al.*, 1990). The wood burning source was estimated using estimates for the production of fuelwood from United Nations (1983). The source is small because of the low nitrogen content of wood and the resulting low emission factors for NO_x (Dasch, 1982; Clements and McMahon, 1980). Agricultural waste burning is estimated to add an additional 1.5 Tg N/yr.

A third major source of reactive nitrogen is soil microbial activity. The magnitude of the source of NO_x from soils, like the magnitude of the source of N_2O from soils, depends on factors such as vegetation type, soil moisture, temperature, fertilization history, recent burn history, and ambient NO concentration (see, for example, Galbally, 1989; Williams and Fehsenfeld, 1991; and Slemr and Seiler, 1991). Estimates for this source have ranged from 1 to 20 Tg N/yr (Johansson, 1984; Kaplan *et al.*, 1988; Slemr and Seiler, 1984; Williams *et al.*, 1987; Galbally, 1989). Penner *et al.* (1991a) developed a global distribution for soil emissions based on scaling the area in grasslands and tropical forests by the observations from Williams *et al.* (1987) and Kaplan *et al.* (1988). Considering the range in measured emission factors for different biomes (see. e.g., Bakwin *et al.*, 1990; Kaplan *et al.*, 1988; Davidson *et al.*, 1991), an average of 10 Tg N/yr remains a reasonable global estimate for the total emissions from soils.

Four of the five surface-based emission sources (wood burning, biomass burning in forests and savannas, agricultural waste burning, and soil microbial activity) are associated with land use or land cover. According to the estimates in Table 10, then, almost 50% of the surface-based source of reactive nitrogen is associated with land use or land cover. Sources in the upper troposphere are also important, however. They include lightning, aircraft emissions, and production in the stratosphere. Although these are small relative to the other sources listed in Table 10, they are input directly to the upper troposphere. The production of O_3 in this region is especially important because upper tropospheric ozone is far more efficient as a greenhouse gas than ozone near the surface (Lacis *et al.*, 1990). Quantifying the relative contributions of the surface sources of NO_x and those sources that directly contribute to NO_x concentrations in the upper troposphere is an important unsolved problem in atmospheric chemistry because it involves correctly estimating the influence of stratosphere-troposphere exchange and the influence of the convection of boundary layer air to the upper troposphere. These dynamical processes are poorly treated in models because their representation requires higher resolution than can be reached on today's computers.

Reactive Sulfur

The major species that contribute as source components to reactive sulfur are SO_2, DMS, H_2S, and CS_2. These species all undergo chemical reactions leading to their oxidation and the production of H_2SO_4 vapor. H_2SO_4 vapor may either condense on preexisting particles or nucleate to form new particles if the appropriate conditions of humidity, temperature, preexisting particle concentrations, and acid vapor production rates are present.

Interest in the reactive sulfur cycle in the atmosphere derives from three concerns. First, SO_2 and sulfate contribute to acid rain and acid deposition. Second, the formation of sulfate increases the aerosol burden in the troposphere. Sulfate aerosol reflects solar radiation, tending to cool the climate (Charlson *et al.*, 1990). Third, the formation of new particles through the homogeneous nucleation of gas phase H_2SO_4 can lead to increased cloud condensation nuclei (CCN) concentrations. These particles act as seeds for the condensation of water when cloud droplets form. It has been hypothesized that an increase in CCN concentrations can lead to an increase in cloud droplet number concentration (Twomey, 1977).

Clouds with a higher concentration of droplets and the same liquid water content are able to reflect more solar radiation. Increased sulfur emissions may therefore lead to clouds that reflect more solar radiation and cool the climate. Indeed, decreasing temperatures in certain regions of the Northern Hemisphere may be associated with the increase in sulfur emissions during the last several decades (Engardt and Rodhe, 1993).

The largest source of reactive sulfur in the atmosphere is fossil fuel burning, estimated as ≈77 Tg S/yr for 1980. Fossil fuel and industrial sulfur are released in the form of SO_2. Because this source dominates over land areas, its change over time has led to significant increases in the deposition of sulfate to ecosystems and lakes. Ice-core evidence from Greenland also indicates that sulfate concentrations over remote areas have increased (Neftel *et al.*, 1985). Although emissions from industrial areas in North America and Europe have decreased during the last decade, emissions from less developed regions are probably increasing.

The second largest source of sulfur in the atmosphere is the production of DMS by ocean phytoplankton. Because stratus clouds over the ocean are particularly susceptible to alteration of their albedos by changes in CCN, this source has been hypothesized as important for climate change (Charlson *et al.*, 1987). The observation that increased sulfate is associated with ice-core data during glacial periods lends support to this theory (Legrand *et al.*, 1988). The mechanisms involved in the production of DMS by ocean phytoplankton are poorly understood, however, as are the responses of phytoplankton to climate change. Only certain species are involved in the production of DMS, so that correlations of DMS in sea water with total primary productivity or with chlorophyll abundance are poor. Also, the production of DMS may be activated as zooplankton graze on phytoplankton. Thus, the oceanic abundance of DMS depends on ocean ecosystem dynamics as well as primary productivity. Measurements of DMS in seawater together with estimates of flux rates have placed the total source strength of DMS in the ocean between 12 and 40 Tg S/yr (Bates *et al.*, 1987; Andreae and Raemdonck, 1983; Erickson *et al.*, 1990; Spiro *et al.*, 1992). Oceanic emissions of H_2S and CS_2 are much smaller than the emissions of DMS.

Volcanic emissions of sulfur are also important. About 10 Tg S/yr are emitted into the atmosphere on a long-term average basis (Stoiber *et al.*, 1987). On an episodic basis, of course, the emissions from highly explosive volcanoes can be significantly higher than the long-term average and can lead to significant increases in the stratospheric aerosol burden. The 1991 eruption of Mount Pinatubo is a case in point, with as much as 10 Tg S emitted directly into the stratosphere during its eruptive stage. The majority of the sulfur emitted from volcanoes is emitted as SO_2.

Terrestrial emissions of sulfur were estimated as 10 Tg S/yr by Watson *et al.* (1990). However, a new estimate, based on combining measured fluxes for different soils and land-cover classes with the areal extent of these soils and land-cover classes, gives a total of only 0.35 Tg S/yr (Bates *et al.*, 1992). Recent measurements indicate that even this estimate may be too high because it is based on mea-

Table 11: Sources of reactive sulfur (Tg S/yr)

Source	Total	Range
Anthropogenic: fossil fuel or industrial		
Fossil fuel and industry	77	60–100
Anthropogenic: land use by man		
Biomass burning (forest and savannas, wood fuel and agricultural wastes)	2.2	1–4
Land cover related (and subject to possible change):		
Terrestrial soils, forests, crops	1.0	0.1–10
Volcanoes	10	3–20
Oceans	25	12–40
Total source strength	115	76–174

surements made with enclosure techniques that used sulfur-free air (Castro and Galloway, 1991). Any estimate of terrestrial sulfur fluxes is uncertain because measured fluxes can vary by over an order of magnitude, depending on temperature, soil type, moisture content, and land cover (Bates *et al.*, 1992). Approximately one-third to one-half of the terrestrial emissions are estimated to be H_2S, while most of the rest is DMS.

Biomass burning also represents a minor source of reactive sulfur. These emissions have been estimated as between 1 and 4 Tg S/yr (Crutzen and Andreae, 1990). I have adopted a figure of 2.2 Tg S/yr based on Bates *et al.* (1992) and Spiro *et al.* (1992).

Table 11 summarizes the global emissions of reactive sulfur. The terrestrial source and biomass burning sources are both associated with land use or land cover. They constitute about 3% of the total reactive sulfur source.

Tropospheric Aerosols

Tropospheric aerosols are important because they scatter and absorb (primarily solar) radiation, thereby changing the amount of radiation absorbed by the earth. The composition of the aerosol is important to its radiative effect because the composition determines whether the aerosol primarily scatters or absorbs solar radiation. A scattering aerosol with an optical depth of 0.125 and a single scattering albedo of 0.95 could cool the earth by as much as 1.6 °C, whereas the same total abundance of aerosol with a single scattering albedo of 0.75 could warm the earth by 0.5 °C (Charlock and Sellers, 1980). The albedo of the aerosol is controlled primarily by the amount of soot (also called black carbon or elemental carbon) in the aerosol (although dark soil aerosols may also absorb solar radiation). Quantification of the amount of soot present in the atmospheric aerosol is therefore of importance in order to quantify the overall effect of aerosols on climate. Recent studies have estimated that the amount of radiation scattered as a result of the anthropogenic sulfate in atmospheric aerosols may be a large fraction of the total amount of radiation trapped by increases in greenhouse gases (Charlson *et al.*, 1990, 1991, 1992). Because emissions of soot are associated with sulfur emissions, it is important to quantify the warming effects of soot as well (Penner *et al.*, 1993).

Aerosols may also impact the earth's radiation balance through their effects on clouds. Aerosols alter clouds in two ways. First, they act as cloud condensation nuclei (CCN). Changes in the concentrations of aerosols that act as CCN can lead to changes in the number of drops present in clouds. Indeed, clouds that form over continents, where CCN concentrations are high, typically have smaller but more numerous drops than similar clouds that form over the oceans. Other things being equal, these clouds are also more highly reflective of solar radiation (Twomey, 1977). Estimates of the effects of anthropogenic sulfur aerosols on cloud droplet concentrations have shown that increases in these may be a substantial source of atmospheric cooling (Charlson *et al.*, 1987; Wigley, 1989). The effects of aerosols produced in biomass burning may also be substantial (Penner *et al.*, 1991b, 1992).

The second way in which aerosols affect clouds is by increasing the number of small drops within a cloud. Warm water clouds with more numerous, but smaller, drops do not form precipitation through the coalescence mechanism as easily as do those with large drops. (Cold clouds initiate precipitation through ice formation, and thus would not be affected by this aerosol mechanism.) As a result, these clouds may be longer-lived and hold more liquid water on average than clouds that form with larger drops (Albrecht, 1989; Radke *et al.*, 1989). Thus the area of the earth covered by clouds may be increased, as well as the amount of liquid water present in clouds. Other things being equal, clouds with more liquid water are also more highly reflective of solar radiation.

Atmospheric aerosol concentrations and optical depths are highly variable, but on average, urban concentrations may equal $10^5/cm^3$ while pristine ocean areas often have concentrations of less than a few hundred per cm^3. The corresponding optical depth for these aerosol concentrations varies from a high of about 1.5 to a low of perhaps 0.05. The largest optical depths occur on humid days. This is because above about 80% relative humidity aerosols with soluble components deliquesce and form larger haze particles. When particles form a haze, they scatter radiation even more efficiently.

Evidence for a trend in aerosol concentrations and optical depths is sketchy at best. However, concentrations in certain industrial regions appear to have increased (Husar *et al.*, 1981; Winkler and Kaminski, 1988). Furthermore, the concentrations of sulfate, nitrate, and trace metals in Greenland ice cores indicate that the trend in aerosol concentrations may extend beyond local source regions (Neftel *et al.*, 1985; Mayewsky *et al.*, 1986). These indications of increasing abundance, however, are not observed at all locations. For example, aerosol concentra-

Table 12. Sources of the tropospheric aerosol (fine particle fraction)

Type	Source strength (Tg/yr)	
	Anthropogenic	Natural
Gas-to-particle conversion:		
Sulfates	120	55
	(90–160)	(25–110)
Ammonium	35	15
	(15–60)	(5–40)
Nitrates	70	70
	(35–100)	(10–270)
NMHCs	–	50
		(20–120)
Primary emissions:		
Biomass burning (savannas, forests,	67	3.0
and agricultural wastes)	(10–110)	(2–5)
Fossil and woodfuel burning	70	–
	(24–120)	
Soot component from biomass and	18	0.2
fossil fuel burning	(5–30)	(0.1–0.4)
Dust	–	150
		(60–300)
Sea salt	–	140
		(100–180)
Biogenic		100
		(0–200)
Total	362	583
	(174–550)	(222–1225)

tions at the South Pole appear to have decreased between 1977 and 1985 (Samson *et al.*, 1990).

Aerosols in the fine particle fraction are most effective at scattering solar radiation. This portion of the aerosol spectrum also determines the CCN number concentrations. Therefore, these aerosols are the most important to quantify in order to determine the climate effects of aerosols. Except in desert dust outbreaks and near the ocean surface where sea salt aerosols are prevalent, the fine particle fraction also contains most of the aerosol mass. Here, I define fine particles as aerosols with a diameter less than about 2.0 μm, because emissions have been defined this way (Gray, 1982; Hildemann *et al.*, 1991a, 1991b). However, to relate aerosol number concentrations (which often peak at 0.05–0.1 μm diameter) to the concentration of particles that act as CCN, a better understanding of the factors that determine the chemically-resolved size distribution is needed.

Particles may be composed of sulfate, nitrate, organics, silicates and other compounds usually associated with soils and dust, sea salt and compounds associated with the sea salt aerosol, ammonium, soot, and trace metals. (The total trace metal mass is usually small and is therefore ignored in the following analysis.) Below, I present estimates for the global source strengths for each of the aerosol components. These estimates are made on a global basis, but one should keep in mind that the composition of the fine particle fraction of the aerosol varies seasonally and regionally. The relative magnitudes of the global source strengths for each component, therefore, may not necessarily correlate with the magnitudes of the aerosol components at any given location.

Table 12 summarizes the estimates for the sources of the atmospheric aerosol. Aerosol sources may be divided into those which contribute to the formation of 'secondary' aerosols through gas to particle conversion and those which are injected directly as 'primary' aerosols. Below, I first estimate the secondary aerosol sources. These include sulfate, nitrate, ammonium, and condensed organics.

Secondary Aerosol Sources

Over many continental areas a major part of the fine particle fraction of the aerosol mass in the boundary layer is sulfate and is largely of anthropogenic origin. The source of fine particle aerosol sulfate listed in Table 12 is assumed to be proportional to the reactive sulfur source (Table 11). For this estimate, 50% of the reactive sulfur source was assumed to be deposited before becoming particulate sulfate.

Nitrates form an additional important fraction of the atmospheric fine particle aerosol mass, though most of the NO_3^- mass tends to be associated with larger particle sizes (Warneck, 1988; Savoie and Prospero, 1982). Therefore, the sources listed for reactive nitrogen may not contribute proportionally with sulfur to the aerosol number and the fine particle mass. In the estimate for Table 12, I assumed that 50% of the reactive nitrogen source from Table 10 is deposited in the gas phase, while the remaining 50% becomes aerosol NO_3^-.

Some NMHC compounds, particularly those formed during the photo-oxidation of terpenes, can condense to form particles. The yield of aerosol from terpene oxidation is variable but may range from 10% to 60% (Pandis *et al.*, 1991; Hatakeyama, 1991) and is estimated from this range in Table 12. The source of aerosols from the photo-oxidation of isoprene is apparently negligible (Paulson *et al.*, 1990; Pandis *et al.*, 1991). The photo-oxidation of anthropogenic NMHCs can also lead to the formation of species which may condense to the aerosol form, but this source is expected to be small based on Table 9 and is ignored in the current estimate.

Finally, NH_3 is also an important precursor of tropospheric aerosols. It is emitted as a gas, but under the right conditions of humidity and temperature readily condenses to the ionic form, NH_4^+, found in aerosols. The oceanic source of NH_3 is largely natural (Quinn *et al.*, 1987), while the continental sources derive mainly from human activity (Bowden, 1986). Over continents, NH_3 is produced as part of animal excrement and as a byproduct of soil microbial activity. It is common to estimate the NH_4^+ in aerosols by assuming that the sulfate that is present in the atmospheric aerosol is present as ammonium sulfate, $(NH_4)_2SO_4$. However, this procedure appears to produce an overestimate when compared with estimates of the NH_3 source over Europe (Ferm, 1990; Buijsman *et al.*, 1987; Thomas and Erisman, 1990). For Table 12, I assumed that a lower bound for the source strength for NH_4^+ in aerosol particles is roughly one-half that derived from assuming an ammonium source sufficient to produce $(NH_4)_2SO_4$, while the upper bound assumed that all sulfate is present as $(NH_4)_2SO_4$. I also assumed that the anthropogenic fraction of the ammonium source scales with the anthropogenic fraction of the sulfate source.

Primary Aerosol Sources

Primary particles are directly injected into the atmosphere. They consist of dust particles, sea salt particles, biogenic particles (fungi, spores, epicuticular plant waxes, etc.); particles from biomass burning of forests, savannas, and agricultural wastes; and particles from burning fossil and wood fuels. Because of the procedure I used for estimating sources, I have separated biomass burning of forests, savannas, and agricultural wastes from biomass burning of wood fuels. As noted above, the fraction of the aerosol that is present as soot is also important to quantify. Soot is formed solely during combustion.

The source of fine particles from the burning of biomass in tropical forests and savannas was estimated as 25–80 Tg/yr (Penner *et al.*, 1991b), with about 6 Tg C/yr of soot (compare Crutzen and Andreae, 1990). This estimate was based on the inventory of CO_2 emissions by Hao *et al.* (1990) together with estimated emission factors for fine particle mass and soot. In Table 12, this estimate was increased to reflect the additional source of biomass aerosols from agricultural wastes. Also, a small source from natural forest fires should be considered.

The source of fine particle emissions from fossil fuel and fuelwood burning is estimated as follows. Penner *et al.* (1993) have estimated the source of soot or black carbon from measured ratios of SO_2 and black carbon from around the world together with the estimated source of SO_2 from fossil fuel burning (e.g., Hameed and Dignon, 1988). They obtained a total of approximately 24 Tg C/yr, which is considered an upper bound because it is a factor of two larger than an estimate based on fuel use and emission factors (Penner *et al.*, 1993). Ghan and Penner (1991) estimated a lower bound for the amount of soot emitted during fossil fuel and fuelwood burning of ≈5 Tg/yr, using emission factors from a study of soot emissions in Los Angeles. Ghan and Penner (1991) also included a lower bound of ≈24 Tg/yr for the source for fine particle emissions. Here, I estimate an upper bound to the fine particle source by scaling this estimate of fine particle emissions by the ratio of soot carbon mass in the two estimates given above. This yields an upper bound for the total emissions of fine particle mass of 115 Tg fine particles/yr for a total range from fossil fuels and wood burning of 24–120 Tg/yr.

The amount of fine particle dust and sea salt particles lofted into the atmosphere was estimated in the Study of Man's Impact on Climate, quoted in Prospero *et al.* (1983). Defining fine particles as having a diameter less than 6 μm, they estimated that the source of fine particle dust was 150 Tg/yr with a range of 60–300 Tg/yr. The source of sea salt particles was estimated as 180 Tg/yr, but could be as small as 100 Tg/yr if one assumes that a smaller fraction of the total source has a diameter less than 2 μm (based on Eriksson, 1959, and Junge, 1972, as quoted by Duce, 1978).

The final category of primary particulate sources are the biogenic particles. The fraction of these as fine particles is not known (Duce, 1978) but is certain to be of importance in some circumstances (Jaenicke and Matthias, 1988; Talbot *et al.*, 1990; Artaxo *et al.*, 1990; Mazurek *et al.*, 1991). The source estimate in Table 12 is highly uncertain. It was derived by difference, assuming a total source strength of fine particle organic carbon that ranged from 100 to 300 Tg/yr in the budgets of Duce (1978) and Jaenicke (1978). Subtracting the portions of this source that are accounted for in other categories (condensed secondary organic and organic aerosols from biomass and fossil fuel burning, yields a total source for primary biogenic particles of 0–200 Tg/yr.

Anthropogenic Sources

Table 12 presents the estimated fraction of aerosol sources from anthropogenic causes. Nearly 40% of the total emissions of fine aerosol particles appears to derive from human activities. This estimate is a factor of two to five times larger than previous estimates of the role of anthropogenic aerosols in the aerosol particle budget (Prospero *et al.*, 1983), in part because I have emphasized the fine particle fraction of the aerosol mass which is the most important fraction for determining light scattering and absorption in the visible spectrum and for determining the number of CCN. All the estimates in Table 12 clearly demonstrate that atmospheric aerosol sources have a large anthropogenic component and may therefore have changed over time. The role of land use and land cover may be judged from the derivation of these fractions in previous budgets for NO_x, SO_x, and the NMHCs and from the forest, savanna, agricultural waste, and fuelwood burning, and primary biogenic fractions of the primary source estimates. I estimate a total fraction of aerosol source related to land use and land cover of approximately 30%. Any change in land use or land cover will lead to aerosol changes that are regional in character. Because these estimated sources are so uncertain, the magnitude of the expected change in aerosol abundance and the climate impact of these changes cannot be determined rigorously but may be expected to be important.

Conclusion

Estimates of the global emissions of trace gases and aerosols show that anthropogenic activities are responsible for a large fraction of total emissions and that emissions related to land use and land cover play a significant role in the budgets of many of these species. Changes in land use and land cover would therefore be expected to lead to significant changes in emissions. The source budgets presented here represent a first step toward quantifying the magnitude of trace gas and aerosol emissions from different land-use categories and land-cover types. Many of the sources as estimated here, however, are poorly quantified. This is especially true for those associated with biological processes. Work toward refining these estimates is needed to better quantify the role of land-use and land-cover change in atmospheric chemistry.

Acknowledgments

I am pleased to acknowledge useful conversations with Jane Dignon, Pamela Matson, Cindy Nevison, and T. Novakov. Daniel Jacob and Hanwant Singh provided preprints of their work. This work was funded in part by the U.S. Environmental Protection Agency under Interagency Agreement DW89932676-01-0, by the National Aeronautics and Space Administration under grant NAGW-1827, by the U.S. Department of Energy Quantitative Links Program, and by the Lawrence Livermore National Laboratory (LLNL) Laboratory Directed Research and Development program. Work at LLNL is performed under the auspices of the U.S. Department of Energy under Contract W-7405-Eng-48.

References

Albrecht, B. A. 1989. Aerosols, cloud microphysics, and fractional cloudiness. *Science 245,* 1227–1230.

Andreae, M. O., and H. Raemdonck. 1983. Dimethylsulfide in the surface ocean and the marine atmosphere: A global view. *Science 221,* 744–747.

Artaxo, P., W. Maenhaut, H. Storms, and R. Van Grieken. 1990. Aerosol characteristics and sources for the Amazon Basin during the wet season. *Journal of Geophysical Research 95,* 16971–16985.

Aselmann, I., and P. J. Crutzen. 1989. Global distribution of natural freshwater wetlands and rice paddies, their net primary productivity, seasonality and possible methane emissions. *Journal of Atmospheric Chemistry 8,* 307–358.

Atherton, C. A., and J. E. Penner. 1988. The transformation of nitrogen oxides in the polluted troposphere. *Tellus 40B,* 380–392.

Bakwin, P. S., S. C. Wofsy, S. M. Fan, M. Keller, S.E. Trumbore, and J.M. Da Costa. 1990. Emission of nitric oxide (NO) from tropical forest soils and exchange of NO between the forest canopy and atmospheric boundary layers. *Journal of Geophysical Research 95,* 16755–16764.

Bartlett, K. B., P. M. Crill, J. A. Bonassi, J.E. Richey, and R.C. Harriss. 1990. Methane flux from the Amazon river floodplain: Emissions during rising water. *Journal of Geophysical Research 95,* 16773–16788.

Bartlett, K. B., P. M. Crill, R. L. Sass, R.C. Harriss, and N. B. Dise. 1992. Methane emissions from tundra environments in the Yukon–Kushkokwim Delta, Alaska. *Journal of Geophysical Research 97,* 16645–16660.

Bates, T. S., J. D. Cline, R. H. Gammon, and S. R. Kelly-Hansen. 1987. Regional and seasonal variations in the flux of oceanic dimethylsulfide to the atmosphere. *Journal of Geophysical Research 92,* 2930–2938.

Bates, T. S., B. K. Lamb, A. Guenther, J. Dignon, and R.E. Stoiber. 1992. Sulfur emissions to the atmosphere from natural sources. *Journal of Atmospheric Chemistry 14,* 315–337.

Bekki, S., and J. A. Pyle. 1992. Two-dimensional assessment of the impact of aircraft sulphur emissions on the stratospheric sulphate aerosol layer. *Journal of Geophysical Research 97*, 15839–15847.

Bingemer, H. G., S. Burgermeister, R. L. Zimmermann, and H.-W. Georgii. 1990. Atmospheric OCS: Evidence for a contribution of anthropogenic sources? *Journal of Geophysical Research 95*, 20617–20622.

Born, M., H. Dorr, and I. Levin. 1990. Methane consumption in aerated soils of the temperate zone. *Tellus 42B*, 2–8.

Bowden, W. B. 1986. Gaseous N emissions from undisturbed terrestrial ecosystems. *Biogeochemistry 2*, 249–279.

Bowden, R. D., P.A. Steudler, J.M. Melillo, and J.D. Aber. 1990. Annual nitrous oxide fluxes from temperate forest soils in the northeastern United States. *Journal of Geophysical Research 95*, 13997–14005.

Bowden, R. D., J. M. Melillo, P.A. Steudler, and J.D. Aber. 1991. Effects of nitrogen additions on annual nitrous oxides fluxes from temperate forest soils in the northeastern United States. *Journal of Geophysical Research 96*, 9321–9328.

Brasseur, G. P., C. Granier, and S. Walters. 1990. Future changes in stratospheric ozone and the role of heterogeneous chemistry. *Nature 348*, 626–628.

Brown, K. A., and J. N. B. Bell. 1986. Vegetation—The missing sink in the global cycle of carbonyl sulfide (COS). *Atmospheric Environment 20*, 537–540.

Brown, S., and A. E. Lugo. 1984. Biomass of tropical forests: A new estimate based on volumes. *Science 223*, 1290–1293.

Buijsman, E., H. F. M. Maas, and W.A.H. Asman. 1987. Anthropogenic NH_3 emissions in Europe. *Atmospheric Environment 21*, 1009–1022.

Calvert, J. G., and S. Madronich. 1987. Theoretical study of the initial products of the atmospheric oxidation of hydrocarbons. *Journal of Geophysical Research 92*, 2211–2220.

Castro, M. S., and J. N. Galloway. 1991. A comparison of sulfur-free and ambient air enclosure techniques for measuring the exchange of reduced sulfur gases between soils and the atmosphere. *Journal of Geophysical Research 96*, 15427–15437.

Charlock, T. P., and W. D. Sellers. 1980. Aerosol effects on climate: Calculations with time-dependent and steady-state radiative-convective models. *Journal of the Atmospheric Sciences 37*, 1327–1341.

Charlson, R. J., J. E. Lovelock, M. O. Andreae, and S. G. Warren. 1987. Oceanic phytoplankton, atmospheric sulphur, cloud albedo and climate. *Nature 326*, 655–661.

Charlson, R. J., J. Langner, and H. Rodhe. 1990. Sulfate aerosol and climate. *Nature 348*, 22.

Charlson, R. J., J. Langner, H. Rodhe, C. B. Leovy, and S. G. Warren. 1991. Perturbation of the Northern Hemisphere radiative balance by backscattering from anthropogenic sulfate aerosols. *Tellus 43AB*, 152–163.

Charlson, R. J., S. E. Schwartz, J. M. Hales, R. D. Cess, J. A. Coakley, Jr., J. E. Hansen, and D. J. Hofmann. 1992. Climate forcing by anthropogenic aerosols. *Science 255*, 423–430.

Chin, M., and D. D. Davis. 1993. Global sources and sinks of OCS and CS_2 and their distributions. *Global Biogeochemical Cycles 7*, 321–337.

Cicerone, R. J. 1988. How has the atmospheric concentration of CO changed? In *The Changing Atmosphere* (F. S. Rowland and I. S. A. Isaksen, eds.), John Wiley and Sons, New York, 49–61.

Cicerone, R. J., and R. S. Oremland. 1988. Biogeochemical aspects of atmospheric methane. *Global Biogeochemical Cycles 2,* 299–327.

Clements, H. B., and C. K. McMahon. 1980. Nitrogen oxides from burning forest fuels examined by thermogravimetry and evolved gas analysis. *Thermochemica Acta 35,* 133–139.

Crutzen, P. J. 1991. Methane's sinks and sources. *Nature 350,* 380–381.

Crutzen, P. J., and M. O. Andreae. 1990. Biomass burning in the tropics: Impact on atmospheric chemistry and biogeochemical cycles. *Science 250,* 1669–1678.

Crutzen, P. J., I. Aselmann, and W. Seiler. 1986. Methane production by domestic animals, wild ruminants, other herbivorous fauna, and humans. *Tellus 38B,* 271–284.

Dasch, J. M. 1982. Particulate and gaseous emissions from wood-burning fireplaces. *Environmental Science and Technology 6,* 638–645.

Davidson, E. A., P. M. Vitousek, P. A. Matson, R. Riley, G. Garcia-Mendez, and J. M. Maass. 1991. Soil emissions of nitric oxide in a seasonally dry tropical forest of Mexico. *Journal of Geophysical Research 96,* 15439–15445.

Dignon, J. 1992. NO_x and SO_x emissions from fossil fuels: A global distribution. *Atmospheric Environment 26A,* 1157–1163.

Dignon, J., and J. E. Penner. 1991. Biomass burning: A source of nitrogen oxides in the atmosphere. In *Global Biomass Burning* (J. Levine, ed.), MIT Press, Cambridge, Massachusetts, 370–375.

Duce, R. A. 1978. Particulate and vapor phase organic carbon in the global troposphere: Budget considerations. *Pure and Applied Geophysics 116,* 244–273.

Engardt, M., and H. Rodhe. 1993. A comparison between patterns of temperature trends and sulfate aerosol pollution. *Geophysical Research Letters 20,* 117–120.

Erickson III, D. J., S. J. Ghan, and J. E. Penner. 1990. Global ocean-to-atmosphere dimethylsulfide flux. *Journal of Geophysical Research 95,* 7543–7552.

Erickson III, D. J., J. J. Walton, S. J. Ghan, and J. E. Penner. 1991. Three-dimensional modeling of the global atmospheric sulfur cycle: A first step. *Atmospheric Environment 25A,* 2513–2520.

Eriksson, E. 1959. The yearly circulation of chloride and sulfur in nature: Meteorological, geochemical, and pedological implications, Part 1. *Tellus 11,* 375–403.

Ferm, M. 1990. Ammonia—A recent atmospheric problem. *Acid Magazine 9,* 12.

Finlayson-Pitts, B. J., and J. N. Pitts, Jr. 1986. Atmospheric chemistry: Fundamentals and Experimental Techniques. John Wiley and Sons, New York, 1098.

Fung, I., J. John, J. Lerner, E. Matthews, M. Prather, L.P. Steele, and P.J. Fraser. 1991. Three-dimensional model synthesis of the global methane cycle. *Journal of Geophysical Research 96,* 13033–13065.

Galasyn, J.F., K.L. Tschudy, and B.J. Huebert. 1987. Seasonal and diurnal variability of nitric acid vapor and ionic aerosol species in the remote free troposphere at Mauna Loa, Hawaii. *Journal of Geophysical Research 92,* 3105–3113.

Galbally, I.E., 1989. Factors controlling NO_x emissions from soils. In *Exchange of Trace Gases between Terrestrial Ecosystems and the Atmosphere* (M.O. Andreae and D.S. Schimel, eds.), John Wiley and Sons, New York, 23–37.

Ghan, S.J., and J.E. Penner. 1991. Smoke, effects on climate. In *Encyclopedia of Earth System Science,* Vol. 4 (W.A. Nierenberg, ed.), Academic Press, San Diego, California, 191–198.

Goldan, P.D., R. Fall, W.C. Kuster, and F.C. Fehsenfeld. 1988. Uptake of COS by growing vegetation: A major tropospheric sink. *Journal of Geophysical Research 93,* 14186–14192.

Gray, H.A. 1982. *Control of Atmospheric Fine Primary Carbon Particle Concentrations.* Report No. 23, Environmental Quality Laboratory, California Institute of Technology, Pasadena, California.

Hameed, S., and J. Dignon. 1988. Changes in the geographical distributions of global emissions of NO_x and SO_x from fossil-fuel combustion between 1966 and 1980. *Atmospheric Environment 22,* 441–449.

Hao, W.M., S.C. Wofsy, M.B. McElroy, J.M. Beer, and M.A. Togan. 1987. Sources of atmospheric nitrous oxide from combustion. *Journal of Geophysical Research 92,* 3098–3104.

Hao, W.M., M.H. Liu, and P.J. Crutzen. 1990. Estimates of annual and regional releases of CO_2 and other trace gases to the atmosphere from fires in the tropics, based on the FAO statistics for the period 1975–1980. In *Fire in the Tropical Biota* (J.G. Goldammer, ed.), Springer-Verlag, Berlin, 440–486.

Hatakeyama, S., K. Izumi, T. Fukuyama, H. Akimoto, and N. Washida. 1991. Reactions of OH with a-pinene and b-pinene in air: Estimate of global CO production from the atmospheric oxidation of terpenes. *Journal of Geophysical Research 96,* 947–958.

Hegg, D. A., L. R. Radke, P. V. Hobbs, R. A. Rasmussen, and P. J. Riggan. 1990. Emissions of some trace gases from biomass fires. *Journal of Geophysical Research 95,* 5669–5675.

Hildemann, L. M., M. A. Mazurek, and G. R. Cass. 1991a. Quantitative characterization of urban sources of organic aerosol by high-resolution gas chromatography. *Environmental Science and Technology 25,* 1311–1325.

Hildemann, L. M., G. R. Markowski, and G. R. Cass. 1991b. Chemical composition of emissions from urban sources of fine organic aerosol. *Environmental Science and Technology 23,* 744–759.

Hofmann, D. J. 1990. Increase in the stratospheric background sulfuric acid aerosol mass in the past 10 years. *Science 248,* 996–1000.

Houghton, R. A. 1991. Releases of carbon to the atmosphere from degradation of forests in tropical Asia. *Canadian Journal of Forest Research 21,* 132–142.

Houghton, R. A., and D. L. Skole. 1990. Changes in the global carbon cycle between 1700 and 1985. In *The Earth Transformed by Human Action: Global and Regional Changes in the Biosphere over the Past 300 Years* (B.L. Turner II, W.C. Clark, R.W. Kates, J.F. Richards, J.T. Mathews, and W.B. Meyer, eds.), Cambridge University Press, Cambridge, U.K.

Houghton, R. A., W. H. Schlesinger, S. Brown, and J. F. Richards. 1985. Carbon dioxide exchange between the atmosphere and terrestrial ecosystems. In *Atmospheric Carbon Dioxide and the Global Carbon Cycle* (J. R. Trabalka, ed.), DOE/ER-0239, U.S. Department of Energy, Washington, D.C., 113–140.

Husar, R. B., J. Holloway, and D. E. Patterson. 1981. Spatial and temporal pattern of eastern U.S. haziness: A summary. *Atmospheric Environment 15,* 1919–1928.

Jaenicke, R. 1978. The role of organic material in atmospheric aerosols. *Pure and Applied Geophysics 116,* 283–292.

Jaenicke, R., and S. Matthias. 1988. The primary biogenic fraction of the atmospheric aerosol. In *Aerosols and Climate* (P.V. Hobbs and M.P. McCormick, eds.), A. Deepak Publishing, Hampton, Virginia, 31–38.

Johansson, C. 1984. Field measurements of emission of nitric oxide from fertilized and unfertilized soils in Sweden. *Journal of Atmospheric Chemistry 1,* 429–442.

Johnson, J. E., A. Bandy, and D. Thornton. 1990. The interhemispheric gradient of carbonyl sulfide as observed during CITE-3. *Eos, Transactions of the American Geophysical Union 71,* 1255.

Junge, C. E. 1972. Our knowledge of the physico-chemistry of aerosols in the undisturbed marine environment. *Journal of Geophysical Research 77,* 5183–5200.

Kaplan, W. A., S. C. Wofsy, M. Keller, and J.M. DaCosta. 1988. Emissions of NO and deposition of O_3 in a tropical forest system. *Journal of Geophysical Research 93,* 1389–1395.

Keller, M., W. A. Kaplan, S. C. Wofsy, and J. M. Da Costa. 1988. Emissions of N_2O from tropical forest soils: Response to fertilization with NH_4^+, NO_3^-, PO_4^{3-}. *Journal of Geophysical Research 93,* 1600–1604.

Khalil, M. A. K., and R. A. Rasmussen. 1983. Increase and seasonal cycles of nitrous oxide in the Earth's atmosphere. *Tellus 35B,* 161–169.

Khalil, M. A. K., and R. A. Rasmussen. 1984. Global sources, lifetimes and mass balances of carbonyl sulfide (COS) and carbon disulfide (CS_2) in the Earth's atmosphere. *Atmospheric Environment 18,* 1805–1813.

Khalil, M. A. K., and R. A. Rasmussen. 1992. The global sources of nitrous oxide. *Journal of Geophysical Research 97,* 14651–14660.

Ko, M. K. W., N. K. Sze, and D. K. Weisenstein. 1991. Use of satellite data to constrain the model-calculated atmospheric lifetime for N_2O: Implications for other trace gases. *Journal of Geophysical Research 96,* 7547–7552.

Lacis, A. A., D. J. Wuebbles, and J. A. Logan. 1990. Radiative forcing of climate by changes in the vertical distribution of ozone. *Journal of Geophysical Research 95,* 9971–9982.

Langner, J., and H. Rodhe. 1990. Anthropogenic impact on the global distribution of atmospheric sulphate. In *Proceedings of the International Conference on Global and Regional Environmental Atmospheric Chemistry* (L. Newman, W. Wang, and C.S. Kiang, eds.), Beijing, China, 3–10 May 1991, 106–117. Available from NTIS, Springfield, Virginia.

Lassey, K. R., D. C. Lowe, M. R. Manning, and G. C. Waghorn. 1992. A source inventory for atmospheric methane in New Zealand and its global perspective. *Journal of Geophysical Research 97,* 3751–3766.

Legrand, M. R., R. J. Delmas, and R. J. Charlson. 1988. Climate forcing implications from Vostok ice-core sulfate data. *Nature 334,* 418–420.

Lerner, J., E. Matthews, and I. Fung. 1988. Methane emission from animals: A global high-resolution database. *Global Biogeochemical Cycles 2,* 139–156.

Levy II, H., and W. J. Moxim. 1989. Simulated global distribution and deposition of reactive nitrogen emitted by fossil fuel combustion. *Tellus 41B,* 256–271.

Lobert, J. M., D. H. Scharffe, W. M. Hao, and P. J. Crutzen. 1990. Importance of biomass burning in the atmospheric budgets of nitrogen containing gases. *Nature 346,* 552–554.

Logan, J. A., M. J. Prather, S. C. Wofsy, and M. B. McElroy. 1981. Tropospheric chemistry: A global perspective. *Journal of Geophysical Research 86,* 7210–7254.

Marland, G. 1989. Fossil fuels CO_2 emissions: Three countries account for 50% in 1988. *CDIAC Communications (Winter),* 1–4.

Matson, P. A., and P. M. Vitousek. 1990. Ecosystem approach to a global nitrous oxide budget. *BioScience 40,* 667–672.

Matson, P. A., P. M. Vitousek, G. P. Livingston, and N. A. Swanberg. 1990. Sources of variation in nitrous oxide flux from Amazonian ecosystems. *Journal of Geophysical Research 95,* 16789–16798.

Matthews, E., and I. Fung. 1987. Methane emission from natural wetlands: Global distribution, area, and environmental characteristics of sources. *Global Biogeochemical Cycles 1,* 61–88.

Mayewsky, P. A., W. B. Lyons, M. Twickler, W. Dansgaard, B. Koci, C.I. Davidson, and R.E. Honrath. 1986. Sulfate and nitrate concentrations from a South Greenland ice core. *Science 232,* 975–977.

Mazurek, M. A., G. R. Cass, and B. R. T. Simoneit. 1991. Biological input to visibility-reducing aerosol particles in the remote arid southwestern United States. *Environmental Science and Technology 25,* 684–694.

Moxim, W. J. 1990. Simulated transport of NO_y to Hawaii during August: A synoptic study. *Journal of Geophysical Research 95,* 5717–5729.

Muzio, L. J., and J. C. Kramlich. 1988. An artifact in the measurement of N_2O from combustion sources. *Geophysical Research Letters 15,* 1369–1372.

Neftel, A., J. Beer, H. Oeschger, F. Zürcher, and R.C. Finkel. 1985. Sulphate and nitrate concentrations in snow from South Greenland 1895–1978. *Nature 314,* 611–613.

Pandis, S. N., S. E. Paulson, J. H. Seinfeld, and R.C. Flagan. 1991. Aerosol formation in the photooxidation of isoprene and β-pinene. *Atmospheric Environment 25A,* 997–1008.

Paulson, S. E., S. N. Pandis, U. Baltensperger, J. H. Seinfeld, R. C. Flagan, E. J. Palen, D. T. Allen, C. Schaffner, W. Giger, and A. Portmann. 1990. Characterization of photochemical aerosols from biogenic hydrocarbons. *Journal of Aerosol Science 21,* S245–S248.

Penner, J. E. 1980. Increases in CO_2 and chlorofluoromethanes: Coupled effects on stratospheric ozone. In *Proceedings of the Quadrennial International Ozone Symposium,* Boulder, Colorado, 4–9 August 1980. International Ozone Commission, published at the National Center for Atmospheric Research (NCAR), Boulder, Colorado, 918–925.

Penner, J. E. 1990. Cloud albedo, greenhouse effects, atmospheric chemistry, and climate change. *Journal of the Air & Waste Management Association 40,* 456–461.

Penner, J. E., M. B. McElroy, and S. C. Wofsy. 1977. Sources and sinks for atmospheric H_2: A current analysis with projections for the influence of anthropogenic activity. *Planetary and Space Science 25,* 521–540.

Penner, J. E., P. S. Connell, D. J. Wuebbles, and C. C. Covey. 1989. Climate change and its interactions with air chemistry: Perspectives and research needs. In *The Potential Effects of Global Climate Change on the United States* (J.B. Smith and D.A. Tirpak, eds.), EPA-230-05-89-056, U.S. Environmental Protection Agency, Washington, D.C.

Penner, J. E., C. S. Atherton, J. Dignon, S. J. Ghan, J. J. Walton, and S. Hameed. 1991a. Tropospheric nitrogen: A three-dimensional study of sources, distribution, and deposition. *Journal of Geophysical Research 96,* 959–990.

Penner, J. E., S. J. Ghan, and J. J. Walton. 1991b. The role of biomass burning in the budget and cycle of carbonaceous soot aerosols and their climate impact. In *Global Biomass Burning* (J. Levine, ed.), MIT Press, Cambridge, Massachusetts.

Penner, J. E., R. E. Dickinson, and C. A. O'Neill. 1992. Effects of aerosol from biomass burning on the global radiation budget. *Science 256* (5062), 1432–34.

Penner, J. E., H. Eddleman, and T. Novakov. 1993. Towards the development of a global inventory of black carbon emissions. *Atmospheric Environment 27A,* 1277–1295.

Pollack, J. B., O. B. Toon, and D. Wiedman. 1981. Radiative properties of the background stratospheric aerosols and implications for perturbed conditions. *Geophysical Research Letters 8,* 26–28.

Prinn, R. D., R. Cunnold, R. Rasmussen, F. Alyea, A. Crawford, P. Fraser, and R. Rosen. 1990. Atmospheric emissions and trends of nitrous oxide deduced from 10 years of ALE/GAGE data. *Journal of Geophysical Research 95,* 18369–18385.

Prospero, J. M., R. J. Charlson, V. Mohnen, R. Jaenicke, A. C. Delany, J. Moyers, W. Zoller, and K. Rahn. 1983. The atmospheric aerosol system: An overview. *Reviews of Geophysics and Space Physics 21,* 1607–1629.

Quinn, P. K., R. J. Charlson, and W. H. Zoller. 1987. Ammonia, the dominant base in the remote marine troposphere: A review. *Tellus 39B,* 413–425.

Radke, L. F., J. A. Coakley, and M. D. King. 1989. Direct and remote sensing observations of the effects of ships on clouds. *Science 246,* 1146–1148.

Rinsland, C. P., R. Zander, E. Mahieu, P. Demoulin, A. Goldman, D. H. Ehhalt, and J. Rudolph. 1992. Ground-based infrared measurements of carbonyl sulfide total column abundances: Long-term trends and variability. *Journal of Geophysical Research 97,* 5995–6002.

Rodriguez, J. M., M. K. W. Ko, and N. D. Sze. 1991. Role of heterogeneous conversion of N_2O_5 on sulfate aerosols in global ozone losses. *Nature 352,* 134–137.

Samson, J. A., S. C. Barnard, J. S. Obremski, D. C. Riley, J. J. Black, and A. W. Hogan. 1990. On the systematic variation in surface aerosol concentration at the South Pole. *Atmospheric Research 25,* 385–396.

Sanhueza, E., W.M. Hao, D. Scharffe, L. Donoso, and P.J. Crutzen. 1990. N_2O and NO emissions from soils of the northern part of the Guayana Shield, Venezuela. *Journal of Geophysical Research 95,* 22481–22488.

Savoie, D. L., and J. M. Prospero. 1982. Particle size distribution of nitrate and sulfate in the marine atmosphere. *Geophysical Research Letters 9,* 1207–1210.

Savoie, D. L., J. M. Prospero, J. T. Merrill, and M. Nematsu. 1989. Nitrate in the atmospheric boundary layer of the tropical South Pacific: Implications regarding sources and transport. *Journal of Atmospheric Chemistry 8,* 391–415.

Seiler, W., and R. Conrad. 1987. Contribution of tropical ecosystems to the global budgets of trace gases, especially CH_4, H_2, CO, and N_2O. In *The Geophysiology of Amazonia* (R. Dickinson, ed.), John Wiley and Sons, New York, 133–160.

Shine, K. P., R. G. Derwent, D. J. Wuebbles, and J.-J. Morcrette. 1990. Radiative forcing of climate. In *Climate Change, The IPCC Scientific Assessment* (G.J. Jenkins, J.T. Houghton, and J.J. Ephraums, eds.), Cambridge University Press, Cambridge, U.K., 41–68.

Singh, H. B., and P. L. Hanst. 1981. Peroxyacetyl nitrate (PAN) in the unpolluted atmosphere: An important reservoir for nitrogen oxides. *Geophysical Research Letters 8,* 941–944.

Singh, H. B., and P. B. Zimmerman. 1992. Atmospheric distribution and sources of nonmethane hydrocarbons. In *Gaseous Pollutants: Characterization and Cycling* (J.O. Nriagu, ed.) John Wiley and Sons, 177–235.

Singh, H. B., L. J. Sales, and W. Viezee. 1986. Global distribution of peroxyacetyl nitrate. *Nature 321,* 588–591.

Slemr, F., and W. Seiler. 1984. Field measurements of NO and NO_2 emissions from fertilized and unfertilized soils. *Journal of Atmospheric Chemistry 2,* 1–24.

Slemr, F., and W. Seiler. 1991. Field study of environmental variables controlling the NO emissions from soil and the NO compensation point. *Journal of Geophysical Research 96,* 13017–13032.

Spiro, P. A., D. J. Jacob, and J. A. Logan. 1992. Global inventory of sulfur emissions with $1° \times 1°$ resolution. *Journal of Geophysical Research 97,* 6023–6036.

Stoiber, R. E., S. N. Williams, and B. Huebert. 1987. Annual contribution of sulfur dioxide to the atmosphere by volcanoes. *Journal of Volcanology and Geothermal Research 33,* 1–8.

Talbot, R. W., M. O. Andreae, H. Berresheim, P. Artaxo, M. Garstang, R.C. Harris, K. M. Beecher, and S. M. Li. 1990. Aerosol chemistry during the wet season in Central Amazonia: The influence of long-range transport. *Journal of Geophysical Research 95,* 16955–16970.

Thomas, R., and J.-W. Erisman. 1990. Ammonia emissions in the Netherlands: Emissions inventory and policy plans. In *Proceedings of the Workshop on International Emission Inventories*, Regensburg, Federal Republic of Germany, 3–6 July 1990. Norwegian Institute for Air Research, Lillestrom, Norway, 257–264.

Turco, R. P., R. C. Whitten, O. B. Toon, J. B. Pollack, and P. Hamill. 1980. OCS, stratospheric aerosols and climate. *Nature 283*, 283–286.

Twomey, S. 1977. *Atmospheric Aerosols,* Elsevier, New York, 302 pp.

United Nations. 1983. *1981 Yearbook of World Energy Statistics.* United Nations, New York, 784 pp.

Vaghjianai, G. L., and A. R. Ravishankara. 1991. Rate coefficient for the reaction of OH with CH_4: Implications to the atmospheric lifetime and budget of methane. *Nature 350,* 406–409.

Wahlen, M., N. Takata, R. Henry, B. Deck, J. Zeglen, J. S. Vogel, J. Southon, A. Shemesh, R. Fairbanks, and W. Broecker. 1989. Carbon-14 in methane sources and in atmospheric methane: The contribution from fossil carbon. *Science 245*, 286–290.

Warneck, P. 1988. *Chemistry of the Natural Atmosphere.* Academic Press, Harcourt Brace Jovanovich, San Diego, California, 757 pp.

Watson, R. T., H. Rodhe, H. Oeschger, and U. Siegenthaler. 1990. Greenhouse gases and aerosols. In *Climate Change, The IPCC Scientific Assessment* (J.T. Houghton, G.J. Jenkins, and J.J. Ephraums, eds.), Cambridge University Press, Cambridge, U.K., 1–40.

Wigley, T. M. L. 1989. Possible climatic change due to SO_2-derived cloud condensation nuclei. *Nature 339*, 365–367.

Williams, E. J., and F. C. Fehsenfeld. 1991. Measurement of soil nitrogen oxide emissions at three North American ecosystems. *Journal of Geophysical Research 96*, 1033–1042.

Williams, E. J., D. D. Parrish, and F. C. Fehsenfeld. 1987. Determination of nitrogen oxide emissions from soils: Results from a grassland site in Colorado, United States. *Journal of Geophysical Research 92*, 2173–2179.

Winkler, P., and U. Kaminski. 1988. Increasing submicron particle mass concentration at Hamburg, 1. Observations. *Atmospheric Environment 22*, 2871–2878.

9

Soils

S. W. Buol

Land-use and land-cover changes have notable impacts on the world's soils, the 'dirt' wherein water combines with the mineral elements necessary to support all vegetation on the earth. In this paper, the major soil processes affected by human action are described and the incidence of the globally significant effects—erosion and physical and chemical degradation—is reviewed. Particular attention is devoted to soil chemical fertility as affected by human land use because of its importance not only in supporting the growth of useful crops but in maintaining a protective cover of vegetation on the soil.

Soil is often discussed as a single entity, but it should be kept in mind that soils differ greatly in their ability to fulfill the task assigned them as the media for the growth of land plants. Not only do soils take on characteristics from the different geologic materials from which they acquire their mineral components, but they also have distinct dynamics of wetness and temperature derived from local climatic conditions (Soil Survey Staff, 1975). Finally, as the major medium of intercourse between the organic and inorganic domains, soils have features that vary in response to the organic domain of the organisms that root in them and extend their aerial appendages into the atmosphere of this planet.

Key Processes of the Soil System

The vital processes that take place in the soil are represented in Figure 1. For the purposes of this discussion, the intrasoil translocations, although of vital interest to the understanding of soil genesis, will not be discussed. The discussion will consider the transfers between the soil and the air, between the soil and the groundwater beneath, and between one soil and another. The primary focus will be on the flow of economically valuable food and fiber in the biocycling loop and the entry of nutrients essential for plant growth via mineral weathering.

In the context of discussing the relation of soil to land-cover and land-use change, the following roles characterize each pathway in Figure 1.

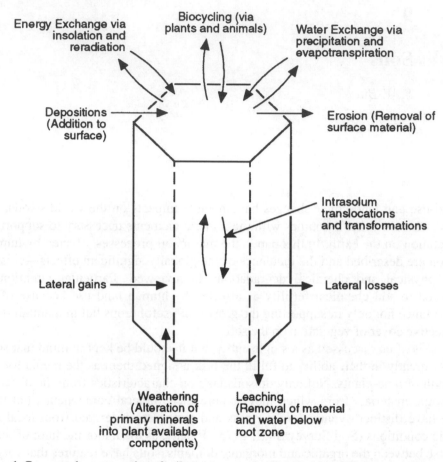

Figure 1. Conceptual representation of soil as an open system (adapted from Buol *et al.*, 1989).

Energy Exchange

The radiation flows from the sun to the earth and from the earth to outer space depend on the absorptive and reflective qualities of the soil cover or the properties of the soil itself in the absence of any cover. Energy exchanges vary considerably in quantity and quality, but an aggregation of scenarios can be defined in association with particular land-cover and land-use categories. For example, humid tropical rainforests of isohyperthermic soil temperature regimes and udic and perudic soil moisture regimes have year-long exposure of photosynthetically active leaves. When this forest gives way to cropland, there are periods of the year when bare soil, incomplete foliar canopy, and ripe nonphotosynthetic vegetation, each of a different albedo, interact with radiation. Each type of crop and/or management system has its own distinct influence on radiative exchanges, as does urban land, covered to a large degree with buildings and pavement.

Water Exchange

All precipitation that falls on the land interacts with soil. As direct precipitation or flow from snowmelt, water either runs off the soil surface, is temporarily incorporated into the soil only to be returned to the air via evaporation or transpiration, or percolates to a depth not reached by plant roots and enters the groundwater. The impacts of land use and other human activities on the quality of water are well known. Particularly notable are irrigation practices that direct water into soils containing relatively soluble minerals. As the irrigation water dissolves, the osmotic value of the water is increased, and it becomes too salty to be useful for normal plant growth. If this practice is allowed to proceed for some years without the salts' being flushed to the groundwater, the soil becomes unusable for plant growth. This process is termed salinization. Irrigation with naturally salty water can also cause salinization. Although the techniques of reclaiming salinized soil are well known to agronomic practices, they may not be economically feasible in many cases. In many places, moreover, unacceptable problems are created in adjacent areas when the salt is flushed to the groundwater.

Erosion and Deposition

Although represented in Figure 1 as separate processes, erosion and deposition are only two segments of the same process of redistributing mineral material across the earth's surface. Although some solid mineral particles do flow to the oceans onto the continental shelves, most particles eroded from one site or kind of soil within a landscape are deposited on another soil at a lower elevation within the same landscape (Ruhe, 1969). Most particles suspended by wind too are deposited on a soil surface at another, often adjacent, site.

Erosion and deposition are natural processes that have shaped the land surface through the history of the earth, but humans dramatically influence their rate and spatial distribution. Human influence takes place via the manipulation of land cover and vegetation and the mechanical loosening (through cultivation) of the surface soil. Major patterns of human habitation on the earth are related to the distribution of erosion and deposition. Ancient civilization developed along the major floodplains (areas of deposition) of such erosive watersheds as the Nile, Mekong, Ganges, and Euphrates river systems. These depositional landscapes are enriched by the vegetative byproducts included in the topsoil of the hills, which are eroded in the headwaters of these river systems. This same interdependence of erosion and deposition is a consideration in land-use and land-cover manipulations at microscales involving fractions of hectares throughout the world. If erosion is reduced, deposition on adjacent soils is reduced.

Not all soils in a landscape hold the same potential for erosion. The same amount of accelerated erosion may severely damage some soils and limit future production potential, but have little effect on some other kinds of soil (Larson *et al.*, 1984).

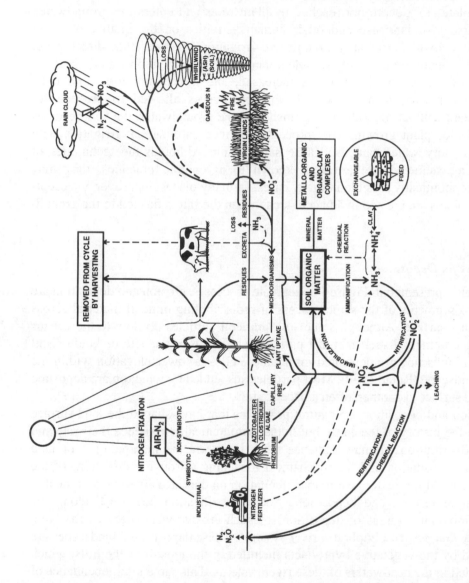

Figure 2. Representation of nitrogen in the ecosystem (adapted from Stevenson, 1982, copyright © 1982, American Society of Agronomy, Inc.).

Leaching

Soluble or suspendable materials that are moved beyond the depth of plant roots can be considered leached. Historically, a major distinction has been drawn between Pedocals, soils with unleached calcium carbonate in the profile, and Pedalfers, or soils from which most of the carbonate has been leached.

The term leaching is also frequently used to describe the movement of material only from the surface layer of soil. This process, which leaves the leached material within the rooting depth, is more properly termed translocation.

Biocycling and Crop Production

All plant and animal life within soil has a short life expectancy relative to the time scale of soil formation. During this growing period, plants accumulate carbon from the atmosphere and the mineral constituents necessary for their metabolism and structure from water in the soil. When plants are allowed to complete their life cycle without being harvested and moved from the site, they concentrate the elements essential to plant growth in or near the top of the soil. The plant thus extracts nutrients from an entire root volume, which may extend to a depth of several meters, and concentrates these elements at the soil surface—through leaves and other plant parts shed during the life of the plant and through the entire plant upon its death. It is this accumulation that accounts for the relative fertility of topsoil vs. subsoil in most soils.

In addition to concentrating elements essential for plant growth and limited by the mineral constraints of the soil, such as phosphorus, potassium, calcium, magnesium, zinc, and copper, some organisms are able to capture nitrogen from the air and convert it into nitrate and ammonia necessary for biological use. Nitrogen, which is acquired by the plant, is taken into the plant through the root system but does not originate in the soil. Nitrogen is obtained from the air and deposited in the soil both by nitrogen-fixing organisms that capture N_2 from the air and directly from rain and/or snow, as shown in Figure 2.

Weathering

Processes of weathering include an almost infinite array of reactions that break down various minerals to release plant-essential elements and place them in a soluble ionic form in the soil water. Not all soils are equally endowed with minerals containing plant-essential nutrients. The differences can be traced to differences in the origin and composition of the geologic material in which the soil has formed.

In general, the richest supplies of weatherable minerals are found in mixtures of rock materials recently (in geologic time) exposed to the earth's surface by processes associated with continental glaciation. They include the wind-deposited loess that occupies large areas equatorward from the great continental ice sheets of recent glaciation in Europe, Asia, and North America. The mineralogically rich

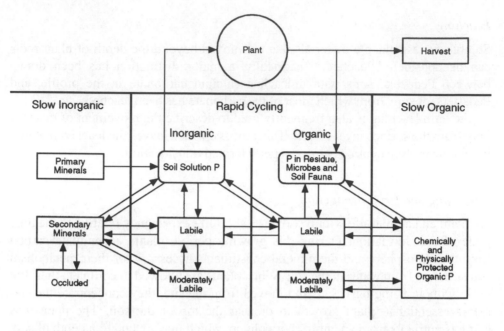

Figure 3. Representation of phosphorus transformations in cropped soils (adapted from Tiessen *et al*., 1984).

materials also include many volcanic extruda and alluvial deposits originating from rapidly eroding mountain ranges. Some of the soils that are least well endowed mineralogically form in sediments of sandstone and other water-reworked sediments on coastal plains. Many igneous rocks weathering on nearly level landscapes have most of the plant-essential elements removed from the rock deep below the root zone; thus the saprolite from which the soils form is quite low in potential fertility.

It must be remembered that only a small fraction of most mineral-borne elements essential to plant growth is brought into solution and thus becomes available to plants. Thus, the biomass growing in poorly endowed soils competes for the ions and uses them in the various tissues—often the seed. When the plant, or its seed, is used as a food and consumed at another location, the site of growth is chemically degraded as compared to a site where the vegetation completes its life cycle *in situ* and returns all its parts to the soil surfaces.

Figure 3 illustrates the phosphorus cycle, indicating several secondary organic and inorganic sinks found within the soil. Note, however, that the only source of P is the primary minerals, mainly apatite, and that the other forms of P in the soil are all secondary. In many soils formed from redeposited materials, such as sandstone, the slowly dissolved secondary minerals are the only primary source of phosphate. Although the other mineral-derived plant-essential elements have somewhat different cycles because of their chemical affinities, each has only a primary mineral source.

Human Impact on Soil

Words used to discuss human impacts on soils are subject to their author's values. Is any change from the natural prehuman state bad, and thus a degradation of the soil? Is a human alteration of the soil that increases the yield of a desired product good in that it reduces the area needed to satisfy human needs and allows more area to be devoted to 'natural' habitat? No less difficult are questions involving total destruction of soil by pavement in major metropolitan areas. This discussion of the changes in soil initiated by human intervention will refrain from using terms like 'degradation' or 'improvement,' which lie outside its scope.

Within the familiar conceptual representation of soil formation $S = F(c,o,r,p,t)$, where the properties of soil (S) are a function (F) of climate (c), organisms (o), relief or slope (r), parent material (p), and time (t), humans have been observed to have a predictable influence on soil properties. Most human impact occurs through the alteration of the plant cover naturally associated with each kind of soil. Humans manipulate natural soil cover for very specific reasons:

- Humans harvest plants and consume them as food and fiber at locations distant from where they grew. The simple act of hunting and gathering thus produces a chemical export from one site to another, chemically degrading the site of growth and enriching the site where the food constituents are returned to the soil. (They are lost to all soil if deposited in the ocean.)
- Being selective in the type of vegetation they desire as food, humans alter the natural vegetative cover to favor the growth of those species that satisfy their needs and desires. In this process, they seek to eliminate crop competition by destroying other plant and animal forms by cultivation, which leaves the soil surface exposed for various periods of time.
- Humans are impatient. They seek to grow their desired crops as rapidly as possible to reduce the time between harvests and also to reduce the time they need to protect their crops from competitive species.
- Humans also have the ability to alter climate and moisture conditions through irrigation and drainage, to alter slope by terraces, and to augment the nutrients provided by the parent material by adding plant-essential elements from distant deposits via fertilization.
- In extreme cases, humans completely remove and replace soils by forming new soils, as seen in the restoration of mine spoils or in extreme forms of urban landscaping.

As previously outlined, the transportation of any organism, or part thereof, from the site where it is grown, such that the residue is deposited elsewhere, subtracts from the chemical balance in the soil at the growing site.

When the soil surface is not protected by a vegetative cover, the rate of erosion increases. The two major agents of erosion are water and wind. In many, if not most, of the accelerated erosion scenarios associated with cropland use, erosion occurs when fields are cleared of vegetation to prepare for crop planting or when

Environmental Consequences

Table 1: Soil evaluation criteria used in Oldeman et al. (1990)

Type of influence
Water erosion
Wind erosion
Chemical deterioration (including pollution)
Physical deterioration
Degree of change
Evaluate the proportion of the area within each mapped unit as
1–5%, 5–10%, 10–25%, 25–50%, or 50–100% of the area
Severity of impact
1. Light impact: restoration possible by local management
2. Moderate impact: major changes required by local management
3. Strong impact: major engineering work required to improve
4. Extreme impact: unreclaimable

vegetation is removed by grazing animals. Physical erosion is often made more severe than it might otherwise have been by previous cropping that has reduced the chemical fertility of the soil, thereby slowing vegetative growth and exposing the soil surface to direct raindrop impact and wind.

Another very visible effect of human activity is registered where irrigation waters, usually in arid areas, concentrate soluble salts in the soil. Salty water has a high osmotic value, and plants unable to utilize such salty soil water die. Extra water to leach the salts is not always readily available to irrigators.

Human activity in the form of animal, machinery, or even human foot traffic on the surface of the soil tends to compact the soil surface, destroys or severely weakens the vegetative cover, and thereby reduces the rate at which rainwater will infiltrate into the soil. This surface compaction can usually be loosened by cultivation, but the energy needed is not always available to the cultivator. Increased runoff and aggravated erosion result. Finally, humans can contaminate soil with industrial waste and thereby alter its ability to grow plants.

These five general categories—chemical change, erosion, salinization, compaction, and pollution—were used by a group of soil scientists to evaluate the present status of the world's soil resources as affected by human impact (Oldeman *et al.*, 1990). Their method was to enlist nearly 200 soil experts from all areas of the world to use the world soil map (FAO/UNESCO, 1971–79) as a base map and evaluate, from their personal knowledge of the areas, human impact on soil resources. Although the project continues with the goal of providing a digital data base, their preliminary report will be used for the following discussion (Oldeman *et al.*, 1990).

The soil scientists were asked to evaluate several categories of human influence within each mapped unit, shown in Table 1.

Human Influence on Soil Chemical Properties

Oldeman *et al.* (1990) considered chemical deterioration in the soil due to human

218

activity when insufficient manure and fertilizer were returned to cropped areas, salinization was created by unwise irrigation or drainage practices, soil acidification was not corrected by liming, and pollution was caused by industrial or other urban uses.

No soil chemical deterioration is identified in the United States, Canada, Argentina, South Africa, or China due to insufficient return of fertilizer or manure. Only small areas in Europe, the United Kingdom, Sweden, Greece, and Finland are identified as having low soil nutrient supply resulting from human activity. India and the Middle East also have only very small areas slightly affected by low fertility.

The major areas of low soil fertility and inadequate fertilizer use are in the humid tropics of Africa, South America, and southeastern Asia. The problem can be traced to a combination of low native soil fertility and a lack of the agrobusiness infrastructure needed to deliver sufficient fertilizer to meet the requirements of the cropland.

Extensive areas of industrial and urban pollution are identified only in eastern and northern Europe.

Soil alteration by human-induced salinization is most extensive in arid regions of the Middle East, southeastern Europe, northeastern Spain, and some of the major delta areas of southeastern Asia, western China, and northern India. Severe salinization is identified in the arid regions of Mexico, and less severe problems are noted in the southwestern United States. In Central and South America, no salinization problems are identified. Outside of Egypt, Senegal, Algeria, and Somalia, no areas of salinization are indicated in Africa. No doubt small areas of salinization do exist in many of these regions, but their identification is precluded by the map scale used.

Acidification from drainage of pyrite-containing soils is most severe in southern Vietnam and the southern part of Indonesian Borneo (Kalimantan). This problem is directly associated with Sulfaquent and Sulfaquept soils, often identified as acid sulfate clays, and can be a severe localized problem whenever coastal marshland areas containing pyrite are drained. More general but less severe acidification is often associated with chemical nutrient depletion without sufficient application of liming materials with the fertilizer. In many of the areas discussed, this problem is identified as chemical nutrient depletion.

Erosion by Water

Water erosion is identified as extensive in all areas of the world. Most of the areas identified as being adversely affected by water erosion are rated by the soil scientists as occupying less than 10% of any mapping unit and as being affected to only a light or moderate degree, such that the areas are still suitable for use in local farming systems. Changes in local farming systems that incorporate conservation practices are seen as the necessary remedial measures.

Some areas, however, are identified as having undergone strong or extreme erosion and as reclaimable only through major engineering practices. Most notable among these areas are northern and western India, parts of Sri Lanka, and large areas in Turkey. In Africa, several eroded areas are located in the Sahel of West Africa, South Africa, Angola, Zimbabwe, Zambia, Zaire, Kenya, Tanzania, Ethiopia, Uganda, Somalia, Sudan, and Morocco. In Europe, Yugoslavia, Romania, and the southwestern parts of the USSR display the most severe erosion by water.

Although high proportions (over 50%) of several mapping units in China appear eroded, the impact is considered light to moderate, except in the loess areas of central China, where it is judged to be more severe. Extensive areas of strong erosion are identified in the Philippines, Vietnam, Malaysia, and Thailand.

Large areas of strong water erosion are identified in central Mexico and several other Central American countries and in Haiti. In South America, strong erosion is found in western Venezuela, Colombia, Ecuador, Bolivia, Chile, northwestern Argentina, and eastern central Brazil.

No strongly water-eroded areas are identified in the United States, Canada, Australia, New Zealand, or western Europe.

Erosion by Wind

Soil erosion by wind is most common in the more arid cropland areas of the world. As with water erosion, light and moderate degrees of erosion considered manageable by local management practices are found in all continents. In Africa, areas with strong or extreme wind erosion are identified only on the southern edge of the Sahara Desert in Mauritania, Niger, and Algeria; in an area in northern Kenya; and in northern Botswana. An area being strongly damaged by overblowing is identified along the coast of Somalia. Several areas of strong wind erosion with deformation of landscape are identified in northern China. No areas of strong wind erosion are identified in the Americas, Europe, or Australia, but extensive areas of light and moderate wind erosion exist.

Physical Deterioration of Soils

Three types of physical deterioration are identified: compaction, waterlogging, and organic soil subsidence. Strong compaction is identified in western Venezuela, Germany, Belgium, Sudan, South Africa, and southeastern Australia. Significant organic subsidence is noted only in Brunei and Malaysia. Waterlogging is identified as a strong limitation only in several areas of Mexico.

Management Systems Used to Mitigate Human Influence

Most, if not all, of the changes in soil properties by human activities have been studied and alteration made in the practices people have used to inhabit the land and obtain the necessities of life from it. Changes in soil management systems are not

Table 2: Nutrient removal by crop

Crop	Plant Part	Yield (tons/ha)	Nutrient (kg/ha)				
			N	P	K	Ca	Mg
Cereals							
Corn (low yield)	grain	1.0	25	6	15	3.0	2.0
	stover[1]	1.5	15	3	18	4.5	3.0
	Total	2.5	40	9	33	7.5	5.0
Corn (high yield)	grain	7.0	128	20	37	14.0	11.0
	stover	7.0	72	14	93	17.0	13.0
	Total	14.0	200	34	130	31.0	24.0
Rice (low yield)	grain	1.5	35	7	10	1.4	0.3
	stover	1.5	7	1	18	2.6	2.2
	Total	3.0	42	8	28	4.0	2.5
Rice (high yield)	grain	8.0	106	32	20	4	1.0
	stover	8.0	35	5	70	24	13.0
	Total	16.0	141	37	90	28	14.0
Wheat	grain	5.0	80	22	20	2.5	8.0
	stover	5.0	38	5	60	10.0	10.0
	Total	10.0	118	27	80	12.5	18.0
Sorghum	grain	1.0	20	0.9	4	4.0	2.4
	straw	1.2	6	0.4	2	4.6	3.2
	Total	2.2	26	1.3	6	8.6	5.6
Root crops							
Cassava	roots	16	64	21	100	41	21
Potatoes	tubers	14	77	14	224	4	9
Sweet potatoes	roots	16.5	72	8	88	–	–
Grain legumes							
Beans	beans	1.0	31	3.5	6.6	–	–
Soybeans	beans	1.0	49	7.2	21	–	–
Peanuts	unhulled weight	1.0	49	5.2	27	–	–
Other							
Sugar cane	all above ground	200	149	29	316	55	58
Oil palm	fruit	15	90	8.8	112	28	–
Bananas	bunches	10	19	2.0	54	23	30
	stems & leaves	–	20	1.3	22	1	3
	Total	–	39	3.3	76	24	33
Pineapple	fruit	12.5	9	2.3	29	3	3

[1]Cured stalks from which the grain has been removed.
Adapted from Sanchez, 1976.

Table 3: Four-year average corn yield (grain kg/ha/yr) on the Morrow plots at the University of Illinois

Soil Treatment	No Rotation Continuous Corn	2-Yr Rotation Corn–Soybeans	3-Yr Rotation Corn–Oats–Clover
None (control)	2684	3638	4949
MLP 1904 →	5544	7439	8040
LNPK 1955 →	7056	7953	8492
MLP 1904 → 1955, LNPK 1955 →	7758	8141	8379

M = manure; L = lime; P = phosphate fertilizer; N = nitrogen fertilizer; K = potassium fertilizer; rates of fertilizer applied as predicted by soil test evaluation procedure.
Note: The four years averaged were 1955, 1961, 1967, and 1973 because all rotations had corn planted in those years.

always possible if only the local resources of society are utilized. Humans have frequently had to move away from certain soil problems. Often, however, areas or uses are abandoned or the management system changed not because of soil problems as such but because other areas are able to compete more effectively in the marketplace or because the commodity produced is replaced by a synthetic product.

This section presents examples of shifts in soil management to counter the major adverse influences that human manipulation has had on soil properties.

Combating Chemical Degradation

All plants take nutrients from the soil during growth. When grain crops are grown, for example, a proportion of the nutrients goes to the grain, which is usually harvested and sold or consumed at a site from which the residue is not returned to the soil where it was grown. Part of the nutrient uptake is in the stem and leaves and may or may not be returned to the site where it was grown. Because the nutrient content of all the crops varies in response to cultivar type, limitations in the soil, water supply during growth, and stage of maturity at harvest, no exact numbers can be assigned to these proportions. Table 2 gives representative values that can be expected from common crops—some grown worldwide and some only in the tropics.

The nitrogen removed by crop harvest can be returned to the soil by rainfall in an amount known to vary greatly (but approximated at 20 kg N/ha/yr) and by including legumes in the cropping system. All the other elements have to originate from minerals in the soil or have to be replaced by fertilizers, mined and concentrated to reduce transportation costs, or manures obtained from animals that obtained their food at another location. In effect, manure, or compost, simply transports nutrients via an organic carrier from one site to another.

Combating this export of nutrients with application of manure or fertilizer is a

Table 4: Prorated kg of grain corn per ha per year over all treatments and rotations (for average yield of 1955, 1961, 1967, and 1973)

Soil Treatment	No Rotation	2-Yr Rotation	3-Yr Rotation
None (control)	2684	1825	1650
MLP 1904→	5544	3719	2678
LNPK 1955→	7056	3976	2829
MLP 1904→1955	7758	4071	2791
LNPK 1955→			

See Table 3 for symbol identification.

common and necessary practice in all parts of the world. The following two examples are used to illustrate how soils of very high and very low natural fertility can be expected to respond to such treatment.

At the University of Illinois, plots have been maintained for nearly 100 years in which corn (*Zea mays*) has been grown continuously and in rotation (Odell *et al.*, 1982). The soil at Illinois is representative of the Midwest—formed in loess from the Wisconsin age glacial period and naturally vegetated by tall prairie grasses (Aquic Argiudoll, fine-silty, mixed, mesic, Flanagan series). Table 3 shows the corn grain yields from 12 subplots for four years. The years 1955, 1961, 1967, and 1973 were selected for averaging because in those years corn was grown on both of the rotation plots and, of course, on the continuous corn plot.

It should first be pointed out that more than 40 bushels of corn per acre (2500 kg/ha) can be produced on such soils with continuous planting and no added chemicals for nearly 100 years. This yield is doubled with lime, phosphate, and manure, but it is nearly tripled when complete fertilization is used. In the two- and three-year rotations where legumes (soybeans or clover) are used as a green manure, less than a doubling of the yield was observed compared to continuous corn, indicating that the nitrogen incorporated from the air via the legumes is not the only limiting nutrient.

A consideration of greater significance to farmers' decisions on how to manage their land, however, becomes apparent if corn is considered as the objective of the farming operation. If the yield of corn per year is calculated by dividing the yields in Table 3 by the number of years in the rotation (Table 4), a very interesting result is obtained. Any of the rotations costs farmers in corn production per year in the long term unless they are able to obtain income from the crop grown in rotation. One could also extrapolate these results to answer a question regarding land use: What if the United States were to meet its requirements for corn without fertilizer or by requiring the farmer to use a rotation for soil conservation practice? The clear answer is that considerably more land area would have to be devoted to producing corn.

The example of long-term farming on one of the best natively fertile soils in Illinois shows several options regarding the use or nonuse of fertility

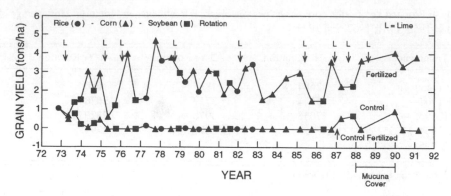

Figure 4. Fertilized and unfertilized crop yields in Yurimaguas, Peru (adapted from TROPSOILS, 1991).

replenishment. Let us now consider a rather poor soil. In 1972, North Carolina State University started long-term soil studies at Yurimaguas, Peru. The soil, formed in transported sediments in the humid Amazon jungle, is quite representative of many in the humid tropics and very similar to those on the Southern Atlantic Coastal Plains of the United States or southeastern China. The soil is classified as a Typic Paleudult; fine-loamy, siliceous, isohyperthermic, Yurimaguas series.

At the time of initial planting, a 20-year-old regrowth jungle was cut and burned following slash-and-burn practices common to the area. Fertilizer was applied to each crop in such amounts and compositional proportion as could best be gauged by soil test and agronomic experience. Lime was added when deemed necessary by determination of exchangeable aluminum content. Control plots were maintained, planted, and harvested the same as the fertilized plots. Weeds were removed manually from both control and fertilized plots.

In the humid tropical climate, three crops per year can be grown, a practice that was maintained until 1985, when mechanical tillage by tractor was adopted on all plots. The work was further altered in 1987 when 80 kg/ha phosphorus was mistakenly added to the check plots. In 1988–90, cropping was interrupted and a legume cover crop of mucuna (*Stizolobium* sp.) was planted over both fertilized and control plots.

As can be seen in Figure 4, the initial fertility in the soil and the ash obtained upon burning the vegetation permitted low yields of 2 tons/ha or less for the first few crops. For the next 20 years, practically no yields were obtained from the control plots despite careful planting, until the accidental incorporation of phosphorus in 1987. The resting period from 1988 to 1990 under a legume cover and the accidental phosphate from two years before combined for one crop of only about one ton per hectare in 1990, but no yield was obtained from the second crop that year. It is easy to see why fields are abandoned and new areas burned in the traditional slash-and-burn management common in the area. Over the same period, plots receiving the lime and fertilizer inputs deemed adequate for crop production pro-

Table 5: Fertilizer and yield comparisons on Norfolk soils in North Carolina and Muscatine soils in Iowa (in kg/ha)

	Norfolk Soils		Muscatine Soils	
	1923	1983	1919	1979
Corn yield	1568–2509	6899	2634	8091–8216
Fertilizer rate				
N	36–63	132–177	0	168–202
P	2–6	20	0	34–54
K	6–11	75	0	75–110

Adapted from Buol et al., 1990.

duced consistent yields. The slow and erratic increase of yields near the beginning of the period can be attributed to several factors, but primarily to biological changes that acidified the soil more rapidly than expected. Too little lime was applied. Lack of experience on the part of the researchers to select correct fertilizer formulations also contributed to low yield. The total result points out that continuous cropping is possible on such low-fertility tropical soils, and efforts to understand sustainable crop production on them require several years at one location.

The long-term experience in the United States on low native fertility soils like the Yurimaguas soils compared to high native fertility soils like the Flanagan soils (Illinois) can be seen in Table 5. The data in Table 5 represent averages or ranges obtained on farmers' fields in Wilson County, North Carolina (Norfolk soils), and Cedar County, Iowa (Muscatine soils), in comparable time periods of the 1920s and 1970s.

From historical records, it is safe to assume that the Norfolk soils (Typic Kandiudults, fine-loamy, siliceous, thermic) had been farmed for several years prior to 1923. The earliest farming was slash-and-burn–type cultivation. Farming on the Muscatine soils (Typic Argiudolls; fine-silty, mixed, mesic) probably started about 50 years prior to 1919. The most significant aspect of the data, considered to be representative of many soils in each area, is that yields in both areas have increased considerably over the approximately 60-year period. At both periods, yields were higher in Iowa than in North Carolina. Even using fertilizer, farmers in North Carolina in the 1920s obtained lower yields than those obtained by Iowa farmers without fertilizer. By 1980, however, the Iowa farmers were using higher rates of fertilizer and obtaining higher yields than the North Carolina farmers. The conclusions drawn from these data are that fertilization must begin earlier in natively poor soils and that, after fertility is cropped out of even the most natively fertile soils, fertilizer is necessary to obtain economical production. Although no data exist, it is very plausible to assume that without fertilizer use, the yields on the Norfolk soil would follow much the same pattern as seen in the Yurimaguas control plots (Figure 4).

Combating Soil Erosion and Compaction

The most effective method of reducing erosion is to keep a vegetative cover on the soil. When soils are used for crop production this is only possible if exported nutrients are replaced. It is the author's postulate that much of the severe erosion and compaction seen in association with cropping has its underlying cause in insufficient soil fertility, which leads to poor plant cover and, in turn, to accelerated erosion.

The erosion process is not uniform on the landscape. As can be seen on the evaluations reported by Oldeman *et al.* (1990), seldom is more than 50% of an area affected, and more often less than 10% of the land in areas recognized as impacted by erosion is affected. Daniels *et al.* (1989) point out that many estimates of the potential impact of erosion have been made by taking the yields or yield estimates on sloping parts of the landscape, those most often identified as eroded phases of a soil, and extending them to more level areas of the same soil. In their detailed evaluation of this spatial relationship of landscape position, yield, and degree of erosion reported in naturally infertile Hapludults of the Piedmont province in North Carolina, they found that when the reduced yields associated with sideslopes (due to the loss of precipitation related to natural runoff) were accounted for, the degree of erosion had little influence on the yields obtained. After comparing eroded and slightly eroded soils by landscape position, they concluded that lower yields, often seen to occur on parts of the landscape most susceptible to erosion, were largely the result of reducing water for crop production and not of the eroded nature of the soils. In another study of similar types of soil, McCracken *et al.* (1989) found that the more steeply sloping areas of Hapludults in the North Carolina Piedmont that had never been cultivated or had any timber harvests would be classified as eroded because of the relative thinness of their topsoil horizons. This finding should come as no surprise considering the shallower nature of soils attributable to a steeper slope in the classic soil formation concept advanced by Jenny (1941).

The conditions of the observations made by Daniels *et al.* (1989) need to be pointed out before extrapolation. Although the sites studied have been cropped for perhaps more than 100 years, common farming practices have returned fertilizer and manure to all of them. Thus, any fertility degradation from loss of nutrients in the naturally more fertile topsoil has long since been replaced and continues to be annually replaced. In areas where cropping depends on native fertility, such erosion would probably more severely affect yield.

Just as the process of erosion is not uniformly active within an area, not all kinds of soils are equally affected by erosion. Larson *et al.* (1984: 244) point out that 'if the nutrients are replaced and lime added and the soil receives reasonable management, the soil may continue to produce crops at a high level for an indefinite period,' unless root-restricting hard layers or bedrock are exposed at a shallow depth by the erosion process.

Several technologies are available to combat the adverse effects of erosion. Common to most of these technologies is to keep the soil surface protected from the impact of rain and wind by the use of plant cover. A fertile chemical condition and adequate water are necessary to all of these technologies. It should be kept in mind that erosion is almost always most severe on only small portions of any given area. Combating erosion requires the active cooperation of all land managers, and practices need to be applied to specific locations.

Salinization and Other Soil Pollution

The technology of preventing or correcting salinization of soils is well understood (U.S. Salinity Laboratory Staff, 1954). Basically, it requires the flushing of relatively salt-free water through the soil to dissolve the soluble salts and remove them from the area. The two constraints on this technology are an inadequate supply of freshwater and/or inadequate drainage structures to remove the flushing water. Overcoming these constraints is often beyond the power of individual farmers because the regional water supply is limited or the removal of the flushing water would encroach on the lands of others. Action usually requires wider cooperation.

Pollution from industry or other urban activities, seen by Oldeman *et al.* (1990) as limiting in a surprising number of areas, especially in Europe, is an obvious concern for larger aggregates of society. Alleviation at the source of the pollution appears the only acceptable solution, although some damaged sites will have to be repaired locally.

Summary

Soil should always be considered in the plural, i.e., soils. Although identifiable processes are common to all soils, the magnitude of each process within different soils differs greatly. These differences are seldom identifiable on large-scale maps but require evaluation and attention at a life-size scale on the land. The unique blend of active processes taking place in each soil requires that the individuality of soils be considered in evaluating their susceptibility to human influence.

Human activities require extractions from soil. Nutrients extracted in crop production are replaceable. Of the essential nutrient elements, nitrogen is unique in that air is its primary source, and the soil serves only to hold small quantities placed there by rainfall or by organisms capable of chemically converting molecular gaseous N into organic forms. The other essential nutrients (notably P, K, Ca, Mg, Zn, and Mn) are derived only from the primary minerals contained in soil. Soils are very unequally endowed with these minerals. When these elements are removed in the harvest of vegetation, they must be replaced if vigorous plant growth is to continue.

Erosion removes soil material. In some soils, this eroded material can be replaced by incorporating underlying materials into the root zone. This is only possible if the underlying material is friable and easily manipulated by tillage. Soils that have undesirable subsoil material, or materials not easily formed into soil, suffer irreparable damage from erosion. Such soils demand protection from any activity that accelerates natural erosive processes. Protection is best accomplished by maintaining continuous vegetation cover. Human needs for food and fiber should be met through the use of highly productive soils that are not subject to erosion.

In most respects, productive soil is a renewable resource, but several strategies are required and technically correct practices are needed to address the individual character of the different soils that collectively form the soil of the planet. Global assessments of human impact on soils reveal chemical deterioration primarily in tropical areas where soils with low native fertility are abundant and human institutions have not developed to replace the nutrients removed in the food. This reduced chemical ability to support adequate vegetative cover invariably leads to accelerated physical soil damage via erosion.

References

Buol, S. W., F. D. Hole, and R. J. McCracken. 1989. *Soil Genesis and Classification*, 3rd edn. Iowa State University Press, Ames, Iowa.

Buol, S. W., P. A. Sanchez, J. M. Kimble, and S. B. Weed. 1990. Predicted impact of climatic warming on soil properties and use. In *Impact of Carbon Dioxide, Trace Gases and Climate Change on Global Agriculture*, Special Publication No. 52, American Society of Agronomy, Madison, Wisconsin, 71–82.

Daniels, R. B., J. W. Gilliam, D. K. Cassel, and L. A. Nelson. 1989. Soil erosion has limited effect on field scale crop productivity in the Southern Piedmont. *Soil Science Society of America Journal* 53, 917–920.

FAO/UNESCO (U.N. Food and Agriculture Organization/U.N. Educational, Scientific, and Cultural Organization). 1971–1979. *Soil Map of the World*, 1:5 million, Vol. I–IX. UNESCO, Paris, France.

Jenny, H. 1941. *Factors of Soil Formation*. McGraw-Hill Book Co., New York.

Larson, G. A., F. J. Pierce, and L. J. Winkelman. 1984. Soil productivity and vulnerability indices for erosion control programs. In *Erosion and Soil Productivity*, Publication No. 8–85, American Society of Agricultural Engineers, St. Joseph, Michigan, 243–253.

McCracken, R. J., R. B. Daniels, and W. E. Fulcher. 1989. Undisturbed soils, landscapes, and vegetation in a North Carolina Piedmont virgin forest. *Soil Science Society of America Journal 53*, 1146–1152.

Odell, T. T., W. M. Walker, L. V. Boone, and M. G. Oldham. 1982. *The Morrow Plots: A Century of Learning*. Bulletin 775, Agricultural Experiment Station, College of Agriculture, University of Illinois, Urbana-Champaign, Illinois.

Oldeman, L. R., R. T. A. Hakkeling, and W. G. Sombroek. 1990. *World Map of the Status of Human-Induced Soil Degradation*. International Soil Reference and Information Centre, Wageningen, The Netherlands, and U.N. Environment Program, Nairobi, Kenya.

Ruhe, R. V. 1969. *Quaternary Landscapes in Iowa*. Iowa State University Press, Ames, Iowa.

Sanchez, P. A. 1976. *Properties and Management of Soils in the Tropics*. John Wiley and Sons, New York, 618 pp.

Soil Survey Staff. 1975. *Soil Taxonomy*. USDA Handbook 436, U.S. Department of Agriculture, Washington, D.C.

Stevenson, J. F. 1982. Origin and distribution of nitrogen in soil. In *Nitrogen in Agricultural Soils* (F.

J. Stevenson, ed.), Agronomy Monograph 22, American Society of Agronomy, Madison, Wisconsin, 1–42.

Tiessen, H., J. W. B. Stewart, and C. V. Cole. 1984. Pathway transformations in soils of differing pedogenesis. *Soil Science Society of America Journal 48*, 853–858.

TROPSOILS. 1991. *Annual Report*. Soil Science Department, North Carolina State University, Raleigh, North Carolina.

U.S. Salinity Laboratory Staff. 1954. *Diagnosis and Improvement of Salina and Alkali Soils*. USDA Handbook 60, U.S. Department of Agriculture, Washington, D.C.

10

Hydrology and Water Quality

Peter Rogers

The purpose of this paper is to summarize what we know about the effects of land-cover changes upon the hydrological cycle and upon water quality. This paper focuses upon natural and anthropogenic land-use changes and their quantifiable impacts on hydrology and water quality. The paper deals with the major biomes of grasslands, forests, croplands, and wetlands; it also examines land uses such as urban developments and transportation corridors.

In addition to the direct effects of land-cover change upon water, there are also the indirect impacts upon climate and the altered climate's subsequent impact upon water. One can go a step further and consider the effect of changed climate upon land cover and the land cover's subsequent impact upon water (see Figure 1). How far one should pursue this cycle depends upon the magnitude of the impacts and also upon the time scale of concern. For most practical decisions, consideration of only the direct (first-order) effects is sufficient. For intermediate-term decisions (between 10 and 50 years), consideration of the primary and secondary effects will suffice. For long-term considerations, on the same order as the predicted CO_2 doubling, consideration of all three pathways is in order. In this paper, effects along each of these pathways are indicated wherever possible.

The Hydrological Cycle

In an ecosystem, everything is connected to everything else. Freshwater ecosystems tend to be more difficult to analyze and understand than terrestrial ecosystems because chemicals and nutrients are transported by water through highly complex processes. The hydrological cycle accounts for the passage of water through the lithosphere and troposphere. Aquatic ecosystems exist wherever water is concentrated in this cycle—in lakes, rivers, marshes, oceans, soil moisture, and groundwater (see Figure 2). In a sense, the freshwater ecosystem is a natural machine: a constantly running distillation and pumping engine. The sun provides thermal energy, which, together with gravity, keeps the water machine moving. Although this water cycle has no beginning or end, the oceans are the major source, the atmosphere is the delivery agent, and the land is the user. In this system, no water is lost or gained, save for that small amount removed from the water

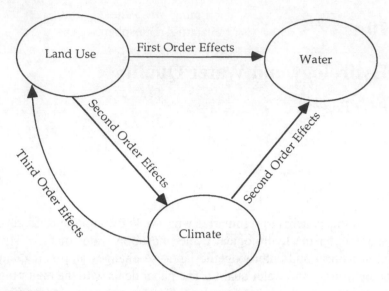

Figure 1. Effect of land-use change on hydrology and water quality.

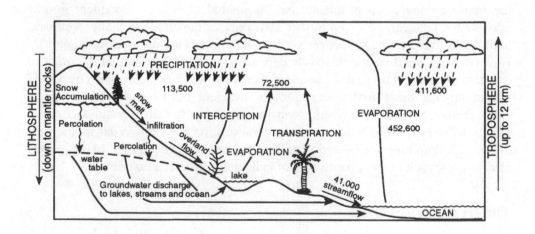

Figure 2. Hydrological cycle showing approximate magnitudes of components in km^3 (from Repetto, 1985: 258; reprinted with permisssion from Yale University Press, copyright © 1985, Yale University Press).

cycle as either hydrogen or oxygen during chemical reactions. But the amount of water available to the user may fluctuate on account of variations in either the source or, more usually, the delivery agent. Large fluctuations in the atmosphere and the oceans produced deserts and ice ages, and even now, smaller, local alterations in the hydrological cycle produce floods and droughts. Weather and climate are primarily influenced by the hydrological cycle (or vice versa).

This water machine has been functioning since the beginning of the earth's atmosphere and will continue until the earth's atmosphere ceases to exist. This is not to imply that the machine has reproduced the same weather and climate over the millennia. On the contrary, we know that the weather and climate have fluctuated over time. What has remained constant, however, is the basic functioning of the machine itself. The machine evaporates water from the oceans and other water bodies, as well as the moist surfaces of the land. This water vapor is transported through the atmosphere by the simple physical principles of convection and radiation of heat until it precipitates as rainfall. In returning to the oceans and other water bodies, the precipitation seeps into the ground and replenishes groundwater reservoirs; is absorbed into permanent ice fields; picks up minerals and organic materials from natural processes such as weathering, or because of anthropogenic activities; and returns contaminated to the oceans, where it continues the endless cycle.

Figure 2 shows the approximate annual magnitudes of amounts of water in each of the major compartments of the global hydrological cycle under current climatic conditions. Evaporation, condensation, and precipitation make up the three most important stages of the hydrological cycle. The atmosphere acquires moisture by evaporation from oceans, lakes, rivers, snow and ice fields, and damp soils, and by transpiration from plants and vegetation in processes referred to jointly as evapotranspiration. Air currents induced by temperature and pressure differences transport water vapor over large distances. Condensation, cloud formation, and precipitation may then occur. During precipitation, some evaporation takes place in the air itself, but much of the water reaches the ground, water surfaces, or vegetation. Of the precipitation that reaches the vegetation, some is held by the canopy and eventually evaporates again, and some eventually reaches the ground. Of the water that reaches the ground, some will infiltrate the surface, some will accumulate in surface depressions, some will begin to move over the surface, and some may percolate through deeper layers and enter the groundwater system, where it may be held for long periods. Water in some of the aquifers under the Sahara, for example, has remained in its location for millions of years.

Climatic conditions influence life on earth and shape the physical and biological environment, which in turn influences the state and composition of the atmosphere. In other words, strong interactions and feedback loops exist between the various components of the ecosystem. Human activities are influenced greatly by weather and climatic conditions, and humans can intentionally or inadvertently modify weather, climate, the water cycle, and such geochemical cycles as the carbon cycle.

The atmosphere's vapor content is directly related to air temperature, and both are greatest in summer and at low latitudes. Exceptions include the deserts of the tropics, where there is little rainfall because, along with other mechanisms, the slow descent of air in high-pressure systems hinders the precipitation. The temporal and spatial distributions of precipitation are major determinants of the availability of land for human uses. For example, where rainfall is seasonal, and precipitation also varies greatly from year to year (as in the semiarid tropics), this can pose problems for agriculture and other activities. Spatial variability in precipita-

Table 1: Annual world water balance

Elements of water balance	Volume (km^3)	Depth (mm)
Land draining into the oceans		
(116.8 million km^2)		
Precipitation	106,000	910
Runoff	41,000	350
Evapotranspiration	65,000	560
Land without access to the sea		
(32.1 million km^2)		
Precipitation	7,500[1]	238
Evapotranspiration	7,500	238
Oceans (361.1 million km^2)		
Precipitation	411,600	1,140
Inflow of river water	41,000	114
Evaporation	452,600	1,254
World (510 million km^2)		
Precipitation	525,100	1,030
Evapotranspiration	525,100	1,030

[1]Including 830 km^3 or 26 mm of runoff.
Adapted from L'vovich, 1979: 56; © AGU.

tion also has great significance in ecosystem stability. Worldwide, average annual rainfall varies from almost zero in some deserts to over 10,000 mm in parts of Hawaii. At Iquique in the Chilean desert, no rain fell for a 14-year period; at the other extreme, 22,000 mm has been recorded in a single year at Cherrapunji, India. In semiarid areas, temporal variability can be devastating. In the Sahel region of Africa beginning in the late 1960s, a decrease in rainfall of 20–50% over a six-year period seriously harmed agriculture, livestock, inland fisheries, and national economies, with many social and political ramifications.

Scarcity of Water

Considering the enormous amounts of water in the hydrosphere, it is difficult to understand how there could possibly be a scarcity of water. Nevertheless, the water available for human use on a sustained basis in the earth's populated regions is uncomfortably close to potential 'needs' of the 21st century, an indication that the process of moderating those needs should begin immediately.

  The most comprehensive studies on the world water balance are based upon the work of L'vovich (1979) (see Table 1). He estimates total annual precipitation to be 525,100 km^3 (1 km^3 is equivalent to 264.2×10^9 U.S. gallons). Most precipitation—411,600 km^3, or 78%—is over the oceans and hence is not readily available for human use. Worldwide diversion for human use in 1970 was about 3500 km^3, and based on this level of use, an estimated 5800 km^3 is contaminated by anthropogenic pollution.

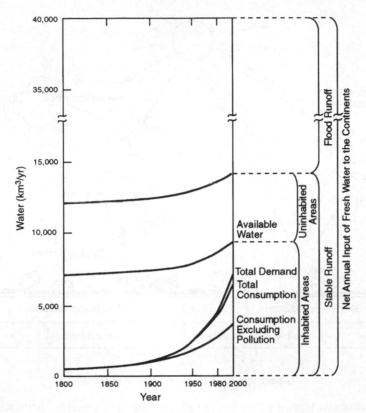

Figure 3. Global water supply and demands (from Ambroggi, 1980; adapted from'Water,' by Robert P. Ambroggi. Copyright © 1980 by Scientific American, Inc. All rights reserved).

The amount of water stored in the global hydrosphere is three orders of magnitude larger than the annual precipitation. Total fresh and saline water (including water vapor) in the hydrosphere is estimated at 1454 million km^3. Of the total 84.4 million km^3 of freshwater, approximately 60 million km^3 is groundwater, 24 million km^3 is in ice sheets, 280,000 km^3 is in lakes and reservoirs, 85,000 km^3 is in soil moisture, 1200 km^3 is in rivers, and only 14,000 km^3 is in the atmosphere at any one time. The entire amount of stored atmospheric moisture is recycled every 10 days, and the storage in rivers is rotated every 11 days.

An estimate of global historical water use and future demands placed upon the annual hydrological cycle is shown in Figure 3 (Ambroggi, 1980). This figure shows that of the amount of precipitation remaining after evapotranspiration and infiltration, only 41,000 km^3 is available annually as surface or groundwater runoff. Of this amount, only 14,000 km^3 is available as stable runoff[1]; the rest

---

[1]Stable runoff is the part of runoff that comes from underground flow into rivers (sometimes called base flow or sustained fair weather runoff). This is a conservative estimate of available water, since it ignores seasonal diversion and storage possibilities. Seasonal storage in impoundments is likely to remain fairly small (1974 estimates range from 1855 to 5000 km^3/yr), while the potential for groundwater storage from seasonal peak flows is potentially larger (3000 to 5000 km^3/yr) and much less expensive per unit of storage.

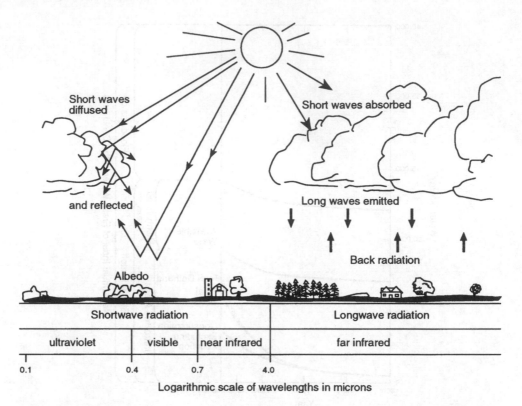

Figure 4. Radiation exchanges at a watershed surface. The land surface receives shortwave radiation both directly from the sun and by diffusion from the sky and clouds. Part is reflected and part is absorbed. Some of the absorbed heat energy is lost as longwave back radiation (from Pereira, 1973: 94).

flows rapidly into the oceans as flood runoff. Of the stable runoff, 5000 km^3 is in sparsely inhabited regions, and thus only 9000 km^3 is readily available for human use. The ratio of use (including dilution of wastes) to readily available resources, then, is in the range of 39–64% instead of the 1% that would be the case if total use were compared to total precipitation.

Hence, although there appears to be no scarcity of the current resource in aggregate, if use were doubled over the next 20 years and nothing were done to increase the stable runoff, water scarcity could become quite serious. As Figure 3 indicates, severe shortages might occur if nothing is done to rationalize water use, curb demands, and improve technology. Moreover, the aggregate resource position disguises many serious instances of scarcity, even at the current level of human use, at national or regional levels. For example, Africa's runoff per unit area is only about one-fifth that of South America and one-half the world average. This is because of lower precipitation and higher evapotranspiration than in other parts of the world. On a per capita basis, however, Asia has only one-half the world's average water availability, and Africa is almost exactly at the world average of 8562 m^3/person/yr. Despite its low runoff per unit area, Oceania has six times the

Table 2: Albedos in Israel and England/Wales

Land Surface	Albedo
Israel[1]	
Open water (lake)	11.3%
Pine forest	12.3%
Evergreen (maquis) scrub	15.9%
Citrus orchard	16.8%
Open oak forest	17.6%
Rough grass hillside	20.3%
Desert	37.3%
England and Wales[2]	
Agricultural grassland	24%
Deciduous woodland	18%
Towns	17%
Conifer plantation	16%
Heather moorland	15%
Peat and moss	12%

[1]Shortwave reflection as measured by instruments mounted on a helicopter by Stanhill et al. (1966).
[2]Airborne measurements from a summary of evidence presented and mapped by Barry and Chambers (1966).
Adapted from Pereira, 1973: 96.

world's per capita level because of its low population. Latin America has almost four times the world's average per capita supply of water.

Water and Energy Budgets

The hydrological cycle is intimately involved in the energy balance of the globe. Evaporating water carries latent heat into the atmosphere. A fundamental insight of H.L. Penman and Thornthwaite (in the late 1940s) was that when water is in free supply, the evaporation and transpiration from a complete canopy of green vegetation can be predicted directly from climatic factors. Figure 4 shows radiation exchanges at watershed surfaces. As the figure shows, the major influence of land-use changes upon the radiation exchanges is due to change of albedo. Open water is about twice as absorptive as grasslands, and forested land lies between the two. (See Table 2 for albedo values of some common land covers.) Land-cover changes that change albedo hence also have a major influence on hydrology, as they alter local energy balances and therefore affect the rate of evaporation and evapotranspiration. This in turn directly influences the amount of water available to other parts of the hydrological system.

The most important way of influencing the heat balance of the globe is by changing the evaporation of water at the surface. Figure 5, which is based on work by V. Ramanathan, demonstrates the relationship between global climate and the hydrological cycle. The diagram shows the relative importance of albedo, long-

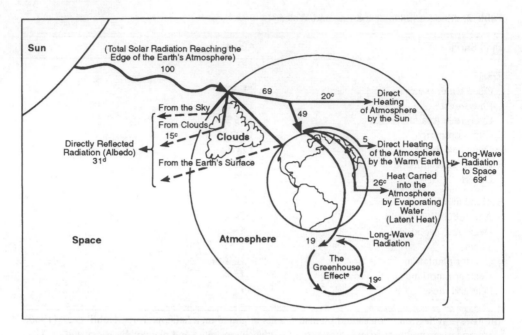

Figure 5. The water cycle and global climate. Notes: (a) Each unit is 3.43 W/m². (b) Numbers may not total due to independent rounding. (c) Processes involving the water cycle. (d) Measured by satellite instruments. Other data are modeled or derived from other measurements. (e) Caused by water vapor, clouds, and carbon dioxide and other gases (from WRI, 1988: 189, reproduced with permission of Oxford University Press).

wave radiative cooling, and evaporation in the heat balance of the globe. The most important of these factors is evaporation, which returns 40% of the radiation actually reaching the earth's surface as latent heat back to the atmosphere. Longwave radiative cooling follows with 29%, and reflection from the earth's surface accounts for about 24% of the total radiation reaching the earth's atmosphere. Because three-quarters of the surface of the globe is covered with water, major changes of land cover can affect at most 10% of the cooling due to evaporation and to changing albedo. One can immediately see why the emphasis in studying global warming has focused upon interference with longwave radiative cooling as the most important mechanism for controlling heat balances of the globe.

Effects on Water Quality and Flows of Land-Cover Change

Land-cover changes can have four major direct (or first-order) impacts upon the hydrological cycle and water quality: they can cause floods, droughts, and changes in river and groundwater regimes, and they can affect water quality. Three of these impacts deal with water quantity; for example, clearcutting of forests can cause flooding in the local streams until the deforested slopes have time to revegetate. Less obviously, the destruction of vegetation by overgrazing in arid and semiarid

Table 3: Point and nonpoint contributions of specific pollutants[1]

Pollutant	% from Point Sources	% from Nonpoint Sources
Chemical oxygen demand (COD)	30	70
Total phosphorus	34	66
Total Kjeldahl nitrogen	10	90
Oil	30	70
Fecal coliform	10	90
Lead	43	57
Copper	59	41
Cadmium	84	16
Chromium	50	50
Zinc	30	70
Arsenic	95	5
Iron	5	95
Mercury	98	2

[1]The data in this table represent the average of individual states' percent contributions, based on average daily loading data for 50 states and the District of Columbia
From EPA, 1984: 1–14.

regions can lead to reduced infiltration into the soil with a subsequent impact upon renewal and replacement of the vegetation and increasing desiccation of the soils. The disturbance of land cover can also make less water available for groundwater recharge, leading to a new regime for the groundwater resource itself. The base flow of perennial rivers in turn could be seriously affected by such a reduction of groundwater returns to the river, with more of the flow being concentrated in the flood or peak periods and less during the dry periods.

A further consequence of these impacts on water quantity is the addition or removal of materials from the rivers, water bodies, and groundwater. For example, clearcutting of trees can lead to large increments of sediment reaching a nearby stream. These sediments can seriously disrupt the natural riverine ecosystems when introduced rapidly and in large quantities. The amount of sediment may be so large as to blanket the river bed, thus disrupting the feeding and breeding habits of many benthic aquatic organisms. Similarly, the change in turbidity induced by suspended sediment may disrupt the feeding and breeding of many surface feeders and also inhibit the action of sunlight in stimulating algae growth.

Human uses of the land introduce many other pollutants into the aquatic ecosystem. In the United States, we have spent over $300 billion since 1970 on controlling 'point source' pollution, only to discover that most of the rivers and water bodies are still heavily polluted by pollutants not emanating from point sources. Agriculture and forestry are the two major sources of nonpoint source (NPS) pollutants, but urban storm runoff and combined sewer overflows are also major contributors. Table 3 shows the estimates of relative contributions to 1984 water pollution conditions in the United States by point and nonpoint sources. The surprise in these data is that the bulk of the pollutants were deemed to be primarily of non-point origin. The implications for water quality management are very important; the time has now come to switch the emphasis of the regulatory programs away from end-of-pipe point source controls to controls of nonpoint sources. These are best achieved by land-use controls. Unfortunately within the U.S. context this is not easily achieved. Without controlling land uses, however, it may not be possible ever to achieve effective control on nonpoint source pollution.

Effects of Forest Change

It is often stated that forests cause rainfall. While this is a plausible inference, there is a paucity of data to resolve whether it is generally true or not, and there is a great controversy among climatologists on this important point. There is evidence that forests influence the occurrence of mists, and the condensation of these mists causes heavy drip from the trees. One study recorded about 10 inches (12 cm) of drip from a pine tree in the Berkeley Hills during each of four rainless summers (Parsons, 1960). Pereira (1973) reviewed the Soviet data on snow trapping by forests and concluded that extra water yield in forested catchments is due to this phenomenon. While it is plausible that evaporation from forests could increase rainfall, Penman (1963) concluded that although vegetation does affect how precipitation is partitioned between runoff and evapotranspiration, there is no evidence that it can affect the amount of precipitation received. Salati and Vose (1984) reviewed the evidence from the Amazon Basin and agreed that, although they reported significant changes in the flood hydrographs of the river, concurrent with deforestation, the results are inconclusive. Richey *et al.* (1989) examined an 83-year reconstructed record of the Amazon lows and showed that there has been no statistically significant change in the discharge. They claim that oscillations of river discharge predate significant human influences in the Amazon Basin and reflect both extrabasin and local factors.

Deforestation does lead to increases of solar radiation received at the ground surface (by as much as 150 times, according to Changnon and Semonin, 1979). This change in radiation causes changes in surface temperature and other low-level weather parameters. Historically, the midlatitudes of the Northern Hemisphere have had their climates changed by deforestation, but the rate of change was so slow that it masked the effects on clouds and rainfall (third-order effects in Figure 1). Deforestation affects albedo, interception by the canopy, and surface roughness, as well as the radiation reaching the ground; all of these have major impacts upon the energy and water balances.

Changing from forests, which use water all year round, to short-season annual crops can result in large reductions in water use. Removing trees in water supply reservoir catchments and replacing them with grasses has become a widespread activity to increase the yields of the catchments. Deep-rooted trees are profligate users of groundwater and soil moisture. Data from Israel show, for example, that Aleppo pine dried out the soil profile to a depth of 5 m during the summer months.

Pereira (1973) reports on the U.S. Tennessee Valley Authority experiences in which afforestation led to a 50% reduction in flows of water, eliminating floods and soil erosion.

Watershed research studies have empirically confirmed the property of forests to absorb heavy rainfall and transmit water to the soil by infiltration through forest litter (Pereira, 1973). These results, however, do not hold for forests with shallow soils that become rapidly saturated. In such cases, after the litter and soil are saturated, the flood control ceases.

Effects of Grassland Change

The hydrological effects of grasslands depend entirely upon their management. Natural grasslands provide satisfactory land cover only when they are managed properly. Improperly managed grazing and burning can lead to removal of vegetative cover and the trampling of soils. The most intensively used agricultural grasslands are typically those derived from cleared forest or woodland (scrub or savanna). In many areas where uncontrolled burning and grazing have been practiced, grazing management has a greater impact upon the hydrology of a watershed than does forest management.

The vegetation of grasslands depends upon the soil depth and rooting depth of the grasses, as well as the local climate. Roots for grasses range from 5 cm to 5 m. The annual water use is heavily dependent on the length of the growing season. Given water to meet the full transpiration rate for the growing season and given artificial fertilizer and intensive cultivation, grasslands can be highly productive. Where such management practices have been carried out in semiarid regions in Australia, storm runoff and soil loss have been reduced to negligible amounts in comparison with unmanaged areas. Two opposing hydrological facts are at work with managed grasslands. In order to control flood flow and soil erosion, control of grazing is essential to preserve the grass cover, to prevent soil exposure, and to prevent excessive trampling. On the other hand, with improved density and productivity of grasslands, the total water yield decreases; the vegetation needs the water for evapotranspiration. This leads to the conclusion that in many cases, rather than looking to afforestation to reduce flood and erosion damage, maintaining grass cover may be more effective without reducing the water yield of the watershed as much as forests (Ives and Messerli, 1989).

Charney (1975) was concerned with the possibility of desertification due to changes in surface albedo in semiarid regions. For example, increased surface albedo caused by overgrazing could generate a net radiative loss that would increase sinking airflow, reduce convection, reduce precipitation, and enhance the aridity of the region. These are the second- and third-order effects of Figure 1. Changnon and Semonin (1979) speculate that these types of effects helped produce the 1.3 million km^2 Rajasthan Desert in India. Recent findings with regard to the actual shrinking of the size of the Sahara Desert (Tucker *et al.*, 1991) cast some doubt on the plausibility of these arguments (see also Melillo, this volume).

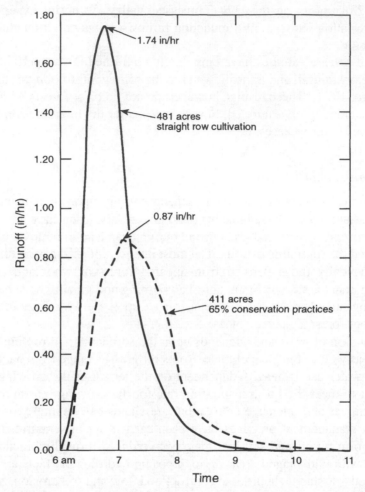

Figure 6. Runoff from a storm of 2.75 in (6.99 cm) on a field with straight row cultivation (solid line) and one of which 65% was prepared using water conservation practices (dashed line). Even with only 65% good farming practices, peak rates of runoff were halved (from Pereira, 1973: 171, reproduced with permission of Oxford University Press).

Effects of Cultivation (Including Irrigation and Agricultural Drainage)

It is important to remember that while arable crops demand water they may demand less than the natural vegetation that they replace. The main hydrological effect of cropped land is a change in the partitioning of the rainfall into evaporation, overland flow, and infiltration.

Worldwide, there is a plethora of literature on farm management practices to reduce soil erosion and flood damage. Indeed, the recent 'no till' movement is credited with making significant reductions in the total soil loss in U.S. agriculture. Even without 'no till,' practices such as terracing, contour plowing, strip cropping, and drainage can lead to very large changes in runoff characteristics. For example, Figure 6 shows a 50% reduction in storm runoff hydrograph for a watershed of over

242

800 acres split between straight-row cultivation on 481 acres and a variety of conservation practices on 65% of the remaining acres. Not only was the magnitude of the peak greatly reduced, but the time to peak runoff was delayed one hour.

Obviously, the greatest impact upon the hydrological cycle occurs when irrigation is practiced. Over 210 million ha of irrigated land is currently under agriculture around the world (the largest shares being held by India, China, the United States, Pakistan, and the USSR; WRI, 1986). In countries where irrigation is practiced, typically 80% to 90% of the water is 'consumed' (not returned to flow) by irrigation. In most arid regions, 10,000 tons of water are consumed per hectare of irrigated crop. Many problems arise from careless use of irrigation; salinization and waterlogging of the soils may account for as much as 30% of the total irrigated area (see Buol, this volume). It has been estimated that more land in the USSR goes out of production due to these effects than is newly brought into production each year; the actual irrigated areas may be decreasing over time rather than increasing. Conflicts between farmers for irrigation water are commonplace in the developing countries, and conflicts between farmers and other sectors of the economy are already widespread in the developed world. The recent five-year drought in California has brought these problems to the attention of a wide audience.

Evidence for the second-order effects of mesoscale climate forcings due to irrigation is available, but conflicting. Irrigation results in a sizeable increase of moisture in the low-level atmosphere which leads to higher humidities, lower albedo, and much lower daytime temperatures, all of which cause changes in wind velocity and cloud and rainfall processes (Changnon and Semonin, 1979). For example, Barnston and Schickedanz (1984) found statistically significant increases (15–25%) in summer rainfall over two large irrigated areas in the Texas Panhandle based upon data from 1931 to 1970. Fowler and Helvey (1974), however, found no significant precipitation effects over irrigated land in the Columbia River Basin.

Effects of Settlement and Other Nonagricultural Land Uses

The clustering of the world population in towns and cities has accelerated greatly in this century (Douglas, this volume). It is likely to be the shift from any land use/cover to an urban one that will have the largest impacts on water quantity and quality. Urban centers create problems that are very different from those of other land uses. Most industries tend to be located in urban areas, adding to the congestion of residential and commercial establishments. To support workers and machines, a better infrastructure is needed: more water is demanded throughout the year and hence reservoirs are created; more protection is demanded from floods and hence dams and embankments are constructed; more municipal and industrial wastewater is produced and hence wastewater treatment plants are built; and the land cover is substantially altered.

Changnon and Semonin (1979) summarized the findings of the Metropolitan Meteorological Experiment (METROMEX), the Regional Atmospheric Pollution Study (RAPS), and other studies of urban climate. They concluded that land-cover

Environmental Consequences

Table 4: Climatic effects of urbanization

Climatic Variable	Ratio of City to Environs
Solar radiation (insolation) in horizontal surfaces	0.85
Ultraviolet radiation, summer	0.95
Ultraviolet radiation, winter	0.70
Mean annual temperature greater in the city by 1° to 1.3°F	–
Annual mean relative humidity	0.94
Annual mean wind speed	0.75
Speed of extreme wind gusts	0.85
Frequency of calms	1.15
Frequency and amount of cloudiness	1.10
Frequency of fog, summer	1.30
Frequency of fog, winter	2.00
Total annual precipitation	1.10
Days with less than 2/10 in. of precipitation	1.10

From Kibler, 1982: 4–5; © AGU.

Table 5: Potential hydrological effects of urbanization

Urbanizing Influence	Potential Hydrologic Response
Removal of trees and vegetation	Decreased evapotranspiration and interception; increased stream sedimentation
Initial construction of houses, streets, and culverts	Decreased infiltration and lowered groundwater table; increased storm flows and decreased base flows during dry periods
Complete development of residential, commercial, and	Increased imperviousness reducing time of runoff concentration thereby increasing peak discharges and compressing the time distribution of flow; greatly increased volume of runoff and flood damage potential
Construction of storm drains and channel improvements	Local relief from flooding; concentration of floodwaters may aggravate flood problems downstream

From Kibler, 1982: 4–5; © AGU.

changes have significantly affected the urban climate and the water resources available to urban regions. For example, they concluded that changes in weather from urbanization encompass all surface weather conditions: contaminants in the air, solar radiation, temperature, visibility, humidity, wind speed and direction, cloudiness, precipitation, atmospheric electricity, severe storms, and fronts. The METROMEX studies showed that daytime urban heat excess often extended ver-

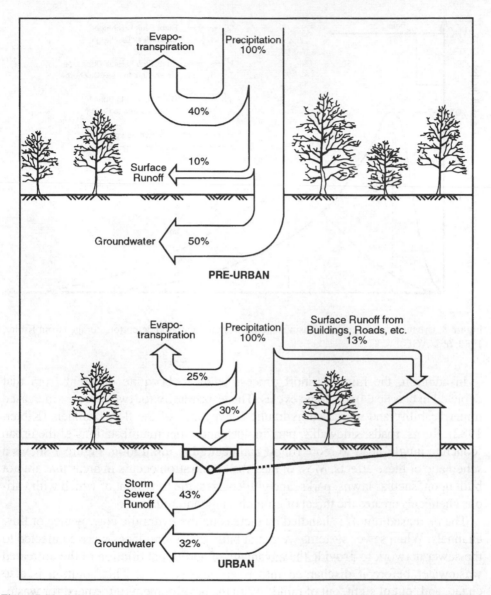

Figure 7. Hydrologic changes in Ontario, Canada, caused by urbanization (from OECD, 1986: 43).

tically for as much as 2000 m, and a thermal and pollutant plume often extended for as much as 50 km downwind.

Tables 4 and 5 show potential climatic and hydrological effects of urbanization. Cities are wetter and warmer than the surrounding countryside. They are also more likely to produce flash floods and increase sediment production than the nearby land uses. What makes the urban area different from similar land uses that are dispersed is that the potential for infiltration into the ground and interception by trees in the intervening nonurban land uses is much lower.

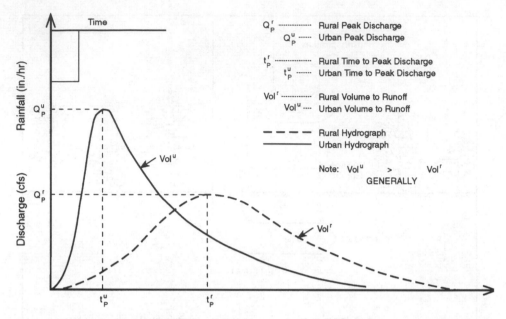

Figure 8. Urbanization impacts on basin response without increased detention storage (from Kibler, 1982: 6; © AGU).

In addition, the rainfall-runoff process tends to be quite different from that depicted in classical hydrologic cycles. This is primarily for two reasons: increased impermeability and increased hydraulic conveyance of the flow channels (Kibler, 1982). Roofs, roads, sidewalks, parking lots, and other manifestations of the urban 'concrete jungle' lead to more runoff and negligible infiltration. Figure 7 shows a schematic of these effects. Most of the urban infiltration occurs in areas that are not built upon, such as lawns, parks, and golf courses, the treatment of which with various chemicals creates the threat of groundwater contamination.

The increased runoff is handled by increasing the hydraulic conveyance of flow channels. When sewer systems were first built, storm drainages were connected to the sewer network to provide the advantage of additional dilution to the untreated wastewater before it discharged into rivers and streams. This solution is also cheap, and 'out of sight, out of mind.' With the development of the need for wastewater treatment, this 'combined sewer' advantage has turned into a serious problem for many cities. During storms, stormwater mixes with sewage and overloads wastewater treatment plants, leading to 'combined sewer overflows' that are usually discharged with no treatment to the receiving body of water. It is best to have separate sewer networks for stormwater and sewage, but it is usually prohibitively expensive to convert existing combined sewers to separate ones.

Also during a storm, there is the 'first flush' effect with increased pollutant loads during the beginning of the storm. Increased levels of suspended solids (SS) and biochemical oxygen demand (BOD) are transmitted downstream, threatening the ecology of the receiving water body. Also, due to increased conveyance of flow channels

Table 6: Comparison of combined sewer overflows and urban stormwater

Characteristic	Range of Values
Combined sewer overflows (l)	
BOD$_5$ (mg/l)	30-600
Total suspended solids (TSS) (mg/l)	20–1700
Total solids (TS) (mg/l)	150–2300
Volatile TS (mg/l)	15–820
pH	4.9–8.7
Settleable solids (mL/l)	2–1550
Organic N (mg/l)	1.5–33.1
NH$_3$N (mg/l)	0.1–12.5
Soluble PO$_4$ (mg/l)	0.1–6.2
Total coliforms (number/100 ml)	$20{,}000–90 \times 10^6$
Fecal coliforms (number/100 ml)	$20{,}000–17 \times 10^6$
Fecal streptococci (number/100 ml)	$20{,}000–2 \times 10^6$
Urban Stormwater	
BOD$_5$ (mg/l)	1–700
Chemical oxygen demand (COD) (mg/l)	5–3100
TSS (mg/l)	2–11,300
TS (mg/l)	450–14,600
Volatile TS (mg/l)	12–1600
Settleable solids (mL/l)	0.5–5400
Organic N (mg/l)	0.1–16
NH$_3$N (mg/l)	0.1–2.5
Soluble PO$_4$ (mg/l)	0.1–10
Total PO$_4$ (mg/l)	0.1–125
Chlorides (mg/l)	2–25,000[*]
Oils (mg/l)	0–110
Phenols (mg/l)	0–0.2
Lead (mg/l)	0–1.9
Total coliforms (number/100 ml)	$200–146 \times 10^6$
Fecal coliforms (number/100 ml	$55–112 \times 10^6$
Fecal streptococci (number/100 ml)	$200–1.2 \times 10^6$

[*]With highway deicing.
From Kibler, 1982: 163; © AGU.

and the absence of natural depressions and storages, the peak of the flood is transmitted fast and almost unattenuated through the system, leading to an increase in both runoff volume and peak flow rates, exacerbated by a decrease in time to peak discharge as compared to undeveloped areas. Comparisons of the effects of urban and rural land-use scenarios on a typical storm hydrograph are shown in Figure 8.

Other than an increase in SS and BOD, water pollution problems are caused by street sweeping, oil and gasoline leaks, application of salt on roads, etc. Urban traffic also generates polluting particulate matter that could contribute to deteriorating water quality by wet and dry deposition. Table 6 compares the characteristics of combined sewer overflows with those of urban stormwater. The surprising conclusion is that urban stormwater is very similar in chemical and biological contaminants to the raw sewage in combined sewer overflows.

Effects on Wetlands

Another category of land type is greatly affected by human development and has fundamental impacts on water and water quality (Foster and Rogers, 1991). This is the wetlands. With respect to size, numbers, and distribution, wetlands are a resource of worldwide importance (Maltby, 1988). By accepted definition, they cover as much as 10% of the land surface of the earth (overlapping, to be sure, with other types of land cover such as grassland and forest and tree cover). In certain countries (e.g., Indonesia), wetlands represent 25% or more of the land area; in others they are extremely rare. Wetlands take many forms: the productive mangroves of the Asian and American tropics, the extensive peatlands adjacent to the South China Sea, the wildlife-rich savanna floodplains of Africa, the interconnected lake and river wetlands in the sheet flood regions of South America, and the fresh- and saltwater marshes of North America, to mention some of the best known. In addition to being diversified, wetlands are unevenly distributed in their occurrence both in the United States and throughout the world.

Despite their extent and significance, wetlands have almost universally been regarded as wastelands. Only recently has a wetland conservation movement arisen. With the exception of the Convention on Wetlands of International Importance, adopted at Ramsar, Iran, in 1971, and now subscribed by 54 nations (World Wide Fund for Nature, 1989), and the continuing missionary work of the World Wildlife Fund and the International Union for the Conservation of Nature and Natural Resources, much of wetland conservation has been restricted to North America.

What is a wetland? It is wet land—that is to say, land covered with (or saturated with) water for appreciable periods of time. These areas bear such designations as swamps, marshes, bogs, fens, and wet meadows. But in recent years, wet land has also become wetland—a natural resource in its own right, duly validated by statute and provided the ultimate distinction of governmentally designated programs and agencies in many developed countries.

Problems in defining wetlands abound. The initial question involves how wet, how often, and for how long. What constitutes a recognizable wetland is another issue. Still a third question, and one currently rife with controversy in the United States, is how one goes about delineating a wetland. The fact is that the definition of a wetland will vary by agency and even by advocate. For example, the scientist's view of a wetland may not be the farmer's. Compounding these perceptual difficulties are those of law and regulation.

Why do wetlands matter? Simply because they perform useful functions and provide important values. The full list can be a long one, but most tend to fall into several discrete categories. The first might be termed *physical and hydrological*. Wetlands, for example, impede wave action and disperse flood flows over area and time. By doing so, they trap and deposit sediments, and reduce erosion. Much has already been discovered about the important natural relationships that exist between wetlands, surface water, and groundwater in particular locations. For example, we know that wetlands can store significant amounts of floodwater and reduce flood damage. In artificial settings, it is no accident that developers often seek to divert urban runoff into existing wetlands. A second functional category is primarily *biological and biochemical*. Wetlands are invaluable as habitat for fish and wildlife, furnishing nutrients, food, and cover and even functioning as nursery grounds for living resources. From the human standpoint, the biological capacity of wetlands to break down and assimilate wastes is becoming increasingly important for point and nonpoint source pollution control. *Economic* benefits represent one of several categories of values. Directly harvestable wetland resources include such items as berries, rice, hay, peat, bottomland forests, fish, and fur. Significant indirect returns can occur when wetlands increase water yields, enhance water quality, or prevent flood damage. But real economic value can show up in other settings as well—for example, the enhanced property values of a second home overlooking a scenic salt marsh. Broad *societal* benefits are an area of increasing importance as our knowledge of wetlands increases. Recreation, research, culture, education, and aesthetics are examples here. Finally, many wetland experts argue that wetlands also have *intrinsic* value. They should simply be allowed to exist as part of the natural landscape. It is important to remember that some wetland systems are more difficult to replace than others. In such cases, destruction or alteration is virtually irreversible.

In the United States, it is generally believed that approximately 10% of the country's nearly 1 billion contiguous land and water hectares—some 89.5 million ha—was once wetland (Dahl, 1990). In the space of 200 years, more than one-half (nearly 47 million ha) has been lost. Since an estimated 90% of the original resource occurred inland, the freshwater wetland component has sustained the highest casualties. One causal factor alone, agricultural drainage, accounts for the bulk of the losses. From the 1780s to the 1980s, on average, over 24 ha of wetlands were lost every hour in the lower 48 states.

The cumulative impacts of land-cover change on wetlands need careful monitoring. While the bulk of wetlands conversions are still to agriculture, many of the best urban sites are in or near wetlands. In many Asian cities the only room for expansion is into nearby wetlands. As indicated above, such land development may affect hydrology and environmental quality far more than development of upland sites.

Prognosis: Prediction and Mitigation of Impacts

In order to predict the impacts on water quantity and quality, three things are necessary: a reliable data base on the current land uses, a model of how the change in land use affects the water cycle, and a reliable way of forecasting future land use. First, as indicated above, in many instances we do not have reliable data bases of land use at the scale needed to show the impacts. Second, there is not always an agreed-upon model that describes the impacts of land use on water quantity and quality at the scale of interest. Finally, models for forecasting future land use at the scale usually required to show the impacts are extremely unreliable.

Having pointed out these limitations, however, there are significant developments in each of these three areas. With the advent of very inexpensive workstations and powerful microcomputers, there has been increased interest in geographical information systems (GIS) (Burroughs, 1986). The goal is to have detailed land-use data for many parts of the world on standardized and compatible data retrieval and manipulation systems. These are already being used to find areas where conflicting land uses in wetlands are occurring, to determine the spread of desertification, to forecast the growth of cities, etc.

The best compendium dealing with modeling the effects of land-use changes on water quantity and quality is that of Basta and Bower (1982). They reviewed 14 major models that were available and had been used in practical applications. Table 7 indicates the range of land uses, temporal and spatial properties, the type of hydrology (ground surface, snowmelt, etc.), the hydraulic characteristics, the quality processes, and the specific pollutants and residuals considered by each of the 14 models. As the table reveals, no single model covers each of the characteristics. The U.S. Environmental Protection Agency's (EPA) SWMM model comes closest to covering all of them. The reader is warned, however, that the result of the generality of the SWMM model is a very large, complex, and data-hungry model that is probably irrelevant to most of the problems that arise in specific instances.

Land-use forecasting models have a long and noble tradition in transportation planning, where some of the largest, most theoretical, and most complex have been constructed (see Ingram *et al.*, 1972, for a good example of this genre). Despite their complexity and elegance, the reliability of their forecasts appear to be little better than those of simple gravity-type models (see Isard, 1960, for a discussion of these models).

Remediation of Impacts

Most of the discussion in the literature of impact remediation deals with water quality. In this review, we follow this tradition while indicating that many of the remediation strategies will also have major effects on quantity impacts.

Water pollution caused by changes in land cover can be dealt with: (1) at the point of generation, either by end-of-pipe treatment or by making a process change to eliminate the production of the waste in the first place; or (2) in the environment itself, either by treating the ambient surface water or groundwater or by treating the water when it is withdrawn for use. Assuming the availability of feasible best management practices, abating runoff changes and pollutant loadings as close to their source of generation as possible is both logistically and philosophically preferable.

Table 7: Applicability of runoff models to various problem characteristics

PROBLEM CHARACTERISTICS		Hydroscience	MRI	SWMM-Level 1	EPARRB	Simple SWMM	ACTMO	ARM	HSP	NPS	QQS	STORM	AGRUN	CAREDAS	SWMM
Applicable Land Area	Urban	●	●	●	●	●			●	●	●	●		●	●
	Agriculture		●	●			●	●	●			●	●		
	Forests		●	●				●	●						
	Wetlands			●					●						
Temporal Properties	Single Storm Events					●		●	●		●	●			●
	Continuous Simulation						●	●	●	●		●			●
	Annual or Seasonal Average	●	●	●	●										
Spatial Properties	Single Catchments	●	●	●	●								●		
	Multiple Catchments					●	●	●	●	●	●	●		●	●
Hydrology	Surface/Total Hydrograph Generation	●	●			●	●	●	●	●	●	●			●
	Subsurface Processes						●	●	●	●			●		
	Snowmelt						●	●	●	●		●			
	Dry-weather/Base Flow		●						●		●	●			●
Hydraulics	Flow Routing in Channels/Pipes								●		●	●		●	●
	Backwater, Surcharging, Pressure Flow								●						●
	Flow Controls and Diversions								●			●			●
	Storage/Reservoir Routing					●			●			●		●	●
Quality Processes	Surface Generation	●	●			●	●	●	●		●	●			●
	Routing in Channels/Pipes								●			●			●
	Sediment/Erosion	●	●	●			●	●	●	●		●	●		●
	Scour/Deposition in Channels/Pipes							●							●
	Parameter Interaction						●								
	Soil/Sediment-Parameter Interaction	●					●	●		●					
	Routing through Storages	●							●			●	●	●	●
	Treatment/Removal in Storages											●	●		●
	Treatment Processes	●					●					●	●	●	●
Residuals	Organics/BOC/COD	●	●	●	●	●			●	●		●			●
	Nitrogen Species	●	●	●	●	●	●	●	●			●			●
	Phosphorus	●	●	●	●	●	●	●	●			●			●
	Suspended Solids	●	●		●			●	●			●	●		●
	Coliforms	●	●			●						●	●		●
	Pesticides		●				●	●							
	Arbitrary or Other Conservative	●	●	●		●		●	●				●	●	●
	Arbitrary or Other Non-Conservative		●				●	●				●			
	Economic Analysis		●												●

From Basta and Bower, 1982: 184.

Environmental Consequences

In traditional discussions of pollution control, it is generally agreed that the closer the intervention is to the production process, the more cost-effective the intervention is. Changes in land-use practices in this context are roughly analogous to process redesign in industry. Unlike the industrial equivalent, however, modifying land-use practices is often relatively inexpensive and, more importantly, well known—requiring no 'risky' technological innovations on the part of the generator. The advantages of early pollution abatement have been described in numerous other generic contexts (Center for Policy Alternatives, 1980).

While generation of agricultural NPS pollution is relatively easily traced to individual farms, the specific origins of urban and silviculture runoff may be too diffuse to allow for source reduction. In such instances, there may be economies of scale in locating best management practices in gullies and waterways that most contribute to the targeted surface water degradation. However, such practices are invariably structural and substantially more costly than on-site reductions.

The choice of whether to use 'natural' processes, suitably improved upon, or to use purely 'human-made' physical, chemical, and biological processes to deal with the pollution has many consequences for the administration and implementation of control systems. In the context of nonpoint sources, the form of intervention is often divided into capital-intensive vs. operation- and management-intensive methods to control pollution.

Controlling Nonpoint Source Pollution from Land Uses

As NPS runoff is increasingly recognized as the primary source of surface water degradation, debates about water quality focus on this issue. While differences of orientation and emphasis arise from the agricultural, industrial, and environmental camps, there is a consensus that the technological means for controlling NPS runoff exists; development of technology is no longer the paramount objective; rather, diffusion and adoption of that technology now takes precedence.

The removal of the waste materials by physical, biological, and chemical processes from sewage and the conversion into sewage sludge, for instance, has been chosen as the major technical approach in the developed countries for wastewater treatment. Recently, many have voiced surprise at the alarming amounts of sewage sludge to be disposed of in 'an environmentally acceptable manner.' Careful consideration of the law of conservation of mass might lead to the choice of cheaper, and more environmentally benign, technologies in the first place.

Because NPS pollutants are produced by widely scattered polluters, they tend to be of widely differing materials. Although typically produced in small amounts, they can give rise to high concentrations when they accumulate in receiving waters. Classification of pollutants is not without its attendant nuances. Nutrients, for example, are often lumped together in NPS pollution schemes. In fact, phosphate P is readily adsorbed on soil and therefore primarily associated with sediment runoff, while nitrate (NO_3) N is mostly lost with subsurface water due to its high solubility and low adsorptivity. Beyond this, variations in amount, persistence, and location within the soil profile suggest different management practices. In nonagricultural areas the materials reflect the nature of the diversity of different mixes of residential, commercial, and industrial uses.

A key element of NPS management involves the centrality of 'transport mode' to assessing the components of NPS impacts on a receiving water body. Even within certain homogeneous agricultural regions, the variety of transport modes (in solution in subsurface drainage, or a combination of solution and association with sediment in surface runoff) may make the control strategies for seemingly similar types of pollutants very different. Thus, the real components of a nonpoint source problem may not always be readily discernible.

Since nonpoint pollutants are not easily measured and are of such diverse composition, methodological problems regarding how and where to measure the pollutants must be overcome. Although sediment is the primary source of adsorbed compounds (such as P compounds), it may not be the critical pollutant to be controlled in a given watershed. Indeed, in some instances, sediment deposition plays a positive role by slowing the photosynthesis/eutrophication process. While control of sediments may reduce runoff of pollutants that are strongly adsorbed into soils, such as chlorinated hydrocarbon insecticides, it will have little effect on pollutants with low soil adsorption coefficients, such as NO_3 nitrogen (Baker and Laflen, 1983).

Since rainfall, a fundamentally stochastic phenomenon, determines the magnitude of NPS loadings, there is a need to deal with uncertainty in the predicted outcomes. The importance and difficulty associated with prediction and modeling should not be understated.

Costs and Benefits of Remediation

One of the things that we learn from economics is that 'if something is worth doing, it is not worth doing well.' In other words, just because some activity is technically feasible, it does not mean that it should be deployed to its fullest extent. There typically is some level of deployment, far short of perfection, that balances the benefits gained from deployment with the economic costs. Indeed, technical feasibility alone is no basis for doing something. In a typical case, recent estimation of government cost-sharing expenses in an Illinois reservoir's watershed revealed that the marginal cost for each 1% reduction of fine particle sediment N and P ranged between $139,000 and $151,000.

When considering mitigation policies, one common fallacy is to focus only on the costs of implementing the policy and understate, if not disregard, the benefits accruing to society, or individuals, as a result of the policy. Doing this undermines the most central of economic imperatives: benefit maximization. Since N. Kaldor and J. Hicks first posed the question, 'Do the benefits exceed the costs?' assuring a balanced and exhaustive accounting of *both* sides of the equation has been a central challenge in policy analysis.

For example, the eutrophication of surface waters due to land-use changes often causes damages that are classified as either recreational or aesthetic. The difficulty in simply putting a price on the fish population of a lake is apparent and well understood by opponents of environmental policies and manifested in their tactics. In an infamous episode in the United States, the Office of Management and Budget successfully thwarted a national acid rain program proposed by the U.S. Environmental Protection Agency (EPA) by simply quantifying the implicit benefits of implementation. A strict monetary accounting of the billion-dollar policy suggested a price of $6000–66,000 per pound of fish saved in the lakes of the Adirondack Mountains.[2]

Although one typically thinks of externalities as a phenomenon arising from the self-interest or myopia of private individuals, activities in the public realm may also impose such costs. In the area of transportation land-use pollution, one case that has had some economic assessment and illustrates the problem of unconsidered externalities is that of highway deicing salts (EPA, 1976). Sodium chloride is generally used because it is the 'cheapest' way of attaining safe highways in the Snow Belt. It has long been known that there are external effects of discharging even a relatively benign chemical in the ambient environment. It is particularly difficult to assess net benefits of salt control because of perceptions of the value of loss of human life due to decreased highway safety (even though solid evidence of increased safety is lacking).

The use of salt and the citizens' demand for 'bare pavement' are relatively recent phenomena, starting in the 1950s. Currently 10% of the global production of salt is spread each year on the highways of the U.S. Snow Belt. The EPA estimated the total damages due to road salting to be $5 billion per year. These damages are caused by 10 million tons of salt use annually at $25 per ton. In other words, when all of the externalities are taken into account, the full economic 'cost' of road salt is $500 per ton. Unfortunately, the alternatives—calcium magnesium acetate, pavement additives, urea, or calcium chloride—are all quite costly, and some have similar external damages. Of course, salt does not have to be used at all. Mechanical snow removal and sanding (which has some environmental side effects) are adequate and are the practices followed in Alaska. So, depending upon how one defines the feasible economic management practices in this case, one ends up with almost a break-even, or with $5 billion a year net benefits. This case shows how externalities can lead to large divergences between the 'financial' price of $25 per ton and the 'economic' price of $500 per ton. How can government departments make the correct choice of management practice when the prices do not reflect the true economic costs? This is a pressing area for research.

Overall Assessment of Land-Cover Change on Water

An assessment of the global impacts of *Homo sapiens* on water quantity and quality was attempted by L'vovich and White (1990). They provided information on

---

[2]Lippincott, 1986. One interesting problem inherent to monetary quantification of wildlife protection was raised by EPA Administrator William Ruckelshaus, when in response he sarcastically suggested: "It all depends on the assumptions you put in there.... You can escalate the cost of the fish even more.... If we kill all these fish, and we only have one left, we could load the whole cost on the fish that's left" (Lippincott, 1986: 18).

water use by activity from 1680 to 1985. Their paper reported that by 1985 there were 2357 storage reservoirs of more than 100 million m³, with a storage volume of 5500 km³, a surface area of 590,000 km², and annual evaporation losses of 130 km³. There was 2.5 million km² of irrigated land withdrawing 2710 km³ of water per year. Domestic livestock withdrawals amounted to 60 km³ per year. The domestic and industrial consumption of water has increased 40-fold over the past 300 years. And evaporation from the land surface was estimated to have increased by 2470 km³ per year over the same period.

All of these estimates add up to the fact that globally we use about 3500 km³ out of the readily available 8000–9000 km³ water (Ambroggi, 1980, arrived at similar estimates shown in Figure 3). In other words, at current rates of use we could double our water consumption. This is *not* a recommendation since long before that happened there would be major water crises in several regions of the world. L'vovich and White (1990) carried out a projection until the year 2080, when, they suggest, with the adoption of drastic pollution prevention methods, the total consumption would be slightly above 5000 km³.

It is hazardous to predict future water use for two reasons: first, the demand for water is heavily influenced by economic and social forces, which are themselves virtually impossible to forecast; second, the supply of the resource itself will be influenced by the types of interactions shown in Figure 1. Land use will affect water availability; land use will influence climate, which in turn will influence water availability; and climate will also influence land use, which influences water availability and so forth.

References

Ambroggi, R. P. 1980. Water. *Scientific American 243(3)*, 101–116.

Baker, J. L., and J. M. Laflen. 1983. Water quality consequences of conservation tillage. *Journal of Soil and Water Conservation (May)*, 186–192.

Barnston, A. G., and P. T. Schickedanz. 1984. The effect of irrigation on warm season precipitation in the southern Great Plains. *Journal of Climate and Applied Meteorology 23*, 865–888.

Barry, R. G., and R. E. Chambers. 1966. A preliminary map of summer albedo over England and Wales. *Quarterly Journal of the Royal Meteorological Society 92*, 543–548.

Basta, J. D., and T. B. Bower. 1982. *Analyzing Natural Systems*. A Research Paper from Resources for the Future, distributed by The Johns Hopkins University Press, Baltimore, Maryland.

Burroughs, P. A. 1986. *Principles of Geographical Information Systems for Land Resources Assessment*. Oxford University Press, Oxford, U.K.

Center for Policy Alternatives. 1980. Biological impact pathway model. In *Evaluating Chemical Regulations: Tradeoff Analysis and Impact Assessment for Environmental Decision Making*, MIT Press, Cambridge, Massachusetts, 32–47.

Changnon, S. A., and R. G. Semonin. 1979. Impact of man upon local and regional weather. In *Reviews of Geophysics and Space Physics 17*, 1891–1900.

Charney, J. 1975. Dynamics of deserts and drought in the Sahel. *Quarterly Journal of the Royal Meteorological Society 101*, 193–202.

Dahl, T. E. 1990. *Wetland Losses in the United States, 1780s to 1980s.* Fish and Wildlife Service, U.S. Department of Interior, Washington, D.C.

EPA (U.S. Environmental Protection Agency). 1976. *Economic Analysis of the Environmental Impact of Highway Deicing.* EPA-600/2-76-105, Environmental Protection Agency, Washington, D.C.

EPA. 1984 *Report to Congress: Nonpoint Source Pollution in the U.S.* Office of Water Policy, Environmental Protection Agency, Washington D.C.

Foster, C. H. W., and P. Rogers. 1991. *Rebuilding the Nation's Wetland Heritage.* Report of the Harvard Wetlands Policy Project, Kennedy School of Government, Harvard University, Cambridge, Massachusetts.

Fowler, W. B., and J. D. Helvey. 1974. Effects of large scale irrigation on climate in the Columbia Basin. *Science 184,* 121–127.

Ingram, G. K., J. F. Kain, and J. R. Ginn. 1972. *The Detroit Prototype of the NBER Urban Simulation Model.* National Bureau of Economic Research, distributed by Columbia University Press, New York.

Isard, W. 1960. *Methods of Regional Analysis: An Introduction to Regional Science.* MIT Press, Cambridge, Massachusetts.

Ives, J. D., and B. Messerli. 1989. *The Himalayan Dilemma: Reconciling Development and Conservation.* Routledge, London, 295pp.

Kibler, D. F., ed. 1982. *Urban Stormwater Hydrology.* American Geophysical Union, Washington, D.C.

Lippincott, D. 1986. *Ruckelshaus and Acid Rain.* Case No. C 16-86-658.0, for the use of John F. Kennedy School of Government, Harvard University, Cambridge, Massachusetts.

L'vovich, M. I. 1979. *World Water Resources and Their Future* (R.L. Nace, ed.), English translation, American Geophysical Society, Washington, D.C., 56.

L'vovich, M. I., and G. F. White. 1990. Use and transformation of terrestrial water systems. In *The Earth as Transformed by Human Action* (B.L. Turner II, W.C. Clark, R.W. Kates, J.F. Richards, J.T. Mathews, and W.B. Meyer, eds.), Cambridge University Press, Cambridge, U.K., 235–252.

Maltby, E. 1988. Wetland resources and future prospects: An international perspective. In *Proceedings: Increasing Our Wetland Resources* (J. Zelazny and J.S. Feirabend, eds.), National Wildlife Federation, Washington, D.C.

OECD (Organization for Economic Cooperation and Development). 1986. *Control of Water Pollution from Urban Runoff.* OECD, Paris, France, 43.

Parsons, J. J. 1960. Fog drip from coastal stratus. *Weather 15,* 58.

Penman, H. L. 1963. *Vegetation and Hydrology.* Technical Communication 53, Commonwealth Bureau of Soil Science, Farnham Royal, Bucks., England.

Pereira, H. C. 1973. *Land-use and Water Resources.* Cambridge University Press, Cambridge, U.K.

Repetto, R., ed. 1985. *The Global Possible.* Yale University Press, New Haven, Connecticut.

Richey, J. E., C. Nobre, and C. Deser. 1989. Amazon river discharge and climate variability: 1903 to 1985. *Science 246,* 101–103.

Salati, E., and P. Vose. 1984. Amazon Basin: A system in equilibrium. *Science 225,* 129–138.

Stanhill, G., G. J. Hofstede, and J. D. Kalma. 1966. Radiation balance of natural agricultural vegetation. *Quarterly Journal of the Royal Meteorological Society 92,* 128–140.

Tucker, C. J., H. E. Dregne, and W. W. Newcomb. 1991. Expansion and contraction of the Sahara

Desert from 1980 to 1990. *Science 253*, 299–301.

WRI (World Resources Institute). 1986. *World Resources 1986–87*. Oxford University Press, Oxford, U.K.

WRI. 1988. *World Resources Institute Report 1988–89*. Oxford University Press, Oxford, U.K.

World Wide Fund for Nature. 1989. *Wetlands in Danger*. Special Report No. 1, World Wide Fund for Nature, Gland, Switzerland.

V

HUMAN DRIVING FORCES

Introduction

What are the underlying human causes of land-cover change? There is no shortage of candidate variables in the literature. Yet much of the discussion has taken the form of dismissing some proposed forces a priori and taking the importance of others for granted, or of extending the results of isolated empirical studies beyond what they can bear—what Rockwell in this section identifies as a reliance on 'received wisdom and selected illustrations' and Sanderson as a process of debatable axioms becoming accepted truths among policy-makers.

One often-used approach to the question of driving forces allocates responsibility for human-induced environmental change among the 'PAT' triad of factors: population, affluence, and technology—or, in some variant forms, population, production, and consumption, or population, resource demand, and pollution emissions. This approach, treated in more detail in the chapter by Sage, has not been widely adopted in the social sciences. For the tutorial papers, the institute adopted a somewhat different and less tightly structured framework. Through a review of the existing literature, four classes of candidate human driving forces were chosen for examination: population/income change, technological change, political-economic institutions, and cultural attitudes. Critical assessments were sought of the theoretical arguments for the importance of each factor as a driving force and the state of knowledge regarding its connection to land transformation.

A complete assessment of the entire relevant literature clearly lay beyond what could be asked of any of the chapter authors, and these contributions should be read in that light. Together they suggest that though much has been asserted, little is known, and yet promising avenues exist for learning more.

Sage and Grübler deal with the more familiar sets of candidate driving forces, similar to the classic 'PAT' variables: population, income, and technology in their varied associations with land transformation. The other papers deal with variables that have received less systematic attention as driving forces and that are less amenable to quantitative analysis, but that represent some of the prevalent social science and social theory perspectives on the nature–society relationship. Sanderson focuses on the literature on institutions in the field of political economy, and Rockwell on the intersection of sociology and anthropology around the theme of societal and cultural beliefs and attitudes affecting environments.

Human Driving Forces

The lesson that emerges from these assessments is that though arguments for the importance of each candidate variable often discount the significance of the others, it appears more likely that the driving forces operate interactively and in a way that is conditioned by socioeconomic and environmental context. Some of these variables are unquestionably important forces of land transformation at the global and lesser scales, and the rest may well be; but none seems to be a fully autonomous driver of land transformation. They operate along different trajectories in different settings and in association with other variables. Researchers should try to identify those situations in which the factor is important, its importance depends on interactions with other factors, and spatial transfers of impact may connect a driving force in one region or situation with an impact elsewhere.

11

Population and Income

Colin Sage

Introduction

The relationship between population numbers and environmental resources has long been recognized as an ideological and polarized area of debate. On the one hand, contributions couched in apocalyptic terms have argued that the growth of human numbers will inevitably outstrip the capacity of the earth to provide food and other vital support services (cf. Ehrlich, 1968). In contrast to such bleak neo-Malthusianism, others have argued that the innate ingenuity of the human species will always provide the technical means to overcome resource constraints (cf. Simon, 1981). Such polarized opinions have been characterized as a quarrel between Cassandras and Pollyannas (Hardin, 1988), and certainly some of the writings of the archprotagonists strike an unusually defensive or spiteful tone (Simon, 1990; Kasun, 1988).[1]

Yet given the importance of the population–resources equation within the context of the debate addressing global environmental change, it is vital to break away from fossilized positions in which population is viewed either as the single cause of impending planetary collapse or as the catalyst for ever higher levels of material development. Whatever attempts are made will be beset by difficulties, not least in the way in which they tackle the contrasting circumstances and causes of environmental degradation in the rich industrialized countries (hereafter the North) as opposed to the poorer and developing economies of the South.[2] For, unfortunately, growing concern about the maintenance of biospheric system integrity has resulted in quite different international policy agendas.

While there is, as yet, little agreement and even less commitment among the rich countries of the North on formulating clear and effective policy responses to global climate change, ozone depletion, and loss of biodiversity, there is some convergence among them about the need to reduce tropical deforestation and population growth in the South. For many of the poorest countries, on the other hand, environmental problems have hitherto largely been treated as second-

---

[1]Simon's collection of writings includes the previously published exchanges with Paul Ehrlich as well as a transcribed debate with Garrett Hardin.

[2]While Kates and Haarmann (1992) are strictly correct to argue that the main differences in national incomes accord to a latitudinal, tropical–temperate divide, the terms North and South are used here because they embody and convey, through conventional usage, a historical and political relationship that "tropical" and "temperate" do not.

263

Table 1: Population indicators for major world regions

Region	Population Millions 1990	Av. Rate Growth % 1990–95	Infant Mortality per 1000 1990	% Urban 1990	Urban Growth % 1990–95
World	5292.2	1.7	63	45	3.0
"North"	1206.6	0.5	12	73	0.8
"South"	4084.6	2.1	70	37	4.2
Africa	642.1	3.0	94	34	4.9
N. America	275.9	0.7	8	75	1.0
L. America	448.1	1.9	48	72	2.6
Asia	3112.7	1.8	64	34	4.2
Europe	498.4	0.2	11	73	0.7
Oceania	26.5	1.4	23	71	1.4
USSR	288.6	0.7	20	66	0.9

Data from WRI, 1990.

order issues that either derive as a necessary consequence, or mark the failures and distortions, of economic development. Their response to the North is that planning for a reduction in rates of population growth can only begin once economic growth has raised standards of living to 'levels compatible with human dignity' (Latin American and Caribbean Commission, 1990). This marked lack of congruence in the perspectives of North and South illustrates the need for innovative and imaginative approaches toward the analysis of population, resources, and development.

It is, nonetheless, necessary to take heed of global demographic dynamics, for world population reached 5.3 billion by mid-1990 and is growing at a rate of 1.7% per year. Table 1 provides a summary of some demographic indicators for the major world regions. Asia accounts for the largest share of world population, with China (1.1 billion) and India (850 million) the most populous countries, although the fastest rate of increase is in Africa, which is growing at an average rate of 3% per year (Sadik, 1990). Meanwhile in the North fertility is generally below replacement levels, and the intrinsic rate of population growth is negative (Alonso, 1987). This bifurcation in the demographic trajectories of North and South has led some commentators to speculate on the international security implications of a world where today's hegemonic powers hold a diminishing proportion of the total global population, but a high proportion of the world's aged population. Mapping such shifts in absolute and relative demographic weights upon a changing resource base and deteriorating environment represents a vital ongoing task, for the demographic effects on international relations cannot be set against a constant environmental backdrop (McNicoll, 1984).

Indeed, one of the few certainties from which to start is with demographic momentum, for, irrespective of the efforts directed at reducing fertility, human numbers are set to grow to between 7.6 and 9.4 billion by 2025. In other words, the world will still have to cope with between 44% and 62% more people by that date, proving that the greatest policy challenge will be coping with the needs and pressures of the inevitable population rather than attempting to avoid the increment (ODA, 1991).

I do not intend to repeat here the main historical–demographic trends and trajectories that characterize the major world regions, a topic that is addressed by Demeny (1990) among others. Rather, the purpose of the chapter is to explore some of the underlying features and dynamics of population change, relating these to aspects of wealth and poverty, and to questions of land-use change and environmental degradation. Nevertheless, it is important from the outset to emphasize that the marked national diversity in demographic indicators is unparalleled in human history. Examples of such diversity include: natural growth rates of countries ranging from +4% to –0.2%; average family size from more than 8 to around 1.4; age structures that encompass the very youthful (50% of the population under 15 years) to the aged (less than 20% under 15); and levels of urbanization that range from 10% to 90% of total national populations. Given this extraordinary national diversity in demographic indicators—which are influenced by the complex interaction of economic activity, historical processes, social structures, and cultural traditions—there is consequently an enormous variety of population–environment relationships. It is the purpose of this chapter to explore some of the features of this variety and to evaluate the place of population within the complex of factors that fashion environmental change.

Conceptualizing Population and Environment

While cutting through the ideological accretions of recent years, it is still useful to embark upon an analysis of population–environment relationships with a juxtaposition of the positions of Thomas Malthus and Esther Boserup. More than 150 years separate their writings, yet they serve as twin pillars of a debate in which many interlocutors still find it necessary to place themselves within, or close to, one or the other camp.

Malthus was concerned with the relationship between population and food supply under conditions where technology and resources in land remained constant. He postulated that human numbers would outstrip the capacity to produce sufficient food and that 'positive checks' such as poverty, disease, famine, and war would impose downward pressure on the rate of population growth in the absence of fertility control. In contrast, Boserup argued that high population densities are in themselves a prerequisite for technological innovation in which agricultural systems continuously evolve into increasingly land-intensive forms. This process, she argued, has given rise historically to five distinct stages stretching from shifting cultivation with long fallow periods to multiple cropping (Boserup, 1981).

It is important to remember that neither author was concerned with the environment other than as a resource for food production, although they shared the assumption of diminishing returns to labor at a given level of technology (Lee, 1986). Yet while technology provides the central element of change in Boserup's model, it is Malthus's explicit preoccupation with the growth of human numbers that has provided the focus for ideological and moral schism, such that many of those involved in the environmental debate adopt either a pro- or an anti-Malthusian perspective on the cause of the world's ills. Thus, on the one side are those such as the Ehrlichs and Garrett Hardin who favor population control, while on the other a range of writers variously attribute environmental problems to inappropriate technology, overconsumption by the affluent, inequality, and exploitation—everything, in fact, but population growth (Harrison, 1990).

Developing a fresh conceptual approach toward understanding the linkages between population and environment means exposing the ideological dimensions of the debate and overcoming the artificial antagonism between the Malthusian and Boserupian models. Lee (1986) has attempted a most elaborate and sophisticated synthesis using graphical representations (phase diagrams) in which levels of technology are plotted against population size. Ultimately, Lee is seeking to understand the behavior of population and technology both as independent elements and in combination. He embarks upon his inquiry by posing a number of penetrating questions, such as: If larger populations encourage rapid technological progress, and if higher technological states induce more population growth, can a steady-state equilibrium exist? And does the Boserupian model of technological progress ultimately depend upon Malthusian features of positive checks? Lee's chapter finally demonstrates a high degree of convergence between the two models, which is bolstered by some historical reflection.

Blaikie and Brookfield (1987) take a similar view in their discussion of population and land degradation, in which they challenge the inevitable outcome implicit in Boserup's model that population growth always produces agricultural innovation. They note that in many instances not innovation but environmental degradation results and that the weakness of Boserup's model is that she isolates population as a single causal variable in a similar way to the neo-Malthusians. Two issues raised by Blaikie and Brookfield are worth noting here. The first is their notion of 'pressure of population on resources' (PPR), which helpfully establishes a linkage without the need to specify critical thresholds embodied in the idea of 'carrying capacity,' a problematic concept that is entirely conditional on the level of technology and climatic factors. The second issue concerns the necessary distinction between intensification and innovation. While the former indicates increased output through the elimination of fallow and the use of inputs, it can also include the conditions of involution exemplified by Geertz (1963), who documented the use of more and more labor to squeeze marginal increments from Javanese rice paddies. Innovation, on the other hand, embodies the qualitatively new ways in which the various factors of production are employed. While intensification may act as a block to innovation, for example, under conditions where high population densities are supported using simple technology, such conditions are also able to contain the consequences by which labor-intensive systems maintain terracing and other forms of landesque capital.[3] If labor is withdrawn from the maintenance of such a system, the consequences, according to Blaikie and Brookfield, can be disastrous if it has become one of high sensitivity. Nevertheless, degradation is not an inevitable outcome of population pressure; it can occur, they argue, 'under rising PPR, under declining PPR, and without PPR' (Blaikie and Brookfield, 1987: 34).

If we treat population as an independent variable, there are four general ways in which it can interact with environment:

- Population growth can result in the expansion of the area under cultivation and lead to resource depletion and ultimately environmental degradation in the absence of institutional and technological change (the Malthusian scenario). Such circumstances may, for example, be found in areas of land settlement and frontier expansion.
- Population growth can result in the intensification of production, involving increasing investments of human, natural, and financial capital, and in innovation, embodying the development of new technical means of production. While this technologically optimistic, Boserupian scenario represents, historically, the evolution of ever more sophisticated land management systems, nonsustainable outcomes may appear in the medium to longer term (e.g., groundwater pollution, declining soil fertility).
- Population growth can be scale neutral in terms of the local resource base, either through the importation of food from elsewhere or as excess population out-migrates, resulting in no demographic pressures for agricultural change. Such migrants must, naturally, be supported by resources elsewhere.
- There may be reverse effects on population or feedback loops when changes in the productive potential of the local environment influence the determinants of population: fertility, mortality, and migration.

These four simplified options are largely covered by Bilsborrow's (1987) three-fold classification of responses to population pressures, which he has labeled demographic, economic, and (rather clumsily) demographic–economic. The first category refers to the variety of social responses available to reduce existing levels of fertility. The second category encompasses production responses in agriculture, and the third largely comprises migration. Grigg (1980) has also generated a typology of responses comprising four categories, and while the first three conform to those listed above, he substitutes off-farm employment for feedback effects.

Nevertheless, although population is certainly one factor that acts to influence land-use changes, it does so in association with two other variables: technological capacity and levels of consumption. Thus, a given environmental impact (I) is derived from the multiplicative interaction of the three variables, population (P), per capita consumption (or affluence, A) and technology (T). This creates a useful

---

[3] The term "landesque capital" is used to describe those investments in land that have an anticipated life beyond that of the present crop cycle and that help ensure the future maintenance of land capability and productivity of labor. Examples include drainage and irrigation systems, land reclamation works (Blaikie and Brookfield, 1987).

shorthand expression, I = PAT, which is now widely used in the literature but originated in the writings of Paul Ehrlich. Its appeal may stem from the way it diffuses the single responsibility of population, enabling it to be applied equally in the industrialized countries of the North and the largely rural economies of the South. A critic of inappropriate technology such as Barry Commoner has used a variant of the I = PAT expression to calculate the total environmental impact of industrial pollution and to argue that it was primarily the technological factor rather than population that needed to be controlled (Commoner, 1972).

Harrison (1992) has recently reformulated Commoner's expression as environmental impact = population × consumption per person × impact per unit of consumption. To determine the 'relative blame' for each factor over a period of time, he uses the following expression:

$$\frac{\text{Population impact}}{\text{(as \% of total change)}} = \frac{\% \text{ change in population} \times 100}{\% \text{ change in use of resources}}$$

Harrison distinguishes between upward and downward pressures on environmental impact by scoring them from +100% to –100% respectively, since one or more of the three factors may be tending to reduce environmental impact. Working through the example of the expansion of arable land in developing countries, Harrison describes the process of calculation:

Between 1961 and 1985, population expanded by 2.3 per cent a year ... [and] agricultural production rose by 3.3 per cent. So production per person, which we can take as our consumption factor, rose by 0.9 per cent. Farmland, meanwhile, expanded by only 0.6 per cent a year. The farmland used per unit of agricultural production—the impact per unit of consumption—actually declined by 2.6 per cent annually, because yields were increasing. Technology in this case exerted a downward pressure on environmental impact, at least in terms of land used. Only population and consumption exerted upward pressure. In this case we can say that population growth accounted for +72 per cent of the growth of farmland, and increase in consumption per person for +28 per cent. Technology gets a score of –100 per cent, since it was the only one of the three factors pushing towards lower use of farmland (Harrison, 1992: 308).

In developed countries, by contrast, population growth accounted for only 46% of the expansion of farmland and increased consumption per person 54%, while technology again provided the only downward pressure. As a final illustration here, Harrison computed the causes of fertilizer use in developing countries and attributed 22% to population growth and 8% to increased consumption (measured by the growth in agricultural production per person), while the technology factor—increased fertilizer per unit of agricultural production—accounted for 70% of the increase.

To summarize, we can say that in the case of developing countries, population appears to be a major driving force of land-use change according to the procedures outlined here. By using the simplest expression,

Farmed area = Population × food consumption per person × area per unit of food production

it is clear that if technology and food consumption do not change, population growth translates directly into land conversion for agriculture. Yet, as we shall see, there are other pressures on land besides its use as a resource for food production. In the North pressures for land conversion derive from qualitative changes as well as quantitative increases in consumption. Some examples include the continuously increasing aspirations for personal mobility, which involve road-building schemes; the ongoing suburbanization of the countryside; leisure interests, most spectacularly illustrated during recent years by the conversion of farm and other land into golf courses (Pleumarom, 1992); and the energy implications of all of these.

Yet one detects a greater tendency in much of the policy literature to isolate population as a factor for attention than to address either consumption or technology. For example, 'For any given type of technology, for any given level of consumption or waste, for any given level of poverty or inequality, the more people there are, the greater is the impact on the environment' (Sadik, 1990: 10).

While this statement is strictly correct, it hardly seems fair or meaningful to hold all factors other than population constant. Indeed, if for a moment we widen the terms of environmental impact beyond changes in land use and consider waste sinks as well as resources, it is clear that in the North, changes in the factors of consumption (A) and technology (T) have the greatest effect. Ehrlich and Ehrlich (1990) conduct an exercise in which they employ per capita use of commercial energy as a surrogate statistic for A and T. They calculate that each baby born in the United States has an impact on the earth's ecosystems 3 times that of one born in Italy, 13 times one born in Brazil, 35 times one in India, 140 times one in Bangladesh, and 280 times one born in Chad, Rwanda, Haiti, or Nepal. Such a range would seem to correlate very well with levels of per capita gross domestic product (GDP).

Yet while the size of the gross world product—the total of goods and services produced throughout the planet—has been growing at a faster rate than world population (the GWP tripled between 1960 and 1989; Porter and Welsh Brown, 1991), it has not been divided any more equally. In 1965 the high-income countries enjoyed 70% of global GDP, while those of low income took 19%. By 1989 the 16% of the world's population living in the rich countries accounted for 73% of global GDP, while the 78% living in the poorer South received less than 16% (Harrison, 1992). The rising affluence of a stable population therefore poses as great a threat to the global environment as does the growth in population itself, if we consider waste sink as well as resource use and the oceans and atmosphere as well as the land. However, although speculating on the relative contributory weight of population, consumption, and technology at a global scale is a sobering exercise, it is not especially amenable to policy formulation, or to remedial action beyond abstract exhortations to reduce environmental impact.

Attributing this or that proportion of the blame to population size and birthrates is by itself a limited exercise. What is required is an understanding of other demographic variables, such as migration flows, the spatial distribution of population, and where people are in relation to the type and resilience of the ecosystem, which may be as important as total size or average density. Moreover, it is indispensable that we understand those factors, or proximate determinants, that give rise to population outcomes. This requires disaggregating the global picture in order to reveal the regional character of demographic change.

Regional Population Trends and Dynamics

The rapid growth of world population in the post–World War II period is the result of a sharp decline in mortality levels due to interventions in public health and disease control. Crude death rates in the developing regions have fallen from 24 per thousand people in 1950–55 to 11 per thousand people in 1980–85, comparable to rates in the industrial market economies (Working Group on Population, 1986). Life expectancy at birth in the low- and middle-income countries has risen from 41 to 62 years over this period (World Bank, 1990). The success of public health, sanitation, and vaccination programs is particularly apparent in reducing rates of infant and child mortality which, overall, fell by 36% between 1950–55 and 1975–80 (Hall, 1989). However, infant mortality rates show much the greatest variation of the basic demographic variables, as Table 1 demonstrates.

While death rates are approaching their anticipated minimum, birthrates in the South are falling but lagging well behind mortality levels. This 'delayed response' to changes in fertility has called into question the applicability of the theory of demographic transition, based upon the experiences of 19th-century Europe, to the developing world today.[4] Unlike mortality, which is highly responsive to improvements in economic development, fertility is a matter of individual and family decision, as responsive to cultural phenomena—gender relations, marriage contracts, inheritance of wealth and property—as to social and economic influences. However, such factors can have complex interactions, depending upon the institutional context. It has been suggested, for example, that the current high fertility rates in sub-Saharan Africa are being sustained by shortened breastfeeding periods and thus reduced lactational amenorrhea, and by a reduction in the custom of post-birth sexual abstinence from around three years to a mere four months (Jones, 1990). Given such complexity, there would appear to be little future in a single, universally applicable theory; instead, we must recognize that there are many pos-

---

[4] Demeny interprets 'demographic transition' at one level as a broader frame of reference for organizing the mass of historical experience, yet, at another level, as 'a set of propositions incorporating generally valid relationships that can be extracted from experience...Such relationships can be used to discern and predict the direction of change in demographic growth and its proximate determinants [or] even quantitative features of such changes...' (Demeny, 1990: 45).

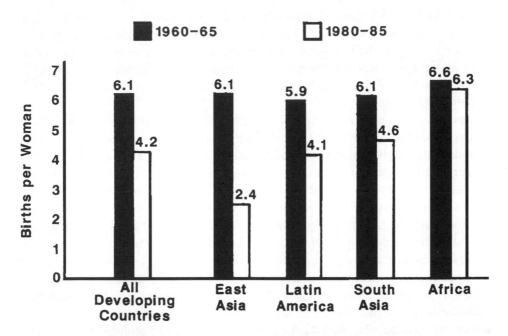

Figure 1. Fertility trends in the developing world, by region (Bongaarts *et al.*, 1990).

sible demographic transitions, each driven by a combination of forces that are institutionally, culturally, and temporally specific (Greenhalgh, 1990a).

While fertility levels have generally stabilized at or below replacement levels throughout the North—the average is now down to 1.9 children per woman, 1.58 in Western Europe (Sadik, 1990)—they have sharply diverged among the major regions of the South during the 1970s and 1980s. As Figure 1 illustrates, during the period 1960–65 there was little deviation from the mean of 6.1 births per woman, but by 1980–85 large regional differentials had appeared. The largest decline occurred in East Asia where fertility fell to 2.4 births per woman, yet there is no single factor to explain such a decline. In China a powerful state apparatus assumed the right to impose fertility control, first through the 'later–longer–fewer' (1971–78) program and then the 'one child' (1979–present) policy (Greenhalgh, 1990b). In Korea, meanwhile, fertility rates fell by 44% between 1965 and 1982 as women married later and later during a period of rapidly growing economic prosperity and full employment. While fertility rates have continued to fall, the decline has slowed as the average age of women at marriage tends toward a limit (Pearce, 1991). In sharp contrast, however, fertility rates in Africa remain high, having fallen from 6.6 births per woman in 1960–65 to 6.2 in 1985–90. Yet this regional average disguises considerable and growing variation among countries.

A comparison of the total fertility rate (TFR) for 1965–70 and 1985–90 for 49 African states indicates a decline in 22 countries, an unchanged rate in 12, and an increase in the average number of births per woman in 15 countries. Included in this latter group are Tanzania, Malawi, and also Rwanda, where the TFR has risen from 8.0 to 8.3 in a country that has the highest population density on the continent. Those countries where the TFR remains unchanged include Kenya (8.1) and Côte d'Ivoire (7.4), the two remaining countries in Africa where average annual population growth rates still exceed 4%. In Nigeria, the most populous country in Africa with 113 million people, the TFR fell from 7.1 in 1965–70 to 7.0 in 1985–90 (WRI, 1990).

Besides a complex raft of factors that underlie a constancy in the birthrate in Africa, population growth averaging 3% per year is also attributable to an infant mortality rate that has been cut by one-third since 1965 and life expectancy that has been raised by seven years. Such improvements derive from the introduction of state-directed health and education services following independence, which have primarily concentrated upon mortality prevention and only recently begun to take some action on fertility reduction. Yet such services have been introduced without reference to traditional systems of education and health care, or to the social, environmental, and political impact they might have. Gradually, the decline in infrastructure, growing bureaucratic costs, and foreign exchange scarcity have made the maintenance and support of these services increasingly difficult. Where institutional decay has led these services to break down altogether, or to be rejected by local people, there has been little with which to replace them as traditional systems have long since been eroded (Spooner and Walsh, 1991).

With the collapse of economic capacity during the 1980s, widespread political insecurity, and the departure of skilled labor, the ability of the state to deliver health and contraceptive services has been undermined. Indeed, it is questionable whether the state or the poor can have any real commitment to birth control under current circumstances of acute deprivation. For example, in parts of Africa children under the age of five years make up 20–25% of the population, yet they account for 50–80% of the deaths. In Europe the mortality rate for the same group is 3% (WRI, 1990). As long as child mortality remains so high and local survival strategies are ineffective, poor people will maintain high fertility rates to ensure child survival and potential income and labor-generating options (Spooner and Walsh, 1991).

The wide variations in fertility levels illustrate the importance of understanding such proximate determinants as marriage, postpartum practices, contraceptive use, and abortion, which are themselves fashioned by the socioeconomic and cultural context. Traditional checks on fertility vary across sub-Saharan Africa; some societies practice controls on the starting pattern of fertility (delayed age at marriage, celibacy), while others employ spacing or stopping patterns (extended postpartum sexual abstinence, breastfeeding, reduced remarriage). Pathological sterility resulting from the spread of sexually transmitted diseases has also suppressed fertility rates in a broad area of Central Africa (Lesthaeghe, 1986). Consequently, it should be noted that there is a highly uneven distribution of population in sub-Saharan Africa, stretching from the high densities of the cities, along the coast, and in the highlands, where population pressures have contributed to environmental degradation, to areas in the interior where low population densities serve as a brake on agricultural development (Hewitt and Smyth, 1992).

Nevertheless, in any of these regions, how far are the proximate determinants of fertility responsive to economic, social, or environmental conditions? It is often noted that Africa, compared to Europe of two centuries and more ago, is characterized by the early and universal marriage of young women (Grigg, 1980). Cain and McNicoll (1988), in an article exploring the institutional determinants of population growth and the consequences for agrarian change, argue that in the case of Africa there are not the same kinds of economic, family, or community controls on fertility that existed in preindustrial Northwest Europe and Japan. They partly attribute the insulation of marital fertility from economic change to the lineage system in which husbands and wives share differential costs and benefits. This echoes a parallel debate in the economic anthropology literature regarding the economic value of children and incentives for higher levels of fertility.[5]

A major factor supporting the differential benefits of children is the importance of male labor migration, which is a feature of many of the Sahelian and southern African economies. With prolonged periods of spousal separation, women rely heavily upon their children as their major productive assets and, consequently, there are low levels of contraception (Lesthaeghe, 1986). The organization of labor and domestic (residential) units is thus a vital area influencing the proximate determinants of fertility. Two other related universal variables play an important part. The first is the character of the specific relations of gender that will determine women's status, the level of participation by women in agriculture and other economic activities, and the degree to which they are able to exert control over their fertility. The second is the social class of the household, or its location according to indicators of economic wealth and control over means of production. Here, there is often a striking correlation between wealth and household size very much in line with Chayanov's notion of demographic differentiation. Lockwood (1991) draws upon three studies in different parts of northern Nigeria which each show that richer households have more dependents but also meet more of their grain needs than poorer households. Under the system of dryland agriculture in the region, larger households (including more wives) mean more labor to engage in soil improvement measures, irrigation, and tree planting. In other words, improved environmental management is more easily achieved among the richer farmers

---

[5] A sense of the issues raised by this debate can be garnered from two extracts that take different perspectives:

> Under general conditions of rural poverty, it may be economically rational for all families to reproduce at a biological maximum due to the economic value of children. But actual family size may vary according to class position, which reflects differences in infant and child mortality rates resulting from variations in household incomes. In addition, the economic value of children may differ according to class position ... (Deere *et al.*, 1982: 101).

In contrast to this position, Youssef proposes an approach to fertility that:

> stresses a linkage to women's workload (both home and productive activities) ... [uses] the mother as the basic unit of analysis (as opposed to the household), and allows for the specification of differential interests in children among household members, since it grounds fertility behaviour in *the women's rational calculation of perceived benefits* (Youssef, 1982: 190).

Youssef's belief that women place a high value on children as substitutes for the workload assigned to them by the labor allocation process within the household is challenged by studies from elsewhere, for example, in Bangladesh where women express a preference for smaller family size (Kabeer, 1985).

273

because they can bear the costs and obtain the benefits in terms of enhanced food security.

These studies illustrate the importance of approaching the issue of population growth through the proximate determinants of fertility at the micro level. But they also highlight the need to consider the range of structural factors that influence population dynamics through mortality rates and mobility. For this reason a political economy of fertility has been proposed that would comprise a hierarchical, multilevel analysis capable of examining household behavior and community institutions in structures and processes operating at regional, national, and global levels, besides the historical roots of those macro/micro linkages (Greenhalgh, 1990a). Such an approach might begin from examining the dynamics and rationale of reproductive behavior and work upward toward the larger forces shaping institutional change. Alternatively, it could start from an analysis of the way the world market shapes local demographic regimes and work down to individual fertility behavior. Direction, however, is less important than integrated, multilevel analysis combining an understanding of both structure and agency (Greenhalgh, 1990a).

Economic Growth, Income Change, and Consumption

A rising level of per capita income, as part of the process of economic development, is a major variable of land-use change, as we have seen in relation to the I = PAT formula. Yet with rising incomes, demand for various goods and services becomes more elastic. In the industrialized countries of the North, for example, the income elasticity of demand for food is low and approaching zero, whereas it is high and positive for such functions as recreation and housing (Pierce, 1990). In the South, by contrast, rising per capita incomes stimulate relatively large increases in demand for basic food goods, although as incomes rise this creates a change in the composition of demand (Crosson, 1986). Thus traditional grains and tubers gradually give way to higher protein sources and acquired dietary tastes, such as livestock products, imported cereals, and processed foods.

Consequently, it is through the changing effective demand for food that rising levels of income exert different pressures upon the agricultural resource base, as shifts to higher protein diets and the increased demand for livestock products require expanded production of animal feeds. This, in turn, leads to an increase in the indirect per capita consumption of plant energy in a developing country with rising incomes. Given that an estimated 40% of net primary production (NPP) currently comprises food production, this has led some authors to express some real Malthusian concerns: 'For an expanding population with a rising income the demands on the agricultural system [will be] nothing short of monumental' (Pierce, 1990: 102).

Rising income also creates other far-reaching changes in relation to land use, for example, by increasing demand for living space, transport, and recreational uses, especially among urban populations. Through rural to urban migration and a growing share of the total population based in urban areas, aggregate demand is also increased, as urban dwellers invariably have higher average levels of income and consumption. The consequences are to produce more food for direct consumption by the urban population and more cash crops for export to pay for imported food, also largely consumed by urban dwellers. The need to engage in export production of land-intensive commodities for high-income countries places a further demand on resources in the South. According to Bilsborrow and Geores (1990), the two main factors responsible for changes in the level of resource depletion are increasing per capita incomes in the North and increasing populations in the South. However, 'with per capita incomes in low-income countries one-tenth those of developed countries, even with four-fifths of the population residing in low-income countries the bulk of the growth in effective demand upon resources in low-income countries in recent decades is attributable to *increases in the high levels of consumption of developed countries*' (Bilsborrow and Geores, 1990: 35; emphasis added). This argument is underpinned by the Ehrlichs, who make reference to the Netherlands in developing their case about worldwide overpopulation. They argue that the Netherlands can support a population density of 1031 people per square mile 'only because the rest of the world does not. In 1984–86, the Netherlands imported almost 4 million tons of cereals, 130,000 tons of oils, and 480,000 tons of pulses. It took some of these relatively inexpensive imports and used them to boost their production of expensive exports—330,000 tons of milk and 1.2 million tons of meat' (Ehrlich and Ehrlich, 1990: 39).

The functional roles of different groups of countries within the world economic system according to the principles of comparative advantage has led many critics to argue that unequal terms of trade prejudice the prospects of the poorest countries to achieve economic growth. Certainly the total size of the world economy has grown at a faster rate than population, increasing from \$5 trillion U.S. to \$17 trillion from 1960 to 1988, at an average annual rate of 3.64%. The relative growth rates of different world regions are, however, quite uneven. While the developed market economies of the North have grown at a rate of 3.5% over this period, rates in the South have ranged from 2.53% in sub-Saharan Africa to 3.84% in Latin America to just over 6% in South and Southeast Asia for the whole period. Yet during the 1980s these differences have widened, for in Africa and Latin America per capita incomes have fallen while they have continued to rise steadily in Asia. Moreover, the overall gap between the rich North and poorer South continues to widen as measured by the ratio of per capita output, which grew from 13.42 in 1980 to 14.7 in 1990 (U.N. General Assembly, 1991).

The relationship between rich and poor countries, as between classes within countries, is ultimately determined by superior economic power in the market place, which provides the necessary signals and incentives to the sphere of production. Supported by structural economic reforms to remove 'distortions,' the market has heralded a growing trend in many developing countries where basic food staples have given way to export crops and feed-grain production for Northern end consumers, resulting in an increasingly segmented structure of agricultural markets in the South (Helwege, 1990).

Export crop production as a means of generating economic growth has a long and often controversial history. Since colonial times, it has been suggested that such crops enhance food security through the exploitation of natural comparative advantage. Critics argue, however, that export crops dominate areas of high agricultural potential and enjoy a disproportionate use of natural resources as well as credit and other inputs. Local food staples are then displaced into areas of low agricultural potential, subject to greater uncertainty of climate and often lacking comparable access to credit and other subsidized inputs. As Leonard (1989) observes, the greatest disparities within rural areas of the South are often those between areas of high agricultural potential that have benefited from agricultural modernization and those areas of low potential that have been neglected.

The displacement of local crops to areas of low viability raises important questions regarding the location of the poor, especially whether the poorest people tend to be found in areas subject to climatic hazards and resource degradation. Before examining this issue, it is necessary to raise some practical considerations regarding the measurement of poverty, especially through the use of income, and whether income constitutes a variable of land-use change.

First, in the poorer countries a large proportion of goods and services is produced and consumed within the household or exchanged through barter or reciprocal arrangements, so that there is no record and little evidence on which to collect income and expenditure data (Anderson, 1991). Second, production and any derived income can fluctuate considerably within and between years, so that figures that might ostensibly refer to annual averages may in reality refer to unspecifiable periods. For this reason consumption may be a better general measure of well-being than income, although by the same token *current* levels of consumption at any given time may not be an accurate measure of a household's typical standard of living (World Bank, 1990).

The idea of 'basic needs' as an index of development and welfare within a society has largely fallen into disrepute, criticized as attempting to universalize Western cultural values (Doyal and Gough, 1991). While efforts are still made to develop a more sophisticated universal measure of well-being, such as the U.N. Development Program's Human Development Index, an often-used rule-of-thumb measure of poverty is the estimates by the U.N. Food and Agriculture Organization (FAO) and World Health Organization (WHO) of daily calorie intake as a percentage of nutritional requirements. Using the first measure of insufficient calories to prevent stunted growth and serious health risks (i.e., below 80% of requirements), some 340 million people fall below the poverty line. A second measure at 90% of requirements, the minimum amount needed for an active working life, produces some 730 million poor people. Of these some 40% are children below the age of ten and 75% are rural dwellers; 80% of income in this group is spent on food (Pearce, 1991). The FAO/WHO figures are very crude, for they do not take account of maldistribution among classes and other groups within a country, and among members of households, which results in an underestimate of the relative malnutrition of women (Doyal and Gough, 1981). As Kates and

Haarmann (1992) demonstrate, the use of different poverty indicators, including both absolute and relative measures, can result in estimates of poverty and hunger that vary by up to 750 million people. Nevertheless, using the incidence of hunger as an indicator of poverty clearly illustrates the extent of food insecurity, which is a problem of access rather than of availability.

Using the incidence and extent of food insecurity as a measure of inadequate incomes—demonstrating the limited entitlements to food by the resource-poor (Sen, 1981)—we might then examine the issue of location. It is widely assumed that a majority of the poor are located in the most ecologically fragile, low-resource areas, marked by limited arable land of low potential and subject to the risks of natural hazards and environmental degradation. The livelihoods of the poor in such areas are generally believed to be highly insecure, and this insecurity is exacerbated by the strongly seasonal demand for agricultural labor. To what degree is this belief correct? Do economically marginal people live in ecologically marginal places?

According to Leonard (1989), 470 million people, or 60% of the developing world's poorest according to his criteria, live in rural or urban areas of high eco-logical vulnerability, such as squatter settlements and regions of low agricultural potential. These are areas where, for Leonard, the poor are 'highly susceptible to the consequences of soil erosion, soil infertility, floods and other ecological disas-ters' (Leonard, 1989: 19). Of this total, some 70% live in Asia, 17% in sub-Saharan Africa, and 13% in Latin America. The 63 million Latin Americans living in ecologically vulnerable areas compose 80% of that region's poorest. Leonard believes that poverty in many developing countries is becoming ever more con-centrated into definable geographical areas that most lack, yet most need, appro-priate infrastructure and technology, making increasing numbers of the poorest people vulnerable to environmental hazards and degradation.

Kates and Haarmann cast doubts on the quality of data employed by Leonard but then observe that, even on his calculations, 43% of the poorest live in areas of high agricultural potential. They argue that it is not appropriate to equate areas of low potential with high ecological vulnerability:

In many parts of the world, land of low agricultural potential is little used, is appropriately used for pastoralism, or is forested. Although subject to erosion, desertification, and deforestation, these areas are not necessarily more vulnerable than are intensively used lands of high agricultural potential that are also subject to erosion, flooding, and, in the case of valuable irrigated lands, salinization and waterlogging (Kates and Haarmann, 1992: 7).

While they believe it is difficult to map poor people directly onto threatened environments, Kates and Haarmann do examine whether poor *countries* have more than their share of drylands, highlands, and rainforests, three major environ-ments that occupy more than one-half the world's land area but support only one-quarter of its people. They find that low-income countries do indeed possess a dis-proportionate share of savanna grasslands, while the very poorest countries, which account for 20% of the land area in the South, possess 63% of its drylands. Moreover, some 71% of tropical moist forests are concentrated in the low-income countries, although not in the very poorest. Highland environments are more equally distributed among poorer and wealthier developing countries.

The main thrust of Kates and Haarmann's article, however, is identifying particular pathways, or spirals, of impoverishment and degradation. Through a review of a number of case studies taken from the literature, they find a remarkable consistency in the analyses of ways poor people lose their entitlements to environmental resources, leading to these downward spirals. Driving these spirals are forces acting in combination, two of which are external to the locality (natural hazards, and development and commercialization) and two of which are internal to the community (population growth and poverty). Although all of the driving forces are implicated in the forms that the spirals assume, Kates and Haarmann suggest that two tend to dominate in each of three major spirals, which they label Displacement, Division, and Degradation. Although the case study environments differ in the particular combinations of driving forces, impoverishment, and degradation sequences, and in the ameliorative responses undertaken, the researchers are nevertheless able to group these studies into the three principal environmental categories of drylands, highlands, and rainforests. This permits some fairly generalized, but useful, descriptions of the way people are displaced, whether by population pressure, resource expropriation, or natural hazards, leading to expansion into areas of lower potential and excessive or inappropriate use, resulting in degradation and impoverishment. This exercise reveals the importance of understanding the different combination of driving forces acting in concert according to local circumstances and dynamics, and the need to develop policy responses across a broad front.

Land-Use Changes: Some Evidence and Conclusions

The caution and circumspection of Kates and Haarmann's paper (1992) describing the combination of driving forces, including population growth, that form the three impoverishment–environmental degradation spirals is most appropriate for a universally applicable approach. Too frequently, efforts to establish a statistically significant relationship between population growth and environmental degradation founder on a priori assumptions about the primacy of population growth and on the extraordinary regional heterogeneity of demographic characteristics, which undermines the relevance of global correlations. Moreover, it has been found that the time scales of population variability are asynchronous with environmental change. Consequently, it has been argued that 'no simple correlations can be established between population and environmental transformations'; rather, there is a 'need for caution in using population as a simple surrogate for environmental transformation' (Arizpe *et al.*, 1992: 62; see also Whitmore *et al.*, 1990).

One of the principal areas where population has been treated as the major driving force of land-use change is deforestation. In some respects this may be because the data for forest clearance are better than for, say, the conversion of pasture or the falling out of production of arable land, and the problem is occurring on a greater scale worldwide.[6] Allen and Barnes (1985) have proposed a statistically significant correlation between population growth and deforestation for a sample of countries from Africa, Asia, and Latin America, yet they avoid any location-specific analysis or explanation. In a wide-ranging review of literature, Bilsborrow and Geores (1990) propose a more circumstantial linkage between population growth and land extensification involving tropical deforestation, although these authors are anxious to incorporate intermediate and explanatory factors such as land fragmentation, patterns of land ownership, and migration.

Myers and Tucker (1987) go one stage further and emphasize the significance of tenure systems and the unequal distribution of land as central to an understanding of the dynamics of deforestation in Central America. They assert that the rise in sheer population numbers is not by itself the cause of rapid environmental degradation in the region. Other authors are even more forthright in deflecting the cause of deforestation away from demographic increase. Barraclough and Ghimire (1990), for example, state that it is a tautology and an overgeneralization to blame growing rural populations, which are characterized as 'hordes of slash-and-burn cultivators.' However, in true agrarian populist style, their argument places the blame on population growth at one remove from the problem by asserting that it is increasing urban populations that stimulate demand for agricultural and forest products, although in many parts of Africa and Latin America these populations may actually be fed by food imports and food aid from abroad.

Barraclough and Ghimire argue that current rates of deforestation 'can be much better comprehended by assuming that they are the outcome of complex historical processes taking place in interacting social and natural systems and sub-systems' (1990: 15). In other words, it is necessary to adopt a regionally specific analysis of the factors creating forces for change. In Africa the authors highlight the expansion of export crops, which have replaced forest either directly or indirectly through the displacement of food crop production, encouraging encroachment by subsistence farmers on to forest land. Meanwhile, in Latin America, it has been the expansion of commercial cattle production that has been the major cause of deforestation: the state of Tabasco, Mexico, for example, has lost 90% of its forests during the last four decades to cattle pastures (Barraclough and Ghimire, 1990). Browder (1989) supports this argument, stating that pasture development has been responsible for 60% of the total tropical forest area converted in the Brazilian Amazon, which until recently was encouraged by substantial government subsidies.

In Indonesia the major dynamic of forest clearance has been the transmigration program designed to relieve high population densities in Java and Bali through resettlement on the outer islands. However, the management of many of these cleared forest areas has proved extremely difficult for farmers accustomed to rich

---

[6] However, data dealing with estimates of rates of deforestation are not without technical difficulties (see Skole, this volume) or political controversy, as the World Resources Institute has discovered in its attempt to calculate a Greenhouse Index (see McCully, 1991; Agarwal and Narain, 1991).

volcanic soils, as they attempt to sustain production on two-hectare farms in the face of low fertility, high acidity, susceptibility to erosion, and the dominance of invasive grasses, most notably *Imperata cylindrica* (Blaikie and Brookfield, 1987). *Imperata* is estimated to occupy 10% of the former extent of forest in the outer islands. It severely retards natural succession and the replenishment of soil fertility under shifting cultivation, while on small farms it is extremely difficult to work because of its extensive rhizomes. Consequently, many small-farm households are forced to choose one of the following options: adopt capital-intensive, high-input solutions; make increasing use of off-farm employment for income generation; or abandon their degraded land and move elsewhere in search of fresh stocks of soil fertility under forest cover. In Lampung Province, southern Sumatra, this has led to extremely high levels of relocation *(transmigrasi lokal),* so that in addition to the more than 200,000 people officially resettled in reserved forest areas from the early to the late 1980s, there are many thousands of spontaneous migrant families occupying land let as logging concessions or retained as public forests and watershed protection in environmentally critical areas (World Bank, 1988; Pain *et al.*, 1989).

Thus, in the context of land settlement in Indonesia and almost certainly elsewhere, the clearance of tropical forests is not being replaced by stable, productive, sustainable, or equitable systems of production, the key indicators of agricultural performance proposed by Conway and Barbier (1988).[7] Rather, forest is giving way to an intermediate and extensive form of land use that is expanding more rapidly by deforestation than it is being reduced by the introduction of successful systems of management (Blaikie and Brookfield, 1987). This, however, demonstrates the vital importance of examining trends in the different categories of land use (which are fraught with definitional and measurement difficulties) and learning how these might correlate with demographic and economic indicators. (See Table 2 for an illustration of the range of data across some relevant categories.) Yet an exercise in complex data analysis or even simple scrutiny should not preclude attention to other variables that mediate between population and resource use: technology most certainly, but also the sociopolitical structures that help us to understand the conditions and dynamics of transmigrant settlement, and the circumstances of poverty that influence household livelihood and demographic strategies. Consequently, 'we need to develop a much deeper understanding of the relationships between human populations, their technologies, cultures, and values, and the natural capital (renewable and non-renewable natural resources) they depend on for life support if we are to achieve sustainability' (Arizpe *et al.*, 1992: 61).

While employing a hierarchical, multilevel framework helps us to remain alert to the ways these components interact at different scales, it is also becoming more widely appreciated that nonlinear behavior might prevent simple aggregation and

---

[7] Conway and Barbier define agricultural sustainability as the ability of a system (field or farm) to maintain productivity in the face of stress or shock. It is clear that where farmers in transmigrant settlements face *Imperata* domination, their agricultural systems are generally unable to sustain production.

Table 2: Population and land use

	Population Density 1989[1]	Average Population 1985–90	% Land Use Change from 1975–77 to 1987				Wilderness Area % of Total Land 1987	Cropland, ha/capita 1989	% Irrigated Land 1985–87	PC Index Agricultural Production 1986 (1979–81=100)
			Cropland	Pasture	Forest	Other				
World	398	1.73	2.7	−0.2	−2.1	1.3	39	0.28	15	102
Africa	212	3.00	4.6	−0.5	−4	1.9	31	0.3	6	94
N&C America	197	1.28	2.1	2.5	−2.2	0.5	42	0.65	9	91
S America	166	2.07	14.1	4.0	−4.5	3.1	24	0.49	6	100
Asia	1139	1.85	0.8	−1.2	−1.5	1.3	14	0.15	31	113
Europe	1050	0.23	−1.0	−3.4	1.3	2.5	4	0.28	12	106
USSR	128	0.78	0	0.2	1.9	−2.6	34	0.81	9	109
Oceania	33	1.44	14.0	−3.8	−6.7	14.3	30	1.87	4	99
Cote d'Ivoire	380	4.12	22.4	0	−42.1	31.2	13	0.3	2	91
Nigeria	1199	3.43	3.8	0.8	−17	7.5	2	0.29	3	100
Costa Rica	576	2.64	6.1	34.1	−23	−16.4		0.18	21	92
Mexico	454	2.2	3.0	0	−12	12.6	2	0.28	21	94
Bolivia	66	2.76	3.0	−1.2	−1.3	4.5	16	0.48	5	94
Brazil	174	2.07	22.7	6.4	−4.2	0.3	24	0.53	3	106
China	1201	1.39	−2.9	0	0	0.2	22	0.09	46	128
India	2811	2.08	0.5	−2.5	−0.4	−3.0		0.20	25	107

Data from WRI, 1990.
[1] People per 1000 ha.

extrapolation. Increasing recognition of the limits of scientific explanation and causation (Mearns, 1991), of problems of indeterminacy and uncertainty, and of plural rationalities (James and Thompson, 1989) should not herald a retreat from rigorous, applied research and the search for solutions to pressing environmental problems. It should, however, make us beware a preoccupation with sheer numbers of people and mathematical 'limits to growth' and rather focus upon the relationships among individuals, regions, and nations, their institutions, technologies, and patterns of resource consumption. The search for sustainable development will require that these become the objects of critical and policy-relevant research.

References

Agarwal, A., and S. Narain. 1991. *Global Warming in an Unequal World: A Case of Environmental Colonialism*. Center for Science and Environment, New Delhi, India.

Allen, J. C., and D. F. Barnes. 1985. The causes of deforestation in developing countries. *Annals of the Association of American Geographers 75*, 163–184.

Alonso, W., ed. 1987. *Population in an Interacting World*. Harvard University Press, London.

Anderson, V. 1991. *Alternative Economic Indicators*. Routledge, London.

Arizpe, L., R. Costanza, and W. Lutz. 1992. Population and natural resource use. In *An Agenda of Science for Environment and Development into the 21st Century* (J.C.I. Dooge, J.W.M. la Rivière, J. Marton-Lefevre, T. O'Riordan, and F. Praderie, eds.), Cambridge University Press, Cambridge, U.K., 61–78.

Barraclough, S., and K. Ghimire. 1990. *The Social Dynamics of Deforestation in Developing Countries: Principal Issues and Research Priorities*. Discussion Paper 16, U.N. Research Institute for Social Development, Geneva, Switzerland.

Bilsborrow, R. E. 1987. Population pressures and agricultural development in developing countries: A conceptual framework and recent evidence. *World Development 15*, 183–203.

Bilsborrow, R. E., and M. E. Geores. 1990. *Population, Environment and Sustainable Agricultural Development*. Draft monograph prepared for U.N. Food and Agriculture Organization, Rome, Italy.

Blaikie, P., and H. Brookfield. 1987. *Land Degradation and Society*. Methuen, London.

Bongaarts, J., W. P. Mauldin, and J.F. Phillips. 1990. *The Demographic Impact of Family Planning Programs*. Research Division Working Paper No. 17, Population Council, New York.

Boserup, E. 1981. *Population and Technology*. Basil Blackwell, Oxford, U.K.

Browder, J. O. 1989. Development alternatives for tropical rain forests. In *Environment and the Poor: Development Strategies for a Common Agenda* (H.J. Leonard, ed.), Overseas Development Council, Washington, D.C., 111–133.

Cain, M., and G. McNicoll. 1988. Population growth and agrarian outcomes. In *Population, Food and Rural Development* (R.E. Lee, W.B. Arthur, A.C. Kelley, G. Rodgers, and T.N. Srinivasan, eds.), Clarendon Press, Oxford, U.K., 101–117.

Commoner, B. 1972. *The Closing Circle: Confronting the Environmental Crisis*. Jonathan Cape, London.

Conway, G., and E. Barbier. 1988. After the Green Revolution: Sustainable and equitable agricultural development. *Futures 20*, 651–670.

Crosson, P. 1986. Agricultural development—Looking to the future. In *Sustainable Development of the Biosphere* (W.C. Clark and R.E. Munn, eds.), Cambridge University Press, Cambridge, U.K.

Deere, C. D., J. Humphries, and M. Leon de Leal. 1982. Class and historical analysis for the study of women and economic change. In *Women's Roles and Population Trends in the Third World* (R. Anker, M. Buvinic, and N. Youssef, eds.), Croon Helm, London, 87–114.

Demeny, P. 1990. Population. In *The Earth as Transformed by Human Action* (B. L. Turner II, W. C. Clark, R. W. Kates, J. F. Richards, J. T. Mathews, and W. B. Meyer, eds.), Cambridge University Press, Cambridge, U.K., 41–54.

Doyal, L., and I. Gough. 1991. *A Theory of Human Need*. Macmillan, Basingstoke, U.K.

Ehrlich, P. 1968. *The Population Bomb*. Ballantine Books, New York.

Ehrlich, P., and A. Ehrlich. 1990. *The Population Explosion*. Hutchinson, London.

Geertz, C. 1963. *Agricultural Involution: The Processes of Ecological Change in Indonesia*. University of California Press, Berkeley, California.

Greenhalgh, S. 1990a. Towards a political economy of fertility: Anthropological contributions. *Population and Development Review 16*, 85–106.

Greenhalgh, S. 1990b. *State-Society Links: Political Dimensions of Population Policies and Programs with Special Reference to China*. Research Division Working Paper No. 18, Population Council, New York.

Grigg, D. 1980. *Population Growth and Agrarian Change: An Historical Perspective*. Cambridge University Press, Cambridge, U.K.

Hall, R. 1989. *Update: World Population Trends*. Cambridge University Press, Cambridge, U.K.

Hardin, G. 1988. Cassandra's role in the population wrangle. In *The Cassandra Conference: Resources and the Human Predicament* (P. Ehrlich and J. Holdren, eds.), Texas A&M University Press, College Station, Texas, 3–16.

Harrison, P. 1990. Too much life on earth? *New Scientist 19*, 28–29.

Harrison, P. 1992. *The Third Revolution: Environment, Population and a Sustainable World*. I.B. Tauris, London.

Helwege, A. 1990. Latin American agricultural performance in the debt crisis: Salvation or stagnation? *Latin American Perspectives 67*, 57–75.

Hewitt, T., and I. Smyth. 1992. Is the world overpopulated? In *Poverty and Development in the 1990s* (T. Allen and A. Thomas, eds.), Oxford University Press, Oxford, U.K., 78–96.

James, P., and M. Thompson. 1989. The plural rationality approach. In *Environmental Threats: Perception, Analysis, and Management* (J. Brown, ed.), Belaven, London, 87–94.

Jones, H. 1990. *Population Geography*. Paul Chapman Publishing, London.

Kabeer, N. 1985. Do women gain from high fertility? In *Women, Work and Ideology in the Third World* (H. Afshar, ed.), Tavistock, London, 83–106.

Kasun, J. 1988. *The War against Population: The Economics and Ideology of World Population Control*. Ignatius Press, San Francisco, California.

Kates, R. W., and V. Haarmann. 1992. Where the poor live: Are the assumptions correct? *Environment 34*, 4–11, 25–28.

Latin American and Caribbean Commission on Development and Environment. 1990. *Our Own Agenda*. Inter-American Development Bank and U.N. Development Program, Washington, D.C.

Lee, R. D. 1986. Malthus and Boserup: A dynamic synthesis. In *The State of Population Theory* (D. Coleman and R. Schofield, eds.), Basil Blackwell, Oxford, U.K., 96–130.

Leonard, H. J. 1989. Overview. In *Environment and the Poor: Development Strategies for a Common Agenda* (H.J. Leonard and contributors), Overseas Development Council, Washington, D.C.

Lesthaeghe, R. 1986. On the adaptation of sub-Saharan systems of reproduction. In *The State of Population Theory* (D. Coleman and R. Schofield, eds.), Basil Blackwell, Oxford, U.K., 212–238.

Lockwood, M. 1991. Food security and environmental degradation in northern Nigeria: Demographic perspectives. *Institute of Development Studies Bulletin 22 (3)*, 12–21.

McCully, P. 1991. Discord in the greenhouse: How WRI is attempting to shift the blame for global warming. *The Ecologist 21*, 157–165.

McNicoll, G. 1984. Consequences of rapid population growth: An overview and assessment. *Population and Development Review 10*, 177–240.

Mearns, R. 1991. *Environmental Implications of Structural Adjustment: Reflections on Scientific Method*. Discussion Paper No. 284, Institute of Development Studies, University of Sussex, U.K.

Myers, N., and R. Tucker. 1987. Deforestation in Central America: Spanish legacy and North American consumers. *Environmental Review 11*, 55–71.

ODA (Overseas Development Administration). 1991. *Population, Environment and Development*. An issues paper for the Third UNCED Preparatory Committee. Overseas Development Administration, London.

Pain, M., B. Benoit, P. Levang, and O. Sevin. 1989. *Transmigrations and Spontaneous Migrations in Indonesia*. Orstom, Paris, France.

Pearce, D. 1991. Population growth. In *Blueprint 2: Greening the World Economy* (D. Pearce, ed.), Earthscan, London, 109–137.

Pierce, J. T. 1990. *The Food Resource*. Longman, Harlow, U.K.

Pleumarom, A. 1992. Course and effect: Golf tourism in Thailand. *The Ecologist 22*, 104–110.

Porter, G., and J. Welsh Brown. 1991. *Global Environmental Politics*. Westview Press, Boulder, Colorado.

Sadik, N. 1990. *The State of World Population 1990*. U.N. Population Fund, New York.

Sen, A. 1981. *Poverty and Famines: An Essay on Entitlement and Deprivation*. Oxford University Press, Oxford, U.K.

Simon, J. L. 1981. *The Ultimate Resource*. Martin Robertson, Oxford, U.K.

Simon, J. L. 1990. *Population Matters: People, Resources, Environment, and Immigration*. Transaction Publishers, New Brunswick, New Jersey.

Spooner, B., and N. Walsh. 1991. *Fighting for Survival: Insecurity, People and the Environment in the Horn of Africa*. Consultants' report prepared for Sahel Program, International Union for Conservation of Nature and Natural Resources (IUCN), Gland, Switzerland.

U.N. General Assembly. 1991. Cross-sectoral issues: The relationship between demographic trends, economic growth, unsustainable consumption patterns and environmental degradation. U.N. General Assembly Document A/CONF. 151/PC/46, United Nations, New York.

Whitmore, T. M., B. L. Turner II, D. L. Johnson, R. W. Kates, and T. R. Gottschang. 1990. Long-term population change. In *The Earth as Transformed by Human Action* (B. L. Turner II, W. C. Clark, R. W. Kates, J. F. Richards, J. T. Mathews, and W. B. Meyer, eds.), Cambridge University Press, Cambridge, U.K., 25–40.

Working Group on Population Growth and Economic Development. 1986. *Population Growth and Economic Development: Policy Questions*. National Academy Press, Washington, D.C.

World Bank. 1988. *Indonesia: The Transmigration Program in Perspective*. International Bank for Reconstruction and Development, Washington, D.C.

World Bank. 1990. *World Development Report 1990*. International Bank for Reconstruction and Development, Washington, D.C.

WRI (World Resources Institute). 1990. *World Resources 1990–91*. Oxford University Press, Oxford, U.K.

Youssef, N. 1982. The interrelationship between the division of labor in the household, women's roles and their impact on fertility. In *Women's Roles and Population Trends in the Third World* (R. Anker, M. Buvinic, and N. Youssef, eds.), Croom Helm, London, 173–201.

12

Technology[1]

Arnulf Grübler

Introduction

This chapter addresses the role of technology (or better, of technological change) in land transformation. The role of technology in changing land-use patterns is usually associated with images of land areas covered by human artifacts like city skylines and sprawling suburbs, but the global quantitative picture contradicts such conceptions. Although detailed statistics are lacking, artifacts of our technological civilization most likely cover less than 1% of the earth's land.[2] In contrast, agriculture and pasture use close to 40% of the global land area (FAO, 1991), while 60% remains largely untouched by human activities, being covered by forests, deserts, or ice sheets.

The role of technology has therefore first to be discussed in its relationship to agriculture, in particular to increases in the productivity of the factor inputs land and labor. The productivity of land determines the land requirements for a given population. The productivity of labor determines the percentage of this population that is required to cultivate the land. The impact of technological change has been most dramatic in raising agricultural labor productivity, but it has also noticeably improved agricultural land productivity. The technologies we use and the pace at which they change therefore matter—not only for the amount of land required to feed people, but even more for allowing ever-increasing fractions of the population to engage in economic activities outside agriculture and to live outside rural (i.e., in urban) areas.

What Is Technology?

Before discussing the relationships between technology and land-use changes, we should ask the question, What is technology? In the most narrow terms, technology is represented by manmade objects. However, such artifacts have to be produced (invented, designed, and manufactured). This requires a larger system: hardware

---

[1]A longer version of this material has been published as Working Paper WP-92-2 by the International Institute for Applied Systems Analysis, A-2361 Laxenburg, Austria.

[2]As explained below, we use a value of around 250 m^2 per capita for the land devoted to building areas (of course not all of it actually covered by buildings) and infrastructures, a value typical of the most densely populated countries like Japan and the Netherlands. As most regions of the world have a significantly lower population density, the actual area covered by human artifacts will be significantly below 1% of the earth's land.

(e.g., machinery, a manufacturing plant), factor inputs (labor, energy, raw materials, and other resources), and finally 'software' (human knowledge and skills), or what the French call *technique*. This last category represents the disembodied nature of technology, which is known in economics as the knowledge base: the information, skills, and procedures (organization) that are required to produce any artifact. Technological change thus not only entails the creation of new and/or the replacement of old artifacts, but also involves changes in the related knowledge base. Finally, *technique* is not only required for the production of artifacts but also for their use (e.g., the *technique* of driving a car or using a bank account), both at the level of the individual and at the level of the whole society. Forms of organization (like the existence of markets), institutions, social attitudes, and beliefs are not only important to understand how systems of production and use of artifacts emerge and function, but even more to understand the origin and choice mechanisms of combinations of particular artifacts and the rates by which these become embedded (if they do) into a given socioeconomic system, i.e., their *diffusion*.

Thus, in this paper the term technology comprises not only manmade artifacts and the (scientific) knowledge, know-how, and skills necessary for their inception, production, and use (the technical knowledge base), but also a large body of evolving social and organizational know-how and techniques. Technology refers to the whole sociotechnical systems of production and use (Kline, 1985) that extend human capabilities and accomplish tasks that humans could not perform otherwise.

From a historical perspective we can conclude that in particular periods, the process of development, stimulated by changing structures of economic activities and technological change, tended to cluster around interrelated sets of artifacts, *techniques*, and organizational/institutional configurations. These mutually interdependent and cross-enhancing sociotechnical systems of production and use cannot be analyzed from the perspective of single technologies, but must be considered in terms of their common (and time-specific) embedding in many other processes of technological, institutional, and social change. Using the concepts of *technology clusters* or *techno-economic paradigms,* we will discuss below their historical evolution and their relationship to technological change in agriculture. We will then relate these technological developments to land-use changes since the beginning of the 18th century.

Technology Clusters since the Industrial Revolution

The cumulative nature of technological change requires us to look back to the Industrial Revolution if we wish to understand the relationship of technology and land-use changes. In the 18th century, a series of innovations (most notably the spinning jenny, the flying shuttle, and the power loom) transformed the manufacture of cotton in England and gave rise to what eventually became a new mode of production: the factory system. Innovations in the fields of energy (stationary steam engines) and metallurgy (replacement of charcoal by coal in the iron industry) were of a similar revolutionary character, and all of these, mutually reinforcing one another, drove an industrial revolution in Britain, making her the world's leading industrial and economic power well into the late 19th century. Technology embodied in machinery, leading to new forms of production, products, and markets, has been, as Mokyr (1990) says, 'the lever of riches.'

It is beyond the scope of this paper to list, much less to discuss, the large number of innovations involved in the takeoff of the Industrial Revolution. Landes (1969) summarizes them under three principles: the substitution of machines for human effort and skill; the substitution of fossil fuels (coal) for animate power, opening for the first time in human history the possibility of unprecedented spatial consumption density and an almost unlimited supply of energy; and the use of new (and more abundant) raw materials in manufacturing. These three principles apply not only to the onset of the Industrial Revolution, but also to later stages in the industrialization process. Today, they also apply to the modernization and economic growth in developing countries.

Important technological innovations can also be identified in earlier periods of human history. What makes the Industrial Revolution unique is a new quality of the innovations introduced: their interdependence and mutual cross-enhancement, and their embedding in profound transformations of the social and organizational fabric of society. The steam engine, coal industry, railroads, and new steel production processes cannot in fact be considered separately: they depended on each other, enhanced each other, and together—via a multitude of what in economics are called forward and backward linkages—contributed to economic growth. The same quality of mutual dependence and enhancement has characterized innovations in the periods since the Industrial Revolution—for example, the internal combustion engine, the oil and petrochemical industries, synthetic fibers, and plastics, to name just a few areas associated with the post–World War II period of economic growth.

Of equal importance were social and organizational changes, which span the whole domain from the generation of (scientific) knowledge to its systematic deployment in the innovation process, incentives for innovation diffusion, new modes of production, new forms of enterprises, the organization of market relations, and so on. Rosenberg and Birdzell (1986) therefore particularly emphasize the decisive role new institutional arrangements, such as the early separation of the political and economic spheres, played in the process of how 'the West grew rich.'

Cameron (1989) cautions against the term 'Industrial Revolution' with its implication of a pronounced discontinuity and emphasis on industrial technology and innovation. He points out that changes were not only industrial but also social, intellectual, commercial, financial, agricultural, and even political. In this 'seamless web' of historical change it is difficult to assign relative weights to different factors, and dangerous to ignore the importance of earlier developments of protoindustrial economies as driving forces and causes of change. Perhaps the intellectual and institutional/organizational changes were indeed the most fundamental, as they provided an environment favorable for the creation and diffusion of

Table 1: Clusters of pervasive technologies

	1750–1820	1800–1870	1850–1940	1920–2000	1980–
Dominant Systems	Water power, sails, turnpikes, iron castings, textiles	Coal, canals, iron, steam power, mechanical equipment	Railways, steamships, heavy industry, steel, coal chemicals, telegraph, urban infrastructure	Electricity, oil, cars, roads, telephone, radio, TV, durables, petrochemicals	Gas, nuclear, aircaraft, telecommunication, information, electronics
Emerging Systems	Mechanical equipment, coal, stationary steam, canals	Steel, city gas, coal chemicals, telegraph, railways, urban infrastructure	Electricity, cars, trucks, roads, radio, telephone, oil, petrochemicals	Nuclear power, computers, gas, telecommunication, aircraft, electronics	Biotechnology, artificial intelligence, space industry & transport
Organizational Style	Manufacture	Factory system	Standardization	Fordism– Taylorism	Quality control

innovations. In this sense, changes in the social context may be seen as the fundamental driving force of change.

From such a perspective, the main characteristic of the period of economic expansion since the 18th century is the interdependence of whole clusters of technological and organizational innovations. Thus, the impact of any individual technology, as important in its own merits as it may be (such as the railways of the 19th century), is necessarily limited. Instead, it is the synergistic interlinkages with other technologies, evolving techniques, and forms of organization that have resulted in the profound transformations of economic, employment, and social structures over the last 300 years.

A particularly convincing description for such technology clusters was developed by Freeman and Perez (1988), who refer to them as 'techno-economic paradigms,' illustrated in Table 1. The top row gives the dominant techno-economic systems for each epoch, and the middle row lists emerging ones. The last row summarizes the predominant organizational and management models during the respective periods. The clusters given in Table 1 are of course not exhaustive, and also the timing is necessarily approximate. However, the table illustrates pervasive technologies and infrastructures, which were to a large extent drivers of the history of economic growth, the spatial division of labor, changes in employment, and to some extent also illustrates the environmental impacts associated with the development of particular technological regimes.

From a historical perspective, we can conclude that the diffusion of such tech-

Table 2: The growth of railway networks

Country	Introduction Year	Length by 1870 (1000 km)	Maximum Length Achieved (1000 km)	Maximum Length (year)[2]	Density[1] in 1923 per 100 km^2	per 10,000 Inhabitants
Austria-Hungary	1837	6.1	23.0[3]	1913[3]	8.0	10.2
France	1828	15.5	42.6	1933	9.7	13.7
Germany	1835	21.5	63.4[3]	1913[3]	12.2	9.6
Russia	1845	10.7	70.2[3]	1913[3]	0.3–1.5	4.8–8.4
USSR	–	–	145.6	1986	0.7	5.5
U.K.	1825	21.6	32.8	1928	16.0	8.8
USA	1829	85.0	482.7	1929	4.3	38.1
Core countries	–	160.4	715.0		2.0	16.7
Rest of the world	–	69.5	540.0		0.6	3.7
World	–	221.9	1255.0	1930	1.0	6.7

[1]Density as calculated by Woytinsky, 1927: 38–39, except Russia and the USSR (own calculation). Range of figures for Russia corresponds to total density and the European part of the territory, respectively. Density figures for USSR are for the 1986 network size.
[2]Data from Mitchell, 1980: 609–616; and Mothes, 1950: 85–104.
[3]Important territorial changes after 1913.

nology clusters is an international phenomenon, but one with great spatial disparities. Particular systems are developed in a number of core countries, from which they spread out via a series of spatial hierarchies to (spatially or economically) peripheral areas. Although adoption starts much later on the periphery, it tends to catch up with the core, albeit at significantly lower levels of adoption intensity.

For instance, the construction of the railway networks of England and the United States spans a period of 100 years (1830–1930), whereas it took typically only about half that time in Scandinavia (1870–1930). Railway networks were also more extensive (in either per capita or unit land area terms) in England and the United States, which led the introduction of this technology, than in follower countries. Altogether, the core countries of railway development (England, Europe, and the United States) had constructed about 60% of the 1.3 million km railway network worldwide by the 1930s (Table 2).

The length of the world railway network has not increased since then. Net additions to the railway network (primarily in developing countries) have been balanced by decommissioning of railway lines (due to the development of newer transport systems) in the core countries.[3] This implies that the pervasive develop-

---

[3]Examples of infrastructure decay can be found in some sectors such as transport (canals, railways) and telecommunications (telegraph), whereas in other sectors (e.g., urban infrastructures) the older system may not lose out at all, continuing to be upgraded and used to the present.

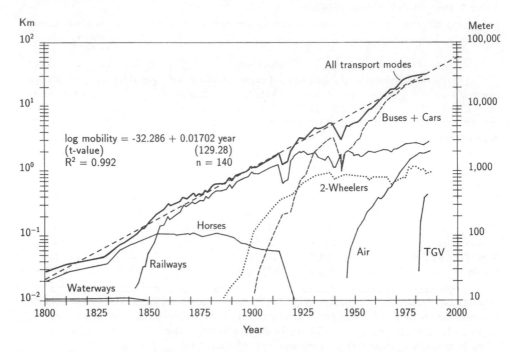

Figure 1. Range covered (average km traveled daily) per capita in France by mode of transport and total since 1800. The succession of transport infrastructures since the onset of the Industrial Revolution expanded the spatial range of human activities by over three orders of magnitude. New transport technologies also enabled increasing spatial division of labor and trade (also in agricultural products) and the growth of cities. Transport infrastructures are land intensive and account for a significant share of built-up land (from Grübler, 1990: 232).

ment of particular infrastructures and technologies is *time dependent*. The densities realized in the leading countries are unlikely to be repeated by follower countries at later periods in history. This is because the 'laggards' often shift to newer systems rather than adopting the older technology at a high level. Thus there is a large heterogeneity among countries in adoption levels of technologies. At the world level, however, there is a broad succession from older to newer technology clusters, as illustrated in Table 1.

Let us take the development of transport systems as an example of this historical process of technological transitions: from canals to railways, to roads, and finally to airways (cf. Figure 1). First, the spread of these transport systems was pervasive in the sense that they were and are important to all branches of the economy, even to nearly every aspect of daily life. In this sense transport (like energy) systems can serve as an indicator of the whole technology cluster they are associated with. Furthermore, one can easily identify the leading countries where the spread of each respective cluster was most important, e.g., for the period up to 1820, England (early canal development); for the period 1820–70, England, France, and the United States (pervasive canal construction, beginning of railway development); for the period 1870–1930, Europe and the United States (pervasive railway development); and finally, for the period up to the present, the countries of the Organization for Economic Cooperation and Development (OECD) (pervasive spread of road infrastructures and of cars) and the USSR and many developing countries (buses). As will be shown below (Figure 14), the dominance of a particular group of countries in the development and resulting application intensity of each of these successive transportation systems is mirrored in their respective intensity of urbanization.

The transition from one technology cluster to the next one can be identified through pronounced discontinuities in the social and economic spheres: increased price volatility, mergers and bankruptcies, and the resulting large-scale disinvestment away from old technologies and infrastructures. The transition, although disruptive, becomes necessary when the older cluster starts showing decreasing marginal returns and generally decreasing improvement possibilities, and the public develops an increasing awareness of adverse social and environmental impacts associated with its further expansion. Its continued intensification in the leading countries and diffusion to peripheral regions become blocked. For the latter areas, a window of opportunity opens in such transitional phases for the introduction of new systems and technologies (Grübler and Nowotny, 1990). On the other hand, countries with pervasive adoption of the previous cluster face considerable transition problems due to the heavy commitments of capital stock and human resources to that cluster. Thus, frequently, the transition from one cluster to another one also changes the 'club' of leading countries.

Technology and Land-Use Changes

How does the succession of various technology clusters relate to changes in land-use patterns?

First, technological change has led to far-reaching transformations in agriculture through increases in land productivity (i.e., decoupling the expansion of agricultural areas from population growth) and in labor productivity (i.e., freeing people for other economic activities and enabling urbanization).

Second, the technological changes involved in the successive 'transport revolutions' increased the spatial division of labor, enabled the expansion of large-scale export-oriented production and trade (in food and agricultural products), and allowed the increasing population concentration in urban areas. Perhaps the most pervasive changes brought about by the Industrial Revolution are the result of the development of transportation systems of increasing spatial density and productivity, allowing people to cover ever larger distances (Figure 1) at lower costs.

Third, new transport technologies increased access to land. Distance can be measured directly or measured in hours of travel time (functional distance). New transport technologies reduced functional distance and connected ever larger territories into systems. This is perhaps best illustrated by the fact that even agriculture today operates as a world system.

Throughout history, the technological level and the dynamics of its change have

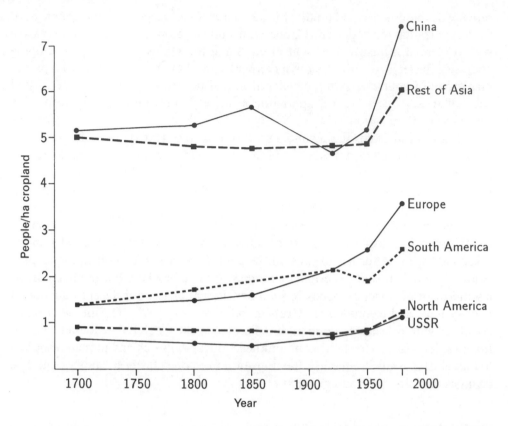

Figure 2. Agricultural land productivity (population per ha cropland). With the exception of Europe, agricultural land productivity did not increase noticeably prior to the 1950s. This implies that population growth resulted in a proportional land-use conversion from forests and grasslands to arable land. Rising land productivity is the result of technological change. Note also the large regional differences and path dependency in the evolution of land productivity (derived from data in Richards, 1990: 164; Durand, 1967: 259; and Demeny, 1990: 49).

been and continue to be spatially heterogeneous. In fact, only over the last 50 years have technologies become truly global. And it is also only during this time that increases in agricultural land productivity have outpaced the rate of population growth. Had the world's population in 1980 been supplied with only the land productivity that prevailed in 1950, the arable land area would have had to be 500 million ha larger than its actual 1980 value (around 1500 million ha). It is our contention that the key factor of such developments is related to technological change.

Impacts on Agriculture and Rural and Urban Populations

Agricultural Land and Labor Productivity Increases

Let us now examine the impacts of the successive technology clusters on agricul-

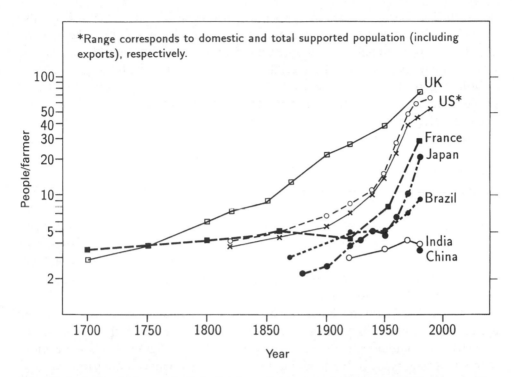

Figure 3. Agricultural labor productivity (total population per agricultural worker). Increases in labor productivity are the result of technological change, encompassing biological, mechanical, and organizational innovations. Note that agricultural labor productivity is even higher when exports are considered. E.g., the two values given for the United States are for domestic (X) and total supported (O) population (derived from data in Durand, 1967: 259; Demeny, 1990: 49; Mitchell, 1980: 161–173; Mitchell, 1982: 84–93; and Mitchell, 1983: 150–160; U.S. data are from Arnold *et al.*, 1990: 72).

ture. Figure 2 presents estimates of agricultural land productivity obtained by dividing the number of people in a given region by the cropland area. The land-use estimates underlying Figure 2 (Richards, 1990) are only first-order approximations. Differences in land productivity reflect different agricultural systems and differences in the stages of agricultural development. More than 200 million people still apply the simplest mode of agricultural production, shifting cultivation, with land requirements of 15–20 ha for feeding one person. On the other extreme, there are areas where three crops per year are grown and less than one-twentieth of a hectare produces enough food for one person (Buringh and Dudal, 1987). Therefore, the regional aggregates of Figure 2 mask persistent differences between and within particular regions. For instance, the land productivity figures of Japan are significantly higher than the Asian average over the whole time period considered. In a similar way, the land productivity figures for France[4] are

---

[4]Land productivity figures for Japan exceeded 8 people per ha arable land even in the 18th and 19th century and currently exceed 20 people/ha (Grigg, 1980). Values for France did not exceed 1.5 people per ha cultivated land (excluding pastures) throughout the 18th century and well into the 1920s, compared to values between 3 and 4 for England and Wales over the same time period (Grigg, 1980).

below the European average throughout the period considered in Figure 2 (cf. also Figure 4). Our crude productivity measure also does also not include the recently significant interregional trade in agricultural products which should increase the land productivity figures of agricultural net export regions (cf. Figure 3).

Still, Figure 2 illustrates clearly the spatial heterogeneity in agricultural land productivity and its evolution since the 18th century. Differences in initial conditions, development paths pursued, and diet and thus the mix of agricultural products produced explain many of the large discrepancies in agricultural land productivity, such as between rice- and grain- (and meat-) oriented agricultural systems. With the exception of modest increases in Europe and perhaps South America (where data are much less certain), agricultural land productivity did not increase in the 18th and 19th centuries, which implies that over this time period there was a one-to-one correlation between population increases and conversion to agricultural land. Increases in agricultural land productivity become noticeable in Europe by the second half of the 19th century, and in all other regions by the second half of the 20th century, primarily in conjunction with the introduction of manmade fertilizers and the diffusion of high-yield crops.

In contrast, agricultural labor productivity, measured in total population per agricultural worker (Figure 3), increased continuously since the onset of the Industrial Revolution (note the semilogarithmic scale of Figure 3). These developments first took place in England, but the other industrialized countries (with the exception of France) followed in the 19th century. Considering the significant export of agricultural products, labor productivity is even higher (as indicated by the two data series for the United States in Figure 3). More recent transformations in the employment structure in the 20th century, such as in the USSR and Japan, were achieved at an even faster pace, so the overall trend is one of convergence in the employment structure with only a few percent of the population employed in the agricultural sector.[5] Although in many developing countries, such as China and India, about 70% of the work force is still employed in agriculture, similar structural shifts are very likely to occur in the future. The experience in the developed countries, with their accelerated rates of change over time, can serve as a guide to derive scenarios about the future pace of this structural transition in developing countries.

The above-outlined increases in agricultural land and labor productivity since 1700 are corroborated by a shorter-term analysis of agricultural productivity increases between 1960 and 1980 from Hayami and Ruttan (1985), illustrated in Figure 4. These authors identify three trajectories of productivity increases, the 'Asian,' 'European,' and 'New Continental' paths, which are related to the relative endowment of land and labor, with starting values around 1000, 10,000, and

---

[5]Keep in mind, however, that many activities previously performed in the agricultural sector have moved to the industrial and service sectors. Hence, the percentage of the work force involved in all food-related activities (farming, production of tractors, food processing and distribution, etc.) is significantly greater than the few percent of the work force that remain on the farms.

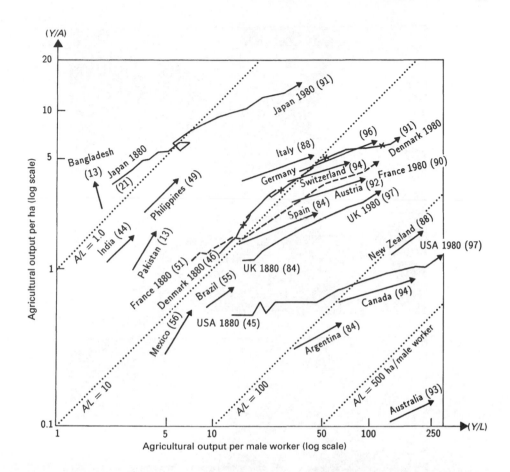

Figure 4. Agricultural output (Y, constant value) per ha land (L) and per agricultural worker (A). Linear vectors indicate changes from 1960 to 1980. Longer trajectories (for Denmark, France, Japan, and the United States) indicate changes from 1880 to 1980. Values in parentheses indicate percent of workforce employed outside agriculture in 1980. Different initial conditions as reflected in the relative availability of the factor inputs land and labor (A/L), and different development paths followed can be grouped into three clusters of trajectories: 'Asian' (e.g., Japan; upper left), 'European' (e.g., Denmark; center), and 'New Continental' (e.g., United States; lower right) (adapted from Hayami and Ruttan, 1985: 121–131).

100,000 ha per agricultural worker, respectively. Thus, the initial conditions and development paths followed in the various regions determine the extent and type of agricultural productivity changes and concomitant changes in land-use patterns. Figure 4 therefore provides yet another illustration of the concept of historical path dependency (Arthur, 1988) developed within the framwork of evolutionary models in economics.

The Importance of Initial Conditions
To understand the large differences between Asia and Europe in agricultural land productivity, we must take an even longer historical perspective. By 1100, China

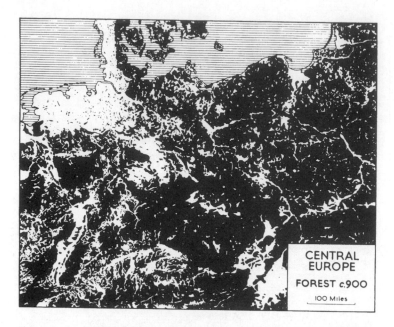

Figure 5. Forests in Europe, A.D. 900 (top) and 1900 (bottom). European deforestation preceded similar transformation processes on other continents in the recent past and even the present (from Darby, 1956: 202–203).

already had a population density of about 25 inhabitants per km² (a population of 100 million people). Europe's population density at that time was less than 7 people per km² with a population of 30 million. Thus Europe could afford a much lower agricultural productivity than China. Typically, European fields did not yield more than three to five (in exceptional harvests, six to seven) times the seed sown (Slicher van Bath, 1963). Although Europe's population increased by over a factor of three to a population density of about 20 inhabitants per km^2 (a population of about 100 million) by the end of the 17th century, agricultural yield per unit of arable land remained modest.

Large areas of virgin forest constituted *the* European resource for increasing agricultural output. Consequently, expanding populations caused large-scale conversion from forests to agricultural areas between the 11th and 15th centuries (Figure 5). By 1400, except in some higher-altitude, swampy, or outlying areas, the original European forest cover had largely disappeared. This large-scale transformation was sometimes halted or reversed by the Black Death or wars, but only temporarily. Throughout this period and also during the Renaissance, the Europeans 'behaved toward their forests in an eminently parasitic and extremely wasteful way' (Cipolla, 1976: 112). Many areas in modern Europe, such as the barren regions of Central Spain and the eroded coastlines of the Adriatic, denuded by the Venetian ship-building industry, testify to the profound land-use changes that accompanied the deforestation of Europe.

By the end of the 17th century, the overall productivity of agricultural land supported just over one person (and two draft animals)[6] per hectare of arable land. Also by this time, the widespread disappearance of forests, particularly in England, resulted in 'timber famines,' with rapidly rising energy (charcoal) prices and many attempts to introduce substitutes (coal). Land seemed finally to become the limiting factor to population growth.

By 1600, China had a population of a similar order of magnitude to Europe at the same time (about 150 million people). However, its agricultural productivity fed 15 people per hectare of cultivated land, which far exceeds even present European land productivity levels.

Although internal colonization opened up additional land areas in China for cultivation at various times (Perkins and Yusuf, 1984), for many centuries agricultural land availability was *the* principal constraint to increases in agricultural output. As a result China developed an agricultural system characterized by labor-intensive, high-intensity rice cultivation with corresponding high yields per area cultivated. In such an agricultural system not only technological innovations are of importance but also social and organizational ones. Wet field rice cultivation required sophisticated civil (terraced fields) and hydraulic (dams, locks, water storage, etc.) engineering techniques to drain and irrigate the land. The scale of these water control projects required elaborate organizational skills. Social organization, in conjunction with an elaborate transport system, enabled effective relief measures in case of food shortages. The related administrative techniques were recorded in legal documents (cf. Yates, 1990).

Agricultural technology was also important. Centuries before Europe the scratch plow was replaced by the iron plow, also adopted for wet field rice cultiva-

---

[6]Inventories of the 16th century in England indicate an average farm size of about 30 sown acres and an average population of 27 draft animals per farm (Langdon, 1986).

Table 3: Global and regional land-use and population change (in million ha and million people, positive or negative sign indicates direction of change)

	1700–1800	1800–1850	1850–1920	1920–1950	1950–1980	Total 1700–1980	% of Global Change	1980 Land Use & Population	% of World
Europe									
Forest	−15	−10	−5	−1	+13	−18	2	212	4
Grassland	−15	−25	−11	−3	+2	−52	–	138	2
Cropland	+30	+35	+15	+5	−15	+70	6	137	9
Population	+53	+63	+105	+79	+92	+392	10	484	11
North America									
Forest	−6	−39	−27	−5	+3	−74	6	942	19
Grassland	0	−1	−103	−22	+1	−125	–	790	12
Cropland	+6	+41	+129	+27	−3	+200	16	203	14
Population	+3	+20	+89	+52	+82	+246	7	248	6
USSR and Oceania									
Forest	−29	−42	−86	−38	−23	−218	19	1187	23
Grassland	+2	+7	−12	−9	−22	−34	–	1673	25
Cropland	+27	+35	+97	+47	+47	+253	20	291	19
Population	+19	+30	+62	+50	+95	+256	7	288	7
Africa & Middle East									
Forest	−11	−15	−68	−96	−118	−308	27	1088	22
Grassland	0	+5	+23	+24	−9	+43	–	2218	33
Cropland	+11	+9	+47	+71	+127	+265	21	329	22
Population	0/+1	+4	+39	+70	+250	+364	10	470	11

Latin America									
Forest	−6	−19	−51	−96	−122	−294	25	1151	23
Grassland	+2	+11	+25	+54	+67	+159	–	767	11
Cropland	+4	+7	+27	+42	+55	+135	11	142	9
Population	+9	+15	+67	+63	+200	+354	9	364	8
Asia									
Forest	−38	−20	−50	−53	−89	−250	22	473	9
Grassland	−1	−8	−11	−12	−31	−63	–	1202	18
Cropland	+38	+29	+61	+65	+120	+313	25	399	27
Population	+195	+171	+216	+372	+1190	+2144	57	2579	58
World									
Forest	−105	−145	−287	−289	−336	−1162	100	5053	100
Grassland	−12	−11	−89	+32	+8	−72	–	6788	100
Cropland	+116	+156	+376	+257	+331	+1236	100	1501	100
Population	+278	+603	+578	+686	+1909	+3755	100	4433	100

Note: Net land conversion may not be equal due to rounding.

Land-use data from Richards, 1990; population data from Demeny, 1990; McEvedy and Jones, 1978.

tion. Seed drills for sowing and many other tools were introduced after the turn of the millennium. The use of a variety of fertilizers (urban refuse, lime, ash) and of insect and pest control measures (e.g., the use of copper sulfates as insecticides) was widespread. Mokyr (1990) highlights yet another feature of Chinese agriculture: the large number of published texts and handbooks, furthering the diffusion of advanced agricultural techniques.

Thus, the long-term historical evolution of agriculture in Europe and China largely explains their different initial conditions at the onset of the Industrial Revolution. These differences in turn determine to a large extent the differences in the development paths followed and the resulting land-use changes that went along with the population growth of the last 300 years. Any analysis of the impacts of technological change on agriculture and the resulting land-use changes has therefore to differentiate between broad categories of agricultural starting conditions and subsequent development trajectories followed.

Agricultural Land-Use Changes

Global Land-Use and Population Changes

Since the Industrial Revolution, it is possible to differentiate three technology clusters that influenced agricultural land productivity and hence land-use patterns. Before discussing these clusters in detail, however, let us consider the magnitude of global land-use and population changes over the last 300 years, summarized in Table 3. The data are of course uncertain, and the numbers should be considered more as indicators of the direction of change than as highly accurate assessments of land use at particular times.

Table 3 indicates the most notable long-term pattern of land-use change since 1700: the large-scale conversion of forests to croplands (see Williams, this volume). Developing countries dominate land-use changes because of their preponderance in population growth. However, the table shows considerable differences among regions with respect to the impacts of population growth on land-use changes. For example, Asia accounts for 57% of the population growth between 1700 and 1980, but it accounts for only 25% of net additions to croplands over this period. On the other extreme, the USSR and Oceania account for only 7% of world population growth but 20% of net additions to croplands.

To compare the land intensiveness of the Asian, European, and New Continental development paths (see Figure 4), we use the data from Table 3 to calculate changes in land use per person of added population for a number of reference periods and regions (Table 4). Such changes help to quantify the impacts of technology on land productivity and the impacts on various regions of entering the international agricultural market (e.g., by highlighting changes due to large-scale export-oriented agricultural production).

As Table 4 shows, for each person added to the world's population since 1700,

Table 4: Land-use change (in ha) per capita population growth

	1700–1800	1800–1850	1850–1920	1920–1950	1950–1980	1700 1980
Europe						
Forests	–0.28	–0.16	–0.05	–0.01	+0.14	–0.05
Cropland	+0.57	+0.56	+0.14	+0.06	–0.16	+0.18
G+C*	+0.28	+0.16	+0.04	+0.03	–0.14	+0.05
North America						
Forests	–2.00	–1.95	–0.30	–0.10	+0.04	–0.30
Cropland	+2.00	+2.05	+1.45	+0.52	–0.04	+0.81
G+C	+2.00	+2.00	+0.29	+0.10	–0.02	+0.30
USSR and Oceania						
Forests	–1.53	–1.40	–1.39	–0.76	–0.24	0.85
Cropland	+1.42	+1.17	+1.56	+0.94	+0.49	+0.99
G+C	+1.53	+1.40	+1.37	+0.76	+0.26	+0.85
Africa and Middle East						
Forests	–11.00?	–3.75	–1.70	–1.37	–0.47	–0.85
Cropland	+11.00?	+2.25	+1.20	+1.01	+0.51	+0.73
G+C	+11.00?	+3.50	+1.80	+1.36	+0.47	+0.85
Latin America						
Forests	–0.67	–1.27	–0.76	–1.52	–0.61	–0.83
Cropland	+0.44	+0.53	+0.40	+0.66	+0.27	+0.38
G+C	+0.67	+1.27	+0.78	+1.52	+0.61	+0.83
Asia						
Forests	–0.19	–0.12	–0.23	–0.14	–0.07	–0.12
Cropland	+0.19	+0.17	+0.28	+0.17	+0.10	+0.15
G+C	+0.19	+0.12	+0.23	+0.14	+0.07	+0.12
World						
Forests	–0.38	–0.24	–0.50	–0.42	–0.18	–0.31
Cropland	+0.42	+0.26	+0.65	+0.37	+0.17	+0.33
G+C	+0.37	+0.24	+0.50	+0.42	+0.18	+0.31

*G+C = grassland and cropland, i.e., including all agricultural area. Note that figures for Africa prior to 1800 are particularly uncertain.

about 0.3 ha of forest has been converted to agricultural land, almost exclusively cropland. However, there is large temporal and spatial variation. For the three development paths described earlier, we assume the following reference values:

Asian	0.2 ha per additional capita
European	0.5 ha
New Continental	1.0–2.0 ha

These values are indicative only. They are typical values from Tables 3 and 4 for before the mid-19th century, i.e., the period prior to large-scale technological changes and international trade in agricultural products. They are yardstick values of what land-use changes would have occurred in the absence of technological advances and external trade. Higher values than these reference values indicate that expansion of agricultural land exceeded population growth; lower values, and especially declining values compared to previous time periods, indicate improving land productivity levels as a result of technological change.

Technology Clusters and Land-Use Changes

Since the Industrial Revolution, three successive technology clusters have influenced changes in agricultural land productivity and the spatial division of agricultural production and hence changing land-use patterns in different regions. Consistent with the larger definition of technology/technique introduced above, we will consider technological/mechanical (tractors, fertilizers, etc.), biological (new crops, high-yield varieties), and social/organizational innovations. As approximate timing, we examine the period until the middle of the 19th century, the period from 1850–70 to the 1930s, and finally from the 1930s to the present. We call these three eras the periods of agricultural innovation, agricultural mercantilism, and agricultural industrialization.

The Period of Agricultural Innovation

The period up to 1850–70 is symbolized by an emerging factory system, particularly in the textile industries; widespread application of stationary steam power; and the development of canals as a new transport infrastructure. At the same time, European agriculture was revolutionized by a combination of biological and organizational innovations, in the form of new crops and new farming practices. None of these innovations was entirely new, all being (at least partly) already in use on a smaller scale in some regions of Europe or imported from the Americas (corn, potatoes). New farming practices and crops were particularly vigorously introduced in England, raising agricultural labor productivity, which enabled drastic shifts in employment patterns toward the newly emerging industries. Particularly important was the widescale introduction of more complex crop rotation patterns in conjunction with new fodder crops (clover and, later, lucerne). This enabled the abandonment of fallow periods and helped to overcome the hitherto limited feed supply for the animal stock (particularly during winter). The better animal husbandry improved fertilizer availability, furthered by imports of guano from Peru (since the 1820s) and later on (after the 1840s) of nitrate from Chile. Grigg (1987) characterizes the new agricultural system as a greater integration of livestock and arable husbandry. Although the new system did not much improve overall land productivity, it enabled the conversion of fallow lands and grasslands to croplands. In fact, Europe appears to be the only region where such conversions (of some estimated 25×10^6 ha in the period 1800–50) took place before the middle of the 19th century, resulting also in a slowdown of the rate of deforestation.

Even more important for feeding larger populations was the introduction of new staple food crops from the Americas: corn (maize) and the potato finally put an end to the frequent famines of Ireland, East Prussia, and some parts of southern Germany. From there the new crops spread fast over the rest of Europe in the 19th century. Additional new crops were tobacco (although of no nutritional value whatsoever) and the sugar beet.

However, it was not only the introduction of new crops and cropping patterns that was important in this time period. Organizational and institutional innovations also played a decisive role. First we must mention the abandonment of peasant serfdom in Europe during the 18th century, as well as a number of land reforms (for an account of Sweden see, e.g., Anderberg, 1991) and the subsequent concentration of farmlands and resulting economies of scale. New fodder crops and the abandonment of fallow lands (used previously for communal pasture) also implied important institutional changes in patterns of land rights and usage. To keep (someone's) grazing animals off (one's) cropland, farmland was becoming increasingly enclosed. In England, between 1760 and 1840, over 6 million acres were redistributed via 'Enclosure Acts,' with the total area being likely even larger (Fussel, 1958). The first horse-powered machines (for threshing) were introduced, but faced opposition (as expressed in the violent Captain Swing movement in England in 1830–31)[7] and diffused slowly. Until the mid-19th century, progress in farming techniques took a similar form in many European countries, but with some time lags (particularly long in France). Yields in England increased slowly from about 16 bushels per acre in the late 16th century to 20–22 bushels 200 years later (Fussel, 1958).

Although land productivity increases were small, the new crops and agricultural practices first introduced in England progressively diffused throughout Europe and enabled considerable increases in food output that sustained population increases between 1700 and 1850, well before industrialization had much effect upon European farming. Although industrialization of agriculture was only modest by the mid-19th century, another development deserves particular attention. In all European countries, centers of agricultural research and education were established by the mid-19th century. In the United States as well, public sector R&D in agriculture became institutionalized with the founding of the U.S. Department of Agriculture in 1862. Institutions and systematic R&D efforts paved the way for even more spectacular future improvements in agriculture through the systematic development of both biological and mechanical agricultural innovations.

The development of textile industries in Europe combined with the income increases of a growing population led to a large demand for cotton and wool, satisfied by imports from abroad. Westward expansion of cotton growing in the United States and, later, more widely in the subtropics, however, appeared to have a large-scale impact on land use only after the 1850s. This is indicated by the relatively modest trade figures (compared to the end of the 19th century) in cotton and wool, which one can derive from available statistics. Therefore, a growing cotton export trade is not the explanation for the fact that land conversion figures in regions outside Europe before 1850 are much higher than would be expected from average

---

[7]For an excellent account of causes, events, and consequences of this manifestation of agricultural 'Luddism' see Hobsbawn and Rudé, 1968.

land productivity figures. Population growth cannot fully explain this phenomenon either. The unexplained residual casts doubts on the estimates of land conversions (and/or population growth estimates), particularly in Africa but to a smaller extent also in North America, that were presented in Tables 3 and 4 above.

The Period of Agricultural Mercantilism

From approximately the mid-19th century to the 1930s, agricultural practices introduced earlier in England and some European countries spread out farther, increasing agricultural productivity in vast peripheral regions of Europe such as Russia. This phase of agricultural expansion was characterized, however, not so much by the diffusion of agricultural techniques as by new developments in transport, manufacturing, and science that accompanied the process of industrialization (Boserup, 1981). These new developments could spread only after the iron and chemical industries were developed and their products became so cheap as to become economical in agriculture. Commercial fertilizer and large-scale imports of food and fodder could not be introduced before a railway network was in place and steamships were widely available. Imports of animal products required refrigeration techniques; increased transport distances for food required new methods of food preservation. It is only after these preconditions were fulfilled that new agricultural methods could be applied and large-scale trade in agricultural products could become possible.

The transport revolution combined tremendous improvements in accessibility with rapidly falling transport costs. This enabled unprecedented regional specialization and the opening of vast new agricultural areas in the Canadian provinces, the American Midwest, the Argentine pampas, the Russian steppes, and the interior of Australia. Thus as the food hinterlands of the industrialized core regions shrank, these regions relied increasingly on external food sources and diversified diets to include products produced only in distant climatic zones.

However, the introduction of mechanical innovations in agriculture was also important, especially for raising labor productivity. Particularly in North America, agricultural labor was scarce relative to land. The introduction of mechanical innovations for stationary applications in agriculture intensified: the mechanical reaper (1831), the transportable threshing machine (1850), and the milking machine (1850), to name just a few examples. However, in the absence of a light, high-output, movable power source (such as the 20th-century tractor), the impact of these innovations, particularly outside North America, remained limited.

Especially important for raising land productivity were the discoveries of man-made fertilizers: superphosphates (invented in 1841 and the only chemical fertilizer of the 19th century), nitric fertilizers (1906), and above all ammonia synthesis for nitrogen fertilizers in 1912 (the Haber-Bosch process). Stimulated by military requirements during World War I, nitrogen fertilizers found widespread application in European agriculture only after the 1920s. Fertilizer, the first pesticides and fungicides, and the breeding of new crop varieties enabled significant expansion of yields per hectare. As a result, land conversions in Europe were reduced to one-half the value (15×10^6 ha additional cropland area) that had prevailed over the previous five decades (cf. Table 3), and land-use changes, especially when compared to population growth (Table 4), were much smaller than would have been expected in the absence of all these technological developments.

New plant varieties were also introduced outside Europe; for instance, new high-yield rice species were introduced in Japan, doubling yields per hectare from 1880 to 1930 (Hayami and Ruttan, 1985). On the whole, however, agricultural land productivity increased mostly in Europe, and this together with large-scale food imports minimized further conversion of forests and grasslands to cropland.

Agricultural land productivity outside Europe did not increase noticeably (with the exception of Japan mentioned above). Particularly in North America, advances in labor productivity were not accompanied by comparable advances in land productivity. As a result, cropland expansion continued vigorously. Wire fencing facilitated conversion from grazing to cropping. An estimated 100 million ha of grassland was converted to cropland in North America from 1850 to 1920 (cf. Table 3, above).

Finally, innovations in food preservation also proved important for agriculture during this time period: tin cans, concentrated milk, and especially refrigeration. (Absorption refrigeration was invented in 1850, and ammonia compression refrigeration in 1876.) Refrigerated steamships enabled the import to Europe of meat from as far away as Australia, New Zealand, and Argentina. All of these developments together, with the drastically decreasing transport costs of the railway and steamship era, enabled an unprecedented expansion of trade in agricultural products. By the 1870s, net imports of agricultural products exceeded the net export value of manufactured goods of the leading economic power, England (Woytinsky, 1927). World trade in agricultural products doubled between the 1870s and 1913. Hence we use the term agricultural mercantilism to characterize this development phase.

What was the impact on land-use changes of this large-scale development of trade in agricultural products? Unfortunately statistical records are scarce, but we have tried in Table 5 to assemble some zero-order estimates of land areas used for export crop production in the mid-1920s. Between about 20 and 50% of new cropland areas in regions outside Europe served export crop production. As the trade in agricultural products by the mid-19th century was rather modest,[8] we can infer rather confidently that nearly all of these areas represent net land-use changes over the period 1850 to 1925. Our crude estimates indicate that about 20% of land-use changes in North America and USSR and Oceania in the 1850 to 1925 period are related to export crop production. The proportion in Asia (excluding China) is estimated as 30%, whereas in Latin America up to one-half of land-use changes can be related to export crop production. In absolute terms North America dominates with some estimated 25 million ha converted to cropland for export (cotton and grains), followed by Asia (mostly India) with some 20 million ha, and USSR/Oceania and Latin America with around 15 million ha each. The available trade statistics

---

[8]Exceptions are, e.g., cotton exports from the United States and Egypt as well as trade in sugar. Areas producing export crops by the 1850s are subtracted from the land-use change figures of the 1850–1925 period given in Table 5.

Table 5: Expansion of cropland[1] for export crop production (zero-order estimates), 1850–1925 ($\times$ 10^6 ha)

	Products				As Percent of Increase in Cropland Area, 1850–1920 (Table 3)
	Luxury[2]	Grain[3]	Industrial Raw Materials[4]	Total	
North America	0.2	17.0	7.7	24.9	19%
USSR and Oceania	–	16.3	–	16.3	17%
Africa	0.2	–	>0.8	>>1.0	?
Latin America	3.5	10.7	>0.1	>14.3	53%
Asia (excl. China)	2.5	5.0	>11.8	>19.3	32%
Total (5 regions)	6.4	49.0	20.4	75.8	21%

[1]Cropland areas in proportion of exports in total production of 15 agricultural commodities by mid-1920s. 1850–1925 expansion assumes world agricultural trade in 1850 was negligible. Exports of cotton (U.S., Egypt, India), wheat (Russia), and sugar (Caribbean) by 1850 are taken into account in the calculations (data from Woytinsky, 1926: 109–220, 265–312).
[2]Sugar (cane), tea, coffee, tobacco.
[3]Barley, corn, oats, rice, rye, wheat.
[4]Cotton, flax, hemp, jute, rubber.

(Woytinsky, 1926; Mitchell, 1982) indicate that export of food and agricultural raw materials from Africa was comparatively modest. This leaves some doubts about the much larger land conversions estimated to have taken place in Africa over the 1850–1920 time period (cf. Table 3 above) than could be expected from the rates of population growth and their additional cropland requirements.

Taking the above land-use changes for export-oriented production into account, the marginal land-use changes of Table 4 are reduced to values of about 1.2 ha cropland expansion per person of additional population in North America, the USSR, and Oceania, and to some 0.2 ha per additional capita in Asia. These results agree well with the estimates given above for the New Continental and Asian development paths under *ceteris paribus* conditions, i.e., in absence of the impacts of technological change and export crop production.

The Period of Agricultural Industrialization

Over the 50 years between the 1930s and the present, world agriculture was transformed from a resource-based to a technology-based industry. Although technology embodied in new farming techniques, new plant varieties, manmade factor inputs, machinery, and equipment is crucial, it is not by itself the primary source of change. Rather, the transformation in agriculture was made possible by a series of institutional innovations furthering the development and diffusion of agricultural technology. Examples include the emergence of public and private sector suppliers of new plant varieties and agricultural technology, institutions and services for transfer of technical knowledge to farmers, public and private sector R&D, and

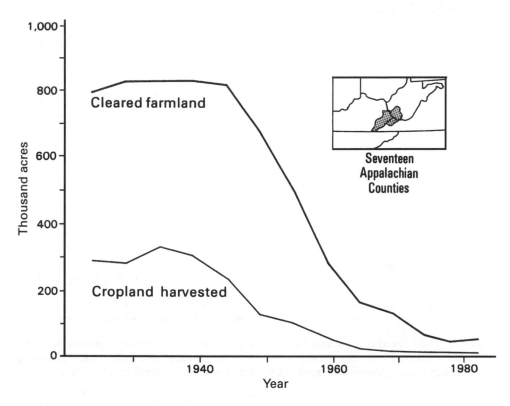

Figure 6. Decrease in cleared farmland and harvested cropland area since the 1930s in 17 counties in Appalachia (United States). Agricultural productivity increases not only have vastly raised output (and surpluses) but also have resulted in significant reconversion of agricultural land to grassland and forests, particularly in North America and Europe (from Hart, 1991: 66; permission of W.W. Norton & Company, Inc., copyright © 1991 by the Commonwealth Fund Book Program).

input supply and marketing organizations, and the development of more efficient labor, credit, and commodity markets. Although we focus below on a quantitative account of some of the most important changes in agricultural techniques and artifacts, the importance of institutions and changing attitudes toward modernization and industrialization of agriculture deserve particular attention.

Agricultural industrialization is characterized by three developments: biological innovations, new cheap factor inputs, and mechanization. All of these areas, mutually interdependent and reinforcing each other, have resulted in spectacular increases in agricultural labor productivity but also, for the first time since the Industrial Revolution, have raised agricultural land productivity throughout the world, i.e., in developing as well as developed countries. The expansion of agricultural land uses became progressively decoupled from the rate of population growth, leading in highly industrialized countries to reconversion of cropland areas to grassland and forest cover (Figure 6).

As the first area of the industrialization of agriculture, we note the introduction of new crops and wide diffusion of new high-yield plant varieties developed

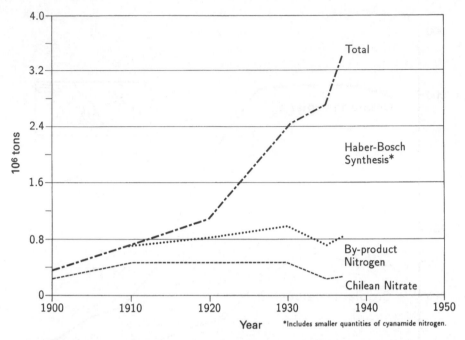

Figure 7. World nitrogen production by process, 1900–38, in million tons (cumulative totals). Ammonia synthesis (by the Haber-Bosch process) was the most important technological innovation for expanding the nitrogen fertilizer supply (derived from data in Zimmermann, 1951: 789).

through systematic agricultural R&D. These are symbolized by, for example, new hybrid corn and rice varieties, perhaps the most important contribution of applied biology in the 20th century. New plant species increased yields, while the further diffusion of crops between continents opened new export markets (e.g., soybeans over the last 30 years in the United States and even more recently in Brazil) or improved and diversified local diets. In the 20th century maize and manioc have become important food supplements in Africa, whereas sweet potatoes, maize, and peanuts have started to diversify rice and wheat diets in Asia.

Industrialization of factor inputs to agriculture in the forms of commercial energy, manmade fertilizers, and pest control substances alleviated most constraints for raising agricultural output. Fertilizer output was no longer dependent on animal production or naturally occurring deposits. Even prior to World War II, ammonia synthesis-based nitrogen fertilizer accounted for over 80% of the global fertilizer output and displaced Chilean nitrate and byproduct nitrogen from coke production (Figure 7). Nitrogen fertilizer output increased globally to close to 80 million tons, with increasing shares for Eastern Europe, the USSR, and especially the developing countries. Total fertilizer application per ha of cropland consequently increased throughout the world, and today it shows, with the exception of Europe and Africa (being significantly above and below the world average, respectively), comparatively small regional disparities (Figure 8).

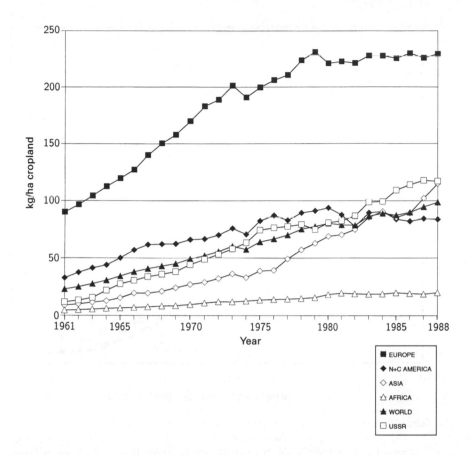

Figure 8. World fertilizer use per ha cropland, 1961–88. Fertilizer application in different world regions (with the exception of Europe and Africa, which are significantly above and below the world average, respectively) has been converging (derived from FAO statistics, various volumes; courtesy of G. Heilig, International Institute for Applied Systems Analysis).

Mechanization, symbolized by the farm tractor, is perhaps the most visible representation of agricultural industrialization. The substitution of inanimate power (and fossil energy) for animal and human power (Figure 9)[9] alleviated yet another constraint for increases in agricultural output: labor.

Mechanization also made available large areas for cropping use that were formerly required to feed working animals. For instance, in the United States, the area required for feeding farm horses and mules amounted to nearly 40 million ha in the 1920s (US

[9]Note that Figure 9 just shows fossil energy inputs to agriculture. Taking into account nonfossil energy consumption (energy from work animals, wind and water power, and fuelwood), estimated to have peaked at around 5×10^{18} J in the 1920s (Fisher, 1974), total energy consumption in U.S. agriculture increased by only about one-third (from about 6×10^{18} J in the 1920s to over 8×10^{18} J in the 1970s), whereas total output more than doubled over the same time period (Hayami and Ruttan, 1985). The resulting improvements in energy consumption per unit of output achieved are the result of better end-use efficiencies of industrial power sources fueled by fossil energy. Recall here that a horse typically converts only 3% of the energy embodied in feed to useful work (kinetic energy) compared to a 30% energy efficiency (kinetic/diesel energy ratio) of a farm tractor.

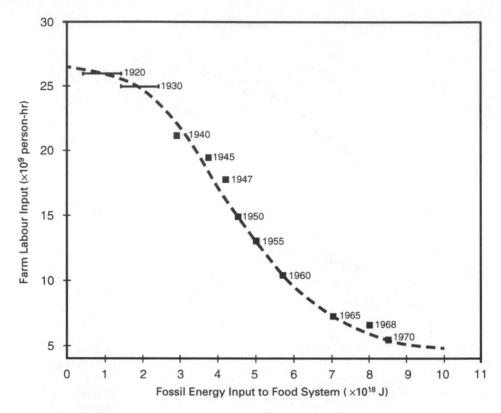

Figure 9. Substitution of human labor in agriculture by mechanization and (fossil) energy inputs, United States, 1920–70. Note that total energy input (including human and animal energy) to U.S. agriculture increased only slightly over this period despite more than a doubling of output. Mechanization does not necessarily imply increasing energy intensity of agriculture due to the much higher energy conversion of commercial energy applications than that of humans and work animals (adapted from Steinhart and Steinhart, 1974: 51; © AAAS).

DOC, 1975), twice as large as the areas devoted to export products and about one-half of the cropland areas used for domestic production. The replacement of farm horses and mules by the tractor thus also minimized further land conversions.

Because of the scarcity of labor relative to other factor inputs in North America, mechanization started in the United States (where the ground was additionally prepared by the 'horse mechanization' over the previous decades) and in some European countries. But since World War II, mechanization has spread to other regions (Figure 10). The mechanization in agriculture is best illustrated by the increasing number of tractors in use worldwide, presently over 26 million. Over the last 20 years, the share of developing countries in the global number of farm tractors has been rising rapidly.

Industrialization resulted not only in new demands for agricultural raw materials but also in the substitution of many raw materials produced by agriculture by manmade products. For instance, with the development of the electrical engineering industry and later motor vehicles, rubber moved from a minor curiosity to a

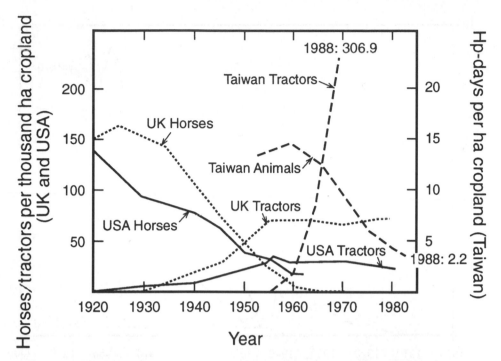

Figure 10. Displacement of animal labor in farming: United Kingdom, United States (number of horses and tractors per ha cropland), and Taiwan (horsepower-days per ha cropland). Mechanization increased energy power available on farms and freed large land areas for agricultural production that were previously required to feed working animals (e.g., 40 million ha in the United States during the 1920s) (adapted from Grigg, 1982: 133, and Jones, 1991: 626).

major raw material; in the early 20th century natural rubber production expanded prodigiously in Southeast Asia, and after World War II synthetic rubber production increased dramatically worldwide.

Production of natural rubber rose rapidly to a level of about 1 million tons in the 1930s, with over 90% of this production concentrated in Southeast Asia. Rubber plantations extended over some 5.6 million hectares in Asia in the 1930s, about equally split between large estates and small holdings (Woytinsky and Woytinsky, 1953). In the late 1980s world rubber production exceeded 14 million tons. It is easy to imagine the land-use impact of this 14-fold increase in rubber production if it were based only on plantation rubber. Fortunately, the actual impact was much smaller due to the introduction of synthetic rubber (yet another outgrowth of the developments in petrochemical industries). Currently two-thirds of the world's rubber output are in the form of synthetic rubber, with a smaller additional quantity of recycled rubber (Figure 11).

As a result of all these developments, output increases could keep abreast of population growth at a global level, and productivity in some regions rose to such levels as to enable large-scale reconversion of marginal agricultural lands to forestry (as in Europe and North America; see Figure 6 and Tables 3 and 4), while still allowing high output levels and even large and costly agricultural surpluses

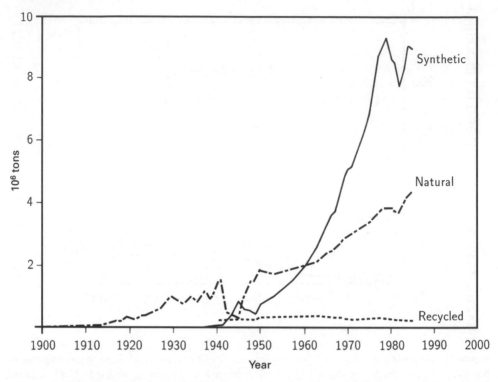

Figure 11. World rubber production (million tons): synthetic, recycled, and natural. Trends in the production of textile fibers have been similar (derived from data in Woytinsky and Woytinsky, 1953: 621–623; and U.N. *Statistical Yearbook*, various volumes).

with the attendant political embarrassments. Agricultural policies in the OECD countries resulted in a total subsidy of agricultural production of about $300 billion U.S. in 1990 (Viatte and Cahel, 1991)—a figure comparable to the total value of the world crude oil trade. This amount was about equally split between direct producer subsidies and transfers away from consumers (considering that consumers have to pay more than the world market prices for agricultural products). On the extreme end (as in Switzerland, Norway, and Japan), subsidies to agricultural producers equal about three-quarters of the value of agricultural output. In addition to vastly raised agricultural output, Western diets have become dominated by the consumption of animal products, and in most Western countries livestock products account for up to two-thirds of the value of output (Grigg, 1987).

Agricultural output and land productivity also increased outside the OECD countries. In the developing world, land productivity grew about 1% annually from 1950 to 1980. Although population in these countries has grown twice as fast (2.1%/yr), this land productivity increase remains a formidable achievement of the Green Revolution. Cropland expansion per person of additional population dropped to a level of about 0.1 ha per capita in Asia, 0.3 ha in Latin America, and to some 0.5 ha in Africa, the USSR, and Oceania. Without the productivity increases of an industrializing agriculture, the cropland area outside Europe and

Table 6: Agriculture: Land, people, and technology, A.D. 1990

	Europe	USSR	North America	JANZ[1]	China	Asia	Africa	Latin America	World
Population (10^6)	497.7	288.0	275.7	143.6	1135.5	1855.9	647.5	448.3	5292.2
Arable land (10^6 ha)	140.1	232.4	235.9	52.2	96.6	351.7	186.7	179.8	1475.4
Irrigated area (10^6 ha)	17.3	20.8	18.9	5.0	44.9	94.9	11.2	15.6	228.7
Farm tractors (10^6)	10.3	2.7	5.4	2.4	0.9	2.2	0.6	1.4	25.9
Fertilizer use (10^6 t)	31.9	26.5	19.9	3.6	18.9	21.4	3.5	8.4	134.1
Food supply (10^9 cal)	1723.9	976.6	994.6	414.3	2946.0	4222.2	1421.0	1204.4	13903.0
Arable land-use intensity:									
people/km^2	355	124	117	275	1175	528	347	249	359
fraction irrigated	0.12	0.09	0.08	0.10	0.47	0.27	0.06	0.09	0.16
tractors/km^2	7.4	1.2	2.3	4.6	0.9	0.6	0.3	0.8	1.8
tons fertilizer/km^2	22.8	11.4	8.4	6.9	19.6	6.1	1.9	4.7	9.1
food output/km^2 (10^6 cal)	1.23	.42	0.42	0.79	3.05	1.20	0.76	0.67	0.94
10^3 cal per person	3.5	3.4	3.6	2.9	2.6	2.3	2.2	2.7	2.6

[1]Japan, Australia, New Zealand

Data from *FAO Production Yearbook*, various volumes.

North America would have had to be expanded by close to 400 million ha above the actual estimated increase of 350 million ha from 1950 to 1980.

Table 6 summarizes the current state of agriculture in selected world regions. Despite large regional variations, industrial innovations have diffused into agriculture on a global scale. Land productivity in terms of food calories per arable ha still shows large disparities among regions as a result of differences in output mix, intensity of cultivation, fertilization, and mechanization. The picture that emerges is that in many regions food production per ha of arable land could be intensified, producing sufficient food for ever-increasing populations.

An open question remains whether in future it will be possible to accelerate agricultural land productivity growth in developing countries to keep pace with population increases. The history of Europe and North America illustrates the potential that technology holds to fulfill such an objective. However, what kind of technologies will be applied, and to what extent, will largely be a function of the economic and social policies adopted. These policies will also have to address the problem of how to solve the large number of constraints (most notably capital shortages) and environmental impacts associated with raising agricultural output and further increasing agricultural land productivity.

Summary

Since the onset of the Industrial Revolution, technological change has been instrumental in raising agricultural productivity. Despite all distributional problems, global agricultural production has kept pace with population growth. Although the

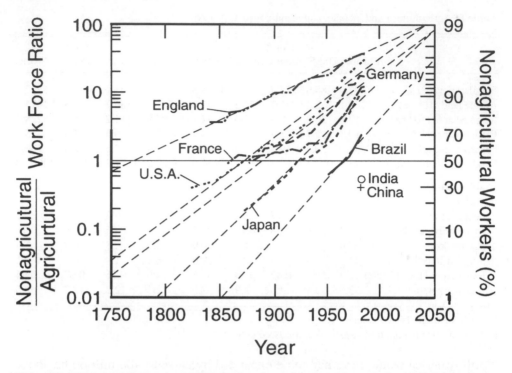

Figure 12. Moving away from agriculture: ratio of nonagricultural to agricultural workforce (logarithmic scale). Countries are converging toward a level at which only a few percent of the work force is employed in agriculture. The rise of urbanization (cf. Figure 13 below) is a perfect mirror image of this pervasive structural transformation (derived from data in Mitchell, 1980: 161–173; Mitchell, 1982: 84–93; and Mitchell, 1983: 150–160).

productivity of land has increased in many developing countries at a slower rate than population, resulting in net expansion of agricultural land uses, land-use transformations would have been significantly (i.e., twice) larger over the last 30 years without technological change. Above all, technology has vastly increased agricultural labor productivity, freeing people from the land to pursue other economic activities.

A further consequence of the industrialization of agriculture was that farming ceased to provide all its own inputs. Farming evolved from a vertically integrated activity (a farmer producing his own inputs like seeds, manure for fertilizer, and livestock for traction power, and also storing and marketing his own production) to horizontal integration with increasing specialization. This shift from vertical to horizontal integration in agriculture also puts a somewhat different perspective on the dramatic shifts in employment patterns away from agriculture. Many activities that previously were performed within the agricultural sector are now performed in the industry and service sectors. Jobs on the farm have moved to industrial manufacturing plants producing seeds, fertilizer, and farm machinery; to food processing industries; and to the service sector (e.g., food retail and restaurants).

With these qualifications in mind, Figure 12 summarizes this transformation in

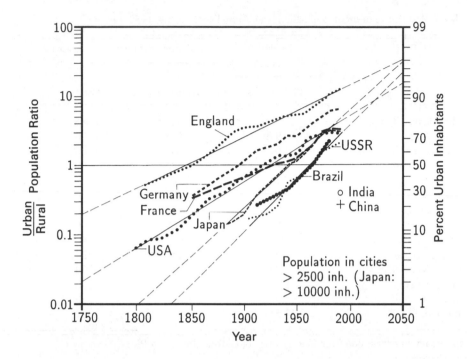

Figure 13. Moving into cities: ratio of urban to rural population (logarithmic scale). 'City' is defined as having more than 2500 inhabitants (in Japan, more than 10,000 inhabitants). German data are for the Federal Republic of Germany alone after 1945 (derived from data in Flora, 1975: 27–56, and U.N. *Statistical Yearbook*, various volumes).

the employment structure away from agriculture. Compared to Figure 3 above (which uses total population per agricultural work force as a productivity indicator), we analyze the ratio of the nonagricultural to the agricultural work force. Plotted on a logarithmic scale, the long-term convergence (although with some lagged developments, as in France) in the employment structure of industrialized countries becomes apparent. The few long-term data we have been able to assemble for developing countries indicate a similar secular trend.

Technological change has thus raised agricultural productivity and permitted an increasing share of the growing rural population to transfer to urban employment, a development most painfully felt today in many rapidly growing megacities of the developing world.

The Urbanization Drive

Urbanization Trends: Catch-Up and Convergence

As we examined the transition from agricultural to nonagricultural employment in Figure 12, let us analyze the shift from rural to urban residence (Figure 13). Despite

317

Table 7: Percentage of urban populations in informal settlements, 1980

	Total Population	Population in Informal Settlements[1]	
	(Thousands)	(Thousands)	(Percent)
Addis Ababa, Ethiopia	1,668	1,418	85
Luanda, Angola	959	671	70
Dar es Salaam, Tanzania	1,075	645	60
Bogota, Colombia	5,493	3,241	59
Ankara, Turkey	2,164	1,104	51
Lusaka, Zambia	791	396	50
Tunis, Tunisia	1,046	471	45
Manila, Philippines	5,664	2,666	40
Mexico City, Mexico	15,032	6,013	40
Karachi, Pakistan	5,005	1,852	37
Caracas, Venezuela	3,093	1,052	34
Nairobi, Kenya	1,275	421	33
Lima, Peru	4,682	1,545	33
São Paulo, Brazil	13,541	4,333	32

Reprinted with permission from WRI, 1991: 76; Oxford University Press.
[1]I.e., shantytowns.

some data consistency problems,[10] the picture is quite consistent and—more noteworthy—also converging in the countries sampled. The similar dynamics in the two structural shift processes (away from agricultural employment and toward cities) point to their close relationship, but with a clear temporal sequence: the shift away from agricultural employment preceded the transition to urban populations in all industrialized countries. On the other hand, in countries that are undergoing the transition in this century, such as the USSR and Brazil, the processes appear synchronized.

The move toward urbanization displays the same dynamic development patterns that were discussed above in other technological and economic structural change processes: a certain convergence in the rates of change and spatial heterogeneity as a function of the time since this transition process was initiated. For example, England has a higher urbanization ratio than Germany or the United States, where this process took off later. In industrialized countries the future growth of urban populations will be comparatively modest since their population growth rates are low and over 80% of their population already lives in urban areas. Conversely, developing countries are in the middle of the transition process, when growth rates are highest. The exceptional growth of urban agglomerations in many

---

[10]The city size definition in Japan is much larger than in other countries due to the absence of more disaggregated statistics. Although this leads to an underestimation of the degree of urbanization in Japan compared to other countries, it should not affect the analysis of the dynamics of this process as reflected in the slope of the curve in Figure 13.

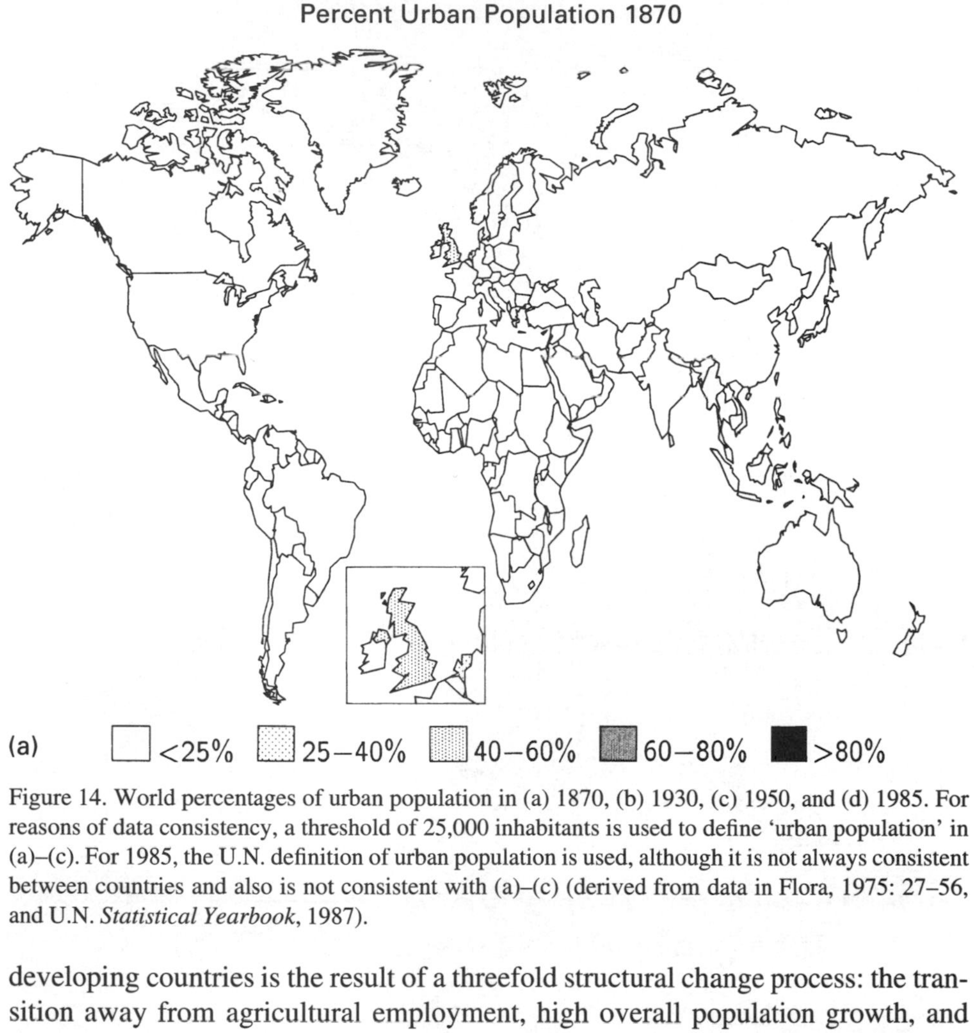

Percent Urban Population 1870

(a) ☐ <25% ░ 25−40% ▒ 40−60% ▓ 60−80% ■ >80%

Figure 14. World percentages of urban population in (a) 1870, (b) 1930, (c) 1950, and (d) 1985. For reasons of data consistency, a threshold of 25,000 inhabitants is used to define 'urban population' in (a)–(c). For 1985, the U.N. definition of urban population is used, although it is not always consistent between countries and also is not consistent with (a)–(c) (derived from data in Flora, 1975: 27–56, and U.N. *Statistical Yearbook*, 1987).

developing countries is the result of a threefold structural change process: the transition away from agricultural employment, high overall population growth, and increasing urbanization rates. Perhaps this is the biggest challenge for technology in the 21st century: how to provide adequate housing, sanitation and health, and transportation services in a habitable urban environment in developing countries. The need for improvement is certainly large, as some estimates of populations in shantytowns (6 million in Mexico City alone) illustrate (Table 7).

With respect to the role of technology in urbanization, Berry (1990) has illustrated anew the linkage between transport infrastructure development cycles and spurts in urbanization in the United States. Increasing the accessibility of cities by improving transport infrastructures (cf. Figure 1, for example) can be considered as the prerequisite to the spread of urbanization from a few countries in the Northern Hemisphere to a global phenomenon. Urbanization ratios at four moments in time are reported in Figure 14. The parallel developments of transport infrastructures—treated in detail elsewhere (Grübler, 1990)—will not be reported here. However, the close relationship between high rates of urbanization and trans-

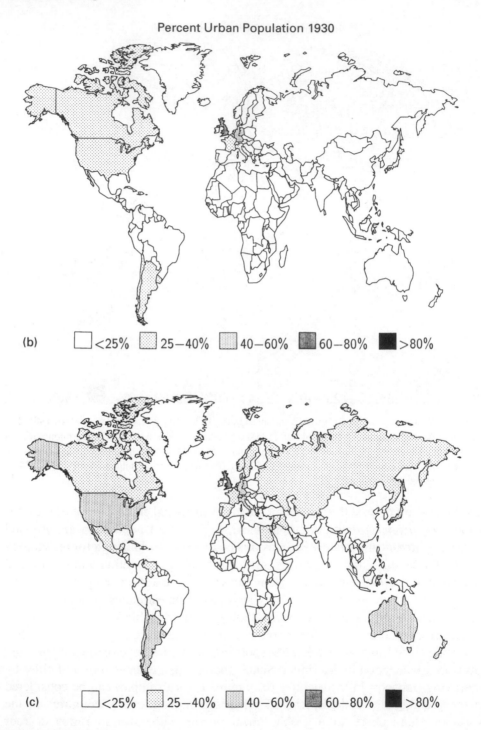

Figure 14. For legend see p. 319.

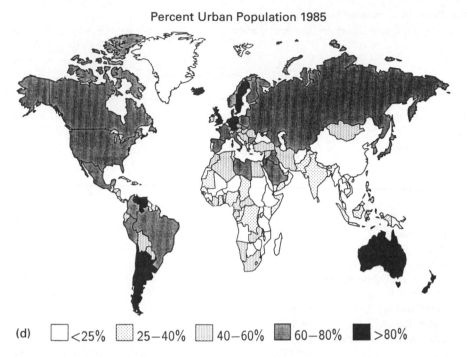

Percent Urban Population 1985

(d) □ <25% ▧ 25–40% ▨ 40–60% ▩ 60–80% ■ >80%

Figure 14. For legend see p. 319.

port infrastructural development cycles, e.g., in the dominance of railway construction in Europe and North America and their related high urbanization rates by 1930, as discussed above, is apparent. Current high transport intensities with either individual vehicles (cars) or public ones (buses and aircraft) based on internal combustion engines also correlate highly with urbanization ratios (Figure 14d), despite wide regional variations. In analyzing land-use patterns it is therefore important to note that land requirements for roads and other transport infrastructure are more a function of population density than of country size, since short to medium-distance trips are the largest part of total travel demand. Total land requirements for transport infrastructures, although small in comparison with agricultural land uses, are significant in relation to other built-up land areas (cf. the following section) and constitute a major human impact on terrestrial and atmospheric environments. Technology could offer vast improvements, either by making current transport technologies more efficient and environmentally benign, or by introducing new high-quality transit systems, such as urban metros or high-speed rail or magnetic levitation trains for intercity transport.

Urban Land Use

Although land-use patterns and their changes both historically and currently are dominated by the impacts of agriculture, let us conclude this section by investigating other land uses, specifically, the ones associated with our technological civi-

Table 8: Green vs. built-up land in densely populated areas

	Bangladesh	Netherlands	Japan	Japan: 3 Largest Metropolitan Areas	Austria: City of Vienna
Population density (people/km^2)	766	388	322	1411	3925
Land use (%)					
(1) Rivers and lakes	9.6	9.0	3.5	3.6	3.4
(2) Forests	13.6	9.7	67.0	52.7	24.6
(3) Grassland	4.2	31.2	1.1	0	
(4) Cultivated land	64.4	22.6	14.3	16.0	26.5
(5) Parks and recreational areas	n.a.	6.3	n.a.	n.a.	9.4
(6) Subtotal (3–5)	68.6	60.1	15.4	16.0	35.9
(7) Infrastructures[2]	n.a.	1.8[3]	0.5	2.8	12.3
(8) Residential buildings	n.a.	7.4	1.9	0.4	14.2[4]
(9) Industry and commerce	n.a.	1.3	0.3	1.1	4.3
(10) Office and public buildings	n.a.	n.a.	0.9	1.1	2.1
(11) Subtotal (8–10)	n.a.	8.7	3.1	3.9	20.6[4]
(12) Built-up land (7+11)	n.a.	10.5	3.6	6.7	32.9[4]
(13) Other uses	8.2	10.8	1.1	7.4	3.9
Per capita land use, m^2/capita					
Forests (2)	170	244	2049	373	63
Green areas (3–5)	854	1513	471	114	91
Infrastructures (7)	n.a.	45[3]	86	37	31
Building area (11)	n.a.	218	121	211	53

[1]Tokyo, Osaka, and Nagoya.
[2]Roads, railways, airports.
[3]Only roads.
[4]Includes private gardens and parks.
Data: Bangladesh: FAO, 1991: 52; Netherlands: van Lier, 1991: 386; Japan: Japan Statistics Bureau, 1987: 7; Vienna: OIR, 1972: I–XXIII

lization beyond agriculture (Table 8).

Leaving aside small islands and city states (such as Hong Kong, with 5400 inhabitants per km^2), the countries with the highest population density in the world are (in decreasing order): Bangladesh, South Korea, the Netherlands, Japan, and Belgium. All of them have population densities of above 300 inhabitants per km^2. Land-use patterns in these countries are of particular interest either because of their high population pressure (Bangladesh) or because of their high population density in combination with a long history and high degree of industrialization (the Netherlands). Japanese statistics (Japan Statistics Bureau, 1987) allow us to investigate the land-use patterns in three metropolitan areas. For comparison, Table 8 also gives land-use patterns in Vienna, Austria, where population densities are obviously much higher (around 4000 inhabitants per km^2) than in the larger administrative regional or national divisions on which aggregate land-use statistics are usually available.

Perhaps the most surprising fact emerging from Table 8 is that even in the most densely populated countries of the world, the dominant land use (typically well above 90%) is still seminatural (forests) and managed ecosystems (water bodies and cultivated land). Built-up areas (which of course are not completely covered with manmade structures) do not account for more than 10% of the land use even in the Netherlands, with its high population density, long industrialization history, and high levels of economic activities. Another surprising aspect is that, even in cities and metropolitan areas, high population densities do not apparently preclude that between 25 and 50% of land is still covered by forests.

Built-up areas (land for buildings and infrastructures) at higher levels of spatial aggregation range typically between 200 and 250 m^2 per capita. The value for the city of Vienna (84 m^2 per capita) is smaller, indicating that the actual ratio of area covered by manmade structures to the total built-up area most likely does not exceed 30–40%. Of the built-up areas, the land requirements for infrastructures are considerable, ranging from 14 to 17% of the national average (Japan and the Netherlands) and from 37 to 42% in urban agglomerations (Vienna and the three largest metropolitan areas in Japan). Land requirements for moving people and goods in densely populated areas thus rival those for housing and for industrial and commercial activities.

Using a value of 250 m^2 per capita for built-up areas and less than 100 m^2 for areas actually covered by manmade structures, the global population of 5.3 billion people is associated with the use of 130 million ha built-up land areas, i.e., only about 1% of the land area of this planet. Actual manmade structures—the physical manifestation of our technological age—most likely do not cover more than 0.4% of the land areas of planet earth.

Conclusion

Land-use patterns have changed over millennia, but the most dramatic transformations have taken place over the last 300 years. With increasing population growth, land transformation patterns accelerated. We have shown that agriculture dominates both land-use patterns and their dynamic transformations. Conversely, the areas covered by artifacts of our technological civilization are small, covering less than 1% of the land area of the earth.

There are many underlying forces of change in land-use patterns: population growth, the rise of an urban society, changes in structures of demand brought about by higher incomes, and the increasing international division of agricultural production, among others. Technological change has been instrumental in these long-term transformation processes. Increases in agricultural productivity, the spatial division of labor, and the accessibility of even the most remote geographical areas were made possible by a succession of technology clusters, which spread first to a limited number of countries but now are global phenomena. We have identified particular transport technologies and infrastructures as appropriate metaphors of the spatial diffusion and intensity of development of particular technology clusters. However, in regard to land-use changes, their effect on agricultural productivity and the spatial division of agricultural production is paramount.

Changes in technology enabled ever larger populations to be supported at higher levels of affluence and consumption (though these are extremely unevenly distributed throughout the world). Technological change has been instrumental in raising agricultural productivity, pushing further away Malthusian 'limits to growth.' In turn, demographic and social changes have also induced technological change. Demographics and technology have therefore to be at the core of any analysis of land-use changes, past and future.

From 1700 to 1980, globally about 1.2 billion ha of forests were converted to arable land. This corresponds to a value of about 0.3 ha per capita additional world population over this time horizon. Large regional variations exist in the relationship between population growth and expansion of arable land due to the specifics of Asian, European, and New Continental systems of agriculture and their respective paths of agricultural productivity increases. Common to them all is that the rate of land-use change compared to population growth has been significantly reduced due to agricultural (land) productivity increases. Technological change has therefore significantly helped to decouple the expansion of agricultural land from population growth. In regions such as Europe and North America, the pervasive adoption of mechanization and of high-productivity agricultural techniques enabled the reconversion of agricultural areas to forests, while at the same time increasing agricultural production (and surpluses) for a rising number of consumers. About 16 million ha of agricultural land has been reconverted to forests since 1950 in Europe and North America, while at the same time population has increased by some 170 million people.

A succession of 'transport revolutions' has allowed people to overcome ever larger distances at lower costs. Increasing spatial division of agricultural production and worldwide trade in agricultural commodities for food and raw materials have been an additional cause of land-use changes. For instance, crude estimates indicate that typically between 20 and 30% of the expansion of arable land outside Europe and China between the 1850s and the 1920s was devoted to export crop production. In some regions and countries (e.g., Latin America and India) this value is likely to have been even higher.

Perhaps the most pervasive impact of technology since the onset of the Industrial Revolution was the tremendous increase in agricultural labor productivity. In industrialized countries today only a few percent of the population is required to supply food for all, compared to 70–80% some 300 years ago. The increases in employment in other sectors of the economy, such as manufacturing and services, combined with tremendous productivity increases enabled the expansion of industrial output and increasing levels of personal consumption and affluence. Changes in agriculture, industry, and urbanization were enabled by the pervasive adoption of new technologies.

Projections indicate an increase of world population by some additional 5 billion people by the second half of the 21st century. If this population increase were accompanied by the magnitude of land conversions that prevailed in the USSR and Oceania over the period 1700–1980, the world's agricultural area would have to expand by some 5 billion ha, equivalent to the total area covered by forests worldwide in 1980. Fortunately, the extent of land-use changes associated with future population growth will be much smaller. Deforestation and land conversion to agriculture per capita additional population in developing countries dropped to 0.2 ha per person in 1950–80. If this figure is multiplied by 5 billion additional world population, arable land use would increase by 1 billion ha, or 20% of the 1980 world forest area. Lowering such figures in the future will to a large degree be dependent on the technologies and agricultural practices adopted to ensure the food supply of future generations.

Malthus considered advances in agricultural productivity to be unlikely to keep pace with the rate of population growth. Consequently, he believed that agriculture and, in particular, land availability would constitute the ultimate constraint to population growth. On the other hand, Boserup (1981) sees increasing population density as a motivation for the development and adoption (diffusion) of more productive technology and social organization, which in turn would allow for increased population and/or rising living standards. The long-term history of population and agricultural productivity increases discussed here clearly supports a Boserupian viewpoint rather than a Malthusian one. As such it perhaps best illustrates the pervasive impacts of the dynamics of technological change. Therefore, the question of what the ultimate carrying capacity of Planet Earth may be (10, 30, or even 1000 billion people, as provocatively argued by Marchetti, 1978) is not the issue. The real question is whether humankind possesses and/or will develop appropriate technologies to feed, house, and employ whatever level of global population will materialize in the 21st century in an adequate, equitable, and environmentally compatible manner.

'Technology' has to be considered in a larger context as comprising not only manmade artifacts, ranging from simple tools to complex technological systems, but also the required knowledge base for the inception, production, and use of artifacts. It is also the social, institutional, and organizational know-how and *techniques* that steer the inception and diffusion of individual or whole clusters of artifacts. Finally, technologies cannot be considered separately; the growth of individual technologies depends on many other technological solutions, giving rise to technology clusters. Their diffusion also depends on a mediating social and institutional framework, ultimately forming (time-specific) regimes of economic expansion which we have referred to as techno-economic paradigms.

Although we consider technology a prime agent of change, it is by itself not the primary cause. Technologies evolve out of social and economic contexts. Policies, institutions, and the social and economic environment shape to a large extent the inception and selection of technologies. The social and economic environment also determines the growth and diffusion (or rejection) of particular (combinations of) technological solutions. However, technology in turn also contributes to create and to shape the social and economic context out of which it has evolved. In this intricate (inter)relationship it appears impossible to arrive at a simple answer to the question of what is the primary driving force of global change. Perhaps it is best to conceptualize technology as a mediator between society at large and its natural environment.

From such a perspective, changes in the technologies we use and in the social and economic context out of which technologies evolve appear necessary. In fact, some[11] argue that we may already be undergoing a transition to a new, environmentally more compatible techno-economic paradigm and a changing social awareness toward environmental change. Although sometimes disruptive, the succession from one dominant techno-economic paradigm to a new one proves essential from a historical perspective for productivity increases and for mitigation of adverse social and environmental impacts associated with the pervasive adoption of particular technological regimes, objectives that present and future technology should aim to fulfill better than in the past.

Acknowledgments

I thank the participants of the OIES 1991 Global Change Institute, in particular Joel Tarr and Vernon Ruttan, for many useful comments and suggestions. Helpful comments by Nebojsa Nakicenovic and editorial assistance by Joanta Green and Marc Clark at the International Institute for Applied Systems Analysis are also gratefully acknowledged.

References

Anderberg, S. 1991. Historical land use changes: Sweden. In *Land Use Changes in Europe* (F. M. Brouwer *et al*., eds.), Kluwer Academic, Dordrecht, The Netherlands, 403–426.

Arnold, R. W., I. Szabolcs, and V. O. Targulian, eds. 1990. *Global Soil Change*. Publication No. CP-90-2, International Institute For Applied Systems Analysis, Laxenburg, Austria.

Arthur, W. B. 1988. Competing technologies: An overview. In *Technical Change and Economic Theory* (G. Dosi, C. Freeman, R. Nelson, G. Silverberg, and L. Soete, eds.), Pinter, London, 590–607.

Ausubel, J. H. 1990. Hydrogen and the Green Wave. *The Bridge 20(1)*, 23–49.

Berry, B. J. L. 1990. Urbanization. In *The Earth as Transformed by Human Action* (B.L. Turner II, W. C. Clark, R.W. Kates, J.F. Richards, J.T. Mathews, and W.B. Meyer, eds.), Cambridge University Press, Cambridge, U.K., 103–119.

Boserup, E. 1981. *Population and Technological Change: A Study of Long-term Trends*. University of Chicago Press, Chicago, Illinois.

Buringh, P., and R. Dudal. 1987. Agricultural land use in space and time. In *Land Transformation in Agriculture* (M.G. Wolman, and F.G.A. Fournier, eds.), SCOPE 32, John Wiley and Sons, Chichester, U.K.

Cameron, R. 1989. *A Concise Economic History of the World*. Oxford University Press, Oxford, U.K.

Cipolla, C. M. 1976. *Before the Industrial Revolution: European Society and Economy, 1000–1700*. Methuen and Co. Ltd., London.

---

[11]E.g., Ausubel, 1990; Freeman and Perez, 1988; Grübler and Nowotny, 1990.

Darby, H. C. 1956. The clearing of the woodland in Europe. In *Man's Role in Changing the Face of the Earth* (W.L. Thomas *et al.*, eds.), University of Chicago Press, Chicago, Illinois.

Demeny, P. 1990. Population. In *The Earth As Transformed by Human Action* (B.L. Turner II, W. C. Clark, R. W. Kates, J. F. Richards, J. T. Mathews, and W. B. Meyer, eds.), Cambridge University Press, Cambridge, U.K., 163–178.

Durand, J. D. 1967. The modern expansion of world population. *Proceedings of the American Philosophical Society 111*, 136–59.

Fisher, J. C. 1974. *Energy Crises in Perspective*. John Wiley and Sons, New York.

Flora, P. 1975. *Indikatoren der Modernisierung*. Westdeutscher Verlag, Opladen, Germany.

FAO (U.N. Food and Agriculture Organization). 1965–1991. *FAO Yearbook: Production*. FAO, Rome, Italy.

Freeman, C., and C. Perez. 1988. Structural crises of adjustment, business cycles and investment behavior. In *Technical Change and Economic Theory* (G. Dosi, C. Freeman, R. Nelson, G. Silverberg, and L. Soete, eds.), Pinter, London, 38–66.

Fussel, G. E. 1958. Agriculture: Techniques of Farming. In *A History of Technology, Vol. IV The Industrial Revolution c.1750–c.1850* (C. Singer, E. J. Holmyard, A. R. Hall, and T.I. Williams, eds.), Clarendon Press, Oxford, U.K.

Grigg, D. B. 1980. *Population Growth and Agrarian Change, a Historical Perspective*. Cambridge University Press, Cambridge, U.K.

Grigg, D. B. 1982. *The Dynamics of Agricultural Change, the Historical Experience*. Hutchinson and Co. Ltd., London.

Grigg, D. B. 1987. The Industrial Revolution and land transformation. In *Land Transformation in Agriculture* (M.G. Wolman and F. G. A. Fournier, eds.), SCOPE 32, John Wiley and Sons, Chichester, 79–109.

Grübler, A. 1990. *The Rise and Fall of Infrastructures*. Physica Verlag, Heidelberg, Germany.

Grübler, A., and H. Nowotny. 1990. Towards the fifth Kondratiev upswing: Elements of an emerging new growth phase and possible development trajectories. *International Journal of Technology Management 5(4)*, 431–471.

Hart, J. F. 1991. *The Land that Feeds Us*. W.W. Norton, New York.

Hayami, Y., and V.W. Ruttan. 1985. *Agricultural Development, an International Perspective* (2nd ed.). Johns Hopkins University Press, Baltimore, Maryland.

Hobsbawn, E. J., and G. Rudé. 1968. *Captain Swing*. Pantheon Books, New York.

Japan Statistics Bureau. 1987. *Japan Statistical Yearbook 1987*. Management and Coordination Agency, Tokyo, Japan.

Jones, D. W. 1991. How urbanization affects energy-use in developing countries. *Energy Policy 19(7)*, 621–630.

Kline, S. J. 1985. What is Technology? *Bulletin of Science, Technology and Society 5(3)*, 215–219.

Landes, D. S. 1969. *The Unbound Prometheus: Technological Change and Industrial Development in Western Europe from 1750 to the Present*. Cambridge University Press, Cambridge, U.K.

Langdon, J. 1986. *Horses, Oxen and Technological Innovation*. Cambridge University Press, Cambridge, U.K.

Marchetti, C. 1978. *On 10^{12}: A Check on the Earth Carrying Capacity for Man*. Publication No. RR-78-7, International Institute For Applied Systems Analysis, Laxenburg, Austria.

McEvedy, C., and R. Jones. 1978. *Atlas of World Population History*. Penguin Books, London.

Mitchell, B. R. 1980. *European Historical Statistics: 1750–1975*. Macmillan Press, London.

Mitchell, B. R. 1982. *International Historical Statistics: Africa and Asia*. Macmillan Press, London.

Mitchell, B. R. 1983. *International Historical Statistics: The Americas and Australia*. Macmillan Press, London.

Mokyr, J. 1990. *The Lever of Riches, Technological Creativity and Economic Progress*. Oxford University Press, Oxford, U.K.

Mothes, F. 1950. Das Wachstum der Eisenbahnen. *Zeitschrift für Ökonometrie 1*, 85–104.

OIR (Österreichisches Institut für Raumplanung). 1972. *Simulationsmodell 'Polis'–Wien*. Arb. Nr. 301.1, ÖIR, Vienna, Austria.

Perkins, D. H., and S. Yusuf. 1984. *Rural Development in China*. Johns Hopkins University Press, Baltimore, Maryland.

Richards, J. F. 1990. Land transformations. In *The Earth As Transformed by Human Action* (B. L. Turner II, W. C. Clark, R. W. Kates, J. F. Richards, J. T. Mathews, and W. B. Meyer, eds.), Cambridge University Press, Cambridge, U.K., 163–178.

Rosenberg, N., and L. E. Birdzell. 1986. *How the West Grew Rich: The Economic Transformation of the Industrial World*. I.B. Tauris, London.

Slicher van Bath, B. H. 1963. Yield ratios 810–1820. *Afdeling Agrarische Geschiedenis Bijdragen 10*.

Steinhart, J. S., and C. E. Steinhart. 1974. Energy use in the U.S. food system. In *Energy: Use, Conservation, and Supply* (P. H. Abelson, ed.), American Association for the Advancement of Science, Washington, D.C., 48–57.

United Nations. 1973–1987. *Statistical Yearbook*. United Nations, New York.

US DOC (U.S. Department of Commerce). 1975. *Historical Statistics of the United States Colonial Times to 1970* (2 vols.). Bureau of the Census, U.S. Department of Commerce, Washington, D.C.

van Lier, H. N. 1991. Historical land use changes: The Netherlands. In *Land Use Changes in Europe* (F. M. Brouwer, A. J. Thomas, and M. J. Chadwick, eds.), Kluwer Academic, Dordrecht, The Netherlands, 379–401.

Viatte, G., and C. Cahel. 1991. The resistance to agricultural reform. The OECD Observer 171, 4–8.

WRI (World Resources Institute). 1991. *World Resources 1991*. WRI, Washington, D.C.

Woytinsky, W. L. 1926. *Die Welt in Zahlen*, Vol. 3, *Die Landwirtschaft*. Rudolf Mosse Verlag, Berlin.

Woytinsky, W. L. 1927. *Die Welt in Zahlen*, Vol. 5, *Handel und Verkehr*. Rudolf Mosse Verlag, Berlin.

Woytinsky, W. L., and E. S. Woytinsky. 1953. *World Population and Production, Trends and Outlook*. The Twentieth Century Fund, New York.

Yates, R. D. S. 1990. War, food shortages, and relief measures in early China. In *Hunger in History* (L. F. Newman, ed.), Basil Blackwell, Cambridge, Massachusetts.

Zimmermann, E. W. 1951. *World Resources and Industries*. Harper, New York.

13

Political-Economic Institutions

Steven Sanderson

Formal analysis of institutional/political factors in land-use change suffers from three inherent problems, two of which are common to other cross-case political analysis: the contingent nature of political intervention, and the tremendous variation in cases—variation that does not follow rigorous laws of behavior and often frustrates efforts to model effectively. The third problem derives from the dynamic impact of landed resources on the nature of institutional intervention and change itself, and our limited knowledge of the complex physical settings in which institutions try to effect change. These problems are compounded by a gross division in the political–economic literature on the role of institutions in social change and a general anemia in treating institutional models of environmental change.

Modern political economy derives from two grand traditions, one emphasizing the constraints of structure, the other stressing choice. Overwhelmingly, the literature describing the political economy of environmental degradation favors choice and gives less attention to structure. Moreover, the literature is undermined by a general imprecision in conceptualizing institutions, as opposed to organizations.

Systematic analysis of institutions and their impacts on land-use/cover change offers a means of creating typologies that define differences among countries and ecological settings and of bringing serious empirical data to bear on these complex phenomena. It is also important in creating frameworks or 'families of models' within which individual cases find their niches (Ostrom, 1990).

But incorporating institutional analysis requires caution. It is not enough to graft a new set of variables onto otherwise unchanged models or systems. Nor is it useful to oversimplify reality by making assumptions that cripple our understanding of the fundamental contributions institutions make to land-use change (Daly and Cobb, 1989). A more insidious danger that has been remarked upon elsewhere (Spooner, 1987) also bears repeating: that the analysis of international and national institutions driving land-use change rarely takes into account the cultural gulf between such institutions and their subjects. Often institutions from outside undertake reforms at the expense of local land-use managers and the institutions that organize them. *Outside* institutions dominate *inside* institutions.

The 'internationalization' of institutionalized decision-making in land-

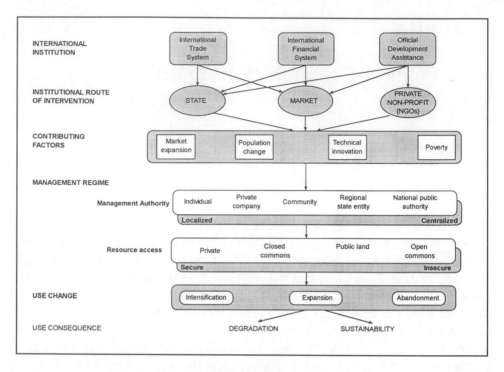

Figure 1. Institutional factors associated with land-use change.

use/cover change parallels the internationalization of poor nations' economies, and the international integration of the system of nation-states itself. Likewise, many more transborder issues affecting land use have appeared on the political agenda—and quickly have been institutionalized outside the traditional decision-making apparatus of the nation-state.

But institutional analysis in political economy has generally derived from a closed system approach. To reflect the internationalization of the economic forces driving land-use change, analysis of the institutional driving forces behind land-use change needs an 'open system' tonic, a means by which the borders of the system are not the nation-state, necessarily, but the nation-state in the context of the most important international change vectors affecting land use/cover. Future analysis needs to take into account the locus of change, institutional routes of intervention, the kind of impact, the management regime, and the degree of consonance between institutions and organizations affecting land-use/cover change and those that depend on the particular ecosystem for survival. The conceptual complexity of such analysis is only suggested by Figure 1.

At the same time, the problems of complexity, contingency, and irregularity are often overdrawn in social science. There is common ground on which institutional analysis and more general models of global change can stand. Many of the methodological problems facing students of institutional driving forces also face ecological theorists, economists, and demographers: issues of scale, rates of change, comparative data sets, and the wrenching difficulties of giving up the pristine case in favor of the global comparison.

Can our understanding of land-use change be improved by looking at institutional driving forces? The tentative conclusion of this paper is yes, but only if institutions are defined with more precision. First tasks include identifying the range and kind of variation we know to exist in institutions affecting land use/cover, finding the appropriate level of analysis, and defining the direction of change, according to the relative weights of different institutions in different countries.

My central argument is that institutional impacts on land-use change must be analyzed by incorporating global-level variables as they affect local level change. Institutional driving forces behind land-use change have been masked to a large extent in the literature of economics and political science by an anachronistic focus on falsely stratified international, national, and subnational levels, in which the international system impinges on the national, and the national on the subnational. This is true of many traditional treatments of trade, finance, investment, and multilateral development assistance. But the weight and direction of various levels and actors in the institutional settings affecting land use have changed radically over the past two decades, as a function of

- The growing dissatisfaction with the state, both for its substantive failures in a wide number of policy areas and for its general fiscal bankruptcy (at least in Latin America and much of Africa) in the wake of the debt crisis
- The urgency of the environmental crisis, which, combined with the 'destatization' of the problem, encourages a policy approach to complex political problems, and movement away from the vested interests of the state toward apparently more 'neutral' organizations whose substantive focus is clearer and less ambiguous or conflictive
- The expansion of nongovernmental organizations (NGOs) and their influence, and the concomitant recognition of such organizations within the traditional context of the international system.

To account for these changes in the way we analyze institutional determinants of land-use change requires that we reconsider some of the main institutions themselves—market, property, and state, among others—from a more international perspective; that we view institutions from both domestic and international angles and disaggregate the international and the domestic side separately; and that we try to treat the dynamics of newer 'transborder' issues from an institutional perspective. So, for example, the study of the impact of agricultural markets and prices on land use cannot focus on a country level or even local market level, without incorporating major international price-makers into the equation, whether they be multinational corporations or agricultural marketing organizations of the Organization for Economic Cooperation and Development (OECD).

Likewise, conventional models of institutional and political change cannot rely on efficiency considerations and property rights alone, but must take into account subnational institutions that may not accept such rationality, or even recognize its rules. Once again, we must take into account differences between 'outside' institutions and their values and 'inside' institutions in a particular land-use setting. In fact, much of the international and national institutional architecture dominating land use today is Western and development oriented, while the targets are largely local and governed by orientations other than development. If the flow of institutional change is multidirectional, we must account for multiple rationalities in the system.

What Are Institutions? A Schematic Review

Institutions are the most obvious and time-worn targets of political and social analysis. Modern political science emerged from classical studies of institutions and processes that guided public sector activity. Much of classical political economy revolved around institutions governing economic exchange, from David Ricardo's focus on the state and Adam Smith's rendition of the market to Karl Marx's critique of political economy, based on an attack on the market and the political and social institutions of bourgeois society.

The post–World War II U.S. behavioral revolution in political science to some extent pushed institutions out of the center of the discipline, in favor of formal analysis of individual political behavior, and in favor of systems analysis and a structural–functional approach that effectively 'black boxed' institutions and processes by focusing on inputs and outputs. As later analysts framed institutions and their dynamics in terms of rational individual behavior instead of institutional architecture, they retained the focus on individual choice in a culturally and politically neutralized framework of choice. Political systems themselves became

choice mechanisms for making economic decisions … [and] economies in their own right making decisions on budgets or the production and distribution of public goods and services. As price theory is a prime focus of private economic choice, so are political institutions the focus of concern for the new political economist. Comparisons of markets and political institutions (usually in terms of efficiency) are, therefore, typical activities (Mitchell, 1968: 77).

Institutional economics has played a role throughout the 20th century, and like its political science counterpart, has been supplanted to some extent by the embrace of one of the 'new institutionalist' perspectives, rational choice, which focuses on the human behavior of the rational individual, and generalizes to questions of collective action. Earlier institutional foci now are resurrected, but in light of the general ascendance of work deriving from individual economic rationality and utility.[1]

The central effort of the choice literature has to do with explaining or encourag-

[1]An important alternative to rational choice is the 'new institutionalism' of March and Olsen (1989), in which institutional routines, inertia, history, power, and other factors enrich the overly simplistic choice frameworks cited above.

ing cooperation from a theoretical standpoint built on individual maximization. In this theoretical tradition, the building block of institutional and organizational analysis is the rational individual decision-maker.[2] In economics, there is wide variation in the extent to which institutions as norms differ conceptually from institutions as 'enabling environments' or organizations (Van Arkadie, 1990; Ruttan and Hayami, 1984; Feeny, 1988).

Efficiency, Power, and Domination

The new institutional literature has been predictably reluctant to deal with questions of power and domination when determining efficient outcomes. It generally views the public sector, for example, as a distorting force that biases economic institutions away from efficiency, in favor of redistribution. That general principle is valid in the sense that the state abides by allocative rules that differ from those of the market. But the premise does not allow for the political power embedded in market institutions, or the power relations embedded in the kind of redistribution undertaken by the state (Cerny, 1990).

In common language, efficiency has come to invoke the same kind of ideological tone as class struggle. But efficiency in the literature of institutional political economy does admit the possibility that imperfect markets and political power and economic domination exist. That permits the possibility that a political–economic system directed toward efficiency will also be directed in favor of a socially regressive distribution of assets. (In fact, treating allocation according to efficiency criteria abandons income distribution, or the distribution of costs from environmental degradation as independent variables; Daly and Cobb, 1989).

Especially in evolutionary models, efficiency may lose its purely economic quality, in favor of 'group dominance at the expense of others' (North, 1990: 21). In most empirical settings, and certainly at most moments in time during the evolution of markets and states, power and domination denominate the institutions of market, state, and property so dramatically that evolutionary interpretations of rational choice correspond no more to understanding everyday life in the market than Marxist teleology does to current Eastern European economic problems.

Institutions vs. Organizations

Despite much valuable conceptualization to the contrary, much literature still fails to distinguish between institutions and organizations (Majone, 1986). Most institutionalists would agree with the distinction, succinctly captured by North.

---

[2] For reasons of time and space, I will not enter into Marxist literature on these questions. Suffice it to say that Marxist and structuralist literature on institutions derives from social class and class-interested organizations as the basic unit of analysis, but has not changed the critical variable of maximization or the question of collective action. In any event, the bulk of the empirical and formal theoretical literature of interest to this institute does not come from a Marxist perspective. I will return to structuralist considerations at appropriate points in the discussion.

Institutions are defined as 'the humanly devised constraints that shape human interaction' (North, 1990: 3). *Organizations* are 'purposive entities designed by their creators to maximize wealth, income, or other objectives defined by the opportunities afforded by the institutional structure of the society' (North, 1990: 73). This distinction mirrors earlier analysis from political science (Uphoff, 1986; Young, 1989) (as well as earlier still from anthropology). It allows customary and informal constraints to play a significant part in the institutional framework of society. It allows us to consider three kinds of institutional–organizational links: institutions that are not organizations (e.g., the law), organizations that are not institutions (e.g., a new law firm), and entities that are both (e.g., courts) (Uphoff, 1986).

It also permits an interactive relationship between institutions and organizations, in which it is entirely possible for the organization to violate the institutional mandate that gave it life. So, for example, the norms and values that created a climate for the Endangered Species Act in the United States are 'undone' by the implementation of the act through its organizational medium, the U.S. Fish and Wildlife Service (Tobin, 1990). The 'perversion' or 'capture' of organizations by special interests has a rich literature and an even richer tradition in politics.

This returns us to the question of power. In real life, if not in models, power in both public and private spheres denominates the distribution of assets. For one thing, unless markets are perfectly competitive (a condition that exists only in theory), market power affects the allocation of resources in ways that may not be consistent with efficiency. Political institutions mirror to some extent the allocation of power in society itself, so that asset distribution by the public sector reflects the values dominant in economic society.

This realization contrasts sharply with general equilibrium political economy, in which policy-makers may even turn out to be inconsequential, because the allocation of values in society is the dynamic reflection of the equilibrium of private interests at play (Magee *et al.*, 1989). This appealingly clever idea shortchanges the role of leadership, and reduces politics to the 'inside game' of political society, rather than to larger and less tidy questions of contesting the control of economic assets in the countryside. Insofar as such analysis is applicable to markets and land-use change in the Third World, it is based on theory, without adequate, empirically verifiable studies. Even in less extreme characterizations, economic theories of choice hold preferences constant and disclaim power or structural inequalities (March and Olsen, 1989).

In general, rational choice models of social change rely on equilibrium. In land-use and land-cover change, one has to ask: What equilibrium? If equilibrium is the product of a well-institutionalized dynamic game, in which *at least* free contention and consensus on the objectives of the game are stipulated, then equilibrium does not define much of the global process of land-use and land-cover change. The magnitude and character of land-use change over the postwar period describe a condition of ecological, social, and economic *dis*equilibrium, *vis-à-vis* a historical equilibrium that may never have existed, except as metaphor. Certainly, political science notions of equilibrium do not carry the weight of physical science models.

Public vs. Private Institutions

The world of policy today has little good to say about the public sector. The contemporary literature of both public choice political economy and ecological economics describes the activities of the public sector as predatory or perverse. In contrast, the private sector and the market are offered up as alternative mechanisms for allocating economic values and biasing social systems toward efficiency. Nurkse's enjoinder that 'the world is not rich enough to despise efficiency' (Nurkse, 1961) still holds true.

At least two problems weaken the complaint against the public sector: first, that redistribution (the defining element of predation) may serve social purposes other than efficiency (such as equity or land conservation) that markets do not; and second, that even though we all embrace efficiency as a goal, its definition has been tied inextricably to maximization of output and the optimization of social organization. Maximization is not necessarily consistent with other policy values, and optimization among levels of institutional play may not be possible or even desirable.

The retreat to the economic foundations of society is insufficient to the set of tasks before us in land-use and land-cover change. In an exceptionally well-tempered treatment, Ostrom (1990) abjures the private vs. public dichotomy in favor of a 'third way,' resting on local managers of common property resources. But even that perspective used elsewhere has been turned to broader ideological goals of privatization.

To some extent, the public vs. private debate is as sterile as the market vs. states debate, in that the dichotomy is too stark to describe the complex interactions between the opposing elements (Cerny, 1990). The question of public vs. private really is a question of what *politically* defines the relevant publics (clientele) for whom land-use policy is designed. The market and property rights generally favor publics who are able to enter the market and to secure property. But the market and property are constitutive rights in bourgeois society, and no one is formally denied access to those institutions.[3]

The state delivers public goods to limited publics. From clean air to public safety, public goods are delivered in reflection of the economic, political, and social stratification of society. To use one pertinent example, the creation of national conservation units for public use generally favors upper-tier consumers,

---

[3] This, of course, was the great ideological claim of land reformers in 19th-century Latin America, or early 20th-century Africa, who wanted to create modern land markets by disentailing Indian 'primitive communism' or tribal communalism and throwing indigenous lands onto the market. Liberal reformers and colonial administrators claimed disingenuously that the Indians had every right to reclaim 'their lands' through purchase, or that rationalization of agricultural property would make its distribution fairer.

and often prejudices the lives of poor residents or ignores them altogether. When confronting state or national public institutions *vis-à-vis* indigenous institutions, one must address the question of whether national-level goals and local institutional norms are consistent, or, alternatively, whether there is some effective translating agent that can make them compatible. This is the age-old institutional question of whether individual preferences and institutional outcomes can converge.

In this quest, is it not likely that the organization charged with implementing institutional values also will become a public? This standard argument of bureaucratic politics is plain evidence of the interaction of institutions and organizations, whereby the values embedded in society become the calling card of the bureaucratic apparatus designed to attend them. It may be that organizational survival goals either eventually displace or conflict with the substantive goals of institutions in society.

The 'Third Way' Redux: NGOs

For years, the three institutional and organizational venues for development have been state, market, and voluntary. The past two decades have seen an increasing reliance on NGOs, a catchall term for nonprofit private voluntary organizations, local community self-help or grassroots organizations, and virtually anything else that falls into the gap between the for-profit sector and the public sector. Enthusiasm for NGOs has accelerated over the 1980s, as faith in the public sector has waned and development assistance specialists have realized that there is an unmet need for complex social services that may be met by the nongovernmental organization (Weisbrod, 1989). NGOs have reached the enviable position of being far more important to the implementation of development assistance (or biodiversity) projects than to the generation of development assistance revenues.

NGOs have been invested with special qualities by the international organizations most involved with land-use change. NGOs have been described as 'efficient and effective alternative[s] to public agencies in the delivery of programs and projects. Moreover they can sometimes reach target groups that public agencies cannot' (World Commission on Environment and Development, 1987: 328) NGOs have the image of 'representing publics' that otherwise would be left out of public policy debate over land use (World Bank, 1990). At the very least, NGOs are nonprofit, meaning that they do not take advantage of consumers' informational handicaps (Weisbrod, 1989).

The result of this enthusiasm and of the proliferation of NGOs themselves has been a fundamental alteration in the way international organizations work to effect land-use change. But the institutional values of NGOs are as murky as their definition, and several problems ensue from substituting NGOs for the public sector, or even allying them with the state (Maniates, 1990). Not the least of these involves the difficult matter of crossing the barriers that separate elite institutions from local ones, especially those in ethnically complex nations.

NGOs complicate the traditional conviction that the state is the proper institutional complex governing the delivery and protection of public goods and the correction of externalities. Whereas welfare liberals might assign a more important role to the state of finding a 'general welfare function,' or at least ameliorating some of the inequities of civil society, individualist political economy focuses on the market and away from the public sector. Some think of politics as a 'non-price rationing device' (de Janvry, 1981); others more categorically restrict public sector regulation to the rare occasions when people must be forced to act against their own economic self-interest (Crosson and Rosenberg, 1989). For obvious reasons, voluntarists would rather see the importance of the state diminished in favor of the private sector, whether for-profit or nonprofit.

Voluntarist and structuralist literatures in political economy also part ways in characterizing the kind of research design required to understand institutional roles in social change, and the importance of research to policies that might promote effective action. For the structuralist agenda, formal analysis is more important to policy formulation, because of the implicitly more 'radical' changes required for the proper impact. If one identifies structure as the problem, policy solutions must take on structural reforms. The choice perspective suggests that policy choices are available to solve the social problems in question, without addressing underlying structural variables. A great deal of environmental literature comes from such a public policy perspective, with a research base that focuses on rationality and efficiency, favoring the *apolitical* aspects of institutional change. Even institutional approaches to environmental change may neglect individual, community, and institutional diversity in favor of a focus on policy 'outputs,' divorced from the social and political fabric into which they are woven. We will return to this theme later in the paper.

What We Know about Institutions and Land Use

Some Methodological Warning Signs

Every analytical enterprise has its methodological minefields, but relating institutional relationships to land-use change is particularly tricky. Let us briefly note some special problems.

Generalizing from the Data

In generalizing from case study material, the uniqueness of individual systems or communities is surrendered to the need to generalize. To the extent that community institutions matter to land use, it may prove extraordinarily difficult. In a paper based on research in the Peruvian Amazon, Padoch and de Jong (1992) observed tremendous diversity, variation, and change in a small, apparently homogeneous community of non-Indian riverine settlers. In a community of 46 households, the authors found 12 distinct agricultural types and 39 variations in resource use strategies. None of these distinctions was overly fine or trivial. No typical strategy or technology characterized the community. And one year later, among 12 households interviewed a second time, only 3 were using the same strategy they had employed only one year before.

This kind of variation, which is certainly replicated elsewhere, should lead not to despair, but to caution. Clearly, there is a mandate in this variation for stronger case analysis, as well as conceptual and theoretical improvement, so that cases can be clustered carefully together with some confidence. And it is apparent that the resource itself must be considered as an independent variable. That suggests that structure informs process at 'both ends of the spectrum,' institutional and resource. The system-wide model that pretends to capture land-use change on the frontier must address the complexity of local variation in its research agenda.

It may be possible to use general aggregate data analysis to target particularly interesting 'outlying cases,' where systemic features of land-use change do not explain the variance, and then to telescope in with careful, replicable case-study methodologies to refine the model.

The Level of Analysis Problem

This raises the level of analysis problem, i.e., to which level is the research design properly pitched. As I have already indicated, it is important as a first step to reject a closed systems approach, which in the case of land use is clearly inappropriate for several reasons:

- Land use is a function of international agricultural prices, national political responses to those prices, and producer strategies to accommodate both.
- World prices in agriculture are themselves a function of government intervention in the 'price-making' countries, generally found in the OECD community. Hence, a direct relationship exists between producer strategies at the local level in the less developed countries (LDCs) and agricultural pricing policies in the developed countries. (It is not an exaggeration to say that the land-use decisions of the Mexican subsistence farmer are a complex function of U.S. agricultural policy supports for maize and the Mexican government's response to them; Sanderson, 1989.)
- The critical institutional values of multilateral and bilateral development assistance focus to a great degree on agricultural land-use change. Such is also true for the conservation agenda of the same communities, plus relevant nongovernmental institutions.
- The paths of development assistance have changed significantly, so that a new consideration of the levels and routes by which land-use change is effected at the international level must be undertaken (Figure 2).
- It is generally conceded (Goodland, 1990) that the critical variable in land-use change in the tropics has to do with rich-country markets for tropical products. This reality has even produced initiatives in the European Parliament about trade in tropical timber, and the creation of an International Tropical Timber Agreement, the ostensible purpose of which is the preservation of tropical

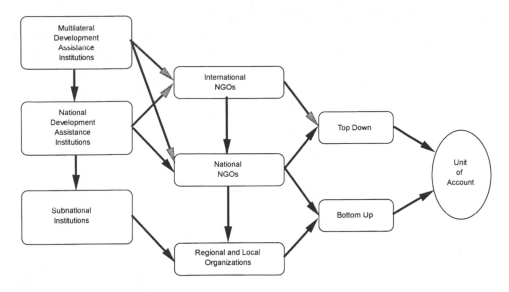

Figure 2. Cascading routes of institutional intervention.

forests through 'sustainable management' (McNeely, 1988).
• One of the most interesting changes in recent years has been the growth in international organizations' claims on land-use policies across national boundaries (e.g., the international movement to establish more protected areas).

Related to the level of analysis problem is the geographic context of analysis. That is, all ecological and geographic settings are not equal. This raises important questions about the 'endemism' of institutions affecting land-use change. Suffice it to say that the greater the degree of endemism, the more difficult it is to model local institutions.

Research in a Data-Poor Environment
Despite the tremendous amount of casework in land-use issues over time, local data have not been generated in consistent, comparable, quantitative formats. Moreover, the character of aggregate data that would permit coherent and replicable cross-national analysis may not even begin to tap the connection between environmental degradation and poverty. That is, it may be clear that poverty and environmental degradation are strong correlates, but the specific dynamics are not revealed in economic data. It may be that greater deforestation, erosion, water contamination, and other practices are taking place without any measurable reflection in output. And the output data are not relevant to the misery-driven changes in land use. (This is particularly true of systems that mine resources, such as semiarid agricultural areas irrigated by groundwater resources; such systems may show high productivity over long periods of time, but in fact may not be sustainable.)

To date, whether demand is managed by government and market working together to 'get prices right,' or by the state alone to 'get policies right,' or in rarer cases to 'get institutions right,' it will be difficult to muster cross-national data that permit us to model international institutions along with land-use change vectors at the grassroots level.

Policy Axioms Becoming Received Wisdom

Due to the urgency and catastrophic potential of environmental degradation in our time, there is a great pressure to shrug off the detached role of the intellectual and to become applied ecologists, anthropologists, etc., and—in the case of political science and economics—to become 'policy relevant.' Unfortunately, much of the policy literature has stipulated a number of arguments about the relationships between institutional forces and land-use change in the LDCs, well in advance of the research required to substantiate them. Over a very short period of time, the relationship between the institutions of international finance, trade, development assistance, and investment, on the one hand, and the institutions of land-use change at the grassroots level, on the other, has been proclaimed as truth and acted on in policy circles. This is contrary to the purpose of science in policy and permits attention to be diverted from fundamental sources of change to more tractable short-term initiatives that may not have sufficient impact.

A number of other cautions are warranted here, but will only be listed. They include the incredible breadth of variables required for proper analysis; the imprecision of lagged effects, given our state of knowledge; and the possibility that institutions may have to change an unknown combination or number of variables to have an effect, but that this effect may express itself as a change in only one variable; and so on. However, the remaining space in this paper will be spent trying to work away from the shortcomings of the data, toward some argument and hypothesis.

Defining What We Know by What We Do Not: The Relationship between Land-Use Change and Markets, Private Property and External Trade

Generally, political economy observes that markets outperform states, a conclusion drawn more from the poor performance of the state than the virtues of markets, particularly in ecologically fragile settings. Markets do allocate economic values more efficiently, when the objective is maximizing output. But the reasons for examining land-use change are more than output-oriented. And they build, not on a simple dichotomy of market and state, but on the interaction of markets, states, and local institutions *in the important context of natural processes and resource endowments* to create a cascade of effects on land use. So, changes in population and labor markets in highland Ecuador create pressures in Ecuadorian Amazonia. That region, in turn, is opened up by oil exploration, which responds directly to global oil markets and the investment preferences of transnational corporations, as well as the realities of the Organization of Petroleum Exporting Countries (OPEC) and U.S. energy consumption patterns. The pattern of land-use change in the Amazonian frontier, however, is a function not only of the market, but of producer strategies and preferences, resource endowments, and a variety of other factors that make the simple assertion of market supremacy oversimple.

At the firm level, when a sawmill or papermill exhausts the supply of raw material locally and has to go farther into the interior to find its timber or pulpwood, it may be seen either as a 'good thing' because the firm is relocating closer to the point of production and returning more value added to local producers, or as a 'bad thing' because the firm is accelerating the degradation of the frontier by expanding its catchment area and making economic returns available to more forest dwellers. Is the market driving resource use? Are poor forest dwellers attracting capital? Is firm decision-making causing deforestation?

Similarly, despite earlier literature to the contrary, the degradation of upland Nepalese hillsides appears now to be mainly a product of natural processes (Blaikie and Brookfield, 1987; de Boer, 1989), rather than human-induced. But certainly natural processes induce certain kinds of human responses, and the resource-use responses of populations affected by natural degradation, in turn, affect future degradation.

The premise that private property is preferable to other forms is found throughout the literature. However, some of the most durable and ecologically sustainable land-use regimes are in common property or usufruct arrangements. The anointment of private property as the most appropriate institutional arrangement has several origins. First, it is a sacrosanct liberal principle, which has been implemented politically in the disentailment of corporate property, the expansion of the agricultural frontier through colonization by small holders, and the redistribution of abandoned, or idle, or counterrevolutionary land assets through progressive land titling schemes. The preference for social redistribution through privatized land assets has been endorsed by a wide political range of governments, and was the watchword of postwar U.S. land reform policy in all the LDCs (Prosterman and Riedinger, 1987; de Onís, 1970).

Private property distribution through land reform has as one of its principal goals the elimination of poverty as a source of land degradation. The preference for private property as the regime of choice is accompanied by the corollary principle that secure land tenure means sound land management (World Bank, 1990). The two core hypotheses here are that private property is more secure than alternative arrangements and that security in land tenure ameliorates poverty-induced environmental damage (OECD, 1989).

One need only turn to Lipton (1985) to see the shortcomings of this theory. First, it is not evident that improving security of land tenure has the net effect of reducing land degradation (although insecure land tenure, especially when accompanied by speculation and frontier expansion, undoubtedly does lead to accelerated degradation). It all depends. What is the impact of land asset redistribution on labor market requirements? What tier of the rural poor does land reform reach? Is land reform accompanied by ecologically responsible technical assistance? Does land reform mean the further impoverishment of claimants who are denied land? More important, perhaps, is there evidence that secure land tenure promotes sound land management in any other dimension than production? And, finally, what is the relationship between land reform and more general resource reform? (For example, the consequences of water redistribution are radically different when the beneficiary population includes the landless as well as the landed; OECD, 1989.)

In fact, in cases throughout the world, usufruct arrangements limiting the rights of reform beneficiaries to alienate their resource rights through mortgage or sale, or to use them in ways that may be profitable in the marketplace, are *de facto* institutional critiques of the supposed security of land tenure under private property and the different values held by the market and the state. Likewise, in many common property arrangements throughout the world, it is not evident that privatization is either necessary or sufficient to the stewardship of common property resources (Ostrom, 1990).

The private property theorem, like market supremacy, is based on a negative proof: that the state is bankrupt and that 'primitive societies' are inefficient or otherwise inviable. The bankruptcy of state policy is hard to gainsay. It is the state that has classified as 'abandoned' any land that is not continuously occupied (Blaikie and Brookfield, 1987) or as 'idle' any land not in production (Sanderson, 1981, 1990). The state as land manager has often forced land out of fallow, to its eventual detriment and to the detriment of the communities that depend on it; government has considered deforestation to be a sign of land improvement (Mahar, 1989). The national government has 'statized' tribal and community rights over forest land in West Africa, to the detriment of forest management (Gillis, 1988). And the institutionalized agricultural and timber subsidies that have governed LDC (and developed country) land policies have been criticized roundly. A general conviction among privatizers is that cost-benefit analysis of alternative uses to those mandated by the state would show the economic inefficiency of the public sector (Deacon and Johnson, 1985).

The other half of the proof rests on generalizations about the inviability of traditional communities. Some of the arguments have to do with the inappropriateness of traditional ways of life in the modern world. As one popular saying in Brazil has it, the future of the country is being sacrificed for the sake of 200,000 Indians. Or, in Mexico, indigenous communities have been annihilated for their rejection of modern government, preferring community-based institutions over the external authority of a central government they had no role in constituting (Spicer, 1962).

More sophisticated treatments of traditional communities stress their inadequacy as stewards of the land, or forest, or water over which they traditionally have held domain. What is not said in this argument is that the failure of common property or other community-based resource management strategies is often a function of their vulnerability before the market. If market supremacy, or private property, were examined from the standpoint of their historical impact on sustainable land-use institutions, they would not enjoy the repute accorded them by policy-makers today.

Another widely accepted theorem is that external sector pressures accelerate land-use change to the detriment of the environment. Defining arguments include several grand-scale hypotheses:

- That natural resources are exploited at a faster rate under conditions of extreme external debt than would be the case without the debt crisis (World Commission on Environment and Development, 1987)
- That trade vulnerability (a function of trade openness, which is touted as the solution to debt problems) and resource exploitation and degradation are closely correlated (Bramble, 1987; Repetto, 1988).

Corollaries include:

- That debt relief would generate an environmental dividend
- That trade insulation would reduce resource exploitation.

Some strong foundations in economic theory underlie these arguments (Capistrano and Sanderson, 1991), but the demonstration of a causal connection between the external sector and land-use change is difficult to make. The process is complicated not only by the weakness of available data, but by shortcomings in analytical frameworks that guide data analysis. Once again, much of the literature is driven by policy, not formal analysis. The one clear conclusion that may be drawn at this point is that country cases vary widely.

Interestingly, the same literature argues the contrary case, that external sector pressures change land use to the benefit of the environment. This argument rests on some rather involved assumptions about the relationship between trade and economic development on the one hand and poverty-induced land degradation on the other. The general premise is that external sector pressure generates more trade, which is good for poverty alleviation. This produces the argument that poverty alleviation yields environmental dividends in land use. The efficiency side of this argument produces the corollary that technological innovation enhances land management. The assumption is that poor people, once elevated from their grinding misery to some better economic state, become better land managers. It is also apparently assumed that poverty-induced land uses are worse than alternative land uses. This case is particularly difficult to defend in the modernization of fragile arid lands, where low-intensity, low-output uses have survived over centuries but modern innovation and intensification have degraded the resources in a generation.

We can see much by examining the United States, Canada, and Mexico in the evolving institutional setting of trilateral free trade. Advocates of free trade argue that the impact overall will be positive, with more efficient allocation of rural resources offsetting potential losses in rural employment, much of which will be offset by alternative opportunities (Levy and van Wijnbergen, 1991). Opponents contend that free trade with the United States will mean wholesale displacement of rural labor in Mexico, with emigration and environmental degradation the natural result. It is also contended, with some justification, that the agriculture that replaces subsistence maize and bean cultivation will be more energy- and input-

intensive, with substantial implications for land use and environmental degradation. In both cases, the evidence is mainly forthcoming.

Rethinking Institutional Approaches to Land-Use Change

Looking at Land-Use Change Itself

The vast range of land-use and land-cover changes is not all cut from the same
bolt of cloth. Land-use/cover change can include both natural and anthropomorphic
degradation, resulting in deforestation, expansion of the agricultural
frontier, modifications in the intensity and purpose of land use, and changes in
the social dynamics associated with land-use decision-making, among others. Each
has its own peculiarities and literatures, compounded by the many natural, agricultural, and social science perspectives that might be employed to examine them.

The first principles of societal models treating land-use change should focus
on 'long and short waves,' meaning some attention to the 'nested' quality of
different cycles that affect land-use change. Land use, like global warming,
must be incorporated into a strong theoretical framework and a deep historical data
base that tell us of the context in which land-use change takes place. Periodicity is
important; so is historicity. For example, 1945 is an obvious choice, not only
because of the creation of the Bretton Woods system and the foundations of the current system of multilateral development assistance, but because it represents the
beginning of decolonization. Other periods suggest themselves, as well, such as the
1890–1914 export commodity expansion in Latin America, or the post-
Independence period in Africa. In any event, time scale is of primary importance.

A second and related principle is that within this time scale, it should be recognized that specific communities have their own time scales. Community cycles
governing ownership or access to assets may or may not coincide with more general time scales in society. Land asset ownership in rural communities is typically
life-cyclical. Land use sometimes varies according to cyclical cropping decisions,
which may be annual or semiannual. Drought cycles and commodity price swings
are but two other examples of possibly relevant periods.

Against this general texture of timed rates of change, there may be important
subperiods that reveal something of institutional pressures for land-use change.
For example, in Mexico since 1982, the two principal support systems for poor
farmer agriculture—producer price subsidies and import restrictions in basic food
grains—have withered in the face of external stabilization and trade liberalization.
Following support prices, targeted agricultural credit, and import levels as a proportion of total food grain supplies (as well as related variables such as per capita
consumption of basic foods) are important surrogate indicators of land-use
change. Time series on competing crops that may indicate cropping changes on
marginal lands (maize to sorghum in Mexico), as well as major shifts in labor market, land asset control, etc., would be important complements to such analysis.

Finally, it may be well to compare across historical epochs. Blaikie and Brookfield (1987) suggest that the historical experience of European land degradation is being repeated during this century at an accelerated rate in tropical countries. This is as precise as saying that because changes in atmospheric carbon dioxide in this century are the same as in some other century, the causes are the same. It is extremely important, especially in the absence of good direct data in history, to compare the leading political, institutional, and economic forces at play during times of high land-use change in the past vs. the present, and to periodize the comparisons carefully.

Key Institutional Settings and Levels of Analysis

Among the many transnational and international institutions and organizations that affect land use, our attention will focus briefly on official development assistance via a wide variety of organizations, and on international trade in the bi-, pluri-, and multilateral system. We will mention briefly, only for reasons of space, international legal conventions, international financial institutions, and direct foreign investment via the transnational corporation. All of these foci of activity correspond to the architecture of the Bretton Woods system that followed World War II and defined U.S. hegemony after 1945. Whether or not the organizational forms and strengths are the same as the original mission of the International Monetary Fund, the objectives of economic stabilization and adjustment are intact. Likewise, the role of the World Bank and its regional affiliates is multilateral development assistance, with the modes and definitions that accompanied its birth. Direct foreign investment has changed remarkably over the same period, but the medium itself and its institutional rules are still inextricably associated with the transnational corporation. And, whatever the tendencies toward regionalism in trade, the principles underlying the General Agreement on Tariffs and Trade— trade liberalism in tension with domestic protectionism—still abide.

At this level, however, new institutional values and organizational forces have emerged via noneconomic institutions with a global mandate: the U.N. World Commission on the Environment and Development (the Brundtland Commission) and its successors in the institutional complex of environment and development issues; the U.N. Environment Program, Development Program, etc.; the Tropical Forestry Action Plan; the International Tropical Timber Organization; and a variety of institutions covering agricultural innovation and research (especially the Consultative Group on International Agricultural Research Institutions and International Service for National Agricultural Research, as well as regional organizations such as the Club du Sahel). And, as already mentioned, the focus of institutional vectors at this level has shifted away from the public sector to NGOs.

In fact, much institutional analysis at this level focuses on a model of international, national, and subnational flows (Uphoff, 1986) that no longer describes appropriately the paths of interaction in land-use change. Figures 1 and 2 offer an alternative view that takes into account the new importance of NGOs in development assistance flows and raises a new set of questions about the impact of the international development assistance community on land-use change. Figure 2 suggests that the 'core modality' of international-national-subnational linkages may still be intact, but that new paths have opened up for the reasons cited above. Those new paths have their own logics and contributions to land-use change, and create a nonlinear cascade of institutional impacts at various levels.

Moreover, a new kind of development assistance now has appeared, oriented not toward traditional output improvements but toward conservation. The multilateral and bilateral development assistance community has 'invested in biodiversity' (Abramovitz, 1991) and in tempering global climate change. An interesting test of the viability of such funding would associate development assistance to agriculture with development assistance to conservation, and make a qualitative examination of the consistency of the two institutional goals in a given country. (Unfortunately, the period is so short, the money so minute, and the target countries so few that such a test would be difficult to execute.) One hypothesis based on the above analysis would suggest that there is a strong association between development assistance for agriculture and deforestation, and that there is a strong association between deforestation and development assistance for conservation. To complete the syllogism, it may be that the multilateral development assistance community is giving money that promotes and retards deforestation at the same time.

In addition to the 'NGOization' of development assistance, and the incorporation of new conservation mandates in development assistance, the international trade system has had tremendous impact on land use throughout the world. As several recent studies have shown (Goldin and Knudsen, 1990; Hathaway, 1987; OECD, 1987), the European Community and the United States both engage in agricultural policies that both support and depress world prices in complex ways. Those countries determine, to some great extent, world prices in coarse grains and livestock, stimulating and depressing investment and production (and therefore land use) in some of the most important agricultural sectors. Moreover, foreign policy instruments indirectly related to land-use policies (such as Soviet grain purchases from the West) affect overall supply and distribution of grains.

All of these policies and variables affect land use in the LDCs, as do other important commodity areas in agricultural trade (coffee, citrus, frozen concentrated orange juice, sugar, bananas, etc.). They suggest two important tacks that enhance our understanding of institutional contributions to land-use change: (1) a clustering of countries according to their primary commodity composition, trade openness, sensitivities to changes in world prices in key commodities, and domestic policy apparatuses capable of moderating such changes; and (2) a thorough treatment of the international agricultural trade system as it affects land use, which would include, for example, the trajectory of various commodity agreements (coffee, wheat, tin, rubber, cocoa) that are in various states of decomposition or renegotiation. There are several global models for estimating the impacts of changes in trade institutions on agricultural production and consumer welfare, each of which distinguishes among country cases according to variables important to land-use change.

At the national and local levels, the characterization of institutions and organizations is likewise inadequate. Relating back to the historical evolution of the international system now being described, it must be acknowledged that through colonization, Northern institutional hegemony, and North–South financial and development assistance flows, the institutions and organizations charged with land-use change in the developing world are products of the international order, and are not endemic institutions or organizations. The irony of postwar institutional change in land-use policies is the substitution of Northern institutions and organizations for indigenous institutions at the national and local levels. This certainly extends to land-use organizations and reforms that shortchange tradition, community, and common property, but also to informal institutions (from apprenticeships to resource exploitation rights that limit harvest) that are poorly expressed in organizational form. So, LDC agricultural research institutions replicate the North's energy-intensive, mechanized agricultural models and its assumptions about effecting agricultural modernization in the LDCs. Cooperative enterprises are often set up on Northern premises and with extracommunity structure and goals, as if no community institutions existed. (This coincides with the general principle that development assistance to communities must be built around the communities' incompetence, rather than their knowledge.) Concessional assistance is built around donor country needs (e.g., agricultural surplus disposal) rather than recipient needs. And so on.

Each of these categorical assertions obviously has its counterargument, but the striking thing about the development assistance institutions of the North is the durability of their programmatic designs, even in the face of strong evidence of environmental and social cost.

Institutions Governing Access to Land

Land reform is an equally durable institutional and organizational value in the West, though its impacts on land-use change and poverty alleviation are less evident than its impact on disenfranchising the rural oligarchy. Three summary difficulties can be pointed out in regard to land tenure reform:

- Its failure to treat the condition of the rural ultrapoor, or those most in need
- Its general inattention to maintaining land cover or limiting aggressive agricultural adaptation of nonagricultural land
- Its preoccupation with titling schemes over access to land assets.

I have treated the last of these, at least briefly. The first is the most serious indictment on the issue of poverty alleviation. If land reform is to contribute to positive land-use regimes, research into land reform must concentrate on the effects of land tenure change on local labor markets, the distribution of benefits of land tenure change among various strata of claimants in the countryside, the sensitivity of 'reform sector' agriculture to macroeconomic changes of the kind described above, and so on. The bias of such research should be directed toward finding the property regime(s) most appropriate to such values as sustainability, putting the least first, and favoring rural over urban distribution values. But a first step toward determining whether there is a systematic land redistribution effect on land use would be to use land reform programs as an attribute when disaggregating country cases. Interestingly, using the models of international agricultural change described above (especially RUNS, which distinguishes between rural and urban benefits), one can discriminate among countries along these lines by using the surrogate of benefits to a given population.

Along the second dimension, land tenure reform generally has been implemented for purposes of redistributing the productive assets of the countryside, without any particular regard for conservation, either of land cover or of flora and fauna. Quite often, as is the case with the western Amazon or the northern Mexican frontier, land reform is undertaken through colonization for the explicit purpose of agricultural frontier expansion. In other cases, land reform fails in its broad ambition to satisfy landless claimants, and new 'empty' areas are opened up for colonization. In still other cases, new areas open up because of the collapse or exhaustion of land reform itself.

This suggests a set of criteria for evaluating land reform experiences in general, and especially the impact of land reform on land-use change:

- Future research should distinguish between countries with an expanding agricultural frontier and those with an essentially closed frontier. (A reasonable surrogate indicator might be average land in cultivation.) That allows the analysis to distinguish between countries whose agricultural potential is a function of increasing or changing output on a given land base and those with the possibility of bringing new land into production. Presumably there is a stronger tendency in the country with a closed frontier to intensify agriculture on existing land and to produce jobs in the nonagricultural sector to reduce pressure on the land.
- The accomplishments of land reform institutions should be compared with the impact of non-land reform agricultural institutions (agricultural credit, employment and income generation projects, etc.; Figure 3).
- The strength of the association between reform agriculture and external economic forces should be taken into account. For example, in Peru, much of the agrarian reform of the 1960s and 1970s was in the coastal river valleys, linked to export agriculture and the domestic commercial market. In contrast, Mexican agrarian reform has been tied overwhelmingly to domestic foodstuff production, and less to the export market and to cash crops (although that is changing over time). The impact of changes in policies in those two cases is radically different, in terms of both alternative land uses and impact on the rural poor.
- The relationship between agrarian reform and rural labor markets must be

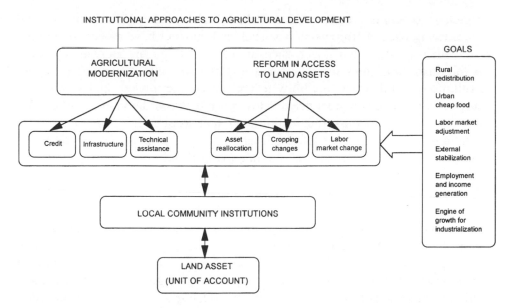

Figure 3. National-level land-use change institutions: institutional approaches to agricultural development.

assessed. On both sides, strong contentions are made that land reform either creates or suppresses potential rural employment for the poor.

The Spatial Aspect of Institutional Impact, or the Level of Analysis Problem Revisited

Urgent issues of degradation often govern analysis of land-use change. Removal of forest cover, expansion of the agricultural frontier, threats to flora and fauna, and the on-site agents of land-use change capture policy analysis, for obvious reasons. However, research on institutional vectors governing land-use change often neglects influences that are geographically remote from the most obvious points of impact. Two examples relate to the analysis in this paper and suggest directions for future research.

• The privatization/destatization literature pushes for removal of the public sector's perverse incentives, with the clear premise that point-of-impact subsidies and transfers pull economic opportunists to fragile areas and motivate environmental degradation through land-use change. In fact, removal of government subsidies may reduce deforestation, though the data are very controversial and so wrapped up in the confounding effects of economic recession, general fiscal austerity, international pressures, etc., that it is virtually impossible to establish a causal relationship between privatization and reduced rates of deforestation. But whether or not the removal of the economic role of the state enhances the prospect of proper land-use change depends on such variables as the 'connectedness' of the land unit in question to external market forces; the role of the

349

regional economic hub around which the land unit revolves; the presence/absence of strong state agencies that may not have an economic mission but may affect the governance of the land (e.g., the Bureau of Indian Affairs and its analogs in other countries); the strength of community-based institutions and their susceptibility to change via the market; and so on. One suspects that there is a great deal of difference between the viability of extractive reserves (or other usufruct institutions) in areas of strong community organization vs. areas of recent settlement and weak community organization; and that such land use will be viable as a function of its remoteness, its importance, the quality of its resources, among other factors.

• Too little attention is given once again to the formal analysis of international influences on local land use. As Fearnside (1988) has observed, in order to slow land-use change in the Amazon, the Brazilian government should look at land reform options, but in the Center South, where demographic and agricultural changes have pushed populations out into the Amazon (see also Skole, this volume). What are the dynamics that govern commercial agricultural land-use change, which in turn has such drastic impacts on the rural poor, the agricultural frontier, and the prospects for land-cover conservation? The answer to date has been to retreat to the closed system approach to land-use issues, in which international—or even important national-level—influences go generally untreated (Pearce and Turner, 1990) or in which it is assumed that international development assistance institutions can counteract the harmful land-use effects of agricultural modernization and development by creating more programs to invest in biological diversity and resource conservation (McNeely, 1988).

The Question of Institutional and Resource Vulnerability

Institutional vulnerability cannot be determined without proper regard for context. Generally, institutions identified with sustainability are weak, compared with those associated with agricultural modernization and land-use intensification. But such a generalization is tempered by the remoteness of the institution from external agents of change, the strength and duration of external attack, the sensitivity and resilience of the resource around which the institution is built, and the permeability of the barrier between local-level institutions and external forces (Figure 4).

A further determinant of institutional and resource vulnerability is how 'change variables' are loaded. For example, top-down vs. bottom-up strategies affecting land use also privilege different institutions, with predictably different outcomes. Local institutional strength or weakness tells much about the proper point of policy intervention; policy intervention also predicts to some extent the survivability of local institutions; the robustness of local institutions are connected to the management prospects of the land asset; and so on.

Continuing with Figure 4, it is useful to discriminate among land-use strategies: expansion with intensification, expansion without intensification, and intensifica-

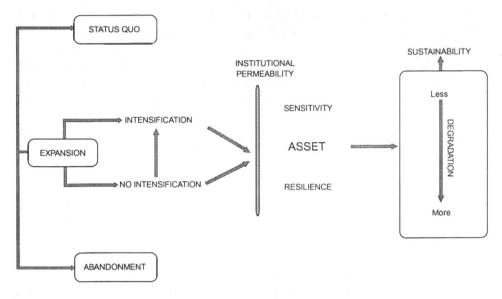

Figure 4. Dynamics of land-use change at the level of the unit of account

tion without expansion, and to link these strategies to the quality of the resource itself and to the affinity of certain institutions with certain kinds of strategies. It is also useful to distinguish between intensification and innovation.

What Is To Be Done?

Research agendas that address the question of institutional driving forces behind land-use change should direct themselves to: (1) finding global indicators of land-use change that correspond to high levels of institutional intervention at the international level; (2) creating a matrix of country or regional cases sensitive to the range of values possible within a given institutional set; (3) determining the sensitivity or 'permeability' of institutions at various scales to macroscale drivers; and (4) selecting out indicative cases to explore the significance of local institutional variation in the face of external mandates for change.

The objective of this agenda is to refine our understanding of the evident economic integration of nation-states through international institutions, and to discriminate among various local responses to that integration. The objective of the second strategy is to determine under what circumstances local laws, norms, and community values 'matter,' in terms of their capacity to affect external pressures to change land use and cover.

The central problem with such a strategy is that the tradition of institutional analysis leaves us with two deficiencies, demonstrated in this paper: (1) a limited understanding of the connection between levels of analysis and the related question of how to join scale-limited research designs, and (2) the unsolved matter of limitations in the data. How can these shortcomings be addressed?

Two strategies are appropriate, one theoretical, one empirical. Theoretically, it is essential to advance our understanding of level-specific or scale-specific institutional determinants of land-use change. This involves several issues:

• What institutional drivers are important at what scales? For example, international market prices drive balance of payments–connected decisions about cropping for export, but they may be rate modifiers at the subnational level and meaningless at the level of the peasant.
• Is there a hierarchy among scales in the social driving forces? That is, are there emergent properties of social systems that govern institutional impacts at different scales?
• Is there continuity among scales, or discontinuity?
• What governing models are most appropriate to each scale, and how many specific scale variations are necessary to reveal the dynamics of institutional driving forces?

These theoretical and methodological concerns, among others, are not well represented in the social science literature, and are crucial to the analysis of institutional driving forces behind land-use change. The task of addressing these theoretical issues is not daunting, but it does involve a systematic reconsideration of the best literature at specific scales and a careful set of propositions about their interconnectedness.

This is certainly linked to evaluating the arguments in this paper and to a mandate for further empirical research with protocols that encourage comparative assessment of field-based inquiry. A good example is international political economy research. A great deal of knowledge already exists in the international agricultural trade literature, but it does not focus on land-use change. First cuts at clustering countries along the variables described in this paper (openness, primary sector dependence, trade concentration, development assistance, public sector involvement in agriculture, etc.) could reveal a set of priorities for country-based research.

Second, these variables must be arrayed against what we know about land-cover change over historical time. This simple observation requires a multi-institutional effort to integrate physical data sets with social science data sets, and to associate variation in land cover at the regional or country level with international institutional interventions, in the form of trade, development assistance, conservation, legal conventions, and so on.

Third, it is important to choose a set of canonical cases, which might offer a template of the most important scale dividers and institutional variables. Using the recommendations of Working Group A (this volume), a multiscale research initiative using a few cases might link the empirical needs of global change research with the theoretical needs outlined above.

Finally, some difficult choices must be made about first-level surrogate indicators of institutional impact on land-use change. A range of possible indicators and attributes is suggested in the findings of Working Group B (this volume), but those must be refined in light of sound economic and political theory, and in light of their value in models.

Conclusion

The institutional dynamics of land-use change have transcended local and even national levels, but the treatment of those dynamics has not. This paper advocates a multipronged approach to understanding institutional driving forces behind modern land-use change: one begins with the international economic institutions governing agricultural output, and the other begins with local institutions that resist, modify, accommodate, retreat, or die before the more systemic processes that impinge on their once-isolated systems. On the macro side, the mandate is daunting but methodologically direct. There must be greater attention to the relationship between international institutions and local impact, first and foremost through agricultural markets and prices. On the micro side, the call is for more careful local case studies, but more importantly, for a coherent, replicable approach to such studies, so that comparable data are drawn and a consistent set of questions asked. Then, both macro and micro considerations must be folded into a new theoretical framework that takes into account the most appropriate models of institutional driving forces at various scales and the data demands of those models. Once these long-term research ambitions are satisfied, it will be interesting to begin to sort out the countries and the cases according to variables discussed in this paper, and to model the results. Whether or not it enhances our predictive power in the short term, greater understanding of institutional driving forces is certainly necessary for any analyst of the human and human-created foundations of global land-use change. And it offers a theoretical research agenda with exciting prospects for joining social science and natural science concerns for cross-scale analysis.

Acknowledgments

The author would like to acknowledge the contribution of Lisa Naughton to this paper. Others who commented on earlier drafts include Kiran Asher, Clyde Kiker, Robin King, Marianne Schmink, and Charles Wood.

References

Abramovitz, J. N. 1991. *Investing in Biological Diversity: U.S. Research and Conservation Efforts in Developing Countries*. World Resources Institute, Washington, D.C.

Blaikie, P., and H. Brookfield. 1987. *Land Degradation and Society*. Methuen & Co., London.

Bramble, B. 1987. The debt crisis: The opportunities. *The Ecologist 17(4–5)*, 192–199.

Capistrano, A. D., and S. Sanderson. 1991. The tyranny of the external: Links between international economic change and natural resource use in Latin America. Paper presented to the 1991 Conference of the Latin American Studies Association, April 1991, Washington, D.C.

Cerny, P. G. 1990. *The Changing Architecture of Politics: Structure, Agency, and the Future of the*

State. Sage Publications, Beverly Hills, California.

Crosson, P., and N. J. Rosenberg. 1989. Strategies for agriculture. *Scientific American (September),* 128–135.

Daly, H. E., and J. B. Cobb, Jr. 1989. *For the Common Good: Redirecting the Economy toward Community, the Environment, and a Sustainable Future.* Beacon Press, Boston, Massachusetts.

Deacon, R. T., and M. B. Johnson, eds. 1985. *Forestlands: Public and Private.* Pacific Research Institute for Public Policy, San Francisco, California.

de Boer, A. John 1989. Sustainable approaches to hillside agricultural development. In *Environment and the Poor: Development Strategies for a Common Agenda* (H.J. Leonard ed.), Overseas Development Council, Washington, D.C., 135–163.

de Janvry, A. 1981. *The Agrarian Question and Reformism in Latin America.* Johns Hopkins University Press, Baltimore, Maryland.

de Onís, J. 1970. *The Alliance That Lost its Way.* Quadrangle, Chicago, Illinois.

Fearnside, P. M. 1988. An ecological analysis of predominant land uses in the Brazilian Amazon. *The Environmentalist 8(4),* 281–300.

Feeny, D. 1988. The demand for and the supply of institutional arrangements. In *Rethinking Institutional Analysis and Development* (V. Ostrum, D. Feeny, and H. Picht, eds.), International Center for Economic Growth, San Francisco, California.

Gillis, M. 1988. Indonesia: Public policies, resource management, and the tropical forest. In *Public Policies and the Misuse of Forest Resources* (R. Repetto and M. Gillis, eds.), Cambridge University Press, Cambridge, U.K.

Goldin, I., and O. Knudsen, eds. 1990. *Agricultural Trade Liberalization: Implications for Developing Countries.* Organization for Economic Cooperation and Development, Paris, France.

Goodland, R., ed. 1990. *Race to Save the Tropics: Ecology and Economics for a Sustainable Future.* Island Press, Washington, D.C.

Hathaway, D. 1987. *Agriculture and the GATT: Rewriting the Rules.* Institute for International Economics, Washington, D.C.

Levy, S., and S. van Wijnbergen. 1991. Agriculture in the Mexico-USA Free Trade Agreement. Paper prepared for the CEPR-OECD Conference on International Dimensions to Structural Adjustment: Implications for Developing Country Agriculture.

Lipton, M. 1985. *Land Assets and Rural Poverty.* Staff Working Paper No. 744, World Bank, Washington, D.C.

Magee, S. P., W. A. Brock, and L. Young. 1989. *Black Hole Tariffs and Endogenous Policy Theory: Political Economy in General Equilibrium.* Cambridge University Press, Cambridge, U.K.

Mahar, D. J. 1989. *Government Policies and Deforestation in Brazil's Amazon Region.* World Bank, Washington, D.C.

Maniates, M. F. 1990. Organizational designs for achieving sustainability: The opportunities, limitations, and dangers of state-local collaboration for common property management. Paper presented at the 1990 Conference of the International Association for the Study of Common Property, October, 1990, Duke University, Durham, North Carolina.

March, J. G., and J. P. Olsen. 1989. *Rediscovering Institutions: The Organizational Basis of Politics.* Free Press, New York.

McNeely, J. A. 1988. *Economics and Biological Diversity: Developing and Using Economic Incentives to Conserve Biological Resources.* International Union for Conservation of Nature and Natural Resources, Gland, Switzerland.

Mitchell, W. C. 1968. The new political economy. *Social Research 35(1),* 76–110.

North, D. 1990. *Institutions, Institutional Change and Economic Performance.* Cambridge University Press, Cambridge, U.K.

Nurkse, R. 1961. International trade theory and development policy. In *Economic Development for Latin America* (H.S. Ellis, ed.), St. Martin's Press, New York, 234–274.

OECD (Organization for Economic Cooperation and Development). 1987. *National Policies and Agricultural Trade.* OECD, Paris, France.

OECD. 1989. *Strengthening Environmental Cooperation with Developing Countries.* OECD, Paris, France.

Ostrom, E. 1990. *Governing the Commons: The Evolution of Institutions for Collective Action.* Cambridge University Press, Cambridge, U.K.

Padoch, C., and W. de Jong. 1992. Diversity, variation, and change in riberéo agriculture. In *Traditional Resource Use in Tropical Forests* (K.H. Redford and C. Padoch, eds.), Columbia University Press, New York, 158–174.

Pearce, D. W., and R. K. Turner. 1990. *Economics of Natural Resources and the Environment.* Johns Hopkins University Press, Baltimore, Maryland.

Prosterman, R., and J. M. Riedinger. 1987. *Land Reform and Democratic Development.* Johns Hopkins University Press, Baltimore, Maryland.

Repetto, R. 1988. Overview. In *Public Policies and the Misuse of Forest Resources* (R. Repetto and M. Gillis, eds.), Cambridge University Press, Cambridge, U.K., 6–41.

Ruttan, V. W., and Y. Hayami. 1984. Towards a theory of induced institutional innovation. *Journal of Development Studies 20(4),* 203–223.

Sanderson, S. E. 1981. *Agrarian Populism and the Mexican State.* University of California Press, Berkeley, California.

Sanderson, S. E. 1989. Mexican agricultural policy in the shadow of the U.S. farm crisis. In *The International Farm Crisis* (D. Goodman and M. Redclift, eds.), Macmillan, London, 205–233.

Sanderson, S. E. 1990. Economy of political violence in the new Republic of Brazil. Working Paper, Centro de Estudos da Violéncia, Universidade de São Paulo, Brazil.

Spicer, E. 1962. *Cycles of Conquest.* University of Arizona Press, Tucson, Arizona.

Spooner, B. 1987. Insiders and outsiders in Baluchistan: Western and indigenous perspectives on ecology and development. In *Lands at Risk in the Third World* (P.D. Little and M. M. Horowitz, eds.), Westview Press, Boulder, Colorado, 58–68.

Tobin, R. 1990. *The Expendable Future: U.S. Politics and the Protection of Biological Diversity.* Duke University Press, Durham, North Carolina.

Uphoff, N. 1986. *Local Institutional Development: An Analytical Sourcebook with Cases.* Kumarian Press, West Hartford, Connecticut.

Van Arkadie, B. 1990. The role of institutions in development. In *Proceedings of the World Bank Annual Conference on Development Economics,* 1989, World Bank, Washington, D.C., 153–175.

Weisbrod, B. A. 1989. Rewarding performance that is hard to measure: The private nonprofit sector. *Science 244,* 541–545.

World Bank. 1990. *World Development Report.* World Bank, Washington, D.C.

World Commission on Environment and Development. 1987. *Our Common Future.* Oxford University Press, New York.

Young, O. 1989. *International Cooperation.* Cornell University Press, Ithaca, New York.

14

Culture and Cultural Change

Richard C. Rockwell

Anthropology and conservation philosophy have taught us that culture either determines or strongly influences how a people use their land. White (1967: 1205) asserted, 'What people do about their ecology depends on what they think about themselves in relation to things around them.' Presumably, the earth itself puts bounds on what a people can do with their land, but after acknowledging such biological and physical constraints, the canonical literature implicitly asserts the strong, direct influence of culture on land use. It illustrates that influence with a selection of case studies. The driving forces of intensification of land use and environmental degradation, such as rapid population growth, dense settlement patterns, excessive consumption, pollution, wasteful technology, and inept political systems, are argued to be cultural in origin, providing indirect paths for culture to influence land use. In the end, culture is widely seen as fundamental in human interactions with the environment.

Although hundreds or perhaps thousands of case studies have touched on the connections between people's ideas and how they use the land, not a single powerful generalization about those connections can be drawn with even minimal confidence. Little is known quantitatively, comparatively, or globally. With few exceptions, the studies reported in this paper have been qualitative explorations without statistical rigor, without an attempt to evaluate quantitatively the relative magnitude of cultural and other driving forces, and without a demonstration of whether the effects of culture are direct or indirect. There has been little effort to produce comparable data; the emphasis is instead upon what is unique about a people and their culture. It is rare for conclusions from field work to be stated so that they can be falsified, and rare for a second researcher to replicate an earlier study. Secondary analysis of data obtained from field work is uncommon. Because there has been no systematic pattern to the choices of places where researchers do field work, the collection of case studies is not representative of the world's peoples, and thus global generalizations cannot be drawn.

I am not alone in drawing these bleak conclusions. Kay (1985: 125) stated, 'This paper does not argue that a culture's environmental beliefs have no influence on its environmental impact, but rather that factors influencing belief and impact need exceedingly careful definition, and that the relationship itself ought to be viewed as a hypothesis to be tested, rather than as an implicit assumption.' Kay emphasized the heterogeneity of ideas in complex societies, questioning whether such a society can be aptly described as having a shared culture about the environment. She extended her questioning of the assumption that beliefs determine environmental behavior to a culture that today might be seen as simple, that of the pioneers on the American frontier.

Tuan (1968) argued that variation in culture does not necessarily lead to variation in environmental outcomes. Although the peoples of preindustrial China and Mediterranean Europe had sharply different attitudes towards nature, the result of human use of the land was the same: deforestation and soil erosion. The reverent Chinese attitude towards the land—humans were seen as part of nature, not its masters—nevertheless led to environmental degradation, a finding also reported in other settings by C.J. Glacken and by R. Dubos.

This paper does *not* defend the proposition that culture and cultural change have little or no influence on land-use/cover changes. To the contrary, the author tends to believe that culture does matter, even that it matters a great deal. The paper argues, instead, that there is not sufficient evidence upon which to decide the question. It poses a challenge for researchers: What kinds of evidence can be collected that would satisfy the skeptic? The paper concludes by sketching a system for comparative observation that would produce the necessary evidence.

This paper is a response to a question from the organizers of the workshop: Does culture have an *independent direct effect* on how a people use their land and how they change its cover, and if so, what is the magnitude of that effect? The operative words in this question are 'direct' and 'independent.' Culture may have *indirect* effects on land-use/cover changes through its influence on such factors as population growth, politics, economic development, and settlement patterns. Whether those indirect effects are large is open to question, because the indirect effects of culture are constrained by such factors as climate, soil type, natural resources, hydrology, native biota, transportation, and technology. The indirect effects of culture might even be inconsequential in the face of such environmental and technological constraints. We will briefly consider some of these indirect effects. However, this paper's major focus is on the direct and independent effects of attitudes, values, norms, and knowledge on land-use/cover changes—effects that are not mediated by other factors and that are additional to the influence of culture through other factors.

This paper does not ask the size of these independent direct effects, nor does it determine where culture ranks relative to other driving forces. That, it turns out, is a matter yet to be researched, and doing so will require substantial improvement in measurement of aspects of culture. Nor does this paper take up a fundamental question: Does land use affect culture more than culture affects land use? The primary direction of causation may, in fact, be the reverse of that implied by much of the literature. A people's beliefs may be strongly shaped by whether they are paddy farmers, nomads, or city dwellers, rather than the reverse. The primary determinant of which of these three lifestyles they pursue may be technology and environment, instead of any cognitive choices they have made, and their lifestyles may shape their belief systems.

The methodological perspective employed here is that of the statistically oriented sociologist, for whom falsifiable empirical generalizations are a central goal and reliable quantitative measurements a necessary step toward objective generalizations. A statistical perspective does not characterize much of the work on culture; indeed, the goal of generalization is often rejected as invalid on the grounds that a phenomenon can be understood only within its cultural context. A once-dominant perspective in cultural studies scorns the quest for objective generalizations, considering that quest to be part of a presumably antiquated positivistic or neopositivistic approach to social science. However, the mandate for this paper was to search for generalizations.

What Is 'Culture'?

In this paper, I have used 'culture' to mean verbal communications by individuals about their values, attitudes, norms, and knowledge. This is a definition that can be operationalized, and it is narrower than many definitions (some of which would encompass virtually the whole of the social science contributions to this volume). Ideas are central to virtually any definition of culture; this definition encompasses four different kinds of ideas, which may be thought of as 'layers of culture.' It omits a deeper layer, that of cognition, and a derivative layer, that of behavior.

Values are ideas about good and bad, about importance and nonimportance, such as 'People should work for their living.' Attitudes are evaluations of people, objects, or situations, such as 'I strongly disapprove of people being on welfare.' Although values are commonly seen as more fundamental than attitudes, the distinction is difficult to maintain in research; accordingly, I will treat the two as a package. Norms are expectations about behavior and sanctions for misbehavior, such as 'If people don't work, they should suffer the consequences of deprivation and low status.' When the force of the state or the ruler is behind them, norms become laws. (Laws can also be imposed on a population without their having a basis in culture.) And knowledge consists of statements about reality, such as 'There are enough jobs for everyone.' Science itself, as a system of knowledge, is a part of culture in this definition, but it shares its place with nonscientific and antiscientific modes of knowing about reality.[1]

---

[1] One could, of course, infer culture from nonverbal behavior, but the tautology is apparent when one then explains that behavior in terms of culture. Tuan (1968: 177) understates the problem when he suggests "It is misleading to derive the one from the other."

Data Collection Techniques

There is no global system of measurement for aspects of culture. As we cannot intrude inside people's minds, we must find ways of measuring their ideas on the basis of things that they say or write. Conventionally we employ informants, survey interviews, and written questionnaires to collect these data. There has never been a survey of the entire human population, and there is no collection of surveys or other instruments that could be cobbled together into a semblance of a worldwide survey. Instead, in the advanced industrial countries there is an unsystematic but somewhat comprehensive set of measurements, some of them in time series extending back 50 or more years. In the less developed countries there is often little or no measurement of anything but the most rudimentary demographic and economic variables, and even that measurement may be of low quality. In some countries, surveys are politically highly sensitive. The dangers in relying on surveys to tell us about the entirety of humanity are obvious.

In sociology and political science, the technology of surveys is highly developed. Sampling techniques are capable of producing estimates subject to small sampling errors, and these errors can be estimated with great precision. Interviewing and questionnaire construction methods are similarly highly developed. There is considerable understanding of the effects of variations in wording of questions, of the interviewer-respondent relationship, and of the effects of question order and question context.

In anthropology there has long been a tradition of conducting local 'case studies,' often focusing on the people of one village or one small region of a country. For many nonindustrial peoples the only information we have comes from anthropological case studies. These studies often rely on less formal methods than are found in survey research, including intensive use of open-ended depth interviews with selected informants and participant observation. Such methods rarely produce quantitative data, but they often yield data that seem to me to be of greater credibility than those gained by surveys. However, while case studies can provide a rich understanding of how and why a particular people live as they do, it is difficult to combine and compare the results of case studies in order to arrive at global generalizations. In addition, they rarely employ sophisticated sampling methods.

Culture as Shared Ideas

Conventionally, culture is conceived as something that exists apart from any one individual's ideas. Conversely, this paper accepts the proposition that in some societies ideas are relatively uniform from person to person, while in other societies there is at a minimum a variety of subcultures and in some cases little sharing of ideas about the human–nature relationship. The degree to which consensus exists is an important attribute of a society. There are, however, many possible meanings of the term 'consensus'; Rossi and Berk (1985) identify five different realistic meanings of the term and propose different models for empirically measuring consensus in a society. All of their models presuppose quantitative measurement of norms and other aspects of culture, and all assume that the data are representative of the society.

Values and Attitudes

This paper devotes most of its attention to values and attitudes, and less to norms and knowledge. This is largely because more is known about the former, but also because values are usually viewed as basic to all the other elements of culture. It is questionable whether that view is correct; arguments can be made that knowledge is fundamental, and that the road to changing values takes the direction of inculcating comprehensive and accurate knowledge—or, even more fundamentally, of changing cognition.

Public Opinion

Viederman reports that 'there has been a sea change in environmental consciousness as reported by myriad public opinion polls' (Viederman, in press: 1). A set of small-sample polls was conducted in 1988 by the Louis Harris Organization for the U.N. Environment Program in 14 countries, including 9 less developed countries. These polls found high levels of concern about the environment in all countries, and everywhere but Saudi Arabia people perceived that their environment was deteriorating. Although few perceived the occurrence of global environmental changes, most saw such problems as desertification and loss of farmland as very serious (Milbrath, 1991). However, Lowenthal (1990) remarks that the results of public opinion polls are inconsistent and not well connected to changes in behavior.

Perhaps the most important series of studies of values and value changes is that done by Inglehart and his colleagues since 1970. In more than 25 countries over the last 20 years, Inglehart's World Values Survey has collected data on people's basic values and societal priorities. He finds that these values are gradually shifting throughout advanced industrial societies. The shift is from 'materialist' values that emphasize economic and physical security to 'postmaterialist' values that *assume* economic and physical security and emphasize individual autonomy and quality of life.

The 1990 World Values Survey was the first survey in history to sample the majority of the human population, people living in 42 countries on six continents. Preliminary analysis of data has been completed for 9 countries: Canada, Chile, the Czech and Slovak Federated Republic, Japan, Mexico, Nigeria, South Korea, Spain, and the United States. These countries span a range of wealth, level of industrial development, economic system, ethnic makeup, national history, and environmental degradation.

In all of these countries, Inglehart finds evidence of a shift toward the postmaterialist values. Less pronounced in Latin America, Eastern Europe, and Africa than in Western Europe, North America, and East Asia, in most countries the shift partly involves a difference between the generations: younger generations, who typically have spent their formative years in greater security than did older groups, are more likely to profess postmaterialist values. And the shift is buttressed by a professed willingness to make financial sacrifices to protect the environment, ranging from a majority of respondents in polluted South Korea to one-third of the respondents in poor Nigeria. Inglehart finds that persons with postmaterialist values are three times more likely than those with materialist values to report that they have voted for environmentalist parties in countries where such parties exist. They are five times more likely to report being members of or being willing to consider membership in environmentalist movements. To the extent that individuals accurately report their behavior, this is evidence that these verbal communications about values correspond with other forms of behavior (Inglehart, 1991).

However, as Lowenthal notes (1990: 126–127), 'The chasm between Green words and deeds is awesome.' The *Wall Street Journal*, reporting on a 1991 poll, observes (Gutfeld, 1991), 'Americans *say* they are willing to make sacrifices for a better environment. But what they *do* is another story.' If the amount of recycling reported in surveys were actually observed in practice, much of our landfill problem would already have been solved. Moreover, California, whose citizens routinely profess environmentally sensitive values on surveys, voted down major (expensive) environmental initiatives in a recent election.

Survey researchers have long known about and taken precautions to avoid eliciting socially acceptable responses. It is, for example, difficult to get people to agree with openly racist items in an interview, because today the social pressures against racism are strong and operate in the interview context as much as in other contexts. Some researchers have attempted to measure the extent of such responses. A study of water conservation efforts in Los Angeles (Berk *et al.*, in press) included questions designed to 'trap' respondents who were inclined to exaggerate their water conservation practices. The questions asked whether respondents had taken a series of general conservation steps, including recycling light bulbs (2% claimed to have done so) and installing ferronic input-output devices on their water heaters (5% claimed this impossible act). More than 20% of the respondents were trapped by at least one of the seven fictitious items. The number of water conservation practices reported by respondents was strongly related to the number of fictitious items that respondents checked; some respondents were apparently struggling to appear 'green.' The collection of this evidence of enthusiasm for 'green' actions permitted the authors to deflate their estimates of actual compliance with water conservation measures, whereas a naive survey would have found too high a level of compliance.

Nevertheless, surveys remain one of the most powerful tools for studying culture and cultural changes, and by no means should we set aside their results in despair that they do not reflect behavioral reality. Instead, we must recognize that there remain major questions about how values and attitudes connect with behavior. This linkage has been one of the abiding problems of the social sciences, but seldom has understanding it had more social and political importance than here.

There is a need for continuation of surveys like the World Values Survey and for devising new ways to test the attitude-behavior linkage. And it is imperative for these survey researchers to employ methodologies that are less subject to artifacts such as social acceptability.

Ideologies

The world view or paradigm through which humans approach nature has been the subject of study by many social scientists. In addition to relying on surveys, they have studied values and attitudes through the examination of religion, philosophy, art, and political statements. Indeed, to many anthropologists, it is in such places, rather than in individual responses to questions, that one should look to find culture.

Lowenthal (1990), in an outstanding review of changes in European ideologies, discerns a transition through three phases of ideology about the relationship of humankind and nature. In the first phase nature is seen as being very much controlled by humanity; natural catastrophes happen because of human sin or human omissions, and nature can be placated by such steps as human sacrifices. In the second phase nature is seen as independent of humans but basically threatening, and it is the job of humankind to subjugate nature by destroying the dark forests and taming the wild rivers. The clearing of forests is thus not simply a means of providing for agriculture and settlement; it is also a means of asserting control over a threat. In the third phase nature is seen as both fragile and the broad base of human life, existing in a continually renegotiated interdependence with humans. These ideologies are represented in religious beliefs, literature, art, folk myths, and popular beliefs. Other researchers (e.g., Bennett and Dahlberg, 1990; Kates *et al.*, 1990) have described variants of these basic beliefs.

A second view is that human behavior is determined by nature—in its most extreme form, that humans cannot act otherwise than in their immediate and selfish interests of procreation and survival. This is a point of view against which the social sciences have fought for many years, although following the publication of E.O. Wilson's *Sociobiology* a sophisticated form of the perspective has again become interesting to social scientists.

The third view is that humans are potent modifiers of nature, dependent upon it but also able to change it in substantial and sometimes irreversible ways. This is the primary perspective underlying today's scientific approaches to problems of global change. Indeed, the concept of culture as a driving force of land-use/cover changes is rooted in this perspective.

Rayner (1990) identifies four variants of 'nature myths': nature is fragile, robust, resilient, or capricious. He connects these myths to differences in 'moral imperatives,' preferences in response strategies, and types of social organizations for dealing with nature. Noting that the typology is a heuristic device, defining pure types rarely found in practice, he remarks that societies are strengthened when they have a diversity of points of view that can provide different kinds of early warnings, diagnoses, and options for response. To Rayner, not only do the myths play a positive role, but their diversity does as well; consensus on one myth might be undesirable. This typology, perhaps more than most, lends itself to quantitative measurement in survey research, and some pilot studies have confirmed that the typology can be effectively used to generate comparable data.

Many Christian denominations in the West are now engaged in a substantial rethinking of the implications of Christian teaching for the human–nature relationship. In doing so, they are exploring alternative conceptions of that relationship, including Eastern religions and the thinking of Native Americans. They are attending to an odd aspect of the creation myth, in Genesis 2: having created plants and then Adam from the dust of the ground, the creator saw that it was not good that Adam was alone and set out to make 'an help meet for him.' That helpmeet was *not* Eve but instead 'every beast of the field, and every fowl of the air.' After some exploration of this arrangement, Eve came along when 'for Adam there was not found an help meet for him.'

Kempton *et al.* (in preparation) note the possible emergence of a land ethic in the United States. Aldo Leopold stated this ethic in 1949 (quoted in Kempton *et al.*): 'A thing is right when it tends to preserve the integrity, community, and beauty of the natural environment. It is wrong when it tends otherwise.' The extent to which this preservationist ideology prevails in U.S. culture (or the extent to which it makes a difference in behavior towards the environment) is not yet documented.

Catton and Dunlap (1978) have identified a 'new ecological paradigm' that takes as fundamental the biological relationship of the human species with the environment. They sketched three elements of this 'paradigm': '(1) Human beings are but one species among the many that are interdependently involved in the biotic communities that shape our social life; (2) Intricate links of cause and effect and feedback in the web of nature produce many unintended consequences from purposive social action; and (3) The world is finite, so there are potential physical and biological limits constraining economic growth, social progress, and other societal phenomena' (cited in Buttel, 1986: 345). Dunlap has measured the extent to which this set of beliefs is held in Western populations, providing evidence that some people do endorse these beliefs. However, Buttel questions whether even a 'paradigm shift' towards this new ecological paradigm would be a significant force for pro-environmental social change (Buttel, 1986).

These slightly different ways of classifying the prevailing ideologies about the relationship of humankind and nature ring true, although their largely Western origin raises the possibility that different kinds of ideologies may be found elsewhere. However, with the exception of Catton and Dunlap's studies, none of them is based on systematic empirical research in general populations. Moreover, none of them demonstrates that these ideologies make a difference in how human beings deal with the land. Intuitively, an ideology *should* make a difference: in the absence of hunger or greed, should not a people who see nature as fragile treat the land more gently than those who see it as robust? The demonstration of this proposition is lacking.

A Dahlem workshop report (Ausubel, in press), in a list of 11 questions, called for studies to answer the following: '(1) Can we describe better and more fully world views and their configuration?; (2) How do diverse world views constrain action at the global level?; (3) At thc national level, how do diverse world views influence choice of policy instruments?; ... (7) How do diverse world views influence the sense of urgency that different groups hold about global environmental change?; ... (11) To what extent can environmental and energy technologies be designed around world views?' Pursuing these questions, as well as the question of whether world views make a difference in such individual behaviors as energy use, recycling, and farming practices, would considerably advance the understanding of how ideology relates to environmental issues. Among the few studies in this area is Glacken (1967), but this study ends with the 18th century.

Norms

Norms specify expectations for how people will behave and sanctions for not conforming to these expectations. Norms are the means by which values are (perhaps imperfectly) translated into behavior, or at least into expectations for behavior. In their institutionalized form, norms become laws and regulations. In this transformation, the power of the state or of the ruler is placed behind norms; without that power, norms have force through social disapproval, banishment, and similar informal sanctioning mechanisms.

Laws

Laws are probably the most potent way in which ideas directly affect land-use/cover change. This occurs primarily through systems of property rights. Property rights norms also exist in prelegal societies, but the critical fact of the modern era has been what Richards (1990) calls the increasing power and scope of the 'centralizing state.' On frontiers, the state asserts and then seizes direct control; within settled areas, the state defines and redefines land-use rights through systems of land tenure, restrictions on land use, and provisions for and restrictions on the exploitation of resources. Richards notes that the modern, centralizing state has assiduously extended its control over every tract of land within its boundaries. He argues that despite speculation as to which driving forces are primary in land-use changes on a global scale, 'Nevertheless, state action repeatedly plays a primary role in exploitation of the world's lands' (1990: 163). In Barnes's (1988) conception of politics as the organization of power around a set of values, laws are therefore either a central or the central linkage between values and land-use/cover changes.[2]

Thus property laws are of enormous importance in understanding land-use/cover changes. On the Amazonian frontier, for example, simply extracting for-

---

[2] To Barnes's definition of politics, I would add that politics can also be organized around self-interest.

est products is not sufficient to establish land tenure rights; as on the American frontier in the 19th century, 'development' is required to claim ownership—and in both cases, destruction of the forest has been one of the clearest ways to signal development intentions. Given this importance of laws concerning property rights, it is surprising that there has never been a comparative study of landed property rights. The first such study, under the direction of Richards, will be initiated by the Social Science Research Council in the fall of 1991.

One particular kind of law may be the single best reflection of values: the tax code (an insight attributable to S. Viederman). In the United States, fallow agricultural land and forests near urban areas are usually taxed so as to encourage or force development. The conservation tax credits of the 1970s are gone, but subsidies are urged for the development of fossil energy sources. There is a push for a cut in the capital gains tax, but no proposed tax on the depletion of natural resources. Tax subsidies are provided for logging and technological development, without regard for whether these processes conserve resources or replenish the earth. Although there is hardly anything consensual about the tax code, it is unquestionably the result of a balancing process in which conflicting values are compromised—and in which disfavored values play little role.

The Informal Regulation of Behavior

At the more informal level of norms concerning land use, there is little information available. We do not systematically know people's expectations for the use of the land nor the sanctions they would seek to impose on people who violate those expectations. (Norms regulating water and pasture use, often embodied in laws, are probably the most widespread and highly developed of environmental norms.) This is an ideal field for investigation by the discipline of human ecology, with its focus on people, organization, the environment, and technology.

There have been studies of norms concerning the use of 'central places.' Firey's study (1947) of the social and cultural sanctification of the Boston Common tells us about norms regulating the use of that land, but this line of research has been little absorbed into Firey's field of human ecology. In general, the study of norms has lagged behind the study of values and laws, even for norms concerning other arenas of human existence than that of the land; J. Gibbs's work of the 1970s attempted but largely failed to revive a field of study focused on norms as an aspect of social control.

Knowledge

The field of cultural ecology has concerned itself with what indigenous people know about the land and its ecology, and about their relationship with it. From these studies it would be easy to infer that tribal peoples of the past treated the land more gently than we do today. There is some evidence of that: Richards (1990) points out that shifting cultivators, hunter-gatherers, and pastoralists created landscapes that tended to be sustainable over extended periods, and that on the whole their environmental impact was restrained. But by no means did these peoples have no impact on their environment, and in some cases their impact was highly destructive. One of the hypotheses concerning the abandonment of the palace city of Masvingo in the 15th century in what is now Zimbabwe is that goats destroyed surrounding vegetation and caused major soil erosion. Farb (1978) describes indigenous methods of slaughtering buffalo on the American plains of the 19th century that were extraordinarily wasteful. Tuan (1968) points out that Plato, in the fourth century B.C., still saw signs of the arable hills, fertile valleys, and forested mountains known to a previous generation of Greeks, but he also saw bare lands and mountains that could provide food only for bees.

Unfortunately, some of the cultural ecology literature has tended towards an 'ecological mystique,' claiming that early hunter-gatherers and subsistence agriculturalists lived in an 'instinctive ecological harmony' with their environments. Bennett and Dahlberg (1990) suggest that cultural ecology tends to describe tribal peoples as having achieved a reasonable balance with nature by becoming part of the natural ecosystem. Such a balance, if it existed, need not have depended upon instinctive knowledge about the environment or even upon trial-and-error experimentation. Instead, these tribal peoples may have maintained a small and slowly growing human population because of high infant mortality, may have placed limited demands on the environment because they had limited knowledge of other possibilities, and may have employed simple technologies that were not capable of making large-scale changes in the environment. No understanding of the human-nature relationship would have been required; the continuation of customary practices would have sufficed.

Richards suggests that indigenous knowledge about dealing with the land, when and if it existed, has largely disappeared in the face of modern patterns of land use. The people have lost or set aside their ecological wisdom and their ways of understanding their environment, and they are even losing the languages in which that knowledge was conveyed. Like the last of the Itza-Maya in the dwindling Peten rain forest of Guatemala, they may be unable to act on the knowledge they have, although they know 'Dying is the forest, dying are the animals, dying also are the people. To live, we need the forest and the forest needs us. We must care for the forest and the forest must care for us' (Declaration of the Itza-Maya, September 1991).

Thus modern land use has 'put an end to a rich source of human understanding about varied ecological systems' (Richards, 1990: 176). Reconstruction of earlier patterns of land use through archaeology may provide clues to this ecological wisdom. There is much to be gained by reconstruction of and experimentation with such technologies as the cultivation of the *ramon* breadfruit that is the remaining hope of the Itza-Maya. However, whether that knowledge will be germane to today's intensified need for resources (generated by higher levels of consumption and a larger population) remains to be explored.

The knowledge of modern, industrialized peoples about the land and their relationship to it has been explored by Kempton and his associates (Kempton, 1991). Their approach is that of cognitive anthropology, relying on 'depth interviews' of a small group of respondents. These interviews begin with open-ended questions and elicit responses that have proved difficult to elicit with standardized question- naires and interview schedules.[3] Kempton finds that a diverse group of U.S. resi- dents is aware of the possibility of global climate change and of other environmen- tal issues, but that there is minimal understanding of the scientific processes. A frequent lay understanding of global warming is that it is part of or an effect of ozone depletion. The sources of ozone depletion are similarly misunderstood (and outdated), with some blaming it on pressurized spray cans but not aware that air conditioners play a role.

It is not surprising that the lay public does not understand the science of envi- ronmental change; many of those who have had the opportunity to learn, from politicians to the scientists themselves, are confused. There is every reason for the public to be bewildered when over a period of a few years the newspapers carry a succession of articles forecasting a cooling trend because of variations in the earth's orbit, then articles forecasting global warming because of the greenhouse effect, and finally articles forecasting global cooling because of a volcano's erup- tion. The fact that all of these forecasts could be correct is incomprehensible to most citizens.

Lack of knowledge can play as large a part in land-use/cover changes as can having knowledge. It can lead to ineffective policy decisions, manipulation of public opinion by politicians, and inability to plan coherently for the future. Arizpe's work (Arizpe and Velazquez, 1992) among the people of the Lancondon rain forest in Chiapas, Mexico, shows that they are aware of the possibility of global warming and understand that cutting down their trees could contribute to it. One informant's suggestion that if Northerners are so worried about this they should rip up their sidewalks and plant trees in their place may seem to miss the point until placed alongside the plan of the present U.S. administration to plant 1 billion trees every year for the next ten years—without, as Kempton notes, taking effective steps on energy policy.

There is no doubt that this aspect of culture, knowledge about the earth and the relationships of humans with it, plays and will continue to play a singularly impor- tant role in how humans use the earth. There is also, I am afraid, not much reason to anticipate a quantum change in the population's level of scientific understand-

---

[3] A serious caveat must be offered before Kempton's data are accepted: his sample consists of 38 persons, 12 of them living near Trenton, New Jersey, and the remainder in rural central Maine (Kempton *et al.*, in preparation). Four of the "informants" in the Trenton area were selected by being approached in parks or shopping centers, and the other eight were selected on the basis of the external appearance of their homes, so as to maximize diversity. The Maine informants were selected from the acquaintances of the interviewer (who was based in Maine), so as to maximize diversity and environmental opinion. With this nonrandom "sampling" technique, the risks of draw- ing a biased sample are real and serious, and without random sampling, there is no way to estimate to what degree sampling error may be a factor. It is reasonable to assume that these informants do not represent the people of the United States, but it is also conceivable that they do not even represent the people of suburban Trenton and rural Maine.

ing. On the evidence of the failure of U.S. public schools to teach elementary science, it is at best an open question whether environmental education of the general public is a realistic possibility. It is, I suspect, the elite—in particular, business leaders, politicians, and the media—for whom education is most important. Elite knowledge will have major direct relationships to human use of the land. The broader public will be affected, as it always has been, by appeals to values and attitudes.

One elite culture could play a singularly important role in how the world deals with global environmental change: the culture of the international scientific community. Based in growing knowledge about the intricate relationships of physical, chemical, biological, and human systems, this culture is also informed by shared values, attitudes, and norms. The culture is not without its schisms: the debate about the possibility of greenhouse warming, in particular, stirs scientists' emotions as well as multiplying their viewgraphs. The debate between population biologists and macroeconomists about the linkages between population growth and environmental degradation begins with radically different assumptions and progresses to an ideological gulf of vast proportions. Despite these divides and differing national allegiances and world views, the international scientific community has far more in common than does virtually any other international community.

Biases

At several points in this paper I have emphasized the danger of relying on studies that primarily come from the Western world. An equally important caution turns upon the fact that there may be a gender bias in much field research: when male social scientists step outside their own societies, they routinely collect data from those whom they perceive to be the leaders of villages, i.e., from other males. Women may have a different relationship to nature than do men. The ideologies of domination and control are arguably more associated with the male psyche. It has been argued that the biological function of childbearing creates in women a different relationship with nature. Some have claimed that many of the environmental movements in the Third World have been formed by women and are based in women's concerns for children and the provision of food, water, and fuel. Others have seen in male behavior towards the land another instance of the dire effects of testosterone poisoning.

The hypothesis of a gender difference merits considerable investigation. Contemporary research more often uncovers new differences between men and women than it undermines the idea that differences exist. Some of these differences are cultural, some socioeconomic, and some biological, but all may be connected in important ways to how humankind uses the land. One organization is conducting notable research on this topic: Development Alternatives with Women for a New Era, a group of women primarily based in the less developed countries.

A similar bias appears in studies of environmental movements and environmentalist ideologies, which are often focused on the middle class in the relatively rich societies of the North Atlantic. This overlooks the environmental movements among the poorest in the poor societies of the world, some of which are of long standing. Martinez-Alier (in preparation) argues that poor people's struggle against poverty is simultaneously a struggle for sound environmental management and against the rich and their exploitation of the resources of the land. Milbrath (1991: 17) says that 'environmental movements in [less developed countries] are best characterized as victims' movements.' The victims of poverty are also the victims of environmental degradation, and their poverty itself leads to environmental change.

There is little empirical evidence of an 'environmentalism of the poor.' Anthropological studies of tribal peoples are largely beside the point, despite the painful stories that they often tell. What is required is research about the majority of the world's peoples: poor people who stand to benefit or suffer from economic growth, whose livelihood is influenced by national and global economies. Such a study is now beginning at the Social Science Research Council under the auspices of the Committee on Latin American Studies.

Studies of People and the Land

The preceding sections have disaggregated the various layers of culture, pulling apart values, attitudes, norms, and knowledge. This section concentrates on several studies of the interactions of people and the land: people of the tropical forest, people at work as farmers and hunters, and people living under the influence of European colonization. Each of these studies encompasses more than one people, so comparative analysis is possible.

People and the Tropical Forest

Few examples of land-use/cover change receive as much scientific, policy, and media attention as does the destruction by humans of the tropical forests. The U.S. Man and the Biosphere Program has sponsored a variety of research projects in this area, 20 of which are summarized in Lugo *et al.* (1987). This research encompassed forests on three continents and involved researchers trained in many disciplines, including the social sciences.

The major finding relevant to the consideration of culture and cultural change as driving forces of land-use/cover change was that indigenous peoples may have sound management methods for their own conditions. 'Isolated groups in Peru, Zaire, and Borneo have developed productive systems adapted to conditions which befuddle modern science. … In some circumstances, stable, intensive agriculture can be practiced with high rates of production.' (Lugo *et al.*, 1987: viii).

The Lun Dayeh tribal people of interior Borneo provide an example. These people have essentially abandoned shifting cultivation in favor of permanent, irrigated wet-rice cultivation. Producing an excess of rice, their system is 'particularly well suited to the low human density conditions existing in the Lun Dayeh homeland' and appears to require far less human labor than usually assumed necessary for paddy cultivation. This is said to stem not from favorable conditions but from 'the intimate knowledge of their varied environment and the well-adapted resource exploitation techniques they employ' (Lugo *et al.*, 1987: 10–11). Perhaps as a result of this success, which includes their continued willingness to explore new products and techniques, the Lun Dayeh clear very little forest each year despite a growing population.

Vayda makes a similar point in his study of the shifting cultivators of an Indonesian forest province with a growing human population. He discusses the usefulness of tapping the people's knowledge of their diverse environments and the rationality of their decision-making processes. However, Vayda observed 'massive' environmental changes and argues that the people 'engaged in environmentally damaging activities not because other means of gaining subsistence were unavailable, but because the activities were more profitable than the perceived alternatives' (Lugo *et al.*, 1987: 15).

A study by Werner and others focused on the tropical dry forest of northern Australia, in particular on the country's largest national park and its aboriginal residents. Among the motivations for setting aside this land for conservation is its spiritual significance to these people. They consider most of the natural features of the park to be 'permanent manifestations of the actions of legendary ancestors' and are obliged by their value system to protect these sacred sites. This brings them into conflict with mining interests, forestry, and cattle ranching. They are also in conflict with tourism, but they share with tourists a preference for maintaining populations of feral water buffalo in the park despite the demonstrated damage that these animals cause to vegetation.

Tribes and the Production of Food

Apart from these studies, Lugo *et al.* includes little that addresses the question of whether values, attitudes, norms, and knowledge matter. Vayda has, however, collected another set of studies (Vayda, 1969). Geertz's study in this volume explicitly attends to these issues. Drawing on Freeman's (1955) observations, Geertz notes that the Iban people of Indonesia are swidden farmers. Although below maximum population density, they have expanded into the areas once occupied by other tribes. The Iban seriously overcultivate their land, using a single plot for three years or more in succession. They have a settlement pattern involving the construction of large villages, which makes shifting between plots an onerous task. But there are cultural reasons for their cultivation practices as well: Geertz thinks that the Iban may have 'a superior indifference toward agricultural proficiency' and 'a warrior's view of natural resources as plunder to be exploited' (Geertz, 1969: 16). He finds a 'historically rooted conviction that there are always other forests to conquer.' In addition to showing that among at least one tribe, values and knowledge do have an effect on land-use/cover changes, Geertz also refutes the universality of any claim that primitive peoples live in an instinctual harmony with their environment.

Geertz finds a very different relationship of humans to the environment in the wet-rice farmers of Java and other areas of inner Indonesia. The wet-rice agricultural regime often results in the support of an ever-increasing number of people within a habitat that suffers little further damage beyond the initial preparation of paddies. Expanding into new terraces requires considerable preparatory labor, including the building of new irrigation works. This makes intensification of cultivation of existing paddies preferable. Values or other aspects of culture, it seems, do not regulate this use of the land; economic rationality does.

These and other cases do not provide consistent evidence that tribal peoples have or act upon an innate, benign knowledge about the environment. Like people in advanced industrial societies, some tribal peoples have found better ways than others to deal with the land, while still others have destroyed it. Moreover, the U.S. Man and the Biosphere study suggests a number of ways in which modern science and engineering can improve the uses tribal people make of the rain forest while reducing the damage they do to it. Some of these include such simple techniques as genetic selection for desirable crop traits; others, changes in the spacing and size of tracts of forests to be cleared.

European Influence on People in the Neo-Europes

Crosby (1986) extensively documents how the expansion of European powers greatly altered the face of the earth. Europeans and their descendants are scattered all over the world. So are the plants Europeans brought with them, the agricultural technologies they favored, the kinds of houses they built, the industries they developed, and the nation-states by which they governed themselves. This expansion sharply altered the relation of people to the land, in many cases eliminating the indigenous cultures and peoples. This process continues today in the Yanomami territories in the rain forests of Brazil and Venezuela.

The land was transformed by the European expansion. For example, the island of Madeira was entirely covered by forest when Europeans discovered it; although some of the trees were cut for export, most were destroyed in massive fires that were set to clear land for settlement, agriculture, and pastures. Folk tales describe fires that burned for seven years, and in at least one case the settlers had to take to the sea in boats to avoid the fires they had set. Madeira became the world's largest producer of sugar, but the great forest was gone (Crosby, 1986).

Dramatic as was this expansion of European ways into the rest of the world, is there evidence that it made any difference for use of the land whether the occupying culture was French, British, Spanish, Dutch, or Portuguese? Crosby provides no evidence that the differing cultures of Europe made a difference in agriculture, forestry, or animal husbandry in the 'neo-Europes.' Although Myers and Tucker (1987) argue that Iberia exported to Latin America the high value its people placed on cattle ownership, the social standing and political power conferred by cattle ownership is found elsewhere among simple peoples who were not subject to Iberian influence; indeed, among some of these people, cattle are the chief monetary unit.

Does Culture Matter?

In the broadest sense of 'culture,' *only* culture matters in human relations with the land, because the broadest definition encompasses virtually all that is human. But in the narrower sense in which the concept has been employed in this paper, there is little evidence for whether culture and cultural change are among the primary driving forces. In fact, land-use change may be a driving force in cultural change: the biological and physical constraints on land use experienced by a people may powerfully shape their knowledge, values, attitudes, and norms regarding land use. One need not adopt an environmental determinist's viewpoint to recognize that there may be powerful feedbacks from the environment to culture. Values may *in general* be as much driven by behavior as vice versa.

This paper has focused on the independent direct effects of culture on land-use/cover changes and has not found strong evidence that they do or do not exist. Yet most of the literature strongly argues that culture has effects of some sort, without distinguishing between direct and indirect effects. Perhaps the more significant effects of culture are not indiscernible direct effects but instead the indirect effects through population growth, politics, laws and other institutional arrangements, social movements, and demand and consumption. Culture is believed to have strong effects on each of these factors, and each of them to have strong effects on the land. We have already considered one such indirect effect, the effect of culture through laws and the tax code. We now briefly expand the discussion of culture's indirect effects to include population growth, settlement patterns, social movements, and politics.

The Indirect Effect of Culture through Population Growth and Urbanization

Values and attitudes have some influence on the size of a population and its rate of growth through their influence on age at marriage, fertility, child spacing, and perhaps mortality. Stott (1969: 102–103) argues that 'the most evident devices for limiting human population are cultural in character.' Ehrlich and Ehrlich (1970) reaffirm this proposition, arguing that population growth—which they see as the primary driving force of environmental degradation—is fundamentally a product of values and attitudes. However, most demographers would see this view as an overly simplistic explanation of why populations have grown and declined.

The great growth spurt of the human population began in the 1700s. It occurred nearly simultaneously in far-distant places, such as western Europe and China, where there is little reason to believe there was a simultaneous change in culture. It reached its most dramatic levels of proportional increase in the New World, where North America's share of the world's population increased eightfold. Despite a catastrophic decline in the indigenous population of South America, that region recovered its pre-Columbian population by 1850 and became for a time the fastest-growing region of the world (Demeny, 1990).

Although the reasons for this human growth spurt will never be fully understood, there is little evidence that a change in values and attitudes precipitated it. A decline in mortality, particularly in infant mortality, was critical, and this was due to more and better food, improved housing and sanitation, and perhaps a decline in the virulence of epidemic diseases such as plague (Demeny, 1990). Cultural change lagged behind these changes: even when it was no longer necessary for a family to have large numbers of children to provide farm labor and to ensure a sufficient number of survivors to provide for parents' security in their old age, women still bore many children.

Values also play a role in the kinds of settlement patterns that people prefer. There are people who consider a neighbor who is within sight as too close for comfort, while others revel in the human density of conurbations. But it is by no means clear that the tropical megacities of the less developed countries are being built upon a foundation of chosen values, unless those choices are defined to include the most basic human needs and wants. People may migrate to the cities not because that is the life they would prefer but because they prefer life.

The Indirect Effect of Culture through Demand and Consumption

Kates *et al.* (1990) hypothesize that values and norms may have a greater impact on the *means* by which transformations of the earth take place and on the feedbacks of these transformations to humans themselves than they directly have on the *kinds* of transformations that occur and on the scales at which they occur. There is, however, little evidence on which to evaluate such an hypothesis.

They point out that values and norms clearly have a major impact on demand (that is, on the expected material standard of living) and on the appropriate means for reaching that standard. Sack (1990: 669–670) argues that consumption allows an individual to create a 'magical sense of context and causality' that draws together 'elements from nature, meaning, and social relations.' Can it be determined whether economic, cultural, or other factors predominate in mass consumption?

Simon (1982) has argued that one of the most significant discoveries of the social sciences is that people judge their present condition by the use of a rubber ruler. The zero point and the length of this ruler are determined by one's own recent experience and by one's perception of the living conditions of one's neighbors. In the modern communications world, one's 'neighbors' may be persons seen on television from half a world away. To the extent that the media convey an image of the satisfying life as one based on consumption and waste, people are likely to value high levels of consumption. And, at least with existing technologies, the desired high levels of consumption exact more from the land and foster land-use/cover changes.

Many writers see bringing about changes in demand to be an essential step for reducing land-use/cover change, environmental degradation, and other forms of environmental change. Our supposedly consumption-oriented Western societies must, in this view, learn a different set of values tied to lowered consumption. During (1991: 162) augments this prescription, suggesting that 'the major determinants of happiness in life are not related to consumption at all' but instead to 'timeless virtues of discipline, hope, allegiance to principle, and character.' He proposes that we might be happier with less, or, more exactly, with an economy built on sufficiency and permanence rather than on excess and discard. Others look for lessons in values to Chief Seneca and to the Hopi, Eskimo, Onondaga, and Northwest Coast peoples. The search is on for an alternative world view.

The Indirect Effect of Culture through Social Movements

Ideologies give rise to environmental movements that attain political notice and sometimes political power. Today's movements include the Green parties of Europe (which some see as of declining importance), feminism, and environmental research and action groups, as well as more focused or fringe movements such as 'deep ecology,' regenerative agriculture, appropriate technology, deindustrialization, and negative population growth. These movements serve, at the least, to place the environment on the national agenda; their acceptance into mainstream ideology may not be essential for them to have considerable political effect. Further study is needed to understand the influence of these environmental movements.

The Indirect Effect of Culture through Politics: The U.S. Example

In the 1970s, the prevailing world view about the environment and the desirable (and possible) society collided with the world view of a new Jeremiah in the publication of the Club of Rome report (Meadows *et al.*, 1972). This report proclaimed a heresy in terms of the prevailing ideology: it would not be possible for the human population to expand indefinitely, and for the scale of economic activity to increase accordingly. Cole *et al.* (1973) mounted an effective counterattack. The concept of limits to growth was briefly revived late in the Carter Administration (Barney, 1980), but the incoming Reagan Administration wanted none of it. There ensued a period in which the behavior of the U.S. government closely corresponded with the world view enunciated by Simon and Kahn (1984). The Brundtland Commission, one of the major ideological influences of the current era, weighed in with the idea of 'sustainable development,' proclaiming, 'Growth has no set limits ... beyond which lie ecological disaster' (U.N. World Commission, 1987: 45). During the first half of Reagan's administration, the world was seen as humankind's oyster, to be opened and devoured at will. Drilling for oil in Yellowstone National Park or in the Alaskan preserve seemed to be on the agenda.

World views—and thus culture—clearly mattered in all of these policies and practices, but it was not the culture of the U.S. public that counted: it was the ideas of the governmental and business elite. It may be that it is primarily elite culture that has direct independent effects on the environment, including on land-use/cover changes. The culture of 'simple people,' the vast majority in all societies, may principally affect the kinds of leaders they have.

How to Discover Whether Culture Directly Influences Land Use

What is required to estimate the independent direct effect of culture on land-use/cover changes is the systematic, worldwide collection of data that are:

• Comparable across cultures, meaning that the same variables are measured in the same ways in different cultures
• Acceptably valid, accurate, precise, and reliable
• Quantitative
• Capable of being replicated by other investigators in the same or in different cultures.

No such collection of data exists, and it seems unlikely that it can be compiled from existing research. The Human Relations Area Files should be ransacked for what it has to offer, but HRAF alone cannot meet the need because it primarily contains data from preliterate peoples. New data must be collected. Some of the kinds of studies reported above—anthropological depth interviews, Inglehart's values-oriented surveys, public opinion polls—should be systematically replicated in new countries and for future time periods.

However, the skeptic in me believes that the most widely used measurement techniques will not prove adequate. Survey interviews and questionnaires will not, by themselves, effectively measure values, attitudes, norms, and knowledge concerning land use. Anthropological depth interviews are unlikely to generate comparable data or, indeed, to generate a data set containing the same measures across cultures. We need something that combines the depth of anthropological interviews with the replicability of survey research.

Mitchell and Carson (1989) have developed a method designed to enable respondents to provide 'contingent valuations' of public goods. Respondents are given detailed information about these goods and then asked how much they would be willing to spend. The questions are complex, and some respondents prove unable to answer them. For those who can answer, the results resemble what is found in laboratory simulations of behavior. Sudman (1991: 244) remarks, 'Clearly, more than noise is being measured.' However, a survey with which some U.S. respondents cannot deal does not promise to be the ideal instrument for producing comparable quantitative data from around the world.

Another technology is needed, one with the capacity to elicit responses that have powerful connections to behavior while assuring the researcher of the quality of the data. Building upon the work of Paul Lazarsfeld in the 1930s, Rossi and his collaborators have developed the method of factorial surveys for study of the evaluations individuals make of social situations (Rossi and Nock, 1982). Vignettes—systematically elaborated descriptions of concrete situations—are provided to the respondents, who are asked their evaluations of the situations. Theoretically important characteristics of the vignettes are varied in an experimental design.

In a 1965 study of normative reactions to deviant behavior, A.L. Clark and J.P. Gibbs told respondents brief stories about such forms of deviant behavior as rape, armed robbery, and adultery. Characteristics of these vignettes were systematically varied along several dimensions believed to be salient for the concrete situation; for example, the rape vignette varied by differences in the respective races of the victim and the rapist, and whether physical injury was done beyond the rape itself. Respondents provided open-ended responses concerning what they thought ought to happen to the deviant and what they thought would happen. These responses were analyzed in terms of characteristics of the respondent such as race and social class. My analysis of these data showed that a high degree of normative consensus existed for crimes such as armed robbery and rape, and a low degree for deviant behavior such as homosexuality and adultery. Furthermore, where consensus was low, a large proportion of respondents was unable to give any evaluative response at all.

The technology for doing such factorial surveys has been greatly elaborated since that time, following upon the precepts of Campbell and Fiske's 'multi-method-multitrait matrix.' It has proved possible to elicit valid and reliable responses in a variety of real-world human judgment processes through the use of carefully constructed and systematically varied vignettes. Respondents seem to approve of this way of asking questions, finding it more realistic than the abstract questions commonly used in surveys. It produces a more manageable interview than that proposed by Mitchell and Carson, which it somewhat resembles. Because of the realism of the vignettes, results tend to speak more readily to policy issues than do conventional opinion polls. Future researchers interested in how culture affects land-use/cover changes ought to consider the vignette as a favored method of data collection.

This vignette survey technique is now being employed as a small part of a large-scale multidisciplinary study of environmental ideas, behavior, and outcomes in the Los Angeles Basin (Berk *et al.*, in preparation). In addition to collecting vignette data and standard public opinion data from members of randomly selected households, the study is assembling public and commercial data on water and power usage, air and water quality, transportation, recycling, health problems, population, economic activity, land use, weather, and consumer behavior. Collected for the most part as a result of the conduct of business or through governmental monitoring of the environment, these data are available at varying levels of spatial resolution throughout the basin. All are repeatedly collected (some data every five minutes, others every week) and are being integrated into a geographic information system. The analytical possibilities of such a data set are enormous.

This study of Los Angeles should be the start of a global pattern of research. A systematic collection of many similar data sets is required for the global study of the relationship of culture with land-use/cover changes—a 'global set of regional situations with which to explore each cluster of land-use changes' (Turner and Meyer, 1991). It is not necessary to monitor the world's land use comprehensively. Instead, a number of large field sites should be selected to be studied as intensively as the Los Angeles Basin is being studied. These sites should encompass a range of environmental and human situations, should girdle the globe, and should permit generalizations to all of the processes believed to underlie land-use/cover changes. There are theoretical advantages to selecting such sites without regard to whether or not researchers have already been there or whether some degree of monitoring is already in place, but there are practical advantages to building on existing studies, such as the Lancondon rain forest, the dry tropical forests of Australia, or the highlands of Indonesia.

Within these sites researchers from many natural and social science disciplines should collect a wide variety of comparable data over a long period of time. The present practice is for the compartmentalization of the university to be sustained in the field: an anthropologist goes to one site and studies culture, a dendrologist goes to another and studies the succession of forests, and an ecologist studies life in the layers of the rain forest on another continent. We cannot afford such dispersal of scientific talent. We need to find a way to organize multidisciplinary teams of researchers to conduct coordinated, detailed studies of selected areas of the world over a long period of time. This research should capture data on all aspects of the environment, not simply the global changes such as greenhouse warming, ozone depletion, and loss of biodiversity. As has been repeatedly recognized by the International Geosphere–Biosphere Program (IGBP), all human, physical, and biological processes may be inextricably linked. A satisfactory research design must reflect that complexity. The consequence of disintegrated science is the inability to make generalizations, as evidenced by this paper.

A large ecological research effort designed (and partially implemented in the Long-Term Ecological Research program) at the National Science Foundation in the late 1970s may offer useful principles for such an effort. Resembling what the social scientist knows as replicated case studies, it was initially focused on intensive observation of natural environments in some 90 specific localities representing most kinds of ecosystems in the United States. Comprehensive measurements were to be made for a prolonged time on plant and animal populations, water, soil, air, microclimates, and human interventions. The object was not only to understand what was happening in these locales at particular moments but also to produce long time series for the comparative study of ecological change at a national level.

Designers of the study adopted several principles: collective but not consensual processes for making decisions, measurements in common across all sites as well as locale-specific measurements, quantitative measurements in all cases, open-ended data collection plans, retention of the capacity to make new measurements of past events, measurements for which the scientific need was not entirely clear, the swift public sharing of data, mandated tests for data quality, calibration standards, a common format for reporting data, and a decades-long commitment to support this research. These principles should enable researchers who doubt the utility of comparative research to collect data alongside those who advocate it, and researchers who want to collect data specifically oriented to a particular people to do so while still contributing other data to a worldwide effort.

Only through such systematically comparative studies will we be able to untangle some of the mare's nests of social science. One of these involves the interactions of population growth, high levels of consumption, and environmental degradation. Another involves the origins of high levels of mass consumption. I suggest that it is through the quasi-experiments offered by comparative research designs that the hard work of untangling such mare's nests can begin.

We are now approaching the ability to mount a worldwide comparative study of humans in relation to the land. The IGBP has outlined a global system of regional research centers around which such a worldwide study could be built. The National Science Foundation is considering establishing a network of regional research centers in the United States. The technology for implementing this study exists in both the natural and the social sciences. In the near future some of the needed measurements will be obtainable by remote sensing from satellites. Other measurements, particularly of human dimensions, will require creating and investing major new resources. In terms of the scientific and social benefits of undertaking this research, it will be a great bargain.

I was asked to conclude by considering how a comparative study of culture and cultural change as driving forces in land-use/cover changes ought to be conducted. What I have offered is far broader than simply the study of culture in relation to land-use/cover change. This is a design for studying *all* of the changes that humans make in their environments and the impacts of those changes on human life. It is a plan for research that is intended to move the study of human-nature interactions from received wisdom and selected illustrations to systematic comparisons and falsifiable generalizations.

References

Arizpe, L., and M. Velazquez. 1992. Population and societies. In *Global Change and the Human Prospect: Issues in Population, Science, Technology and Equity,* Sigma Xi Forum Proceedings, Sigma Xi, Research Triangle Park, North Carolina, 31–64.

Ausubel, J. Group report: What knowledge is required to tackle the principal social and institutional barriers to reducing CO_2 emissions? In *Limiting the Greenhouse Effect: Options for Controlling Atmospheric CO_2 Accumulation* (G.I. Pearman, ed.), John Wiley and Sons, Chichester, U.K., in press.

Barnes, S. H. 1988. *Politics and Culture.* Center for Political Studies, Institute for Social Research, University of Michigan, Ann Arbor, Michigan.

Barney, G. O. 1980. *Global 2000: The Report to the President Entering the Twenty-First Century.* Seven Locks Press, Cabin John, Maryland.

Bennett, J. W., and K. A. Dahlberg. 1990. Institutions, social organizations, and cultural values. In *The Earth as Transformed by Human Action: Global and Regional Changes in the Biosphere over the Past 300 Years* (B. L. Turner II, W. C. Clark, R. W. Kates, J. F. Richards, J. T. Mathews, and W. B. Meyer, eds.), Cambridge University Press, Cambridge, U.K., 69–86.

Berk, R. A., D. Schulman, M. McKeever, and H. E. Freeman. Measuring the impact of water conservation campaigns. *Climate Change,* in press.

Buttel, F. H. 1986. Sociology and the environment: The winding road toward human ecology. *International Social Science Journal 109,* 337–356.

Catton, W. R., and R. E. Dunlap. 1978. Environmental sociology: A new paradigm? *The American Sociologist 13(1),* 41–49.

Cole, H. S. D., C. Freeman, M. Jahoda, and K. L. R. Pavitt, eds. 1973. *Models of Doom: A Critique of The Limits to Growth.* Universe Books, New York.

Crosby, A. W. 1986. *Ecological Imperialism: The Biological Expansion of Europe, 900–1900.* Cambridge University Press, Cambridge, U.K.

Demeny, P. 1990. Population. In *The Earth as Transformed by Human Action: Global and Regional Changes in the Biosphere over the Past 300 Years* (B.L. Turner II, W.C. Clark, R. W. Kates, J. F. Richards, J. T. Mathews, and W. B. Meyer, eds.), Cambridge University Press, Cambridge, U.K., 41–54.

Durning, A. 1991. Asking how much is enough. In *State of the World 1991* (L.R. Brown, project director), W.W. Norton, New York, 153–169.

Ehrlich, P. A., and A. H. Ehrlich. 1970. *Populations, Resources, Environments.* W.H. Freeman, San Francisco, California.

Farb, P. 1978. *Man's Rise to Civilization: The Cultural Ascent of the Indians of North America,* rev. 2nd edn. Dutton, New York.

Firey, W. I. 1947. *Land Use in Central Boston.* Harvard University Press, Cambridge, Massachusetts.

Freeman, J. D. 1955. *Iban Agriculture.* Colonial Research Studies No. 18, Her Majesty's Stationery Office, London.

Geertz, C. 1969. Two types of ecosystems. In *Environment and Cultural Behavior: Ecological Studies in Cultural Anthropology* (A.P. Vayda, ed.), Natural History Press, Garden City, New Jersey, 3–28.

Glacken, C. J. 1967. *Traces on the Rhodian Shore: Nature and Culture in Western Thought from Ancient Times to the End of the Eighteenth Century.* University of California Press, Berkeley, California.

Gutfeld, R. 1991. Shades of Green. *Wall Street Journal CXXV(24),* 1.

Inglehart, R. 1991. *Postmaterialism and Environmentalism: The Human Component of Global Change.* Center for Political Studies, Institute for Social Research, University of Michigan, Ann Arbor, Michigan.

Kates, R. W., W. C. Clark, and B. L. Turner II. 1990. The great transformation. In *The Earth as Transformed by Human Action: Global and Regional Changes in the Biosphere over the Past 300 Years* (B.L. Turner II, W.C. Clark, R.W. Kates, J.F. Richards, J. T. Mathews, and W. B. Meyer, eds.), Cambridge University Press, Cambridge, U.K., 1–18.

Kay, J. 1985. Preconditions of natural resource conservation. *Agricultural History 59,* 127–135.

Kempton, W. 1991. Lay perspectives on global climate change. *Global Environmental Change 1(3),* 183–208.

Kempton, W., J. Hartley, and J. Boster. *American Culture and Global Environmental Change.*

School of Engineering and Applied Science, Princeton University, Princeton, New Jersey, in preparation.

Lowenthal, D. 1990. Awareness of human impacts: Changing attitudes and emphases. In *The Earth as Transformed by Human Action: Global and Regional Changes in the Biosphere over the Past 300 Years* (B.L. Turner II, W.C. Clark, R.W. Kates, J.F. Richards, J. T. Mathews, and W. B. Meyer, eds.), Cambridge University Press, Cambridge, U.K., 121–135.

Lugo, A. E., J. J. Ewel, S. B. Hecht, P. C. Murphy, C. Padoch, M. C. Schmink, and D. Stone. 1987. *People and the Tropical Forest: A Research Report from the United States Man and the Biosphere Program.* U.S. Government Printing Office, Washington, D.C.

Martinez-Alier, J. *Varieties of Environmentalism: Scientific Perceptions and Social Movements.* Proposal of the Social Science Research Council Committee on Latin American Studies, Universidad Autonoma de Barcelona, Spain, in preparation.

Meadows, D. H., D. L. Meadows, J. Randers, and W. W. Behrens. 1972. *The Limits to Growth: A Report for the Club of Rome's Project on the Predicament of Mankind.* Universe Books, New York.

Milbrath, L. W. 1991. The world learns about the environment. *International Studies Notes 16(1),* 13–17, 30.

Mitchell, R. C., and R. T. Carson. 1989. *Using Surveys to Value Public Goods: The Contingent Valuation Method.* Resources for the Future, Washington, D.C.

Myers, N., and R. Tucker. 1987. Deforestation in Central America: Spanish legacy and North American consumers. *Environmental Review 11(1),* 55–71.

Rayner, S. (ed.). 1990. Managing the global commons. *Evaluation Review 15.*

Richards, J. F. 1990. Land transformation. In *The Earth as Transformed by Human Action: Global and Regional Changes in the Biosphere over the Past 300 Years* (B.L. Turner II, W. C. Clark, R. W. Kates, J. F. Richards, J. T. Mathews, and W. B. Meyer, eds.), Cambridge University Press, Cambridge, U.K., 163–178.

Rossi, P. H., and R. A. Berk. 1985. Varieties of normative consensus. *American Sociological Review 50(3),* 333–347.

Rossi, P. H., and S. L. Nock, eds. 1982. *Measuring Social Judgments: The Factorial Survey Approach.* Sage Publications, Beverly Hills, California.

Sack, R. D. 1990. The realm of meaning: The inadequacy of human–nature theory and the view of mass consumption. In *The Earth as Transformed by Human Action: Global and Regional Changes in the Biosphere over the Past 300 Years* (B.L. Turner II, W.C. Clark, R. W. Kates, J. F. Richards, J. T. Mathews, and W. B. Meyer, eds.), Cambridge University Press, Cambridge, U.K., 659–671.

Simon, H. A. 1982. Are social problems problems that social science can solve? In *The Social Sciences: Their Nature and Uses* (W.H. Kruskal, ed.), University of Chicago Press, Chicago, Illinois, 1–20.

Simon, J. L., and H. Kahn. 1984. *The Resourceful Earth: A Response to Global 2000.* Basil Blackwell, New York.

Stott, D. H. 1969. Cultural and natural checks on population growth. In *Environment and Cultural Behavior: Ecological Studies in Cultural Anthropology* (A. P. Vayda, ed.), Natural History Press, Garden City, New Jersey, 90–120.

Sudman, S. 1991. Book review. *Contemporary Sociology 20(2),* 243–244.

Tuan, Y.-F. 1968. Discrepancies between environmental attitude and behaviour: Examples from Europe and China. *Canadian Geographer XII (3),* 176–191.

Turner, B. L., II, and W. B. Meyer. 1991. Land use and land cover in global environmental change: Considerations for study. *International Social Science Journal 43*, 669–79.

U.N. World Commission on Environment and Development. 1987. *Our Common Future.* Oxford University Press, Oxford, U.K.

Vayda, A. P. (ed.). 1969. *Environment and Cultural Behavior: Ecological Studies in Cultural Anthropology.* Natural History Press, Garden City, New Jersey.

Viederman, S. Changing the environmental behavior of individuals and institutions: Questions toward an agenda for funders and others concerned with the future. Jesse Smith Noyes Foundation, New York, in press.

White, L., Jr. 1967. The historical roots of our ecologic crisis. *Science 155,* 1203–1207.

VI

ISSUES IN DATA AND MODELING

Introduction

The last three tutorial paper authors were asked to step back from the job to examine the available tools: to address, from the perspectives of both the social and the natural sciences, problems of modeling and data acquisition and analysis in the area of land-use/land-cover change.

The need for modeling in global change studies is evident given the essential functions that models can serve; they can organize a vast array of data for systematic analysis, uncover gaps in the data, expose discrepancies between theoretical expectation and empirical reality, and provide a laboratory for exploring relationships that cannot be experimentally tested in the real world. The report of Working Group C in this volume explores approaches to the development of a global land model. The papers by Melillo and Lonergan and Prudham provide valuable background reviews.

Melillo offers a view from the natural sciences regarding the modeling of connections within the 1988 Earth System Sciences Committee "wiring diagram": in this case, of interactions between the land cover and the atmosphere, including trace gas contributions to global climatic change, albedo impacts on subglobal-scale climate systems, and the impacts of climatic change on ecosystem dynamics. Lonergan and Prudham explore some of the obstacles to integrating social and natural science models. Many of these obstacles do not seem to differ in kind from those noted by Melillo in trying to link one kind of physical model to another: different units of measurement, for example; variations in the quality of data; and differences in temporal and spatial scales. Though data on land uses and candidate causes (e.g., population) are typically collected according to political boundaries rather than natural zones, similar mismatches also exist between different kinds of physical data. The differences in degree between social and physical data and models, however, are probably far more substantial than between different kinds of physical ones, and they represent a major obstacle to bridging the two realms. Above all, human reflexivity complicates the modeling of human behavior but not of natural forces. Lonergan and Prudham illustrate the existing variety of approaches for modeling the interaction of the two realms with examples drawn from the literature, arguing for the particular usefulness of work done at the regional scale.

Skole's chapter is an illustrative exercise that follows the research chain from data acquisition on land-cover changes to the analysis of their human driving forces. It explores a particular question—deforestation in Rondônia in the Brazilian Amazon—and general problems encountered at each step. As Skole notes, an understanding of the biases and errors introduced by different kinds of remote sensing as sources of data is still emerging. Refinement of such techniques promises in the future to expand our ability to monitor land cover much more rapidly than our ability to monitor human land uses and driving forces, where comparable advances are unlikely. Skole's trial explanation of Rondônian deforestation—which incorporates elements as varied as Brazilian national policy and pricing decisions by the Organization of Petroleum Exporting Countries (OPEC), while rejecting the primacy of simple population growth—usefully illustrates the complexity of driving force–cover change interactions, demonstrating the importance of far-flung spatial linkages within the global system. In doing so, it raises an important question: how can factors so diverse and seemingly unpredictable and unconnected be brought into the modeling and forecasting of global land-cover change?

15

Modeling Land-Atmosphere Interactions: A Short Review

Jerry M. Melillo

Human activities have significantly altered biogeochemical cycling at regional and global scales since the beginning of the Industrial Revolution. Over the past 30 years, the pool of carbon in the atmosphere (in the form of CO_2) has increased from 670 to almost 750 Pg (10^{15}) C as a result of fossil fuel burning and forest clearing (Houghton and Woodwell, 1989; Watson *et al.*, 1990). This increase in atmospheric CO_2, as well as increases in other greenhouse gases, has caused concern about the energy balance of the earth. One recent review suggests that the greenhouse effect will result in a rate of increase of global mean temperature during the next century of about 0.3 °C per decade (Houghton *et al.*, 1990). Such temperature increases will also affect amounts and distribution of precipitation. In addition, human activities have had a major impact on the global nitrogen cycle. Industrial and agricultural activities are responsible for the increased inputs of N to many midlatitude ecosystems in the Northern Hemisphere (Peterson and Melillo, 1985). A recent estimate places dry plus wet deposition of N to the temperate and boreal regions of the Northern Hemisphere at almost 18 Tg (10^{12} g) per year (Melillo *et al.*, 1989).

The prospect of major changes in the global environment—a warmer earth with a modified precipitation regime, an altered precipitation chemistry, and an atmosphere with double the present-day CO_2 concentration—presents the scientific research community with a formidable challenge: to devise ways of analyzing the causes of and projecting the course of these shifts as they are occurring. Observational approaches alone are inadequate for providing the required predictive information. We need models to express our understanding of the complex subsystems of the earth—atmosphere, oceans, and land.

A general structure for an earth system model (Figure 1) has been proposed. It was originally developed as the architecture for a numerical computer model suitable for prediction on the time scales of decades to centuries. The intent was to represent our current understanding of the major workings of the earth system. A process was included if it appeared to be extensive or influential enough to have a global-scale impact comparable to that of the anticipated doubling of atmospheric CO_2. The direct impacts of human activities on the natural system were to be treated as inputs in scenario mode; outputs were to be examined in detail only

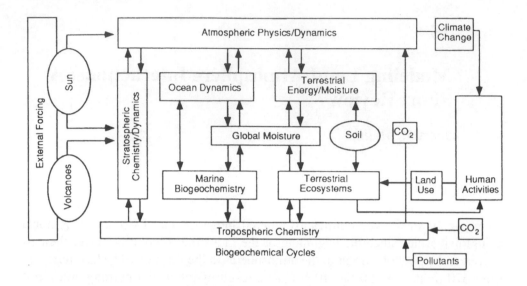

Figure 1. Conceptual model of the earth system (from ESSC, 1988).

insofar as they did not require consideration of economic, social, or political factors.

Particularly important components of the earth system model are the couplings between the atmosphere and the oceans and the atmosphere and the land. The oceans and land are linked to the atmosphere through fluxes of energy, water, and gases such as CO_2, CH_4, and N_2O: the gas fluxes help to regulate atmospheric composition, and so chemical processes, and all of the fluxes affect the global climate system. In this chapter I give an introductory review of a number of different kinds of models that have been or can be used to model the fluxes of energy, water, and gases between land systems and the atmosphere. As part of the review I report on some new research that my colleagues and I are doing on modeling the global carbon cycle.

Land Systems and Climate Modeling

Historically, climate models have evolved out of general circulation models (GCMs) of the atmosphere developed by meteorologists. GCMs are computer models that implicitly or explicitly include the important atmospheric processes that affect climate. Land systems were originally thought of merely as defining lower boundary conditions and were put into the models as simply as possible (Dickinson, 1992). The land properties of interest included fluxes of radiation, water vapor, sensible heat, and momentum. The problem of specifying these fluxes is complex because they can vary over space and time in association with variations in the characteristics of the vegetation, which are caused by many factors, notably human transformations of the landscape.

Land Transformations and Climate: an Early Study

In 1974, Jule Charney introduced the idea that an increase in the earth's surface albedo can change both rainfall and plant cover. As explained by Charney, this mechanism operates because of the dependence of surface albedo on plant cover. Generally, a decrease in plant cover is accompanied by an increase in surface albedo: soil covered by plants has an albedo in the range of 10–25%, while bare soil often has a higher albedo, up to 35–45% in the case of dry, light sands.

An increase in albedo leads to a decrease in the net incoming radiation and an increase in the radiative cooling of the air. This, in turn, causes the air to sink to maintain thermal equilibrium by adiabatic compression, and cumulus convection and its associated rainfall are suppressed. The lower rainfall has an adverse effect on plant growth and tends to cause a further decrease in plant cover. Charney argued that this mechanism offers a possible explanation of the droughts in the southern Sahara, and he suggested that the process could be initiated by overgrazing.

Charney and his colleagues (1975) tested this possible coupling of climate and albedo in the Sahara using a GCM developed at the Goddard Institute for Space Studies (GISS). They made two model runs to determine the effect of surface albedo on rainfall in the Sahara. For one run the surface albedo was set to 14% to simulate a plant-covered Sahara; and for the other run the surface albedo was set to 35% to simulate a plant-free Sahara. The model-predicted mean rainfall over the Sahara during July, the height of the rainy season, was 4.4 mm/day in the low-albedo experiment and 2.5 mm/day in the high-albedo experiment, a decrease of 43%. The GCM that Charney and his co-workers used did not include an adequate biosphere model for calculating changes in vegetation cover resulting from changes in rainfall. They called for a closer coupling of biosphere models and climate system models so that these feedbacks could be captured.

Coupling the Terrestrial Biosphere and the Climate System

During the past two decades, scientists have been working on schemes to more adequately model linkages between the biosphere and the climate system. A recent analysis of the problem (USGCRP, 1990) suggests that the general approach requires the linking of three submodels; a GCM, a land surface parameterization model, and an ecosystems dynamics model. In this section of the paper, I briefly review these types of models and discuss a strategy for linking them.

GCMs

GCMs are based on the physical conservation laws that describe the redistribution of momentum, heat, and water vapor by atmospheric motions (Figure 2). These processes are formulated in a set of general equations that describe the behavior of fluids (air and water) on a rotating body (the earth) under the influence of differential heating (the temperature gradient between equator and pole) caused by an

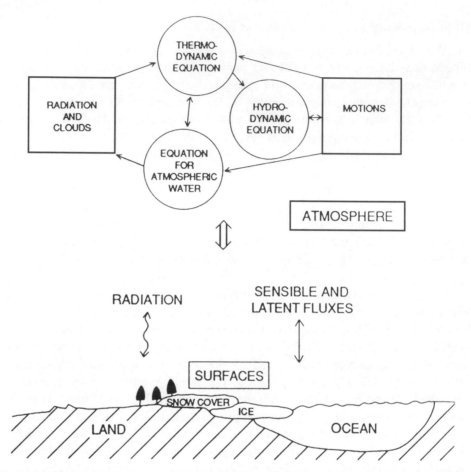

Figure 2. Highly simplified schematic of a general circulation model and linkages to the earth's surface (from Dickinson, 1986).

external heating source (the sun). The equations are nonlinear partial differential equations, and their solutions must be obtained by numerical methods. These numerical methods subdivide the atmosphere both horizontally and vertically. Typical models have a horizontal resolution of 300 to 1000 km and between 2 and 19 vertical layers. For each cell of the model's grid, the general equations predict variables such as wind, temperature, humidity, surface pressure, rainfall, and so on, with a time step between 10 and 30 minutes. The resolutions of the general circulation models are sufficient to represent large-scale features of climate, but allow only a limited interpretation of results on the regional scale. And because of the short time step and the fact that an adequate description of a model's climatology requires many runs, there is an operational limit on the time over which a particular scenario can be explored or the number of different scenarios that can be investigated (Cubasch and Cess, 1990).

The largest uncertainty in estimating the sensitivity of the earth's surface temperature to a given increase in radiative forcing arises from the problem of parameterization (Schneider, 1989). Due to their limited spatial resolution, general circulation models do not resolve several variables important to climate, such as clouds, which control the radiation budget of the earth. However, the statistical effects of these subgrid-scale processes have to be incorporated into the models. Therefore, climate modelers seek to find a parametric representation that relates implicitly the effects of cloudiness and other processes that operate at the subgrid scale but still have effects at the resolution of the GCM. A parameter or proportionality coefficient might be used that describes the average cloudiness in a grid cell in terms of the mean relative humidity in that cell and some other measures of atmospheric stability.

Another important parameterization is the transfer of heat and water within the soil. Issues of interest include the balance between evaporation and precipitation, snow melt, storage of water in the ground, and river runoff. This component of land surface parameterization is important for climate change predictions, since it shows how local climates may change from humid to arid and vice versa depending on global circulation changes.

Land Surface Parameterization Models

The direct model inputs from land processes to a general circulation model are through a land surface parameterization model. These models are designed to cover the relatively short-term biospherically controlled interactions between the land surface, both vegetated and nonvegetated, and the atmosphere. The processes involved in the interactions include albedo (radiative transfer), roughness length (momentum transfer), and the exchanges of sensible and latent heat (biophysical control of evapotranspiration). Until the mid-1980s GCMs treated these processes as independent of each other (Sellers, 1986). The surface fluxes were a function of independently specified surface albedo, an independently specified surface roughness length, and an independently formulated relationship between evapotranspiration and soil moisture.

The relationship between evapotranspiration and soil moisture was originally thought of as a 'bucket' in which the level of water is lowered when evaporation exceeds precipitation, and is raised when precipitation is larger. The water level of the bucket rises up to the point at which the bucket overflows and produces 'runoff.' Over some range in the level of water in the bucket, which varies among models, the rate of evaporation is taken as equal or nearly equal to that from a freely evaporating surface, and is reduced only when the water level is low.

Although this approach places reasonable bounds on surface evaporation rates, it is not the way water vapor is actually transferred from the land to the atmosphere. In reality, plants play an active and regulating role. They are complex organisms that control the passage of water and gas through their systems as a consequence of a variety of factors ranging from the particular biochemical pathway they use to fix carbon in photosynthesis to their canopy architecture. Several research groups interested in climate modeling recognized this fact and began to build models of links between the surface and the atmosphere that introduce some of the influences of the physiology and morphology of vegetation into land surface parameterization models (Dickinson, 1983, 1984; Sellers, 1986).

At present, few land surface parameterization models have been successfully combined with GCMs (Dickinson and Henderson-Sellers, 1988; Sato *et al.*, 1989). Most land surface parameterization models still assume a static ecosystem structure and a prescribed phenology, which in turn defines the albedo and roughness length of a given area and the evapotranspiration response as a function of soil moisture (Sellers, *et al.*, 1988). Generally, the surface vegetation type, and hence albedo and roughness length, is prescribed from data, and the soil moisture is initialized from off-line climatological studies. As a result, these models in their current state have a limited usefulness for the study of global change because they do not yet capture the subtle but important effect that biology can have on the climate system through land–atmosphere interactions. To improve land surface parameterization models they must be coupled with ecosystem dynamics models.

Ecosystem Dynamics Models

The term 'ecosystem dynamics model' refers to a large class of models that includes at least three partially overlapping levels. These start at the plant physiological level and progress through populations/communities and on to ecosystems. At the physiological level, plant processes are described in depth, often including biochemistry. Ecosystem models integrate the physiology to major plant parts or the whole plant and add plant–soil interactions. Population/community models often deal with the 'life-table dynamics,' such as birth and death, of individual plants in the context of their environment.

- *Plant physiology models.* The main objective of most plant physiology models is to predict plant growth and water balance using detailed information on processes such as leaf energy balance, carbon uptake through photosynthesis, respiration, and exchange of water vapor through transpiration. Many of these models exist (to offer a partial listing: for woody vegetation de Wit *et al.*, 1978; Goudriaan *et al.*, 1984; Caldwell *et al.*, 1986; Makela and Hari, 1986; Matyssek, 1986; Arp and McGrath, 1987; Wang, 1988; Running and Coughlan, 1988; McMurtrie *et al.*, 1989; Reynolds *et al.*, 1989; for nonwoody vegetation Detling *et al.*, 1979; Coughenour *et al.*, 1984; Morris *et al.*, 1984; White, 1984; Lauenroth *et al.*, 1986; MacNeil *et al.*, 1989; and references in Hanson *et al.*, 1985; and Joyce and Kickert, 1987). Most plant physiology models have a short time step (one hour to one day), and they share a short list of meteorological driving variables (humidity, precipitation, radiation, temperature, wind speed). Models based on plant physiology are ideal for analyzing and interpreting the detailed short-term reactions of plants to various aspects of climate change, such as plant responses to an increase in the concentration of CO_2 in the atmosphere.
- *Ecosystem models.* Ecosystem models are generally models of system biogeo-

chemistry that consider whole-system carbon and nutrient fluxes. The models are process-based, but relative to the plant physiology models, they have highly aggregated representations of plants and plant processes. Ecosystem models typically lump all green plant biomass into one compartment, whereas the plant physiology models often divide plants up into a number of parts. Details of biochemistry are less evident in ecosystem models than in plant physiology models.

Ågren and his colleagues (1991), in a recent review paper, have recognized two types of ecosystem models, those that use short time steps (1 day or less) and those that use long time steps (1–12 months). The short-time-step models, such as the ELM grassland model (Innis, 1978), generally require input of driving variables on a daily basis and simulate ecosystem dynamics in the years-to-decade time frame. The long-time-step models, such as the Century model (Parton *et al.*, 1988) and the general ecosystem model or GEM (Rastetter *et al.*, 1991), usually require input of driving variables on a monthly basis and simulate ecosystem dynamics over periods ranging from decades to centuries. In considering the effects of climate change on ecosystem structure and function, the long-time-step models are likely to be the most useful.

- *Population/community models*. The most widely used population/community models have been developed for forests (e.g., Botkin *et al.*, 1972; Shugart and West, 1977; Kimmins and Scoullar, 1979; Hagglund, 1981; Belcher *et al.*, 1982). Although these models can differ in certain details, they generally consider birth or recruitment, growth of individuals as modulated by limiting factors, stand structure and spacing, and mortality (Shugart, 1984). The JABOWA/FORET family of population/community models (Botkin *et al.*, 1972; Shugart, 1984) has received the most attention. These models are known as 'gap models,' and they simulate the population/community dynamics of trees on small (ca. 0.1 ha) patches of forest. The trees in a patch are considered individually. The death-birth-growth-death cycle of a small forest patch has both stochastic and deterministic phases (Figure 3). Tree death in these models is stochastic. Tree regeneration is also strongly influenced by stochastic factors. Once the individual trees are established, the growth and competition among them are much more deterministic. Climate has its largest effects during the stochastic phases.

The way tree growth is simulated in these models has changed through time. In the early versions of these models, a maximum rate of tree growth was defined and then reduced to the extent that temperature and light levels during the growing season were less than optimal for the species (Botkin *et al.*, 1972; Shugart and West, 1977). Later, Solomon and his colleagues (1984) factored in nongrowing season temperature. They defined a species-specific winter temperature minimum, or frost tolerance; if the temperature fell below this minimum, growth in the following year was assumed to be entirely allocated to repairing frost damage.

Moisture stress has been simulated in several ways in these models. In the origi-

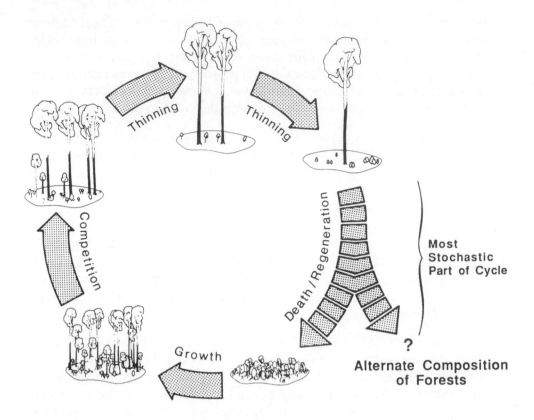

Figure 3. The death-birth-growth-death cycle of a small patch of forest as conceptualized in the JABOWA/FORET family of forest population/community models. Stochastic phases of the cycle are noted (from Shugart *et al.*, 1986).

nal JABOWA model, Botkin and his co-workers (1972) related tree recruitment to actual evapotranspiration; that is, recruitment was reduced in water-stressed environments. More recently, as part of their Linkages model, Pastor and Post (1985, 1986) have developed a soil moisture simulator to calculate water stress. Plant growth is then limited during periods when soil moisture is below the wilting point.

Recent versions of these models also simulate the effects of nutrient limitation on tree growth (Aber and Melillo, 1982; Aber *et al.*, 1982; Weinstein *et al.*, 1982; Pastor and Post, 1985, 1986). Aber and his colleagues used their model, FORT-NITE, to evaluate the effects of forest rotation length, harvest intensity, and fertilization on fiber yield from northern hardwood forests in New England.

Linking Models

The U.S. Global Change Research Program (USGCRP) has set out a plan for linking general circulation models, land surface parameterization models, and ecosystem dynamics models to predict energy and water fluxes between the land and the atmosphere (USGCRP, 1990). This plan supports further work on directly coupling GCMs and land surface parameterization models. It questions, however, the feasibility and desirability of directly coupling ecosystem dynamics models with either land surface parameterization models or GCMs because of the mismatches in space and time scales. The plan does support the construction of 'forcing modules' to convert GCM output into 'forcing' climatologies for ecosystem dynamics models, and 'aggregation modules' to aggregate the effects of ecosystem dynamics into representative land surface parameterizations.

Perhaps the next step in linking dynamic meteorology to ecology at the regional and subregional scales should be pursued with mesoscale models run embedded in a general circulation model (Dickinson *et al.*, 1989). The distillation and integration of information from the three types of ecosystem models to develop dynamic land surface parameterizations might best be considered in the context of hierarchy theory (Allen and Starr, 1982; O'Neill *et al.*, 1986). The three types of ecosystem dynamics models just reviewed can be thought of as levels of a hierarchy. Kittel and Coughenour (1988) have argued that a 'nested approach' should be used for coupling models at various levels in a hierarchy. In this scheme, higher-resolution models are enveloped by lower-resolution models. Model simulations in a hierarchy scheme involve 'zooming in and out' of different levels. High-resolution models are periodically called up to calculate parameter values to be used in low-resolution models, which themselves provide the framework within which the high-resolution models can operate.

With respect to the problem of modeling energy and water balance as part of an earth systems model, one hierarchical approach that might work is as follows. A population/community model such as Linkages would be used to convert annualized indexes of climatic conditions and the current ecosystem state into total leaf area and structure for the next year. Within these total values, an ecosystem model would be used to calculate the phenology of leaf production and loss, the rate of nutrient mobilization and uptake, and so the seasonal pattern of ecosystem dynamics. In the context of the seasonal dynamics established by the ecosystem model, the physiology model would convert climatic data into energy and water balances over very short time steps. In this hierarchical scheme, information from the physiology model would feed back to the ecosystem model, which would, in turn, feed back to the population/community model. As part of the larger earth system modeling exercise, the output of the physiology model could serve as input to a land surface parameterization model.

The scheme outlined above is not without problems. For example, since at least some of the ecosystem dynamics models (e.g., the JABOWA/FORET class of models) include stochastic descriptions of ecological processes, good representations of the mean and variability of a region's climate must be used in the model. This means that before a believable model result can be produced, many variations of a climatology must be applied to an ecosystem dynamics model that incorporates stochasticity. Another problem is that the hierarchical approach outlined above does not consider the possibility of long-term shifts in the global distribution of vegetation units. Some attempts have already been made to predict these shifts with simple biome distribution models. These models could serve as a fourth level in the ecosystem dynamics model hierarchy.

Extant biome distribution models predict the potential equilibrium vegetation (climax vegetation) as a function of climate. Two general types of these models exist. The first is correlation models, which base vegetation distribution on the observed correlation between climate and vegetation type (see Prentice, 1990, for a review). The most well known of these is the Holdridge Life Zone Classification (Holdridge, 1947, 1964). In this scheme the bioclimate is defined by three factors: the annual biotemperature, the annual evapotranspiration ratio, and the annual precipitation. The annual biotemperature is the mean annual temperature excluding temperatures below 0 °C; the annual evapotranspiration ratio is the ratio of annual potential evapotranspiration to annual precipitation. Holdridge arrayed these variables in a triangular coordinate system, divided into zones by the intersection of the variables. Each zone is designated as a vegetation 'life zone' or biome.

Recently, Prentice (1990) tested the Holdridge scheme by using it to predict current vegetation from 1° longitude × 1° latitude climatic data. The original Holdrige scheme performed poorly relative to existing global vegetation maps, with only about 40% of the grid cells predicted correctly. When Prentice reclassified the biome types into fewer classes, she increased the agreement between mapped and predicted vegetation to 77%.

Even if general agreement can be achieved between mapped and model-predicted vegetation distributions, the correlation-modeling approach has several shortcomings that make it of questionable use in climate change research. Systems like the Holdridge scheme infer that climate has a causal relationship to vegetation. The problem is that the climate variables used are not necessarily those to which plants respond, and furthermore, climate is only one of the determinants of vegetation distributions. In addition, to use a system like the Holdridge scheme for climate research, one must assume that today's plant communities will migrate intact into future sites. Based on evidence of past vegetation distributions, it is highly likely that ecosystems will reorder into new compositional and structural configurations in the future (Webb, 1986). Perhaps the most important shortcoming is that correlation models are designed for interpolation rather than extrapolation. Since they provide no mechanistic basis for the correlation, it is not possible to extend them to new conditions.

Dissatisfaction with bioclimatic schemes has led to the second type of biome distribution model, which attempts to predict climax vegetation on the basis of simple mechanistic principles (Woodward, 1987; Woodward and McKee, 1991). These mechanistic approaches use hydrologic balance and temperature-growth relationships to predict vegetation distribution. Woodward and McKee (1991) have used their model to predict the vegetation distribution that would occur in a doubled CO_2 climate predicted by a general circulation model developed at GISS. Mechanistic models are also beginning to be used to model various aspects of

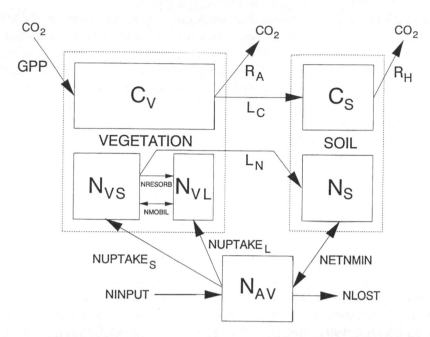

Figure 4. The Terrestrial Ecosystem Model (TEM). Carbon enters the vegetation pool (C_V) as gross primary productivity (GPP). Some is lost in autotrophic respiration (R_A) while the rest transfers to the soil pool (C_S) as litter production (L_C); it leaves the soil as heterotrophic respiration (R_H). Nitrogen inputs from outside the ecosystem (NINPUT) enter the inorganic N pool (N_{AV}); losses leave this pool as the flux NLOST. Nitrogen in the vegetation occurs either in the structural pool (N_{VS}) or the labile pool (N_{VL}). Structural N in vegetation is constructed from N that is derived from either the labile pool as the flux NMOBIL or from soil inorganic N pool as the flux $NUPTAKE_S$. The labile pool is replenished from N that is reabsorbed from senescing tissue (NRESORB), N that is allocated for storage (NMOBIL), or N in uptake that does not enter directly into tissue construction ($NUPTAKE_L$). Nitrogen is transferred from vegetation to the soil organic pool (N_S) as the flux L_N. Net N mineralization (NETNMIN) accounts for N exchanged between the organic and inorganic N pools of the soil (from Melillo *et al.*, 1993; printed with permission from *Nature*, copyright by Macmillan Magazines Limited).

global biogeochemistry. Some of the work that my colleagues and I are doing in this area is discussed next.

Global Carbon Cycle Modeling

In this part of the paper, I summarize a recent global modeling study of net carbon exchange between the land and the atmosphere under current and future climates as predicted by general circulation models. I also discuss plans for new approaches to carbon cycle modeling that include land transformations.

Modeling Net Primary Production

As part of a research program to understand how global change will affect the global cycles of carbon, nitrogen, phosphorus, and sulfur, my research team at the Marine Biological Laboratory, in cooperation with a research team from the University of New Hampshire led by Berrien Moore, have developed a global model of carbon and nitrogen cycling in terrestrial ecosystems. We have just completed a study in which we have combined a global terrestrial ecosystem model (TEM) with the output from four GCMs to estimate the effects of a CO_2 doubling and associated climate changes on net primary production (NPP) for the world's land ecosystems (Melillo *et al.*, 1993). Annual NPP is the net amount of carbon captured by land plants through photosynthesis each year. It is of fundamental importance to humans because it is that portion of solar energy that is available to support life of all components of the biosphere. The largest portion of our food supply is from productivity of plant life on land, as is wood for construction and fuel. Changes in NPP represent changes in sustained carrying capacity of the earth for humans.

Global Extrapolation of TEM for Contemporary Climate

The TEM (Figure 4) is a process-based ecosystem simulation model that uses spatially referenced information on climate, elevation, soils, vegetation, and hydrology to make monthly estimates of important carbon and nitrogen fluxes and pool sizes (Raich *et al.*, 1991; McGuire *et al.*, 1992, 1993 a, b; Melillo *et al.*, 1993). The data sets used to drive TEM are organized at a resolution of 0.5° latitude × 0.5° longitude. The model is grid-cell based, and global extrapolation is accomplished by running the model to equilibrium for each grid cell in the terrestrial biosphere. Because TEM is an equilibrium model, its estimates of carbon and nitrogen dynamics only apply to mature, undisturbed (potential) vegetation; it does not include the effects of land use.

The application of TEM to a grid cell requires the use of monthly climatic/hydrologic data and soil- and vegetation-specific parameters appropriate to the grid cell. We do not make estimates for grid cells defined as ice, open water, or wetland ecosystems, so that our global extrapolation of TEM requires application of the model to 56,090 grid cells in the terrestrial biosphere.

To estimate the carbon dynamics of potential vegetation for 'current conditions' we applied TEM globally at 355 ppmv CO_2 using the long-term or 'contemporary' climate data (Legates and Willmott, 1990a, 1990b; Hahn *et al.*, 1988). Under these conditions, TEM estimates the global annual NPP for potential vegetation to be 53.2 Pg C, with about 60% occurring in the Northern Hemisphere (Figure 5). Mean NPP estimates for ecosystems range from 53 g $C/m^2/yr$ for desert to 1098 g $C/m^2/yr$ for tropical evergreen forest. Estimates for individual grid cells range from 0 g $C/m^2/yr$ in desert, polar desert, and xeromorphic forest to 1422 g $C/m^2/yr$ in tropical deciduous forest (see Plate 14, facing p. 13). The two most productive ecosystems are tropical evergreen forest and tropical savanna. Tropical evergreen forest accounts for 35.9% of the exchange of CO_2 between terrestrial vegetation and the atmosphere, although it covers only 13.7% of the earth's land surface. Tropical savanna, which occupies 12% of the land surface, accounts for 11% of global NPP.

The TEM estimate of global NPP for terrestrial ecosystems, which is a process-

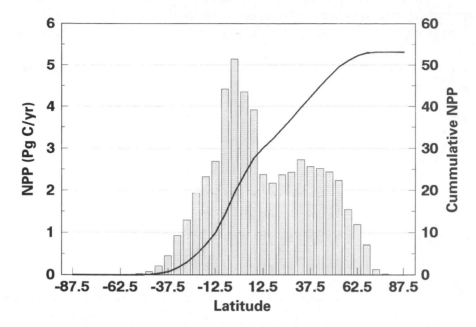

Figure 5. Annual NPP by latitude. The bars represent 5° latitude aggregations, the smooth line represents a cumulative global total. The units are Pg C/yr.

based estimate, is similar to many of the estimates that have appeared in the ecological literature over the past three decades (Table 1). Most of the estimates in Table 1 have been calculated by multiplying the mean NPP for an ecosystem, as determined from a literature survey of field studies, by the area of the ecosystem and then summing across ecosystems. A few of the estimates have been determined by regression approaches involving climate variables, and some consider the effects of land use.

Effects of Predicted Climate Change on Potential NPP
We obtained the output of four GCMs from the National Center for Atmospheric Research (Jenne, in press). The simulations estimate equilibrium climates that correspond to a doubling of the atmospheric CO_2 concentration and include: the GISS GCM (Hansen *et al.*, 1983, 1984), the Oregon State University (OSU) GCM (Schlesinger and Zhao, 1989), and two GCM simulations from the Geophysical Fluid Dynamics Laboratory (GFDL 1 and GFDL Q; Manabe and Wetherald, 1987; Wetherald and Manabe, 1988). Among the GCMs, mean global temperature is predicted to increase between 2.8 °C and 4.2 °C, global precipitation to increase between 7.8% and 11.0%, and global cloudiness to decrease between 0.4% and 3.4% (Table 2).

We generated 'GCM climates' for TEM by using the output variables of surface air temperature, precipitation, and total cloud cover for the current and $2 \times CO_2$ simulations of each GCM to modify the contemporary climate data for TEM. First, we organized each of the output variables of each GCM at the 0.5° × 0.5°

Table 1: Estimates of global terrestrial net primary production

Reference	Estimate $(10^{15}$ g C)
Deevey (1960)	56.4
Bowen (1966)	106
Whittaker and Likens (1969)	49.5
Olson (1970)	54
Bazilevich *et al.* (1970)	78
SCEP (1970)	56
Golley (1972)	40.5
Lieth (1973)	45.5
Whittaker and Likens (1973)	48.2
Whittaker and Likens (1975)	53.4
Ajtay *et al.* (1979)	59.9
Esser (1987)	48.6
Esser (1990)	45
Melillo *et al.* (in press)	53.1

Table 2: Characteristics of the GCMs used in this study

GCM	GFDL 1	GFDL Q	GISS	OSU
Date of model run	1984–85	1987–88	1982	1984–85
Model resolution (lon° × lat°)	7.5 × 4.44	7.5 × 4.44	10.0 × 7.83	5.0 × 4.00
Baseline CO_2 (ppmv)	300	300	315	326
Change in mean global temperature for $2 \times CO_2$	+4.0°	+4.0°	+4.2°	+2.8°
Change in global precipitation for $2 \times CO_2$	+8.7%	+8.3%	+11.0%	+7.8%
Change in mean global cloudiness for $2 \times CO_2$	–0.4%	–0.7%	–3.4%	–3.2%

scale with a spherical interpolation procedure (Willmott *et al.*, 1985). Next, similar to the method of Adams *et al.* (1990), we calculated for each grid cell the ratio of the monthly output of the $2 \times CO_2$ simulation to that of the $1 \times CO_2$ simulation for each of the three output variables; temperature was converted to Kelvin before calculating monthly temperature ratios. We then multiplied each ratio by the corresponding entity in our data for contemporary climate to determine the input data for TEM that represent the $2 \times CO_2$ climate for each GCM.

To help separate the effects of changes in CO_2 concentration from those of the GCM climates on estimates of NPP we performed a factorial experiment with TEM involving two levels of CO_2 (312.5 ppmv and 625.0 ppmv) and five climate scenarios (contemporary and the four GCM climates). We chose the CO_2 level of 312.5 ppmv because it was the average baseline concentration of the four GCMs. For doubled CO_2 with no climate change, TEM predicts a global NPP increase of 16.3%. The responses differ widely among vegetation types and range from no increases for some northern ecosystems to 50% for deserts.

In many northern and temperate ecosystems, NPP is known to be limited by the availability of inorganic nitrogen in the soil (Mitchell and Chandler, 1939; Chapin *et al.*, 1986; Vitousek and Howarth, 1991). Because of nitrogen limitation, TEM predicts that these ecosystems have low capacity to incorporate elevated CO_2 into production (McGuire *et al.*, 1993). In dry regions, on the other hand, TEM generally predicts that water availability limits productivity more than nitrogen availability. Thus, TEM predicts that vegetation in dry regions is able to substantially incorporate elevated CO_2 into production, which results in increased water-use efficiency.

The predicted degree of N limitation of NPP in tropical forests is weak in comparison to that in temperate and boreal forests. Thus, TEM predicts that these systems are able to use a substantial proportion of elevated CO_2 in productivity. However, this result must be treated with caution because TEM does not model phosphorus dynamics and much of the Amazon Basin is covered with phosphorus-deficient soils (Sanchez *et al.*, 1982). A substantial reduction in NPP response for tropical evergreen forest due to phosphorus limitation would result in a much lower estimate of global NPP response to doubled CO_2.

Changes in climate with no change in CO_2 concentration are predicted to have little effect on global estimates of NPP; the changes in NPP range from a decrease of 2% for the OSU climate to essentially no change for the other GCM climates. However, the effects of climate change vary among ecosystems. For northern and temperate ecosystems, increases in NPP are generally predicted, with northern ecosystems showing the greatest response to climate change. For all the GCM climates, predicted temperature increases are greatest toward the poles and least in the tropics (see Mitchell *et al.*, 1990). Productivity in northern and temperate ecosystems is substantially limited by nitrogen availability, and increases in productivity predicted by TEM are primarily driven by the effect of elevated temperature in enhancing the mineralization of nitrogen in the soils of these regions.

With climate change and unchanged CO_2, NPP for tropical evergreen forest is predicted to decrease between 9% and 21% among the GCM climates. These decreases may, in part, be related to increased cloudiness. The largest decrease is associated with the OSU climate which predicts the largest increase in mean annual cloudiness (10%) for the grid cells designated as tropical evergreen forest. Increases in cloudiness decrease photosynthetically active radiation (PAR), and seasonal NPP predicted by TEM for tropical evergreen forest in South America is significantly correlated with PAR (Raich *et al.*, 1991). When both climate and CO_2 are changed, global increases in NPP do not vary greatly among the GCMs and range between 20% and 26% among the GCM climates. The lowest global response, which is predicted for the OSU climate, is associated with the largest predicted increase in cloudiness for tropical evergreen forest. For tropical and dry temperate ecosystems, the responses of NPP are dominated by the effects of elevated CO_2. However, for boreal and moist temperate ecosystems, the responses reflect primarily the effects of elevated temperature in enhancing nitrogen availability.

Although the predicted global and ecosystem-wide responses of NPP at elevated CO_2 are generally similar for all the GCM climates, there are differences in NPP response at smaller spatial scales. For example, in Southeast Asia, NPP predicted for both the GFDL 1 and GFDL Q climates generally increases between 5 and 25%, although increases of between 25 and 50% are predicted for Borneo and southeastern China. The pattern of NPP responses for the GISS climate in Southeast Asia is similar to those for the GFDL climates except for moderate decreases in the vicinity of the Malaysian peninsula. For the OSU climate, NPP is predicted to decrease substantially throughout much of Indonesia. These decreases are associated with a predicted increase in mean annual cloudiness of 13% for tropical evergreen forest in the region.

Summary

While this research does provide some quantitative estimates of the potential effects of climate change on important processes in terrestrial ecosystems, perhaps its main contribution is highlighting uncertainties in current knowledge. For instance, the differences in precipitation and cloudiness patterns predicted by different global climate models result in substantial differences in NPP in some parts of the globe, especially some tropical regions. A second contribution of this study is the demonstration of the value of linking geographically referenced models from various disciplines, here ecosystem science and atmospheric science.

The Effects of Land Transformations on Carbon Storage in Terrestrial Ecosystems

The storage of carbon in terrestrial ecosystems can be changed considerably by land transformations, particularly the conversion of forests to pastures or croplands. This is so because the vegetation and soils of forests contain up to 100 times more carbon per unit area than agricultural systems. Over a 135-year period from 1850 to 1985, it is estimated that land transformations, particularly deforestation, have resulted in a net loss of carbon from the land to the atmosphere of 115 ± 35 Pg C (Houghton and Skole, 1990). In 1980, the estimates of carbon loss from land range from 0.6 to 2.5 Pg C (Houghton *et al.*, 1985, 1987, 1988; Detwiler and Hall, 1988).

The results reported by Houghton and his colleagues summarize a decade-long joint research effort by scientists from the Marine Biological Laboratory and the University of New Hampshire. We made the carbon flux estimates using a simple empirical model (the Marine Biological Laboratory/ Terrestrial Carbon Model, MBL/TCM) linked with geographically referenced information on (1) ecosystem

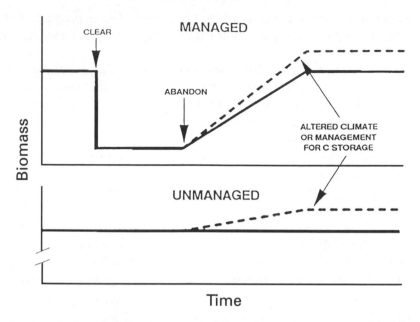

Figure 6. Effect of altered climate, or management for C storage, on recovery of C stocks in managed vegetation (upper), and on C storage in unmanaged vegetation (lower). Although both panels show climate change or management practices resulting in increased C storage, reduction in C storage is also possible in different scenarios. Parameter values for these curves are to be predicted using a detailed process-based ecosystem model and then applied to a more empirical model to calculate carbon storage.

type prior to transformation and (2) type of transformation. The model time step was one year, and the spatial resolution varied from the continental level early in the research effort to the country level later on.

To calculate net carbon flux between the land and the atmosphere following a land transformation using the empirical modeling approach, two general types of information were needed: (1) the area undergoing a specific land transformation within each ecosystem type (initially 12 types—e.g., boreal forest, temperate grassland, tropical woodland), each year; and (2) the response of the carbon pools of each system to the change. A full accounting of the response of carbon pools to land transformation, in turn, required that we consider four factors: (1) the difference between the biomass (vegetation mass) and soil organic matter of natural and managed systems, (2) the amount of biomass removed from the transformed area and the rate at which it burns or decays, (3) the amount of biomass left as 'slash' (woody debris generated from vegetation during a harvest or clearing) on the transformed area and the rate at which it burns or decays, and (4) the rate and maximum level of carbon accumulation in the biomass and soil as the system regrows after harvest or agricultural abandonment.

The dynamics of the various carbon pools and of the ecosystem following transformation were incorporated in the model through a suite or vegetation, soil, and slash response curves. Each ecosystem type had its own characteristic response functions for forest harvest and agricultural clearing. Wood products removed from the transformed system were assumed to decay with rates dependent on the use of the wood.

This purely empirical model has been useful for providing estimates for the effects of past and current land transformations on the global carbon cycle. The approach assumes a constant climate. For this reason it cannot be used to predict how land transformations in a future global climate will affect carbon storage in terrestrial ecosystems. To address this problem, my research group is currently coupling an empirical model of ecosystem response to transformation with a process-based ecosystem model (Figure 6). As the model is currently conceived, land transformation is a specified input; it can come from either a table of possible future land transformation scenarios or from a 'land transformation' model. The latter would be preferable, but at the present time predictive land transformation models do not exist. An early priority for social scientists concerned with the human dimensions of global change should be to outline an approach for building a predictive land transformation model and to set out a strategy for getting the job done.

References

Aber, J. D., and J. M. Melillo. 1982. FORTNITE: A computer model of organic matter and nitrogen dynamics in a forest ecosystem. *University of Wisconsin Research Bulletin R3130*.

Aber, J. D., J. M. Melillo, and C. A. Federer. 1982. Predicting the effects of rotation length, harvest intensity and fertilization on fiber yield from northern hardwood forests in New England. *Forest Science 28(1)*, 31–45.

Adams, R. M., C. Rosenzweig, R. M. Peart, J. T. Ritchie, B. A. McCarl, J. D. Glyer, R. B. Curry, J. W. Jones, K. J. Boote, and L. H. Allen, Jr. 1990. Global climate change and U.S. agriculture. *Nature 345*, 219–224.

Ågren, G. I., R. E. McMurtrie, W. J. Parton, J. Pastor, and H. H. Shugart. 1991. State-of-the-art of models of production-decomposition linkages in conifer and grassland ecosystems. *Ecological Applications 1(2)*, 118–138.

Ajtay, G. L., P. Ketner, and P. Duvigneaud. 1979. Terrestrial primary production and phytomass. In *The Global Carbon Cycle* (B. Bolin, E. T. Degens, S. Kempe, and P. Ketner, eds.), SCOPE 13, John Wiley and Sons, Chichester, U.K., 129–182.

Allen, T. F. H., and T. B. Starr. 1982. *Hierarchy: Perspectives for Ecological Complexity*. University of Chicago Press, Chicago, Illinois.

Arp, P. A., and T. P. McGrath. 1987. A parameter-based method for modeling biomass accumulations in forest stands: Theory. *Ecological Modeling 36*, 29–48.

Bazilevich, N. I., L. E. Rodin, and N. N. Rozov. 1970. *Untersuchungen der Biologischen Produktivitat in Geographischer Sicht*, 5th Tagung Geogr. Ges., Leningrad, USSR.

Belcher, D. M., M. R. Holdaway, and G. J. Brand. 1982. *A Description of STEMS—the Stand and Tree Evaluation and Modeling System*. U.S. Forest Service General Technical Report NC-79, U.S. Forest Service, Washington, D.C.

Botkin, D. B., J. F. Janak, and J. R. Wallis. 1972. Some ecological consequences of a computer model of forest growth. *Journal of Ecology 60*, 849–873.

Bowen, H. J. M. 1966. *Trace Elements in Biochemistry*. Academic Press, New York, 241 pp.

Caldwell, M. M., H.-P. Meister, J. D. Tenhunen, and O. L. Lange. 1986. Canopy structure, light microclimate and leaf gas exchange of *Quercus coccifera* L. in a Portuguese macchia: Measurements in different canopy layers and simulations with a canopy model. *Trees 1*, 25–41.

Chapin, F. S., III, P. M. Vitousek, and K. Van Cleve. 1986. The nature of nutrient limitation in plant communities. *American Naturalist 127*, 48–58.

Charney, J., P. H. Stone, and W.J. Quirk. 1975. Drought in the Sahara: A biogeophysical feedback mechanism. *Science 187*, 434–435.

Coughenour, M. B., S. J. McNaughton, and L. L. Wallace. 1984. Modeling primary production of perennial graminoids—uniting physiological processes and morphometric traits. *Ecological Modeling 23*, 101–134.

Cubasch, U., and R. D. Cess. 1990. Process modeling. In *Climate Change: The IPCC Scientific Assessment* (J. T. Houghton, G. J. Jenkins, and J. J. Ephraums, eds.), Cambridge University Press, Cambridge, U.K.

Deevey, E. S., Jr. 1960. The human population. *Scientific American 203*, 195–204.

Detling, J. K., W. J. Parton, and H. W. Hunt. 1979. A simulated model of *Bouteloua gracilis* biomass dynamics on the North American shortgrass prairie. *Oecologia (Berlin) 38*, 167–191.

Detwiler, R. P., and C. A. S. Hall. 1988. Tropical forests and the global carbon cycle. *Science 239*, 42–47.

de Wit, C. T., J. Goudriaan, H. H. van Laar, F.W.T. Penning de Vries, R. Rabbinge, and H. van Keulen. 1978. *Simulation of Assimilation, Respiration and Transpiration of Crops.* ,PUDOC, Wageningen, The Netherlands.

Dickinson, R. E. 1983. Land surface processes and climate-surface albedos and energy balance. *Advances in Geophysics 25*, 305–353.

Dickinson, R. E. 1984. Modeling evapotranspiration for three-dimensional global climate models. In *Climate Processes and Climate Sensitivity* (J. E. Hansen and T. Takahashi, eds.), Geophysical Monograph 29, Maurice Ewing series 5, American Geophysical Union, Washington, D.C., 58–71.

Dickinson, R. E. 1986. Global climate and its connections to the biosphere. In *Climate-Vegetation Interactions* (C. Rosenzweig, and R. Dickinson, eds.), Proceedings of a Workshop held at NASA/Goddard Space Flight Center, Greenbelt, Maryland, 27–29 January 1986. Report 2, Office of Interdisciplinary Earth Studies, Boulder, Colorado, 87–90.

Dickinson, R. E. 1992. Land surface interaction. In *Modeling the Earth System* (D. Ojima, ed.), Office for Interdisciplinary Earth Studies, Boulder, Colorado, 131–150.

Dickinson, R. E., and A. Henderson-Sellers. 1988. Modeling tropical deforestation: Study of GCM land-surface parameterizations. *Quarterly Journal of the Royal Meteorological Society 114*, 439–462.

Dickinson, R. E., R. M. Errico, F. Giorgi, and G. T. Bates. 1989. A regional climate model for the western United States. *Climate Change 15*, 383–422.

ESSC. 1988. *Earth System Science: A Closer View.* Report of the Earth System Sciences Committee, NASA Advisory Council, National Aeronautics and Space Administration, Washington, D.C., 208 pp.

Esser, G. 1987. Sensitivity of global carbon pools and fluxes to human and potential climatic impacts. *Tellus 39B*, 245–260.

Esser, G. 1990. Modeling global terrestrial sources and sinks of CO_2 with special reference to soil organic matter. In *Soils and the Greenhouse Effect* (A.F. Bouwman, ed.), John Wiley and Sons, Chichester, U.K., 249–261.

Golley, F. B. 1972. Energy flux in ecosystems. In *Ecosystem Structure and Function* (J. A. Wiens, ed.), Oregon State University Annual Biology Colloquium 31, Oregon State University, Corvallis, Oregon, 69–70.

Goudriaan, J., H. H. van Laar, H. van Keulen, and W. Louwerse. 1984. Simulation of the effect of increased atmospheric CO_2 on assimilation and transpiration of a closed crop canopy. *Wissenschaftliche Zeitschrift der Humboldt-Universitat zu Berlin, Mathematisch-Naturwissenschaftliche Reihe XXXII,* 352–356.

Hagglund, B. 1981. *Forecasting Growth and Yield in Established Forests.* Report 31, Department of Forest Survey, Swedish University for Agricultural Sciences, Umea, Sweden.

Hahn, J., S. G. Warren, J. London, and J. L. Roy. 1988. *Climatological Data for Clouds Over the Globe From Surface Observations.* U.S. Department of Energy, Oak Ridge, Tennessee.

Hansen, J., G. Russell, D. Rind, P. Stone, A. Lacis, S. Lebedeff, R. Ruedy, and L. Travis. 1983. Efficient three-dimensional global models for climate studies: Models I and II. *Monthly Weather Review 111,* 609–662.

Hansen, J., A. Lacis, D. Rind, G. Russell, P. Stone, I. Fung, R. Ruedy, and J. Lerner. 1984. Climate sensitivity: Analysis of feedback mechanisms. In *Climate Processes and Climate Sensitivity* (J. E. Hansen and T. Takahashi, eds.), Geophysical Monograph 29, Maurice Ewing series 5, American Geophysical Union, Washington, D.C., 130–163.

Hanson, J. D., W. J. Parton, and G. S. Innis. 1985. Plant growth and production of grassland ecosystems: A comparison of models. *Ecological Modeling 29,* 131–144.

Holdridge, L. R. 1947. Determination of world plant formations from simple climate data. *Science 105,* 367–368.

Holdridge, L. R. 1964. *Life Zone Ecology.* Tropical Science Center, San Jose, Costa Rica.

Houghton, R. A., and D. L. Skole. 1990. Changes in the global carbon cycle between 1700 and 1985. In *The Earth as Transformed by Human Action* (B. L. Turner II, W. C. Clark, R. W. Kates, J. F. Richards, J. T. Mathews and W. B. Meyer, eds.), Cambridge University Press, Cambridge, U.K., 393–408.

Houghton, R. A., and G. M. Woodwell. 1989. Global climate change. *Scientific American 260,* 36–44.

Houghton, R. A., R. D. Boone, J. M. Melillo, C. A. Palm, G. M. Woodwell, N. Myers, B Moore III, and D. L. Skole, 1985. Net flux of carbon dioxide from tropical rainforests in 1980. *Nature* 316, 617–620.

Houghton, R. A., R. D. Boone, J. R. Fruci, J. E. Hobbie, J. M. Melillo, C. A. Palm, B. J. Peterson, G. R. Shaver, G. M. Woodwell, B. Moore, D. L. Skole, and N. Myers., 1987. The flux of carbon from terrestrial ecosystems to the atmosphere in 1980 due to changes in land use: Geographic distribution of the global flux. *Tellus 39B,* 122–139.

Houghton, R. A., G. M. Woodwell, R. A. Sedjo, R. P. Detwiler, C. A. S. Hall, and S. Brown. 1988. The global carbon cycle. *Science 241,* 1736–1739.

Houghton, J. T., G. J. Jenkins, and J.J. Ephraums, eds. 1990. *Climate Change: The IPCC Scientific Assessment.* Cambridge University Press, Cambridge, U.K.

Innis, G. S., ed. 1978. *Grassland Simulation Model.* Springer, New York, New York.

Jenne, R. L. Climate model description and impact on terrestrial climate. In *Global Climate Change: Implications, Challenges and Mitigation Measures* (Majumdar *et al.,* eds.), Pennsylvania Academy of Science, Easton, Pennsylvania, in press.

Joyce, L. A., and R. N. Kickert. 1987. Applied plant growth models for grazinglands, forests and crops. In *Plant Growth Modeling for Resource Management. Volume I. Current Models and Methods* (K. Wisiol and J. D. Hesketh, eds.), CRC, Boca Raton, Florida, 17–55.

Kimmins, J. P., and K. Scoullar. 1979. FORCYTE: A computer simulation approach to evaluating the effects of whole-tree harvesting on the nutrient budget in northwest forests. In *Proceedings, Forest Fertilization Conference* (S. P. Gessel, R. M. Kenady, and W. A. Atkinson, eds.), Contribution No. 40, University of Washington, Seattle, Washington, 266–273.

Kittel, T. G. F., and M. B. Coughenour. 1988. Prediction of regional and local ecological change from global climate model results: A hierarchical modeling approach. In *Monitoring Climate for the Effects of Increasing Greenhouse Gas Concentrations* (R. A. Pielke, and T. G. F. Kittel, eds.), Cooperative Institute for Research in the Atmosphere, Fort Collins, Colorado, 173–193.

Lauenroth, W. K., H. W. Hunt, D. M. Swift, and J. S. Singh. 1986. Estimating aboveground net primary productivity in grasslands: A simulation approach. *Ecological Modeling 33*, 297–314.

Legates, D. R., and C. J. Willmott. 1990a. Mean seasonal and spatial variability in global surface air temperature. *Theoretical and Applied Climatology 41*, 11–21.

Legates, D. R., and C. J. Willmott. 1990b. Mean seasonal and spatial variability in gauge-corrected global precipitation. *International Journal of Climatology 10*, 111–127.

Lieth, H. 1973. Primary production: Terrestrial ecosystems. *Human Ecology 1*, 303–332.

MacNeil, M. D., J. W. Skiles, and J.D. Hanson. 1985. Sensitivity analysis of a general rangeland model. *Ecological Modeling 29*, 57–76.

Makela, A., and P. Hari. 1986. Stand growth model based on carbon uptake and allocation in individual trees. *Ecological Modeling 33*, 205–229.

Manabe, S., and R. T. Wetherald. 1987. Large scale changes in soil wetness induced by an increase in carbon dioxide. *Journal of the Atmospheric Sciences 44*, 1211–1235.

Matyssek, R. 1986. Carbon, water and nitrogen relations in evergreen and deciduous conifers. *Tree Physiology 2*, 177–187.

McGuire, A. D., J. M. Melillo, L. A. Joyce, D. W. Kicklighter, A. L. Grace, B. Moore III, and C. J. Vorosmarty. 1992. Interactions between carbon and nitrogen dynamics in estimating net primary productivity for potential vegetation in North America. *Global Biogeochemical Cycles, 6(2)*, 101–124.

McGuire, A. D., L. A. Joyce, D. W. Kicklighter, J. M. Melillo, G. Esser, and C. J. Vorosmarty. 1993. Productivity response of climax temperate forests to elevated temperature and carbon dioxide: A North American comparison between two global models. *Climatic Change, 24(4)*, 287–310.

McMurtrie, R. E., J. J. Landsberg, and S. Linder. 1989. Research priorities in field experiments on fast-growing tree plantations: Implications of a mathematical model. In *Biomass Production by Fast-Growing Trees* (J. S. Pereira and J. J. Landsberg, eds.), Kluwer Academic, Dordrecht, The Netherlands, 181–207.

Melillo, J. M., P. A. Steudler, J. D. Aber, and R. D. Bowden. 1989. Atmospheric decomposition and nutrient cycling. In *Exchange of Trace Gases Between Terrestrial Ecosystems and the Atmosphere* (M. O. Andreae and D. S. Schimel, eds.), John Wiley and Sons, New York.

Melillo, J. M., D.W. Kicklighter, A.D. McGuire, B. Moore III, C.J. Vorosmarty, and A.L. Grace. Global climate change and terrestrial net primary production. *Nature,* in press.

Mitchell, H. L., and R. F. Chandler Jr. 1939. The nitrogen nutrition and growth of certain deciduous trees of northeastern United States. *Black Rock Forest Bulletin 11*, 1–94.

Mitchell, J. F. B., S. Manabe, V. Meleshko, and T. Tokioka. 1990. Equilibrium climate change—and its implications for the future. In *Climate Change: The IPCC Scientific Assessment* (J. T. Houghton, G. J. Jenkins, and J. J. Ephraums, eds.), Cambridge University Press, Cambridge, U.K., 131–172.

Morris, J. T., R. A. Houghton, and D. B. Botkin. 1984. Theoretical limits of below ground production by *Spartina alterniflora*: An analysis through modeling. *Ecological Modeling 26*, 155–175.

Olson, J. S. 1970. Carbon cycles and temperate woodlands. In *Temperate Forest Ecosystems* (D. E. Reichle, ed.), Springer-Verlag, New York, 226–241.

O'Neill, R. V., D. L. DeAngelis, J. B. Waide, and T. F. H. Allen. 1986. *A Hierarchical Concept of Ecosystems*. Monographs in Population Biology No. 23, Princeton University Press, Princeton, New Jersey.

Parton, W. J., J. W. B. Stewart, and C. V. Cole. 1988. Dynamics of C, N, P and S in grassland soils: A model. *Biogeochemistry 5*, 109–131.

Pastor, J., and W. M. Post. 1985. *Development of a Linked Forest Productivity-Soil Carbon and Nitrogen Model*, ORNL/TM-9519, Oak Ridge National Laboratory, Oak Ridge, Tennessee, 162 pp.

Pastor, J., and W. M. Post. 1986. Influence of climate, soil moisture and succession on forest carbon and nitrogen cycles. *Biogeochemistry 2*, 3–27.

Peterson, B. J., and J. M. Melillo. 1985. The potential storage of carbon by eutrophication of the biosphere. *Tellus 37B*, 117–127.

Prentice, K. C. 1990. Bioclimatic distribution of vegetation for general circulation models. *Journal of Geophysical Research 95*, 11818–11830.

Raich, J. W., E. B. Rastetter, J. M. Melillo, D. W. Kicklighter, P. A. Steudler, B. J. Peterson, A. L. Grace, B. Moore III, and C.J. Vorosmarty. 1991. Potential net primary productivity in South America: Application of a global model. *Ecological Applications 1*, 399–429.

Rastetter, E. B., M. G. Ryan, G. R. Shaver, J. M. Melillo, K. J. Nadelhoffer, J. E. Hobbie, and J. D. Aber. 1991. A general biogeochemical model describing the responses of the C and N cycles in terrestrial ecosystems to changes in CO_2, climate and N deposition. *Tree Physiology 9*, 101–126.

Reynolds, J. F., B. Acock, R. L. Dougherty, and J. D. Tenhunen. 1989. A modular structure for plant growth simulation models. In *Biomass Production by Fast-Growing Trees* (J. S. Pereira, and J. J. Landsberg, eds.), Kluwer Academic, Dordrecht, The Netherlands, 123–134.

Running, S. W., and J. C. Coughlan. 1988. A general model of forest ecosystem processes for regional applications. I. Hydrologic balance, canopy gas exchange and primary production processes. *Ecological Modeling 42*, 125–154.

Sanchez, P. A., E. E. Bandy, J. H. Villachica, and J. J. Nicholaides. 1982. Amazon Basin soils: Management for continuous crop production. *Science 216*, 821–827.

Sato, N., P. J. Sellers, D. A. Randall, E. K. Schneider, J. Shukla, J. L. Kinter III, Y.-T. Hou, and E. Albertazzi. 1989. Effects of implementing the Simple Biosphere Model in a general circulation model. *Journal of Atmospheric Science 46*, 2757–2782.

SCEP (Study of Critical Environmental Problems). 1970. *Man's Impact on the Global Environment*. Report of the Study of Critical Environmental Problems, MIT Press, Cambridge, Massachusetts.

Schlesinger, M. E., and Z. Zhao. 1989. Seasonal climatic changes induced by doubled CO_2 as simulated by the OSU atmospheric GCM/mixed-layer ocean model. *Journal of Climate 2*, 459–495.

Schneider, S. H. 1989. The greenhouse effect: Science and policy. *Science 243*, 771–781.

Sellers, P. J. 1986. The simple biosphere model (SiB). In *Climate-Vegetation Interactions* (C. Rosenzweig, and R. Dickinson, eds.), Proceedings of a Workshop held at NASA/Goddard Space Flight Center, Greenbelt, Maryland, 27–29 January 1986. Report 2, Office of Interdisciplinary Earth Studies, Boulder, Colorado, 87–90.

Shugart, H. H. 1984. *A Theory of Forest Dynamics*. Springer-Verlag, New York.

Shugart, H. H., and D. C. West. 1977. Development of an Appalachian deciduous forest succession model and its application to assessment of the impact of the chestnut blight. *Journal of Environmental Management 5*, 161–179.

Sellers, P. J., Y. Mintz, Y. C. Sud, and A. Dalcher. 1986. A simple biosphere model (SiB) for use within general circulation models. *Journal of Atmospheric Science 43*, 505–531.

Shugart, H. H., M. Y. Antonovsky, P. G. Jarvis, and A. P. Sandford. 1986. CO_2, climate change and forest ecosystem. In *The Greenhouse Effect, Climatic Change and Ecosystems* (B. Bolin, B. R. Döös, J. Jäger, and R. A. Warrick, eds.), SCOPE 28, John Wiley and Sons, Chichester, U.K., 475–501.

Solomon, A. M., M. L. Tharp, D. C. West, G. E. Taylor, J. W. Webb, and J. L. Trimble. 1984. *Response of Unmanaged Forests to CO_2-Induced Climate Change: Available Information, Initial Tests and Data Requirements.* DOE/NBB-0053, U.S. Department of Energy, Washington, D.C.

USGCRP. 1990. *Research Strategies for the U.S. Global Change Research Program.* National Academy Press, Washington, D.C., 291 pp.

Vitousek, P. M., and R. W. Howarth. 1991. Nitrogen limitation on land and in the sea: How can it occur? *Biogeochemistry 13*, 87–115.

Wang, Y. P. 1988. *Crown Structure, Radiation Absorption, Photosynthesis and Transpiration.* Dissertation, University of Edinburgh, Edinburgh, Scotland.

Watson, R. T., H. Rodhe, H. Oeschger, and U. Siegenthaler. 1990. Greenhouse gases and aerosols. In *Climate Change: The IPCC Scientific Assessment* (J.T. Houghton, G. J. Jenkins, and J. J. Ephraums, eds.), Cambridge University Press, Cambridge, U.K., 1–40.

Webb, T., III. 1986. Vegetational change in Eastern North America from 18,000 to 500 yr B.P. In *Climate-Vegetation Interactions* (C. Rosenzweig, and R. Dickinson, eds.), Proceedings of a Workshop held at NASA/Goddard Space Flight Center, Greenbelt, Maryland, 27–29 January 1986. Report 2, Office of Interdisciplinary Earth Studies, Boulder, Colorado, 63–69.

Weinstein, D. A., H. H. Shugart, and D. C. West. 1982. *The Long-Term Nutrient Retention Properties of Forest Ecosystems: A Simulation Investigation.* ORNL/RM-8472, Oak Ridge National Laboratory, Oak Ridge, Tennessee, USA.

Wetherald, R. T., and S. Manabe. 1988. Cloud feedback processes in a general circulation model. *Journal of the Atmospheric Sciences 45*, 1397–1415.

White, E. G. 1984. A multispecies simulation model of grassland producers and consumers. II. Producers. *Ecological Modeling 24*, 241–262.

Whittaker, R. H., and G. E. Likens. 1969. Net primary production and plant biomass for major ecosystems and for the earth's surface. Table presented at Brussels Symp. 1969. In *Communities and Ecosystems* (R.H. Whittaker, ed.), Macmillan, New York, 162 pp.

Whittaker, R. H., and G. E. Likens. 1973. Primary production: The biosphere and man. *Human Ecology 1*, 357–369.

Whittaker, R. H., and G. E. Likens. 1975. The biosphere and man. In *Primary Productivity of the Biosphere* (H. Lieth and R. H. Whittaker, eds.), Springer-Verlag, New York, 305–328.

Willmott, C. J., C. M. Rowe, and W. D. Philpot. 1985. Small-scale climate maps: A sensitivity analysis to some common assumptions associated with a grid-point interpolation and contouring. *Journal of the American Cartographer 12*, 5–16.

Woodward, F. I. 1987. *Climate and Plant Distribution.* Cambridge University Press, New York, 165 pp.

Woodward, F. I., and I. F. McKee. 1991. Vegetation and climate. *Environment International 17*, 535–546.

16

Modeling Global Change in an Integrated Framework: A View from the Social Sciences

Steve Lonergan and Scott Prudham

There have been numerous attempts over the past two decades to incorporate biophysical elements into social science models (cf. Braat and van Lierop, 1987; Leontief *et al.*, 1977; and Pearce *et al.*, 1989) or to 'link' or 'merge' models of biophysical and social processes (cf. Lonergan, 1981; Lonergan and Woo, 1990). The emphasis in this research has primarily been on describing the relationships that exist between social and biological systems and on attempting to 'value' ecological systems to ensure consistency with economic models. As the linkages between biophysical changes and human impacts have become more apparent, there has been a resurgence of interest in methods of linking or merging models, particularly to address questions on the types of social and economic impacts that can be expected from global environmental change. This interest has been frustrated by both the difficulties inherent in modeling social systems and the problems one must confront when attempting to link systems that are, seemingly, quite disparate. One fundamental problem is that there are many different types of models. The focus of this paper is on *symbolic* models, i.e., those where the subject matter is conceptualized in mathematical terms, where the symbols are subjected to standardized transformations, and where measurement is important and statistical data are processed to be used in the model. These models are often termed 'operational' or 'empirical,' to distinguish them from more formal, theoretical models. The aim of this chapter, then, is to discuss how we might use information provided by the physical and biological sciences (often obtained from symbolic models) to model the behavior of social systems and what difficulties exist in adopting such a mechanism.

The chapter has four main sections. The first of these examines some generic methodological difficulties associated with linking or merging biophysical and social science models. These difficulties can generally be classified into four categories: (1) spatial scale, (2) temporal scale, (3) unit of measurement, and (4) quantitative vs. qualitative information [a variation of (3)]. The remaining three sections discuss the three options available in applying global models. These three possibilities are described by Robinson (1985), who provides an excellent review of several models simulating global resources, population, and economic systems and discusses the application of these and similar models to climate change impact analysis. The issues facing the application of global models in this context are not unlike those facing the application of such models in the social sciences in general. According to Robinson, the three possibilities are: (1) to use existing models, (2) not to use formal models, and (3) to develop new models.

Methodological Concerns in Linking Models

Spatial Scale

Two questions arise when addressing the spatial scale question in modeling. First, what is the most appropriate spatial scale for the problem at hand? Second, is there spatial consistency in the information needed to calibrate the models? The first question can generally be handled by examining the question the modeler, or decision-maker, is attempting to answer. There have been global models, such as the Limits to Growth model (Meadows and Meadows, 1972; Meadows *et al.*, 1982), that have served to heighten the awareness of the linkages between the natural environment and society. There have also been global models that have been 'regionalized' or disaggregated by region, such as Leontief's (1977) model of the world economy and the global systems models developed at the International Institute for Applied Systems Analysis (IIASA; Hafele, 1981). The major contribution of these world or world-regional models has been to identify the important interactions between the economy, resource use, and pollution. Evolving from these global models were a set of integrated models at the national level, many related to the interactions between energy and the economy. The most notable example is the Strategic Environmental Assessment System (SEAS) model developed at the U.S. Environmental Protection Agency (EPA; Lakshmanan and Ratick, 1980). SEAS was actually a series of economic, resources (the main one being energy), and pollution models linked together. Although the interest in national models waned during the 1980s, there has been a resurgence of research centered around national resource accounting models (cf. Repetto *et al.*, 1989; Weber, 1983). Resource accounting approaches have attempted to link resource use directly into the systems of national accounts of countries and, in some cases, to provide an integrated framework within which to assess social/ecological interactions. Most operational modeling in the social sciences that attempts to integrate environmental and resource variables, however, has been undertaken at the regional (subnational) level.

A more onerous problem of spatial scale exists when trying to ensure consistency in the level of spatial aggregation of biophysical and socioeconomic models. Socioeconomic data are generally available by political jurisdiction, be it municipal, regional, state/provincial, or national. Biophysical data, on the other hand, are typically available by biophysical region, such as watershed, airshed, or ecosystem. Three alternatives are then available to the modeler. The appropriate spatial scale for the model becomes the smallest region for which there are consistent biophysical and socioeconomic data. This is often at the provincial/state or even national level, foreclosing a regional orientation. In some respects, global or national models are more easily integrated than those with a regional focus, since the data (albeit very aggregate) are spatially consistent. A second option is to 'force' socioeconomic data to adhere to biophysical boundaries by assuming a constant (or a specific) spatial density in the socioeconomic data and generating a multiplier to assign data to a biophysical region. A third method, and one that is gaining in popularity, is to geocode the socioeconomic data so that they can be used at whatever spatial level is desired. There has been considerable interest in the development of global data bases and in establishing digital data bases at the subnational level (e.g., the digital atlas being developed in British Columbia). Attempts are also being made to geocode the data from input–output tables, which can then be integrated with biophysical data (VHB Consultants, 1990).

Temporal Scale

One of the more contentious issues in modeling socioeconomic systems is how to incorporate dynamic components into the models. The implications of global environmental change for human systems may be considerable; a ssessing the magnitude of these implications over the long term in dynamic systems is difficult at best. Scenario-based modeling has attempted to confront the dynamic aspects of social systems by assuming certain feedbacks in the system. An example is the interactive impacts models discussed below. Such considerations are necessary if the model will be used to develop mitigation or adaptation strategies to deal with global environmental change. In addition to the concern with a dynamic system, however, the temporal scale problem manifests itself in two other ways. First is the debate over discounting the future. Discounting has become an integral component in all economic modeling, but has been criticized as economic models have increasingly incorporated ecological systems. The concerns with discounting are twofold: it shifts the burden of costs to future generations, and it precludes future generations from inheriting accumulated natural wealth (Pearce et al., 1989). (There are also concerns with the size of the discount rate used.) This has resulted in the demand by some groups for an environmental compensation fund that would accumulate interest and compensate future generations for the loss of natural capital (caused by climate warming, for example). Depreciating *natural* assets has also been problematic, with two approaches being used, the first a straight depreciation approach (Repetto et al., 1989) and the second a user cost approach to calculate the 'true resource value' (El Serafy, 1989).

The last concern with temporal scale to be discussed here is cumulative change. There has been recent interest in the existence of cumulative environmental effects

Issues in Data and Modeling

(Cocklin *et al.*, 1988). These are effects that result from multiple perturbations of a system, often over time. It is important to note that there are also cumulative socioeconomic effects. These result from the complexity of interrelationships that determine the state of the system and are unique at any given point in time. The 'system' we see, and attempt to model, is one that has been derived from specific conditions and relationships that have evolved over time. This conditionality argument, which has been adopted in the sciences as well as the social sciences (Gould, 1990), essentially forecloses the possibility of predicting what may happen in the future. Social scientists advocating structural explanations of social phenomena reject the symbolic modeling approach altogether. This argument is not trivial, and it effectively sets strict bounds on the utility of modeling the impacts of global environmental change.

Units of Measurement

Possibly the major constraint to effectively linking biophysical and socioeconomic models in the past has been the difficulty in developing a unit measure to assess all systems. Although attempts have been made using energy as a basis for 'man [sic] and nature' (Odum and Odum, 1976) or setting information as a universal metric, valuing social, economic, and environmental systems with a single metric has been controversial. At one extreme are models that value all elements in a single unit, such as dollars. Environmental cost/benefit analysis is the most notable example. At the other extreme are models that keep valuation separate from the model and can handle different variables, regardless of the units. A good example is lexicographic goal programming, which can effectively deal with different unit valuations. There are mechanisms for inferring the socioeconomic implications of biophysical change, however. A straightforward example is the work on the implications of climate warming for agricultural productivity, discussed in more detail below. In the simple formulation, climate variation directly affects the productivity of the system; this can be translated into changed yields and, by assuming an appropriate price structure, specific economic impacts on the farmer. Less straightforward is the assessment of the socioeconomic impacts of climate warming on other systems, for example, transportation. Modeling these impacts requires making a strong assumption of not only the extent to which warming will affect the physical operation of the transportation system, but the way this effect will be translated into a change in personal income and, in turn, to the regional economy and society.

Quantitative and Qualitative Data

Another aspect of the units of measurement problem that deserves separate treatment is the difficulty in dealing with qualitative information in the context of quantitative models. There has been considerable research on qualitative data analysis in recent years, and multicriteria analysis and multidimensional scaling have been used to transfer qualitative information into quantitative models. At present we are working on linking multicriteria analysis with goal programming for use in assessing the tradeoffs implicit in sustainable development strategies. The importance of qualitative information in modeling socioeconomic systems poses one of the major barriers to effectively linking biophysical and socioeconomic models. This is a major concern when modeling for sustainable economic development, for example. The key to sustainability lies in the interrelationships among three systems: the biophysical, the economic, and the sociocultural. This last system historically has been the most difficult to incorporate into symbolic models, since the relationships are not well understood and the units of measurement are quite varied. Models of global environmental change, however, must include social systems in their frameworks. The challenge to modelers in the future will be not only how to link ecological and economic models, but how to include social systems in their analysis as well.

Existing Models

Since Robinson (1985) has discussed the use of existing models at some length, this section is limited to a brief discussion of the models presented in her paper, along with a more detailed examination of two models familiar to the authors and devised since 1985. The application of these or similar models under different problem specifications or with new user-defined attributes remains an option, particularly when one considers the costs and effort associated with building new global models (Robinson, 1985).

Table 1 is adapted from a similar table in Robinson, and provides a brief overview of several models, including their respective time horizons, general methodological approaches, and focuses. There is considerable breadth with respect to each characteristic. In general, temporal resolution is defined somewhat by the method, i.e., the use of continuous or discrete mathematical functions. Although temporal resolution may be a critical variable in climate change impact analysis, this is less likely to be the case in modeling global land-use change. Time horizons of different models are most certainly important inasmuch as predictive power is significantly reduced over longer time frames. However, the utility of extremely short term models (1–2 years) is questionable.

Robinson (1985) further compares the models' treatment of intersectoral flows, agriculture, energy, demography, and political ramifications. Little more is to be added in this discussion. Modeling intersectoral flows may be a particularly critical component of global modeling in that it involves explicit identification of indirect effects. Input–output analysis is particularly well suited in this regard and is considered at some length in the methodology section below. There may also be particular concern about existing global models and their capacity to represent political issues and implications. Here, game theory and simulation may be the most promising approaches. However, political and social issues are notoriously difficult to include in formal symbolic models. The discussion of lexicographic

Issues in Data and Modeling

Table 1: Summary of various global models

Model	Time Horizon	Method	Focus
Coevolution model	Hundreds of years	Dynamic simulation	Society, atmosphere, biogeochemical balances
World 2	200 years	System dynamics	Population, food, soils, industry, pollution
World 3	200 years	System dynamics	Population, food, soils, industry, pollution
Latin American world model	100 years	Dynamic optimization	Allocation of labor and capital to meet basic needs
SARUM	90 years	Dynamic simulation, input-output, econometric	Food and mineral resource adequacy
MOIRA 1	45 years	Algorithmic, optimization, econometric	Hunger, food production, food trade, trade policies
World integrated model	25–50 years	Dynamic simulation, input-output	Population, capital, energy, food, trade, intersectoral flows
International futures simulation	25+ years	Dynamic simulation	Population, economic development, energy, agriculture
Grain buffer stock	Roughly 25 years	Dynamic stochastic simulation	Rules for managing grain buffer stocks
U.N. world model	25 years	Input-output (static)	Requirements for pollution generated by U.N. development targets
Interactive agricultural model	20 years	Cross-impact, interactive projection	Global food problem, grain trade
Optimal grain reserves	Roughly 20 years	Dynamic stochastic optimization	Management of grain reserves
FUGI	Roughly 10–15 years	Econometric input-output (dynamic)	Macro-economic detail, energy and resources
USDA Grains, oils and livestock	10–20 years	Econometric (static)	Production, exports, imports, trade of oils, grain and livestock
Input-output	Roughly 10 years	Input-output (static)	International interdependence
FAO price equilibrium	Roughly 10 years	Econometric (static)	World agricultural market prices, trade flows
World food economy model	1–2 years	Econometric, quadratic programming	Global agricultural markets and trade
UNFPA/FAO/IIASA	1975–2000	Agroecological analysis, linear programming	Population support, land resources, food production

Adapted from Robinson, 1985.

goal programming below is applicable to this issue since explicit tradeoffs between social objectives may be assessed.

The Global 2100 Model

Two additional models have been selected for more detailed review, owing to our familiarity with these models and to their emergence since the publication of Robinson's paper. The first of these is the Global 2100 model, developed to facilitate the analysis of long-term energy–economy linkages, specifically addressing the implications to the global economy of limiting carbon dioxide emissions (see Manne and Richels, 1990, 1991).

According to Beaver and Huntington (1992), the Global 2100 model can be categorized as an energy–economy model as opposed to an energy model proper. That is, the model includes a considerable macroeconomic component in order to present a detailed picture of the interplay among economic sectors and the production and consumption of energy. The cost of such an approach, relative to true energy models, is usually a loss of detail in modeling energy supply and demand. The advantage is that macroeconomic modeling is a component of the framework as opposed to energy models, where economic parameters are largely determined exogenously.

Global 2100 operates on the basis of five global regions: the United States, other members of the Organization for Economic Cooperation and Development (OECD), the USSR and Eastern Europe, China, and the rest of the world. Each region is subject to parallel modeling under the ETA-MACRO framework, a hybrid of an energy model and a macroeconomic growth model employing a dynamic nonlinear optimization approach. Five main features of the model are listed in Manne and Richels (1990). These are:

- Energy costs affect economic growth.
- Energy conservation is driven by energy prices.
- Energy conservation and increased energy efficiency can result from policy-induced changes in the energy sectors and by economic structural alterations. This is called autonomous energy efficiency improvement (AEEI).
- Fuel substitution is driven by relative price disparities.
- Supply technology constraints are built into the model (e.g., time frames required for technological changes).

As is the case for all modeling of this nature, several potentially controversial specifications must be provided. These include a future economic growth rate, the elasticity of substitution between energy and capital-labor, and the rate of AEEI. These depend upon user bias and preference, and the potential costs of carbon dioxide limitations determined by the model depend in large part on the initial specifications of the user.

The MESSAGE II Model

A second model chosen for more detailed review is the Modeling of Energy Supply Systems and Their General Environmental Impact (MESSAGE) II model, an aggregate energy model that examines the future development of energy supply and demand under user-specified scenarios. A more detailed description of the model as well as a presentation of some results can be found in EcoPlan International (1990). The description below is based upon this source as well as on experience with the model as applied to the Canadian energy system by researchers at the Institute for Integrated Energy Systems at the University of Victoria. Although these two applications are national rather than global, the model itself is quite flexible in this regard and is best suited for aggregate level analysis.

This model is a true energy model, with macroeconomic variables specified exogenously, thus serving as a contrast to the Global 2100 model. The MESSAGE II model employs a dynamic linear programming method to produce future specifications of technology-based user-defined scenarios of energy supply and demand. It draws upon a wealth of experience with similar approaches to modeling at IIASA. The model requires four main inputs:

- Energy reserves and resources and their specific extraction costs
- Detailed technical information regarding the supply and demand of energy, tracing energy from its sources, through various transformations to end uses
- Information regarding the development of future demand for energy
- Import prices for energy (EcoPlan International, 1990).

This information is contained in a detailed techno-economic data base describing the various facets of the energy sectors. Model outputs include detailed breakdowns of energy use by type, allocated to the various sources of energy demand, calculated over predetermined time intervals up to a given terminal year. In the application of the model to both Norway and Canada, the model reports on the basis of five-year intervals from 1990 to 2010, with one ten-year interval from 2010 to 2020, even though the actual time horizon is 2040. This is in order to avoid complications involved in terminating the model within the planning horizon of interest (EcoPlan International, 1990).

The linear programming (LP) and optimization approach to modeling is widely familiar and is not reviewed in detail here. However, the use of LP carries with it significant conceptual baggage, as does the overall MESSAGE framework. The limitations of LP are mainly that linear models are insensitive to nonlinear effects, such as changes in economic and energy efficiency with changes in scale, and are not well suited for spatial analysis. On the other hand, linear programming is conceptually simple relative to nonlinear approaches and under certain circumstances is warranted given that improvements in model accuracy with increased complexity may be small relative to the required effort. One particular problem with the MESSAGE model is that it is very data intensive, as it requires compilation of a detailed data base for baseline conditions of energy supply and demand. Also, the techno-economic specifications of energy technologies may be difficult to obtain, particularly for developmental technologies. Finally, as with all models, there are the problems of oversimplification and of developing the model with sufficient foresight to allow for a realistic array of possible future developments.

Conclusion

Aggregate, long-term models of energy–economy linkages are enjoying somewhat of a renaissance due in no small part to the emergence of global climate change as a dominant policy issue (Beaver and Huntington, 1992). Certain facets of these models and the issues that they face are common to all global-scale models, including those examining land-use change. It is worth reiterating that time is an important variable. The emergence of the global change theme has precipitated an emphasis on longer-term modeling. This is a challenge inasmuch as accuracy becomes more and more difficult to maintain over these longer time scales. In addition, the sociocultural conditions under which these models operate may change dramatically over their time horizon. This must be a consideration in interpreting results. Another particularly critical issue is that of model spatial resolution (see 'Spatial Scale,' above). All global models are inherently aggregate in their approach. Nevertheless, the position taken in this chapter is that, for global modeling, there is always considerable capacity to aggregate regions (being careful not to assume too much in the process); however, it is extraordinarily difficult to disaggregate from the global to the regional scale. The inclusion of some regional analysis can be considered critical to social science modeling in that its exclusion precludes, among other things, consideration of equity issues. An example of the critical nature of the spatial resolution question is found in the Global 2100 model. Its five world regions may not be reasonably coherent even five years from now, let alone in the year 2100. In fact, one could argue that the inclusion of a USSR and Eastern Europe region is already obsolete.

Avoiding the Use of Formal Models

Robinson (1985) discusses several areas of concern in the field of global modeling, cautioning users not to overly 'inflate' their expectations. Some of these problems have been mentioned above, including appropriate spatial resolution and oversimplification. In addition, Robinson presents the relative youth of global-level modeling as a problem in that the scientific standards against which these efforts have been held lack somewhat in rigor. Robinson (1985: 470) points out that this is also partly a result of the development of global models from disciplines 'such as economics, demography, ecology and political science...,' academic fields that have lagged behind the so-called hard or natural sciences in their theoretical development and predictive capacity. However, the lack of rigor in these models and the failure of the social sciences in achieving explicative power may be largely attributable to what Kuhn (1970) would call the lack of unifying paradigms in these disciplines. The lack of general methodological and epistemological agreement in the social sciences forces one to consider seriously Robinson's second option for the use of global models in these disciplines, i.e., do not model.

Kuhn (1970) is widely recognized as a seminal influence on the way scientific progress is interpreted. His conception of the development of scientific disciplines is one in which there are three basic phases. The first is pre-science. In this phase there is no unifying theory governing the beliefs of the practitioners in a discipline; community coherence is lacking. Because this unifying effect is absent, there is no agreement on either the rules that prescribe the types of activities that researchers should undertake or the rules proscribing other avenues. Mature science is the second phase, during which a unifying paradigm develops around a critical theoretical framework. The paradigm tells the scientists what subjects they should consider relevant to study and what subjects they should not. The final stage is that of revolution in the discipline(s). At this stage, theories may compete and if the urgency of the inaccuracies perceived in old theories is great enough, a period of upheaval occurs during which an old paradigm is replaced by a new one. Following a chaotic transition period, a new theoretical framework emerges and guides scientists according to its own heuristic agenda (Kuhn, 1970).

These interpretations can be important in attempting to understand the hesitancy with which social science disciplines have adopted symbolic modeling, since one might argue that under Kuhnian theory the social sciences are not characterized by the existence of unifying paradigms. It seems clear that social sciences are more closely associated with the pre-science stage of Kuhn's system. Arguably, this is a contributing factor to the methodological divisions in many of the social science disciplines. For example, Macmillan (1989) identifies four criticisms that have been leveled at modeling in geography: (1) modelers tend to ignore explanation and concentrate on emulation; (2) any theorizing that is done by the modelers is restricted to hypothetical systems; (3) questions of 'real importance' are not considered; and (4) theories must be tied to their societal context. These are fundamental problems of modeling in the social sciences; they are also problems with which most modelers are familiar and the significance of which most modelers appreciate.

The disagreement in social science disciplines over the use of symbolic models is not likely soon to disappear; neither is this disagreement necessarily unhealthy. However, one must keep in mind two points when evaluating models. The first is that each model has a specific purpose that establishes its context and that determines the ultimate usefulness of the model results. Models should be judged according to how well they fulfill these specific purposes, rather than in the abstract. Second, even those loath to model will admit readily that it is important to explore the complex interrelationships in intricate systems. Modelers explore these mathematically. Whether one chooses to follow suit depends upon one's belief in the ability of mathematical models to emulate human behavior and on the ability of modelers to place results in proper context. These questions raise the issues of oversimplification and overemphasis on form.

In the use of symbolic models, oversimplification and preoccupation with form can become particularly problematic. While all models necessarily embody considerable simplification from the real world, there is generally a limit of simplification beyond which the model becomes too different from its intended subject to lend insight. Modeling in the social sciences has been severely criticized as being too simplistic to capture the complexity of human experience and action (some criticisms stating that modeling can *never* capture this complexity). As a result, there has been a significant rift in recent years, particularly in geography, between the quantitative modelers on the one hand and the behavioral/structural analysts on the other. In addition, standards of how accurately models replicate reality are difficult to identify and are subject to constant revision by practitioners; there are many who claim that symbolic modeling has little to contribute to a better understanding of how social systems operate. A second major concern is that undue emphasis on the form of the model can obscure the original and valid intent, which is always to investigate properties of the model subject. There has been a tendency to become absorbed with the development of a model while ignoring the importance of model refinement as a key element to understanding the actual subject matter (Kaplan, 1964).

Building New Models: Methodology

The third option identified by Robinson (1985) when considering the application of global models to new problems is to build new ones. Following a brief discussion of some conceptual aspects of model development, we discuss some particular methodologies. Rather than attempt an exhaustive description of the methodologies available to would-be modelers, this section presents three particular methodologies applicable to modeling at the global or national scale. These are input-output analysis, resource accounting, and lexicographic goal programming. These three have been selected in part because we believe them to be particularly promising approaches to modeling socioeconomic systems, because they are very different and present users with distinct strengths and weaknesses, and because we are familiar with their application to specific problems.

Linking Biophysical and Socioeconomic Systems: Conceptual Models

Most of the recent work on integrating biophysical and socioeconomic models has concentrated on assessing the impacts of global environmental change on these systems. Most of the studies have used the basic cause-and-effect impacts model illustrated in Figure 1a. This has been typical of the climate warming impacts studies conducted by the EPA and the Canadian Climate Center, for example. Outputs

a) the basic impact model

b) the basic model with indirect effects added

c) interactive model with feedbacks

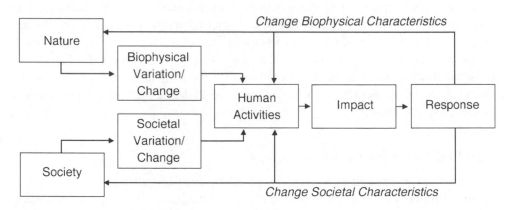

Figure 1. Types of impact models (adapted from Kates, 1985).

from large numerical simulation models (general circulation models, or GCMs) of the earth's atmosphere, typically under a scenario of a doubling of carbon dioxide levels, have been used as inputs that affect physical, biological, and social systems. In some cases these climate changes have been translated directly into human effects, while in other cases they have been used to project biophysical changes and, eventually, socioeconomic impacts. An example of the latter is the work on assessing the implications of climate warming on agriculture. Baseline climate conditions are used to project agricultural yields and, in turn, societal effects (Figure 2a). These simple conceptual diagrams belie the complex nature of the climate–agriculture–society linkages, however, as Figure 2b illustrates.

There are two major problems with the basic impacts approach, however complex the linkages that are described. First, the model does not include impacts beyond the first order. The best example is that of a decline in agricultural output, which would then affect (positively or negatively) the demand for other inputs to

a) the basic impacts model

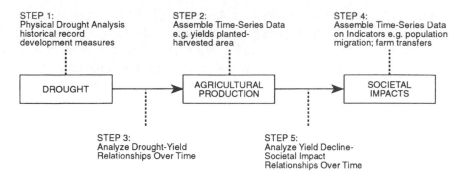

STEP 1:
Physical Drought Analysis
historical record
development measures

STEP 2:
Assemble Time-Series Data
e.g. yields planted-
harvested area

STEP 4:
Assemble Time-Series Data
on Indicators e.g. population
migration; farm transfers

DROUGHT → AGRICULTURAL PRODUCTION → SOCIETAL IMPACTS

STEP 3:
Analyze Drought-Yield
Relationships Over Time

STEP 5:
Analyze Yield Decline-
Societal Impact
Relationships Over Time

b) the basic model with indirect effects added

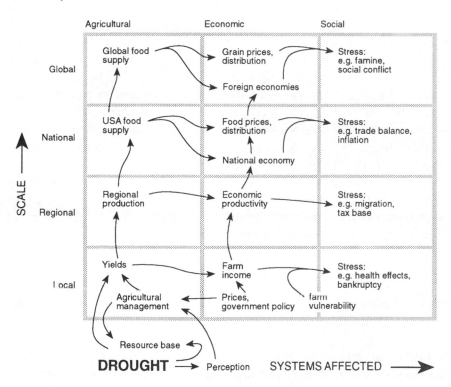

Figure 2. Modeling the impacts of drought on society (from [a] Kates, 1985; [b] Warrick and Bowden, 1981).

agricultural production, such as water, fertilizer, and so on. This changed level of demand for direct inputs would then affect the demand for second-order inputs (used to produce irrigation systems and fertilizer), which would affect third-order inputs, and so on down the line (Figure 1b). Some of these inputs are captured in the process described by Figure 2b; others can only be picked up through a detailed process analysis or, in some cases, an input–output analysis (discussed below).

A second problem with the basic impacts approach is more disturbing, however. Most models do not consider the dynamic nature of the systems involved, due to a combination of uncertainty and mathematical complexity in modeling socioeconomic systems. This is a particular concern when modeling long-term change, such as climate warming. Figure 1c illustrates an interactive impacts model that incorporates both biophysical and socioeconomic feedbacks into the process. Note that the model also takes into consideration the human responses associated with the impacts (mitigation and adaptation strategies). Attempts are now being made to incorporate possible societal changes into the basic impacts models through scenario analysis, but one of the main reasons that many social scientists have been loath to work on models of global environmental change has been the component of uncertainty in dealing with long-term changes.

Scenario analysis, however, cannot incorporate all of the possible feedbacks inherent in social systems. Demographic and structural economic changes will occur as a result of, and in spite of, global environmental change. Mitigation and adaptation strategies will be adopted that may significantly alter the expected impacts from global change. Geopolitical forces will have a major role in determining human migration, the vulnerability of certain regions to environmental change, and the contribution of human systems to global change. We can only begin to incorporate such changes into modeling frameworks; in their absence, however, modeling outputs can be viewed as only general indicators of the implications of global change for human systems.

The conceptual frameworks discussed above provide a general structure to which most socioeconomic models (particularly impact models) that try to incorporate biophysical changes adhere. The discussion below presents examples of models that explicitly attempt to link these systems.

Input–Output Analysis

Input–output analysis, since its initial development by Leontief (1936), has become an invaluable tool for economists and others to estimate the impacts of exogenous changes in the economy. The basic structure of an input–output table is simply an accounting framework of interindustry dollar flows, with additional columns added to represent final demand sectors and additional rows to represent payments to government, labor, and value added. The literature in input–output analysis is quite extensive, and texts describing the basic method have been written by Miernyk (1965), Richardson (1972), and Miller and Blair (1985). The standard industry-by-industry input-output table is a framework for listing the activities in a regional economy. The table can be transformed to allow one to estimate all direct and indirect impacts (nth-order impacts) of an exogenous change in the economy. The model can also incorporate various types of multipliers, including pollution, so that one can calculate the total pollution in a region (or what-

		Agriculture	Textile		Petroleum refining		Sport fishing		Plankton production	Herring production	Cod production
Economic commodities	Wheat										
	Cloth										
				2 Economic system: Intersector coefficients			**1** Ecological processes: Their input and output coefficients re: economic commodities				
Ecological commodities	Crude oil										
	Water intake				−						
	Alkalinity				+						
				3 Economic sectors: Their input and output coefficients re: ecological commodities			**4** Ecological system: Interprocess coefficients				
	Detritus										
	Plankton								+	−	
	Herring									+	−
	Cod									+	

Figure 3. The Isard economic–ecological input–output model (from Isard, 1969). Reprinted with permission from the *Papers of the Regional Science Association*.

ever spatial scale is used) resulting from a change in economic structure (e.g., a new firm moving to the region). In the late 1960s, however, a few researchers began including environmental variables. Models were developed by Cumberland (1966), Daly (1968), Isard (1969, 1972), and Victor (1972), and a complete review of economic–ecological input–output models can be found in Lonergan and Cocklin (1985). For this paper, the Isard model will be explained briefly.

Extension of the traditional interindustry input–output model structure to account for ecological and pollution impacts, other than the simple multiplier approach mentioned above, is inherently difficult because the model assumes single-product industries and must assign market prices to all industry outputs. These difficulties are minimized by the use of a commodity-by-industry model, where there are more commodities than industries. Commodities are listed in rows and industries or activities are listed in columns.

Using this structure, Isard (1969, 1972) developed a model that included environmental variables (Figure 3). The upper left quadrant accounts for the flows between economic sectors, while the lower right quadrant represents interactions within the ecological system. The upper right quadrant describes the use of economic products by environmental processes, pollution being a prominent example. Coefficients in the lower left quadrant record the use of ecological commodities by economic sectors and the export of commodities from economic sectors to the ecological system. The entries of this quadrant include pollutants produced by economic activities and exported to the ecological system.

The entries in each of the quadrants of this accounting framework are in the form of coefficients. That is, the entries in the cells of the economic system (the upper left quadrant) are typical input–output coefficients, specifying the dollar worth of inputs from one sector needed to produce one dollar of output from another sector (or the same sector). In a similar manner, a set of ecological interprocess coefficients, the entries of the lower right quadrant, can also be defined. The coefficients in the off-diagonal matrices of Figure 3 are expressed in terms of the ecological inputs to and outputs from the economic sector per dollar of economic output. Thus, in the lower left quadrant, a coefficient might refer to pounds of sulfur dioxide produced per dollar output of the pulp and paper industry.

Although Isard was able to utilize the ecological–economic input–output model for assessing regional development proposals, there are major technical limitations to this type of model. Most significant are the difficulties associated with obtaining appropriate data. Richardson (1972) and Victor (1972) questioned the ability to specify the ecological interprocess matrix, a difficulty also recognized by Isard. The assumption of fixed coefficients is a limitation of input-output models in general, but is considered to have more serious implications with the addition of the ecological system.

The interest in economic-ecological input–output models diminished in the 1970s as the difficulty in applying the models became apparent. Nevertheless, they represent one of the first attempts by social scientists to integrate economic and ecological systems and were the precursor to the resource accounting models discussed below.

Natural Resource Accounting

There are essentially two types of resource accounting models: those that are strictly derivative of the systems of national accounts (SNA) and those that are not. The former result from broad recognition of the shortcomings in the SNA structure for reflecting the contribution of environmental systems to economic productivity. In particular, the aggregates of the SNA (gross and net domestic product, or GDP and NDP) include income from so-called defensive expenditures and from resource depletion. Resource accounting models not derivative of the

SNA have grown from a perspective that recognizes the immense potential of a comprehensive system of resource information for multiple applications.

Hicks (1946) is widely credited with recognizing that income should be defined and measured as the capacity to consume without reducing future capacity to consume *ad infinitum*. Inasmuch as excess consumption is capital depletion, this makes the concept of sustainable income redundant (Daly, 1989). It is clear that the SNA does not reflect this perception of income.

There are two main problems with the SNA where environmental systems are concerned (El Serafy and Lutz, 1989). These are (1) in the treatment of expenditures on environmental protection and (2) in the treatment of resource depletion. Both problems can be addressed by making the macroaccounting system more consistent, i.e., by treating the environment as a capital asset of the economy.

Correcting the inclusion of expenditures on environmental improvement is simple in principle. Once the error is recognized, these can be deducted from GDP.[1] The difficulty is in defining these expenditures in a functionally acceptable fashion. The term is generally applied to outlays by industries or households that are directed at improving environmental quality or guarding against environmental deterioration. An obvious example is the expenditure by industry on pollution abatement equipment.

The inclusion of the net proceeds from resource depletion in income measures is a second perceived flaw in the SNA. The recommended correction once again is to treat the environment as a capital asset, with all the conceptual baggage this entails, and reduce NDP by an additional amount for nonsustainable resource use.[2]

In both instances, the environment is classified as another form of capital asset. Resource accounting models that emphasize this perspective tend to be structurally derivative of the SNA, in some cases the accounting system being a series of satellite accounts to the main macroaccounts. This presupposes that the contribution of environmental systems can be described in an accounting format and, in particular, in one that is devoted only to market transactions (with limited exceptions).

There are resource accounting systems that have been developed autonomous to the SNA. These also presuppose that environmental systems can be described in an accounting format. However, monetary valuation is not strictly applied, and nonmarket resources may fall within the scope of the accounts. Gilbert and Hafkamp (1986) have convincingly argued that resource accounting must be multiobjective in order to address the current lack of understanding vis-a-vis economic/ecological linkages.

The Indonesian Accounts
Repetto *et al.* (1989) have examined the issue of resource depletion and the SNA in Indonesia. Resource accounts were established for timber, petroleum, and soil

---

[1]While some argue that the correction should be to NDP, this debate is of little consequence to the current discussion.

[2]Here too there has been some disagreement over the locus of the correction, i.e., GDP or NDP.

resources. Each account contained a measure of opening stock, closing stock, unit value of the resource, and various adjustments to the stock. Depletion was valued at the average unit value of the resource over the accounting period. The authors recommend that the resulting total for all resources be deducted from Indonesian NDP. The magnitude of this adjustment is significant to an economy like that of Indonesia. Applying the methodology as outlined reduces the average rate of economic growth from 7% to 4% over the 1971–84 period.

The case of Indonesia is an example of resource accounting as satellite to the SNA. Physical account entries are necessary only as precursors to valuation. There is no mention of the application of the resource accounts in other contexts demanding detailed systematic environmental statistics.

The French Accounts

The French patrimony[3] accounts are, by contrast with Indonesia, structurally independent and advanced. This system is intended to supplement the SNA with information on the contribution of environmental inputs to economic production; however, the accounting structure is independent of the SNA. In addition, the system is set up more as an independent environmental statistical system (Theys, 1989). It is thus designed to track the status of all natural resources in France and to relate this information to existing information systems, including the macroeconomic accounts.

The system is organized around three themes. The first of these divides the natural resource base into constituent components. These components include nonrenewable, physical, and living resources. The second theme organizes the environment as linked ecosystems. The third theme organizes information on resources by resource users or agents. Each of the three themes can be described by two distinct types of accounts: central accounts and peripheral accounts. In addition, the French have outlined plans for developing territorial accounts for regions of France, one of the exceptions to the generally macro level orientation of resource accounting systems.

The French system is unique in its complexity and pioneering in the formulation of resource accounts as a source of environmental statistics *per se*. The patrimony system has been influential in the development of autonomous resource accounting and environmental statistical systems elsewhere, including the resource accounting system devised at the Institute on Environment and Economy in Ottawa, Canada (Friend, 1990), and the regional resource accounting framework developed at the Center for Sustainable Regional Development at the University of Victoria (Prudham, 1992).

Conclusion

The significance of this discussion to the topic at hand is apparent in the emerging focus of resource accounting on the provision of environmental statistics in sys-

---

[3]The French have used this term to denote all natural products and systems that are subject to human influence to a significant degree, i.e., relative to natural cycles.

tematic, account-style format. Structurally, resource accounts are remarkably similar to input–output transaction matrices, describing origins and destinations for resources and associated services. One of the key problems in modeling linkages between different facets of the economic and environmental systems relates to the use of disparate metrics. This is also an obstacle to modeling in general. A resource accounting system offers a forum for reporting in numerous metrics, yet also allows for the subsequent application of single metric analysis, such as dollars or joules.

Certainly the problem of single vs. multiple metrics cannot be resolved through the use of resource accounts alone. However, the possibility of investigating relationships between facets of economic and ecological systems is robust. Hannon (1991) has devised a hybrid accounting and input–output system that is set up to describe ecosystem flows of energy and matter, facilitating economic/ecological modeling. Developments in this direction are important for modeling at the global and regional levels, since they provide conceptual models for the compilation of data bases, as well as providing a forum for the linkage of more specific modeling exercises.

Goal Programming Models

One of the more interesting models that allows an assessment of the tradeoffs among various goals that need not be specified in the same units is lexicographic goal programming (Lonergan and Cocklin, 1988). The linkage of economic/ecological systems represents one area where lexicographic goal programming can be applied to aid decision-makers. Goal programming models have found extensive application in business management, where goals or objectives can be clearly specified, but their application to public sector planning has not been without criticism. These criticisms focus on whether the development targets required in the goal programming framework can be satisfactorily identified (Cohon and Marks, 1975; Dyer, 1979). The technique has been applied to many public sector planning problems, however, including forest management (Dane et al., 1977; Schuler et al., 1977), agricultural and recreational resource planning (McGrew, 1975; Cortes, 1981), and industrial (Werczberger, 1976; Charnes et al., 1977) and residential (Courtney et al., 1972) location problems. One example of the richness of these models is the multiple land-use allocation model developed by Dane et al. (1977) to assist with planning decisions for the Mount Hood National Forest in Oregon. The model was able to provide information on the sensitivity of land allocations to combinations of planning goals, the goal constraints that had the greatest effect on model solutions, the sensitivity of the allocations to goal priorities, and the tradeoffs between goals. These types of information are crucial in identifying and resolving economic/ecological conflicts. With the aim of providing similar types of information, the resource management model described below was developed.

This goal programming model was developed to analyze resource options relating to forest energy plantation development in eastern Ontario, Canada.

This is an economically and socially depressed rural region with a declining agricultural base. In the late 1970s, the provincial and federal governments initiated a cooperative development program, titled the Eastern Ontario Development Program, the purpose of which was to stimulate the regional economy through the promotion of industry, agriculture, and forestry. An important component of the forestry development initiative was the New Forests in Eastern Ontario program, centering on the promotion and development of hybrid poplar plantations.

Six planning goals were identified as relevant to forest energy plantations in eastern Ontario following discussions with the research and planning agencies involved with the biomass program. These were included in the model, along with a set of technical and resource constraints. All can be categorized under one of the following general headings:

- Technical/resource constraints, including constraints on total land availability, total biomass produced, and the size of conversion facility available
- Economic efficiency in biomass production, including the net returns to biomass production under alternative management systems
- Economic efficiency in energy conversion, incorporating net annual returns to three end-use facilities: electricity cogeneration, heating plants, and methanol production
- Regional employment generation, i.e., total employment created at both the production and conversion sites
- Regional income generation, i.e., total income generated within the region as a function of biomass production and energy conversion operations
- Energy efficiency, i.e., total energy produced minus inputs to biomass production and energy conversion
- Environmental quality, i.e., erosion and nutrient loss from the plantations.

The objective of the model was to select the management alternative and end-use combination that minimized the deviations from the target levels that were established for each goal listed above. The output of the model assigned a management system to each of the 27 townships in the region, and simultaneously selected an optimum combination of end-use facilities. In the context of our study, there was little concern if the target level was exceeded; only a failure to meet a particular target level was considered a deviation. The only exception was the environmental quality goal, for which levels of soil erosion greater than the specified target were considered undesirable.

Target levels for all goals except environmental quality were initially set by maximizing the level of achievement for each goal; this value was termed the resource potential relative to the particular goal. The resource potential is particularly important because it identifies the maximum level for each goal that can be achieved given available resources. If this maximum level is desired, or mandated, by decision-makers, the achievement levels of the other goals can be easily determined, as discussed above. The question emerges, though, as to what the effect

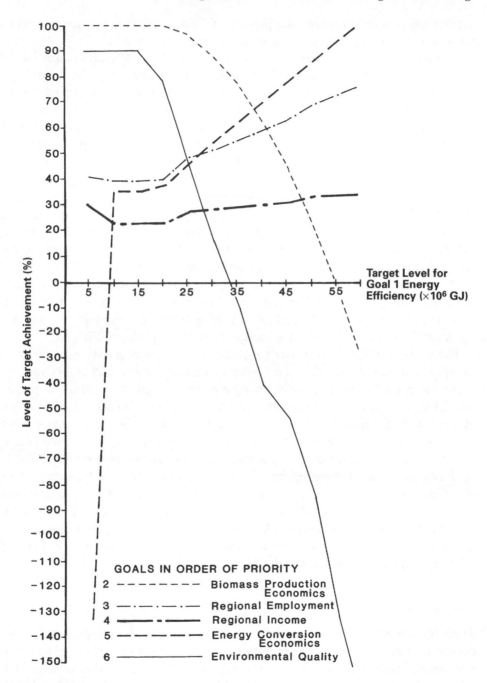

Figure 4. Assessing economic/ecological/energy tradeoffs with a goal programming model.

will be of specifying an achievement level of less than maximum for a particular goal. The programming model can also be used to estimate the sensitivity of the solutions to a lowering of the target level for a specific goal. This is relevant in sit-

431

uations where decision-makers are prepared to accept less than maximum achievement of a particular goal in favor of greater achievement of other goals.

Similar to the assessment of goal relationships that could be undertaken on 720 scenarios given the six goals, the evaluation of the sensitivity of the target levels can also be conducted on as many scenarios. However, given the nature of the conflicts among the goals (in problems with many conflicts, the analysis will rarely proceed past the second goal), far fewer runs are necessary to completely analyze the tradeoffs. By varying the target level of the highest priority goal between a specified minimum and its maximum value, the relative levels of achievement for the other lower-priority goals can be estimated. Figure 4 depicts the results of the analysis when conversion economics is assigned the highest priority; other goals in order of decreasing priority are biomass production economics, regional employment, regional income, energy efficiency, and environmental quality. Increasing the absolute level of achievement for the conversion economics goal results in increasing levels, in terms of percentage of target level achieved, for all goals except environmental quality and production economics, which decrease quite dramatically as economic returns to biomass conversion rise. This can be accounted for by the type of management system and the resultant end uses that accompany the varying levels of achievement of the highest priority goal.

The study outlined above shows the value of using lexicographic goal programming to analyze conflicts between economic and ecological (and other) variables present in natural resource planning problems. The explicit recognition of the multiple objectives evident in the planning process and their representation in the goal programming framework allows for a determination of the relationship among goals as well as the effect of varying the level of achievement of any one goal on the rest. The benefits for natural resource planning are obvious; no longer is one subject to the rigidity of an 'optimal' solution or simulating system behavior, but the tradeoffs between economic and ecological factors can be both qualitatively and quantitatively assessed. Applications at the global resolution are hindered by added complexity in specifying realistic objectives. However, the model design is appealing in its capacity to integrate and examine quantitatively the tradeoffs between biophysical and socioeconomic objectives.

Conclusion

There have been numerous attempts to 'link' or 'integrate' biophysical and socioeconomic systems in the context of social science models. These range from cost/benefit analysis to linking simulation models with optimization models, to the integrated resource accounting framework developed by the French. Models have been developed at all levels of spatial aggregation, from local to global, and for various temporal frameworks. Which of these is most appropriate for modeling land-use change? If we do not choose to use one of the existing model frameworks, can we develop new frameworks to specifically address land-use change? Or should we model at all?

We take the perspective that to have an impact on policy, models must be developed at the regional (subnational) scale. Regardless of the scale question, however, there remain inherent difficulties in any attempt to link socioeconomic and biophysical systems in a structured model. In some cases, for example, to ensure spatial consistency, attempts are being made to geocode social, economic, and biophysical data, which will alleviate one of the problems. In other cases, such as the valuation problem, solutions may not be forthcoming, and ways to circumvent the problem must be used. There is also a set of fundamental criticisms of any models in use in the social sciences. Although these criticisms have not been countered, it is important to reiterate that models must be evaluated with respect to the purpose for which they are designed. Despite these difficulties, attempts to link models have achieved some success and do force modelers and decision-makers to address the fundamental question of how biophysical and social systems are related. The present interest in assessing the implications of global environmental change again focuses attention on linking these systems through symbolic modeling; there must be continued attempts to model these systems in an integrated fashion.

References

Beaver, R. D., and H. G. Huntington. 1992. A comparison of aggregate energy demand models for global warming policy analysis. *Energy Policy 20(5)*, 568-574.

Braat, L. C., and W. F. J. van Lierop, eds. 1987. *Economic-Ecological Modeling*. Elsevier, Amsterdam, The Netherlands.

Charnes, A., K. Haynes, J. Hazleton, and M. Ryan. 1977. A hierarchical goal programming approach to environmental land-use management. *Geographical Analysis 7(2)*, 121–130.

Cocklin, C., H. Hay, and A. Fenn. 1988. *The Concept of Cumulative Environmental Change.* Environmental Science Occasional Publication No. CEC-01, University of Auckland, Auckland, New Zealand.

Cohon, J., and D. Marks. 1975. A review and evaluation of multiobjective programming techniques. *Water Resources Research 11(2)*, 208–222.

Cortes, A. 1981. A multiobjective model for analysis and planning of an agricultural area. Paper presented to the European Congress of the Regional Science Association, Barcelona, Spain, August.

Courtney, J., T. Klastorin, and T. Ruefli. 1972. A goal programming approach to urban suburban location preferences. *Management Science 18(6)*, B258–B268.

Cumberland, J. H. 1966. A regional inter-industry model for analysis of development objectives. *Papers of the Regional Science Association 17*, 65–94.

Daly, H. 1968. On economics as a life science. *Journal of Political Economy 76*, 392–406.

Daly, H. 1989. Toward a measure of sustainable net national product. In *Environmental and Resource Accounting for Sustainable Development* (Y. J. Ahmad, S. El Serafy, and E. Lutz, eds.), World Bank, Washington, D.C.

Dane, C., N. Meador, and J. White. 1977. Goal programming in land-use planning. *Journal of Forestry 75(6)*, 325–329.

Dyer, A. 1979. Implications of goal programming in forest resource allocation. *Forest Science 24(4)*, 535–543.

EcoPlan International. 1990. *Choosing Europe's Energy Future*. EcoPlan International, Paris, France.

El Serafy, S. 1989. The proper calculation of income from depletable natural resources. In *Environmental and Resource Accounting for Sustainable Development* (Y. J. Ahmad, S. El Serafy, and E. Lutz, eds.), World Bank, Washington, D.C.

El Serafy, S., and E. Lutz. 1989. Environmental and resource accounting: An overview. In *Environmental and Resource Accounting for Sustainable Development* (Y.J. Ahmad, S. El Serafy, and E. Lutz, eds.), World Bank, Washington, D.C.

Friend, A. M. 1990. *Economics, Ecology and Sustainable Development: Are They Compatible?* Occasional Paper Series No. 1, Institute for Research on Environment and Economy, University of Ottawa, Canada.

Gilbert, A., and W. Hafkamp. 1986. Natural resource accounting in a multi-objective context. *The Annals of Regional Science 20(3)*, 10–37.

Gould, S. J. 1990. *A Wonderful Life*. Prentice-Hall, New York.

Hafele, W. 1981. *Energy in a Finite World*. Ballinger, New York.

Hannon, B. 1991. Accounting in ecological systems. In *Ecological Economics* (R. Costanza, ed.), Columbia University Press, New York.

Hicks, J. R. 1946. *Value and Capital*. Oxford University Press, Oxford, U.K.

Isard, W. 1969. Some notes on the linkage of ecological and economic systems. *Papers of the Regional Science Association 22*, 85–96.

Isard, W. 1972. *Ecologic-Environment Analysis for Regional Development*. Free Press, New York.

Kaplan, A. 1964. *The Conduct of Inquiry: Methodology for Behavioral Science*. Thomas Y. Crowell, New York.

Kates, R. W. 1985. The interaction of climate and society. In *Climate Impact Assessment* (R.W. Kates, J.H. Ausubel and M. Berberian, eds.), SCOPE 32, John Wiley and Sons, New York.

Kuhn, T. S. 1970. *The Structure of Scientific Revolutions*, 2nd edn. University of Chicago Press, Chicago, Illinois.

Lakshmanan, T. R., and S. Ratick. 1980. Integrated models for economic-energy-environmental analysis. In *Economic-Environmental-Energy Interactions* (T.R. Lakshmanan and P. Nijkamp, eds.), Martinus Nijhoff, Boston, Massachusetts.

Leontief, W. 1936. Quantitative input-output relations in the economic system of the United States. *Review of Economics and Statistics 18(3)*, 105–25.

Leontief, W., *et al.* 1977. *The Future of the World Economy*. Oxford University Press, New York.

Lonergan, S. C. 1981. A methodological framework for resolving ecologic/economic problems. *Papers of the Regional Science Association 48*, 117–133.

Lonergan, S. C., and C. Cocklin. 1985. The use of input-output analysis in environmental planning. *Journal of Environmental Management 20*, 129–47.

Lonergan, S. C., and C. Cocklin. 1988. The use of lexicographic goal programming in economic/ecological conflict analysis. *Socio-Economic Planning Sciences 22(2)*, 83–92.

Lonergan, S. C., and M.-K. Woo. 1990. *Climate Change and Transportation in Northern Canada: An Integrated Impact Assessment*. Final report to the Atmospheric Environment Service, Environment Canada, Downsview, Ontario, Canada.

Macmillan, B., ed. 1989. *Remodelling Geography*. Basil Blackwell, New York.

Manne, A. S., and R. G. Richels. 1990. Global CO_2 emissions reductions: An economic cost analysis for the USA. *The Energy Journal 11(2)*.

Manne, A. S., and R. G. Richels. 1991. Global CO_2 emission reductions: The impacts of rising energy costs. *The Energy Journal 12(1)*, 87–107.

McGrew, J. 1975. *Goal Programming and Complex Problem Solving in Geography*. Papers in Geography 12, Pennsylvania State University, University Park, Pennsylvania.

Meadows, D., and D. L. Meadows. 1972. *Limits to Growth*. Universe Books, New York.

Meadows, D., J. Richardson, and G. Bruckmann. 1982. *Groping in the Dark*. John Wiley and Sons, New York.

Miernyk, W. H. 1965. *Elements of Input-Output Analysis*. Random House, New York.

Miller, R.E., and P.D. Blair 1985. *Input-Output Analysis: Foundations and Extensions*. Prentice-Hall, Englewood Cliffs, New Jersey.

Odum, H.T. and E.C. Odum. 1976. *Energy Basis for Man and Nature*. McGraw Hill, New York.

Pearce, D., A Markandya, and E. B. Barbier. 1989. *Blueprint for a Green Economy*. Earthscan, London.

Prudham, W. S. 1992. *A Regional Resource Accounting Framework*. Master's Thesis, Department of Geography, University of Victoria, British Columbia, Canada.

Repetto, R., W. Magrath, M. Wells, C. Beer, and F. Rossini. 1989. *Wasting Assets: Natural Resources in the National Income Accounts*. World Resources Institute, Washington, D.C.

Richardson, H. W. 1972. *Input-Output and Regional Economics*. Redwood Press, Trowbridge, U.K.

Robinson, J. 1985. Global modeling and simulations. In *Climate Impact Assessment* (R. W. Kates, J. H. Ausubel and M. Berberian, eds.), SCOPE 32, John Wiley and Sons, New York.

Schuler, A., H. Webster, and J. Meadows. 1977. Goal programming in forest management. *Journal of Forestry 75(6)*, 320–324.

Theys, J. 1989. Environmental accounting in development policy: The French experience. In *Environmental and Resource Accounting for Sustainable Development* (Y.J. Ahmad, S. El Serafy, and E. Lutz, eds.), World Bank, Washington, D.C.

VHB Consultants. 1990. *Economy-Environment Linkages and Sustainable Development in Ontario*. Interim report to the Ontario Ministry of Environment, Toronto, Ontario, Canada.

Victor, P. 1972. *Pollution: Economy and Environment*. Allen and Unwin, London.

Warrick, R., and M. Bowden. 1981. Changing impacts of drought in the Great Plains. In *The Great Plains, Perspectives and Prospects* (M. Lawson and M. Baker, eds.), Center for Great Plains Study, University of Nebraska, Lincoln, Nebraska.

Weber, J. L. 1983. The French natural patrimony accounts. *Statistical Journal of the United Nations Economic Commission for Europe 1(4)*, 419–444.

Werczberger, E. 1976. A goal programming model for industrial location involving environmental considerations. *Environment and Planning A 8(2)*, 173–188.

17

Data on Global Land-Cover Change: Acquisition, Assessment, and Analysis

David L. Skole

The contemporary state of the world's land cover is a constantly shifting mosaic of cover types determined by both the physical environment and human impact. The distribution of vegetation cover reflects large-scale variations in the distribution of temperature, precipitation, and various edaphic factors. This array of environmental factors translates into a heterogeneous but fairly straightforward description of potential global biotic cover (cf. Holdridge, 1947). Climate and soils alone, however, do not account for the rich and changing set of land-cover types seen today. Humans have continually modified the landscape. Often, human transformation of land cover is obvious, as when tropical forests are cut, burned, and converted to pastures. But human land-cover transformations also take more subtle forms. The phenomenon of forest decline, found in many industrialized regions of both developed and developing countries, is an example of modification of cover through chronic deposition of atmospheric pollutants or excess nutrients (Aber *et al.*, 1989). Whether directly or indirectly, humans now play a major role in determining the state of land cover at a global scale.

Over the next 20 to 50 years the global effects of land-cover conversion on ecosystems and on human wealth and well-being may well be much larger than those arising from climate change. And yet very little is known about land cover and its alteration. Few data sets exist, and they manifest numerous technical and interpretive problems. Three areas where much better documentation and analysis are needed are: (1) the state (i.e., information on biomass, net primary production, etc.) and distribution of existing land cover; (2) the rate and distribution of land-cover change, both historically and currently; and (3) the human driving forces that determine the rate and extent of land-cover change. Some combination of historical reconstruction and remote sensing is needed to refine the first two areas. The third area of uncertainty will require the closer linkage of physical and social analysis.

In this chapter I review current approaches to acquiring and analyzing data on land cover and land-cover conversion, particularly deforestation. The review is not exhaustive. I focus on approaches to acquiring data on vegetation cover conversion as a prologue to a case study of deforestation and its human causes in Amazonia.

Land Cover, Land Use, and Vegetation

The conceptual framework used throughout this paper contains four overlapping components: (1) vegetation, (2) land cover, (3) land-cover conversion and modification, and (4) land use. All of these terms may mean different things in different disciplines.

When we think about what covers the earth's land surface, we think first of vegetation. Indeed, most of the land is covered by vegetation of one type or another, just as most of the living biomass is plant matter. Land surface classification has typically focused on mapping vegetation, and the remainder of this chapter retains that emphasis. It should be remembered, however, that vegetation is only one category of land cover. Land cover encompasses three categories: (1) natural vegetation; (2) water, desert, ice, and other natural features that do not have a predominant vegetative cover; and (3) vegetative and nonvegetative landscapes created by human activities, such as agricultural and urban land.

Land-cover *conversion* involves a change from one cover type to another: for instance, the conversion of forests to pasture, an important process in the tropics. *Modification* involves alterations of structure or function without a wholesale change from one type to another; it could involve changes in biomass, productivity, or phenology. The line between conversion and modification is not always sharp. Long-term and chronic modification can eventually result in complete conversion. The transformation of a closed canopy forest to a palm forest or woodland as a result of long-term degradation by humans is one example.

Land-cover conversion and modification alike operate through specific processes, sometimes termed proximate sources of change. For instance, deforestation is a process that converts forest to pasture or some other cover type. Site abandonment is a process that results in the conversion of pasture to forest through regrowth. These processes or proximate sources of change can be envisioned as forcing functions, which have both magnitude (rate of deforestation) and direction (forest becoming pasture or vice versa).

Land use itself is the human employment of a land-cover type, the means by which human activity appropriates the results of net primary production (NPP) as determined by a complex set of socioeconomic factors. Forests, for instance, can be used by foresters (selective logging), by rubber tappers, or not at all. Grasslands can be used as grazing land. Land-use change frequently brings about land-cover conversion. For instance, forest is converted to pasture in an effort to convert the land use from rubber tapping to cattle ranching. Characterizing land use often involves questions of scale, intensity, and tenure. These properties are not necessarily inherent in the land-cover properties. Small-scale, subsistence agriculture is not synonymous with large-scale commercial enterprises, even if they create similar types of land cover.

Land-Cover Data: Some General Needs

A serious problem in the study of global land-use/cover change has been the paucity of spatial data sets or maps. Mueller-Dombois (1984) notes three important, if general, reasons for mapping land cover: (1) to outline land-cover patterns for the purpose of developing inventories, (2) to extrapolate field observations and measurements to an appropriate level of geographic and ecological generalization, and (3) as a step in developing predictive explanations of land-cover distributions in terms of past, present, and future environments. These three objectives generally apply for all disciplines and at all scales.

Specific requirements for global change research are now being identified. For instance, the spatial arrangement of land transformation is an important factor in estimating its influence on biogeochemistry and climate. One component of the current uncertainty in modeling the net flux of carbon between terrestrial ecosystems and the atmosphere is lack of information on the types of ecosystems that have been cleared for agriculture. This information is important since terrestrial ecosystems vary considerably in biomass, soil organic matter, moisture regimes, rates of recovery, and other characteristics. For example, the conversion of high-biomass humid forest would produce a significantly greater total net flux of carbon than conversion of tropical dry savanna, and the resultant biomass burning would likely result in greater releases of trace gases such as carbon monoxide and methane than in the dry savanna systems. The same information lack also hampers attempts to develop global budgets of important trace gases (see Penner, this volume).

The spatial arrangement of land-cover conversion, particularly deforestation, will also influence results of model simulations of continental-scale climate and energy balance. Deforestation distributed as a few large blocks may have greater influence on sensible and latent heat flux than the same area distributed as many, widely scattered small patches (Henderson-Sellers, 1987; Henderson-Sellers and Gornitz, 1984).

Finally, there is increasing concern that land-cover conversion in humid tropical forests will result in the loss of a significant number of species (Ehrlich and Wilson, 1991). The impact on biodiversity is related to the total area of forest conversion and the amount of forest fragmentation. Quantifying fragmentation requires an understanding of the spatial arrangement of cleared areas (Soule, 1991; Wilson and Peter, 1988; Skole and Tucker, 1993).

More could be said about other areas where better knowledge is needed: about land-cover modification, for instance, and about change in nonvegetational land cover. Even so, these few examples demonstrate the importance of knowing more than just areas of land-cover types or the magnitude of land-cover conversion; for many purposes, we must know where conversion occurs and with what spatial

Table 1: Methods for classification of vegetation and land cover

Properties of the vegetation
Physiognomic: life form, structure, periodicity
Floristic: dominant species, groups of species
Properties of the environment
Environmental: potential vegetation, related to climate, topography, soil, human use
Geographical location
Combination of vegetation and environment
Correlation: by map overlay
Integration: into combined units: ecosystems, landscape units
Function: groups related by processes

Source: Mueller-Dombois, 1984.

characteristics. Yet, at this time virtually nothing is known about the coincident distribution of land cover, land-cover attributes, and land-cover conversion. New initiatives must involve strategies that focus on spatial and temporal characteristics of both natural land cover and conversion activities.

Mapping and Documenting Land-Cover Patterns

Vegetation Classification Approaches

Generally speaking, the aim of vegetation mapping has been to delineate vegetation distribution using a system of classification based on observed structural or environmental factors. The focus has been on vegetation in its natural condition or potential vegetation. It is important to note that where natural vegetation exists, it is not always the climatic potential vegetation. For instance, large areas of natural savanna exist in the center of the Amazon Basin surrounded by extensive closed canopy rainforest. Their existence is due largely to edaphic factors, principally soil conditions (Marden dos Santos, 1987).

A typology of vegetation classification approaches is shown in Table 1. This table presents the canonical approach, which has been to describe or map vegetation based on any one of three general criteria: (1) physiognomic and floristic characteristics of natural vegetation; (2) properties of the environment that influence the type and distribution of potential natural vegetation; and (3) a combination of vegetation characteristics and environmental influences.

Physiognomic classifications relate heavily to structure and morphology. This approach might also include characteristics of periodicity, such as phenological development of vegetation between dry and wet seasons. In contrast, floristic systems use species composition as the basis for classification and delineation (Whittaker, 1978). Either physiognomic or floristic classification might be amenable to large-scale delineation of vegetation, but physiognomic systems tend to be broader in scope and more suited to remote sensing, particularly when the remote sensing data capture some characteristics of phenology. Environmentally determined classifications, such as those by Holdridge (1967) and others, predict the distribution of vegetation types in the broadest sense, often without consideration of species, over a range of climatic regimes. They are useful for characterizing past and future potential vegetation at the climatic equilibrium, but they do not capture edaphic factors that may determine actual vegetation.

Systems combining environmental and physiognomic approaches have been widely used recently. One approach has been to array vegetation by empirical relations between physiognomic or floristic characteristics and features of the physical environment. One way to make such a determination would be through map overlays. The U.N. Educational, Scientific, and Cultural Organization (UNESCO, 1973) classification was derived in a similar way. A recent derivation of this approach has been what Mueller-Dombois calls the ecological landscape map, which takes advantage of principles of ecology, incorporating climatic and edaphic gradients. This view is comprehensive and tends to be oriented to land cover rather than simply vegetation. An additional factor well known to ecologists is that of successional state. The incorporation of successional state carries with it two assumptions: (1) the potential or natural vegetation of a region is determined and predicted in part by an understanding of what constitutes and what controls climax vegetation, and (2) preclimax conditions can be determined at any point in time based on an understanding of primary and secondary succession. To some extent an understanding of secondary succession implies consideration of disturbance and vegetation response to disturbance.

Few systems consider characteristics of vegetation that are not readily characterized by morphology, climate, or edaphic factors. A functional classification would integrate vegetation and environmental factors, taking into consideration variation along gradients and the dynamics of succession. Instead of focusing on vegetation itself, however, a functional classification would be based on ecosystem processes. Field and Mooney (1986) demonstrated a strong relationship between net photosynthesis and leaf nitrogen concentration across a range of plant genera. Similarly, Vitousek (1982) has demonstrated a functional relationship between nitrogen in litterfall and leaf nutrient content across a range of forest types around the world. Melillo *et al.* (1982) have shown a strong relationship between the lignin-to-nitrogen ratio in litter and first-year decomposition rate and nutrient cycling. In the first year climate has an important influence on the rate (Meentemeyer, 1978), but long-term rates may be more simple functions of lignin (Aber *et al.*, in press). These studies suggest functional differences between groups and raise possibilities for delineating vegetation into groups that illuminate processes.

Historical Approaches Using Tabular Data

Recent historiographic analyses have demonstrated new techniques for documenting earlier land cover and its long-term change. Most of these studies have been qualitative or limited in geographic scope; some have only focused on certain subcategories of land cover, such as forests. These kinds of case study analyses are important for elucidating the fundamental causes of land transformation, but as foundations for large-scale data sets they are somewhat limited.

Flint and Richards (1991) have demonstrated a data-intensive approach in South and Southeast Asia. It relies on historical documents such as revenue records, gazetteers, land assessments, or forestry reports. In some cases direct evidence of the type of land cover present or predominant in a region can be found in these documents, particularly in former colonies of the British Empire. In most cases, however, records of land use or agricultural area are more readily available than direct records of vegetation and land cover. In these cases, estimates of land cover are indirectly derived from data on changes in agricultural area. A baseline delineation of natural land cover is established, and these areas are reduced in proportion to the increase in agricultural area.

Changes in the area of cultivated land can be used to approximate land-cover conversion rates since the most important form of land-cover conversion has historically been agricultural expansion (Richards, 1984; Tucker and Richards, 1983). This assumption is generally accurate, but it must be noted that in some instances the omission of logging and other nonagricultural causes of land-cover conversion could be significant. Certain areas in the tropics, such as the Ivory Coast, Nigeria, Malaysia, and Thailand, have experienced large-scale losses of forest in this way (Myers, 1980; Lanly, 1982; WRI, 1990).

This approach to historical reconstruction of land-cover conversion presents a number of other difficulties. A review can be found in Houghton (1986). There are two important problems:

- Sparse sampling in space and time can result in a variety of interpretations from a single data set. Figure 1 shows three possible time series that can be derived from data points used by Richards *et al.* (1983) to estimate historical changes in carbon stocks on land as a result of agricultural expansion. These series in turn would result in different histories of carbon flux when the data were coupled to a numerical model (Houghton, 1986). To reduce this type of data aliasing, sampling at a finer temporal resolution is required.
- Data from the same source may vary considerably from year to year due to changes in methodology or terminology. As Figure 2 shows, time series derived from later editions of the U.N. Food and Agriculture Organization (FAO) Production Yearbooks, an important source of this kind of data for recent history, differ from the same time series derived from earlier editions.

It must also be noted that historical reconstructions rarely document land-cover modification, but only actual conversion. Moreover, land-cover conversion estimates obtained this way are based on *net* changes in the amount of agricultural land. With this basis, it is not possible to know how much abandonment there is at any point in time. For analyses of net fluxes of carbon, for instance, this could be

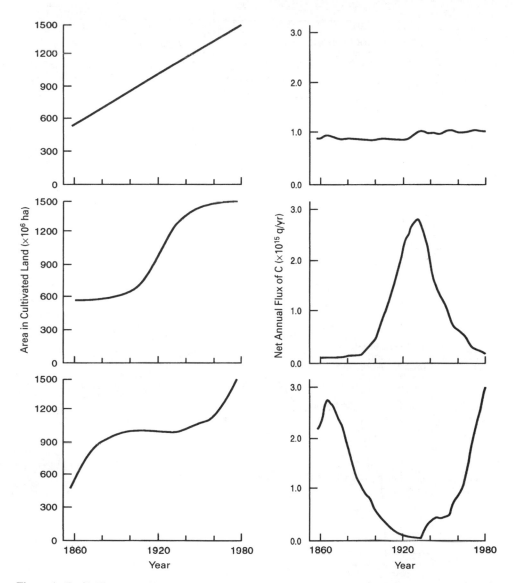

Figure 1. (Left) Three possible time series of land-cover conversion derived from the same tabular historical data on cultivated land. (Right) The net flux of carbon resulting from each series (from Houghton, 1986).

important since a large amount of abandonment would result in lower net releases to the atmosphere.

Historical reconstructions use tabular data only. In addition, a number of important sources of information come in tabular form (e.g., the FAO/U.N. Environment Program Forest Assessment, which provides tabular data at the country level on forest area). There is no inherent spatial organization to the data. One way to develop digital maps of historical land-cover conversion is to use geographic information systems (GIS). The GIS approach permits accurate spatial representa-

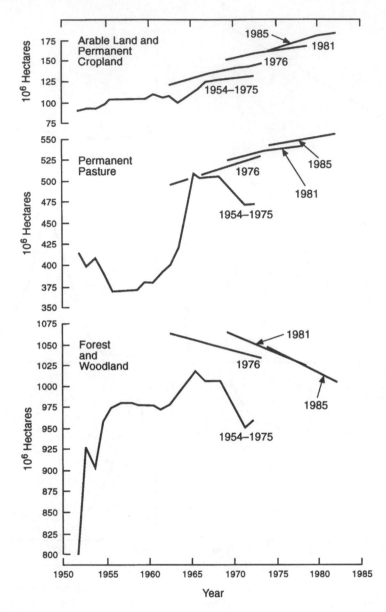

Figure 2. Different editions of the FAO *Production Yearbook* (years shown on figure) give different time series due to changes in methodology or terminology (from Houghton *et al.*, 1991).

tion of data on agricultural area, which can then be merged with digital maps of land cover to estimate conversion rates by land-cover type. The GIS also makes it possible to integrate a wide variety of other data, such as roads and other human-use features, hydrology, soils, and the like. Moreover the GIS approach makes it possible to link historical data to remote sensing data.

I have been developing a technique to study land-cover conversion in Brazil using geographic information systems in conjunction with reconstructions based

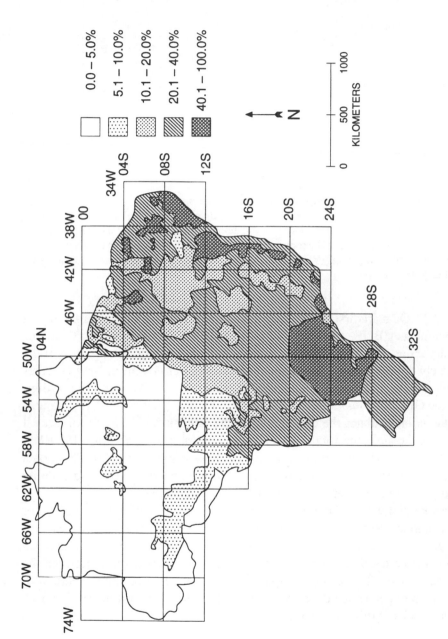

Figure 3. The results of mapping tabular data in a geographic information system to estimate the geographic extent of land-cover conversion in Brazil. Legend shows the fraction (%) of natural land cover converted to human use as of 1980.

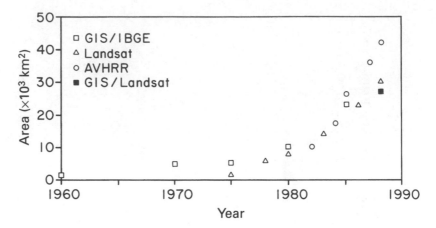

Figure 4. A comparison of different methods for estimating deforestation in the state of Rondônia, Brazil. The open squares represent data derived from agricultural statistics, which could be used to obtain long-term deforestation rates prior to the availability of remote sensing data. The triangles show results from Landsat analyses. The triangle for 1988 is from the Brazilian space agency study. Circles show published results from AVHRR analyses. The closed square represents my analysis for 1988 using Landsat photoproducts.

on tabular data. Data are first tabulated by administrative district from agricultural census documents (IBGE, 1960, 1970a, 1980a). These documents report the area in temporary crops, permanent crops, pasture, and other categories for each of the approximately 4000 *municipios* (the smallest administrative unit in Brazil). Maps showing the borders of each of these *municipios* are also available and can be digitized into the GIS. Each *municipio* in the GIS can have associated with it a numerical identifier, which relates the digital map to the tabular data base. In this way, I have been able to map the total area of land converted to agriculture for 1960, 1970, 1975, 1980, and 1985. Figure 3 presents these results for 1980. These data can then be compared with data derived from direct observations using satellite data during the period of overlap (Figure 4).

To determine the type of land cover that was converted, a digital vegetation or land-cover map is created. Generally, sources of these data are maps or satellite imagery, which are encoded as another layer in the GIS. One conversion map can be subtracted from another to create a map showing areas converted between the two successive dates. This change map can then be overlaid on the vegetation or land-cover map to determine the rate and type of land-cover conversion. Figure 5 presents the results of this type of analysis.

Digital Map Data Bases of Land Cover

Although land-cover mapping and classification is a fundamental approach to understanding the global environment, there are very few sources compiled at the global level. Even fewer sources exist in digital form readily usable for quantita-

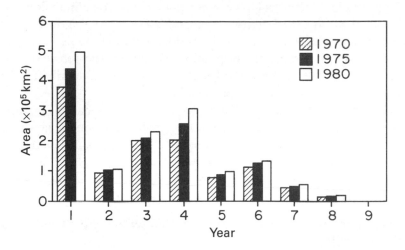

Figure 5. The total land-cover conversion in Brazil distributed across cover types for different time periods (see also Figure 3). The cover types are (1) forest, (2) forest steppe, (3) bush forest, (4) cerrado, (5) wet tallgrass savanna, (6) dry lowgrass, (7) true steppe, (8) wetland, and (9) desert.

tive analysis. Mueller-Dombois (1984) has argued that the two basic features of land cover that could be classified and mapped are its variation in space and its variation over time. If the former is sparse, the latter has been virtually nonexistent.

The few global-scale land-cover maps that do exist have been derived at very coarse scales. Some of the earliest formulations of digital data sets have been developed to support general circulation and atmospheric tracer models (e.g., Matthews, 1983; Wilson and Henderson-Sellers, 1985). The purpose of these data sets has been to map broad ecosystem classes as a way of defining surface roughness, albedo, and other modeling parameters mediated by vegetation. These data have also provided a means to estimate global distributions of NPP and water/energy balance. The formulation by Matthews (1983) is probably the most often used data set. It was derived by combining approximately 100 individual map sources into a single, UNESCO-based vegetation classification scheme. The major vegetation type is defined for each $1° \times 1°$ (horizontal resolution of approximately 110 km at the equator) land surface grid cell.

Other data sets have been developed to support global ecosystem models, particularly for analyses of the global carbon cycle. Olson *et al.* (1983) have developed a $0.5° \times 0.5°$ grid cell map (50 km horizontal resolution) of vegetation types and associated carbon contents. Like the Matthews map, this data set delineates both natural and disturbed land uses. Researchers at the University of New Hampshire have prepared a digital data set of actual and natural land cover in vector form at a horizontal resolution of approximately 10 km. It differs from the previous data sets in its finer resolution, somewhat comparable to the global area coverage (GAC) scale of the advanced very high resolution radiometer (AVHRR) satellite sensor, and in its delineation of predisturbance vegetation.

Issues in Data and Modeling

A third type of digital data set has been developed based on climatological parameters. Emanuel *et al.* (1985) have developed a 0.5° × 0.5° grid cell map of Holdridge Life Zones of the world, an ecoregionalization scheme based on temperature, precipitation, and evapotranspiration. Being climate-sensitive, the data set has been used to make first-order projections of vegetation distribution under a scenario of doubled atmospheric CO_2 concentrations.

These digital data sets have provided the best available global cover maps. The fact that they portray land cover using some predefined nomenclature has its advantages in that it allows assignments of parameters, such as carbon or biomass, to categories with which ecologists are familiar, and because to some degree a type of land cover conveys a general sense of structure and function familiar to ecologists. Nonetheless, all of the data sets suffer from certain critical problems:

- The nomenclature itself may vary from one data set to the next, and mean different things to different scientists.
- Type classifications require a modeler to make (somewhat arbitrary) parameterization assignments, usually from point measurements found scattered in the literature.
- The complete data set may be derived from individual primary sources, each using a different system of nomenclature, and from different points in time.
- The data set represents some interpretation, generalization, or abstraction of vegetation and vegetation boundaries, and thus does not necessarily portray actual distributions.
- Most existing data sets are very coarse resolution (50–100 km).
- None of the existing data sets provides indications of phenology or intra-annual variation.
- All are static generalizations, not able to provide indications of interannual changes.

Clearly, existing sources have proven to be useful first-order delineations of land cover, and they are likely to be useful for a few years to come. The shortcomings of these cartographic approaches, however, suggests the strong need to develop land-cover data sets drawing on remote sensing.

Remote Sensing of Land Cover and Land-Cover Conversion

Satellite remote sensing provides large-area, multitemporal data over a range of spatial and spectral resolutions. Table 2 provides an overview of some of the major satellite sensor systems now in operation. They vary considerably in resolution and coverage. The AVHRR on board the National Oceanic and Atmospheric Administration (NOAA) polar orbiting satellites provides spatially coarse resolution but frequent data (1–4 km, daily). On the other hand, the French Système Probatoire d'Observation de la Terre (SPOT) and the U.S. Landsat satellites provide high-resolution but infrequent data (10–30 m, weekly or biweekly). These characteristics influence the types of applications for which each sensor is suit-

Table 2: A comparison of satellite remote sensing systems in use today

	Landsat-MSS	Landsat-TM	SPOT	AVHRR
Resolution	79 m	30 m	10 m (panchromatic) 20 m (multispectral)	1 km (LAC) 4 km (GAC) 16 km (GVI)
Number of spectral bands	4	7	1 pan, 3 multispectral	5
Spectral range (micrometers)	0.5–0.6 0.6–0.7 0.7–0.8 0.8–1.1	0.45–0.53 0.52–0.60 0.63–0.69 0.76–0.90 1.55–1.75 2.08–2.35 10.4–12.5	0.51–0.73 (pan) 0.50–0.59 0.61–0.68 0.79–0.89	0.58–0.68 0.73–1.1 3.55–3.93 10.3–11.3 11.5–12.5
Frequency of coverage	16 days	16 days	26 days (8 day sequence with variable look angle)	twice daily
Data cost	moderate	high (digital) low (photo)	high (new acquisitions) moderate (archival)	low
Processing burden	low	low	low	high

GAC	=	global area coverage
GVI	=	global vegetation index
LAC	=	local area coverage
MSS	=	multispectral scanner
TM	=	thematic mapper

able. A complete description of the characteristics of the AVHRR, Landsat, and SPOT sensors can be found in the various operations manuals and published literature (e.g., Kidwell, 1991, 1990; CNES, 1988; Colwell, 1983).

The AVHRR's ability to cover large areas with frequent repeat intervals makes it useful for regional land-cover mapping and deforestation detection (Townshend and Tucker, 1984; Malingreau and Tucker, 1987). Justice *et al.* (1985) discuss the use of a greenness index computed from the visible and near-infrared channels of AVHRR data to map the distribution of vegetation and its phenology at very large scales. This greenness index, or normalized difference vegetation index (NDVI), has been correlated with vegetation parameters such as green-leaf biomass, leaf area, and phenology, thereby making it valuable for multitemporal analysis of vegetation dynamics as well as static vegetation classification. The NDVI has also been shown to reproduce the seasonal variation in the global concentration of atmospheric carbon dioxide (Fung *et al.*, 1987). The NDVI also reduces the effect of topography and radiometric variations due to sun and view angle (Holben and Justice, 1981). It has promise as a very important sensor for global vegetation

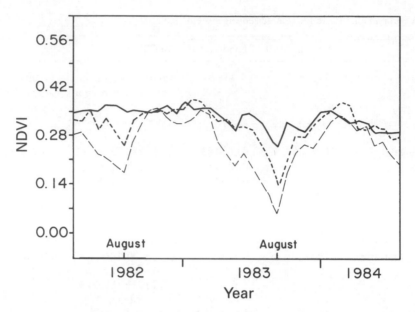

Figure 6. The use of multitemporal AVHRR-derived NDVI data in the Amazon can distinguish two forest types and cerrado (from Malingreau and Tucker, 1987; © IEEE).

monitoring, but much research remains to be done.

Analyzing data from the Brazilian Amazon region, Malingreau and Tucker (1987) have shown that seasonal, multitemporal NDVI data can be used to discriminate among tropical forest, semiseasonal tropical forest, and cerrado (wooded savanna) ecosystems since differences in greenness increase as the season progresses into drier conditions (Figure 6). It is sometimes difficult, however, to separate natural savanna areas from deforested areas using the NDVI. Other radiometric relationships between bands of the AVHRR hold promise for discriminating between types of natural vegetation and deforestation simultaneously (J.P. Malingreau, personal communication).

Tucker *et al*. (1991) have used the AVHRR data to monitor changes in the size and distribution of the Saharan desert in Africa. This case demonstrates the capability of remote sensing data to define inter- and intra-annual transformation of land cover, a kind of characterization that is not possible with maps. Loveland *et al*. (1991) have reported a similar application of AVHRR temporal sequences for a North American land-cover classification.

Recently an International Geosphere Biosphere Program (IGBP) project has been developing 1-km AVHRR data globally to be used for land-cover classification (Townshend *et al*., in press). Such datasets contain high spatial resolution and temporal compositioning at 10-day periods, and are thus very good for developing cover classifications globally using vegetation phenology (Andres *et al*., in preparation).

The potential of AVHRR for deforestation mapping has been noted recently by a number of investigators. Malingreau and Tucker (1989, 1988) and Tucker *et al*.

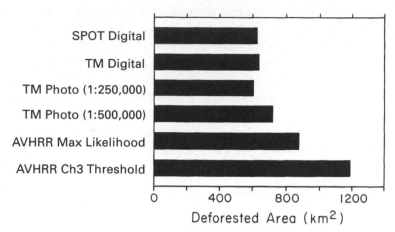

Figure 7. Results of a study in which TM, SPOT, and AVHRR data were acquired during the same period in 1988 in a study site in Rondônia, Brazil. SPOT data were processed at full resolution using supervised digital image processing techniques. TM data were processed at full resolution using supervised digital image processing techniques, and also analyzed at 1:250,000 and 1:500,000 scale using interpretation and GIS digitizing on top of single-channel photoproducts. AVHRR-LAC (local area coverage) data were processed using a multichannel maximum likelihood classifier and Channel 3 brightness temperature threshold.

(1984) observed large-scale deforestation in the southern fringe of the Amazonian forest using AVHRR. Woodwell *et al.* (1984) suggest AVHRR could provide the first stratification of a global sampling scheme. The appeal of AVHRR is that both the Channel 3 signal and the NDVI seem to discriminate deforested areas from intact forest.

In addition to detecting deforested areas, thermal channels on the AVHRR are able to detect the occurrence of fires. The extrapolation of a thermal anomaly to an estimate of area requires several assumptions, however, and much research remains to be done. For example, Setzer and Pereira (1990) used the location and number of thermal anomalies on the AVHRR Channel 3 to estimate the area of deforestation in Brazil in 1987. These factors were applied to the total number of observations to estimate a deforestation rate of 8.0×10^6 ha. This number is now thought to be high. In fact, the rate of deforestation in Rondônia estimated by this method was higher than the *total* area deforested based on high-resolution studies.

There has been considerable success using fine- or medium-scale remote sensing techniques to monitor land-cover conversion and land-cover distribution, mostly in case studies in selected areas (Woodwell *et al.*, 1986; Nelson *et al.*, 1987a, 1987b; Nelson and Holben, 1986). Because a global or regional-scale effort to monitor deforestation with fine-resolution imagery such as SPOT or Landsat would be time consuming and expensive, many investigators have looked favorably at the use of coarse-resolution sensors such as the AVHRR. But it now appears that this gain in efficiency comes with a loss in accuracy. There is evidence that coarse-resolution imagery tends to overestimate the deforested area; for 1-km-resolution data the overestimation might be 50–75%. This point is demonstrated by some of my intersatellite comparisons for the Brazilian Amazon (Figure 7). In a test site in the state of Rondônia I have simultaneously (±4 weeks) acquired digital and photoproduct data from SPOT, Landsat's thematic mapper (TM), and AVHRR. These data were analyzed in conjunction with a field study that verified classification accuracies. A simple classification of forest/nonforest was made using several common methods: digital image processing of SPOT and Landsat TM, analysis of TM single-channel photoproducts using a vector-based GIS at 1:250,000 and 1:500,000 scale, a maximum likelihood classification of AVHRR Channels 1 and 2, and a brightness temperature threshold classification of AVHRR Channel 3. The results indicated that both AVHRR techniques overestimate deforestation. One of the most interesting results is the fact that the use of photoproducts, even at relatively coarse scales, seems accurate.

Townshend and Justice (1988) compared a series of degraded Landsat multispectral scanner (MSS) scenes to analyze the amount of variance in data sets of various resolutions. They were interested in understanding the effect spatial resolution had on locating and quantifying land transformation at relatively coarse scales. Their conclusion was that land transformation could not be assessed with any reasonable accuracy at resolutions above 1 km. They recommended an optimal resolution of 250 m. This important conclusion varies from some earlier conclusions (cf. Woodwell *et al.*, 1986).

Another problem relates to the needs of global carbon models. These analyses need to know the fraction of all deforestation coming from primary forests and secondary forests, since the difference in biomass may be significant. Coarse-resolution data may not be able to discriminate between these two types of conversion and may not provide necessary insight into the dynamic pattern of clearing, regrowth, and reclearing that is characteristic of developing areas.

An optimal approach to deforestation monitoring could be to use high-resolution remote sensing data from the Landsat or SPOT sensors. I have used this approach, as has the Brazilian space agency (Instituto Pesquisas Espaciais, or INPE), to map deforestation in the legal Amazon, that is, the Amazon as defined by the Brazilian government (Tardin *et al.*, 1980; Tardin and Pereira da Cunha, 1990). Results from the state of Rondônia suggest that approximately 25,000 km^2 were deforested as of 1988 (Figure 4). This number compares well with estimates from INPE (Fearnside *et al.*, 1990). It also seems to confirm the possibility that AVHRR estimation of deforestation is biased on the high side and that consequently previous estimates of deforestation in the Amazon based on AVHRR analyses (e.g., Malingreau and Tucker, 1987; Kaufman *et al.*, 1990) are also too high.

It is possible to develop a low-cost, low-technology approach by using photoproducts rather than digital data. In this method, photoproducts are visually interpreted for deforestation. The scale used could range from 1:250,000 to 1:500,000, which appears adequate for most large-scale regional analyses. Data on the rate and pattern of land-cover conversion could be acquired frequently—on the order of two- to four-year intervals. The high-resolution approach requires the acquisition of a large number of individual scenes (approximately 200 in the Brazilian

Amazon). This approach must be coupled to the development of capabilities for data management in a spatial and relational context using GIS. The development of digital data sets using GIS methods has the advantage of making the data interchangeable with numerical models and provides the opportunity for detailed spatial analysis of land-cover conversion.

Coupling Remote Sensing and Geographic Information Systems

It is not the purpose of this paper to review GIS in detail, a task that has been done very well elsewhere (Burrough, 1986; Smith *et al.*, 1987; Bracken and Webster, 1989; Maguire *et al.*, 1991). Berry (1987) provides a simple but functional description of GIS as having three general characteristics: (1) they process and manipulate geographically referenced, spatially coherent data; (2) they internally relate geographical location data and attribute information; and (3) they are computerized. In essence, what distinguishes GIS from other information systems is their ability to use spatial data sets, such as maps and remote sensing imagery, in digital formats. They can perform difficult coregistration and reprojection computations, enabling them to manipulate many kinds of data from different sources in a single system. They store all coordinate and attribute information in the same system as well, thus enabling them to couple multiple layers of geographically referenced data with numeric models. GIS use two general data models: a tessellated or grid model, and a vector model.

Here I provide an example of how large-area GIS can be used to map land-cover conversion. Maps of deforestation for the period 1975–78, which were prepared by INPE from Landsat imagery (Tardin *et al.*, 1980), were digitized in a vector-based GIS. The entire data set consisted of 28 individual map modules. Each module covered one Universal Transverse Mercator (UTM) zone at 1:500,000 scale. Areas deforested as of 1975 and 1978 were vector encoded and joined to form a master data base. This data base was then digitally combined (overlaid) with the map of *municipios* described above to delineate deforestation and deforestation rates by political district. This step created a second-generation data base and permitted the relational linking of the satellite-based data with the tabular land-use data described above. The deforestation data base was also combined with a digital map of forest and cerrado ecosystem distributions, which was derived from a multitemporal analysis of the AVHRR GVI.

The development of geographically referenced land-cover and land-cover change data bases, as described above, enables us to estimate the net flux of carbon between the atmosphere and the biota by coupling these data to a geographically referenced numerical model. Figure 8 shows the GIS-generated map of the net flux of carbon from deforestation between 1975 and 1978. The results are in a form suitable for coupling to atmospheric general circulation models.

The use of GIS also makes it possible to analyze the spatial distribution of land-cover change. Remote sensing can provide the primary source of spatial data, while the GIS provides the computational environment for analysis. The spatial

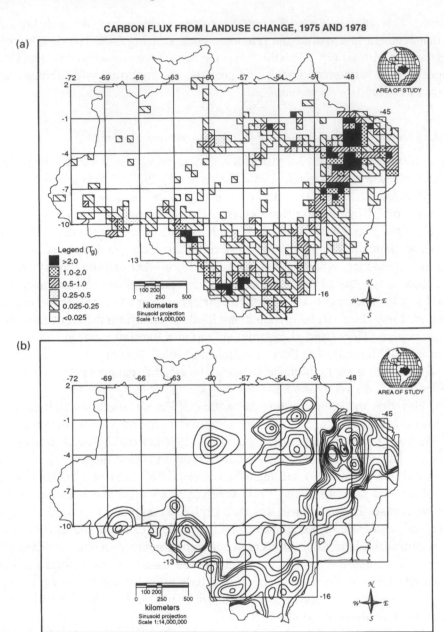

Figure 8. An estimate of the net annual flux of carbon from deforestation between 1975 and 1978 in the legal Amazon of Brazil. (a) The net flux aggregated into 0.5° × 0.5° grid cells. (b) The net flux contoured at intervals of 10^{12} g C.

arrangement of land-cover change provides insights into land use. It has been shown (Lambin, personal communication) that the spatial patterns detected at a certain scale on remote sensing data are related to some key characteristics of farming systems. Knowing something about the spatial dynamics of land-cover

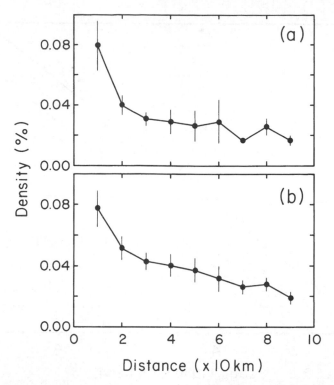

Figure 9. The spatial arrangement of deforestation shown as the density (percent of area) as a function of distance from dense settlements in (a) 1975 and (b) 1978 (Malingreau, 1986).

conversion over time could also be important when developing models. As an example of the latter, I have used the GIS data set on deforestation for 1975 and 1978 described above to analyze the spatial pattern of deforestation. Figure 9 shows the density of deforestation as a function of distance from settlement centers, averaged for the entire Amazon Basin. The figure suggests two points: (1) the difference between 1975 and 1978 reflects the rate and pattern of spatial diffusion over time, and (2) the spatial arrangement is generally quite persistent from one time period to the next.

Deforestation as a Form of Land-Cover Conversion

Tropical deforestation is an important component of global change; it has been shown to have an important influence on regional hydrology, large-scale and long-term climate systems, and global biogeochemical cycles (Houghton and Skole, 1990; Houghton *et al.*, 1991; Shukla *et al.*, 1990; Crutzen and Andreae, 1990; Salati and Vose, 1984). The precise rate and geographic distribution of deforestation, however, are poorly known at continental or global scales (Williams, 1990; Williams, this volume). Remote sensing provides perhaps the best method for obtaining geographically and temporally detailed estimates of changes in land cover.

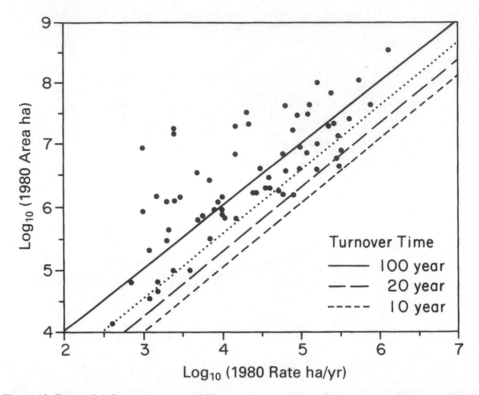

Figure 10. Tropical deforestation rates of 76 tropical countries in 1980 plotted against area of closed forest in 1980 based on FAO/UNEP data. Isolines show turnover times for various time periods, computed assuming constant rates of clearing and no regrowth. The dotted line shows a 20-year isoline assuming an exponential increase in the clearing rate, which doubles in 20 years. Points to the right of a particular isoline represent countries that would be expected to lose all of their closed forests during that time interval.

In 1980 the U.S. National Academy of Sciences released one of the first surveys of tropical deforestation (Myers, 1980). This survey was a broad examination of the best available literature (see also Melillo *et al.*, 1985). Another was conducted by the U.N. Environment Program (UNEP) and FAO for 76 tropical countries (FAO/UNEP, 1981; Lanly, 1982). This study reported deforestation rates for two periods, 1975–80 and 1980–85, the latter being projections. It relied primarily on national reports based on forest inventories, with remote sensing surveys where available. The study estimated an annual deforestation rate of 11.3×10^6 ha/yr for 1975–80, of which 6.9×10^6 ha was in closed forests (Tables 3 and 4). Six countries accounted for 54% of the global total: Brazil, Colombia, Indonesia, Ivory Coast, Mexico, and Thailand. The FAO/UNEP data are represented in Figure 10, which shows the 1980 deforestation rate for each country plotted against the total area reported in closed broad-leaved forest.

Myers (1989) compiled a country-by-country estimate of annual tropical deforestation rates in 1989 for closed forests only (Table 3). He estimates a global rate

Table 3: Estimates of closed forest deforestation (10^3 ha/yr)

	FAO/UNEP, 1981	Myers, 1989	WRI, 1990
Africa			
Cameroon	80	200	100
Congo	22	70	22
Gabon	15	60	15
Ivory Coast	310	250	290
Madagascar	165	200	150
Nigeria	285	400	300
Zaire	165	400	182
Other Africa	277	–	300
Asia			
India	132	400	1500
Indonesia	550	1200	900
Kampuchea	15	50	25
Laos	120	100	100
Malaysia	230	480	255
Myanmar (Burma)	92	800	677
Papua New Guinea	21	350	22
Philippines	100	270	143
Thailand	325	600	397
Vietnam	60	350	173
Other Asia	124	–	84
C.&S. America			
Bolivia	65	150	87
Brazil	1360	5000	8000
Colombia	800	650	820
Ecuador	300	300	340
Guyanas	4	50	5
Mexico	420	700	595
Paraguay	160	–	190
Peru	245	350	270
Venezuela	125	150	125
Other C.&S. America	328	330	477
Total	6893	13,860	16,544

of 13.9×10^6 ha/yr. If this figure is compared with the FAO/UNEP figure for closed forests, 6.9×10^6 ha, the annual rate of deforestation appears to have increased by 100% in the approximately ten years between the two surveys. Six countries (Brazil, Burma, Colombia, Indonesia, Mexico, and Thailand) account for 8.95×10^6 ha, or 65%, of the global rate of deforestation in the late 1980s.

The World Resources Institute (WRI, 1990) has also published country-by-country estimates of deforestation rates in total forest and closed forest for the late

Table 4: Recent estimates of the rate of deforestation and the change in the rate from the late 1970s (10^3 ha/yr)

	Africa	Asia	C.&S. America	Total
FAO/UNEP Late 1970s	1319	1767	3807	6893
Myers Late 1980s (% increase)[1]	1580 (20)	4600 (160)	7680 (101)	13,860 (101)
WRI Late 1980s (% increase)[1]	1359 (3)	4276 (142)	10,909 (187)	16,544 (140)
Our Update[2] Late 1970s			4447	7533
Our Update[3] Late 1980s (% increase)[4]			5509 (24)	11,144 (48)

[1]The percent increase over the FAO/UNEP report.
[2]Our update is computed from our GIS data base on deforestation between 1975 and 1978, as remeasured from Tardin *et al.* (1980). These values were substituted for the Brazilian values reported by Myers and WRI.
[3]Our update based by substituting the rates given in Fearnside *et al.* (1990) and our analysis of the Amazon Basin for the values for Brazil reported by Myers and WRI.
[4]Percent increase over the 1975–78 rate.

1980s (ca. 1988). Their global estimate is 20.7×10^6 ha/yr total, of which 16.5×10^6 ha is in closed forests. Their estimate for closed forest deforestation is a 140% increase over the FAO/UNEP estimate of a decade earlier, but their estimate for total forest is only an 83% increase, that is, somewhat smaller than the estimate of Myers (1989).

Approximately 70% of the WRI (1990) increase in deforestation rate is attributed to Brazil alone. A full 83% of the WRI increase is from just two countries, Brazil and India. (Note that WRI's estimate of deforestation for India is almost fourfold higher than that of Myers.)

The Brazilian estimate given in WRI (1990), 8.0×10^6 ha/yr, is probably too high (see above). Myers's (1989) estimate of 5.0×10^6 ha/yr, which appears to be based on loose accounts and trends rather than objective assessment, is probably also too high. A more conservative figure has been published by Fearnside *et al.* (1990), based on INPE analyses. This estimate is 2×10^6 ha/yr averaged over 1978–88. Fearnside *et al.* (1990) also report the rate of deforestation between 1988 and 1989 to be slightly higher at 2.6×10^6 ha. If these latter estimates are used in place of the figure reported by WRI, we get a new estimate of the deforestation rate in the late 1980s of approximately 11.1×10^6 ha/yr, which yields a 48% increase in the global rate of deforestation for the 1980s (see Table 4). The analysis of deforestation rates made from satellite data by Skole and Tucker (1993) suggest lower rates of deforestation than previously believed ($\sim 1.5 \times 10^6$ ha/yr).

These various estimates highlight some of the problems of measuring deforestation. The work in Brazil represents the state of the art in remote sensing monitoring of tropical deforestation. But, as the earlier sections of this paper have shown, all methods have some problems, and some are still in a trial stage.

Analysis of Deforestation Using Social and Demographic Factors

Demographic Change and Deforestation

In recent years, researchers have analyzed the relationship between deforestation and population growth, mostly at the global level. Allen and Barnes (1985) examined the statistical correlation between population and deforestation data in 76 tropical countries. They also conducted multiple regressions of deforestation against other variables such as arable land, roundwood production, and gross domestic product (GDP). Their analysis suggested a low, but significant, correlation between population growth rates in 1970–78 and deforestation reported for the period 1975–80 in the FAO Forest Assessment (Lanly, 1982). They also showed a low, but significant, correlation between arable land change and deforestation. Their conclusion was that population growth is a major cause of deforestation.

Using data that I have compiled on deforestation from both statistical land-use surveys (IBGE, 1970a, 1980a, 1989) and satellite measurements (Tardin *et al.*, 1980), it is possible to look at deforestation trends in Brazil in relation to population growth, derived from data in IBGE (1970b, 1980b). In the first analysis, I have tabulated remote sensing data reported by Tardin *et al.* (1980). Data on population for each *municipio* for 1970 and 1980, arranged by rural and urban populations, were encoded into a GIS along with deforestation rates. The remote sensing data were for the period 1975 and 1978/9. By simple change detection, an average annual rate can be computed. This rate was regressed against the 1980 rural population for each *municipio* in the legal Amazon. As Figure 11 shows, the apparent relationship found by Allen and Barnes (1985) at the global level is not clear in this analysis. It must be noted that although the correlation was weak, it was significantly different from zero ($p > 0.001$).

Simple regressions of this kind are difficult since various regions could have different, but strong and internally consistent, linear relationships, which would not show up when the regions are viewed collectively. Therefore, I performed a simple rank (Spearman's) correlation. Again, no strong correlation appeared.

Explanations by Understanding Social and Economic Drivers

The regression approaches are straightforward and present a promising

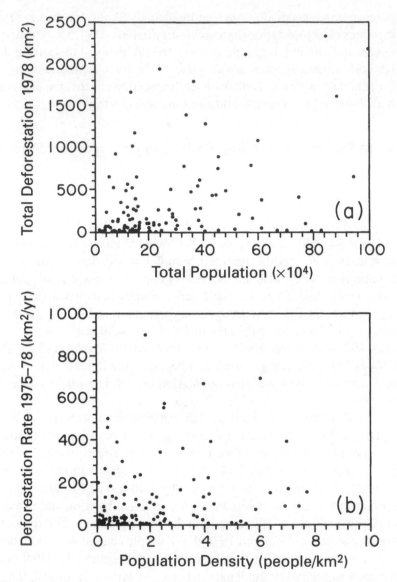

Figure 11. There appears to be no relationship between deforestation and population density. Scatter plots of population against (a) total area deforested in 1978 and (b) population density against the deforestation rate in 1978.

approach for prediction over short time periods, but they do not provide a causal explanation for the explosive rates of deforestation in the Amazon. Hecht (1983; Hecht and Cockburn, 1989) studied cattle ranching in the Amazon and concluded that government policies, fiscal incentives, and the nature of individual farmer decision-making in an inflationary economy (i.e., cattle are a good hedge against uncertain economic conditions) control deforestation rate more than do demographic considerations. She argues, 'It is ludicrous to describe environmental degradation in this situation as only a function of demographics.' A similar view has come out of recent studies of declining wood stocks in sub-Saharan Africa (Anderson, 1986). In these case studies, population growth is seen as part of a multiple feedback system, where it is as much a consequence of poverty and land degradation as it is a cause.

Nonetheless, I believe it is useful to develop an explanatory conceptual framework that links to demographic and 'modernization' factors (e.g., rural to urban migration, increasing substitution of machinery for labor in a rapidly developing industrial economy). The next section presents a conceptual model that relates deforestation to demography and spatial organization, focusing on Rondônia, Brazil, as a case study.

Deforestation in Rondônia, 1970–80

The highest deforestation density in Brazil is in the state of Rondônia, where 13% of the forests had been cleared as of 1987. The state of Rondônia has experienced nearly exponential deforestation rates over the last 30 years as new colonization and settlement programs have opened large tracts of forest. These settlement programs, along with specific fiscal incentives, were established in the 1970s to encourage migration to the region from overpopulated, poverty-stricken, and drought-ridden regions in the south and northeast of the country. The vast Amazon region was seen by many as an empty frontier, which at once could be consolidated under Brazilian national sovereignty and could provide opportunity for millions of the poor and landless (Bunker, 1984a, 1984b).

In my model, this trend can be explained by changing demographic and economic conditions in the south of Brazil, particularly the state of Paraná, during the period. I will focus my discussion on changes in the state of Paraná in the early 1970s and explore how changes in land tenure and land use there directly influenced land use in Rondônia. But first it is important to consider certain international activities that were taking place at the time, particularly related to world oil production, distribution, and price.

The Flood of Petrodollars

After the increase in the price of oil stimulated by the Organization of Petroleum Exporting Countries (OPEC) in the mid-1970s, large amounts of what have been called petrodollars flooded international money markets (Pool and Stamos, 1987). The price of oil went from $1.30 a barrel in 1970 to $28.67 by 1980. Most energy-dependent countries paid OPEC prices, resulting in a large net transfer of wealth from industrial economies to OPEC. OPEC, in turn, deposited these revenues in U.S. and European banks. Since banks must pay dividends or interest to depositors, U.S. commercial institutions were eager to find borrowers.

Developing countries such as Brazil were eager for foreign capital to fund economic development, modernization, and industrialization programs. They also needed dollars to pay for oil, which is bought and traded in dollars. Brazil borrowed heavily to finance domestic economic development programs. One important pro-

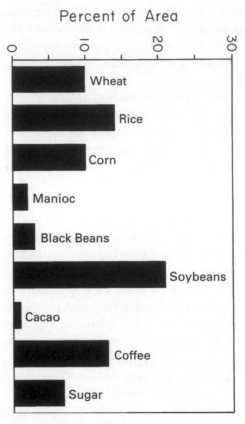

Figure 12. Allocation of crop credits in Brazil in 1978, as a percent of agricultural crop area.

gram in the 1970s was agricultural modernization. It was believed that agricultural products could provide a profitable export industry, which in turn would help pay for loans to modernize agriculture and the rest of the country (Mahar, 1989).

Agricultural Modernization in Brazil in the 1970s

Between 1970 and 1980, there was large-scale investment in agriculture. Crop credits in Brazil experienced an almost fivefold increase (World Bank, 1982). I have no direct data that show the amount of these credits derived from foreign loans. It is probably safe to suppose that the deployment of credits was an effort to build an active agricultural export system to balance foreign debt and offset the rising cost of petroleum (World Bank, 1982). Such investment programs required large amounts of capital from abroad, but they resulted in some degree of success; rising crop credits occurred with increasing crop output; the net value of agricultural output increased 2.68-fold between 1970 and 1980 (IBGE, 1989, 1980a, 1975, 1970c). By 1977, export crops accounted for more than 50% of the total value of principal crops.

Figure 12 shows the distribution of crop credits by crop type. Three general patterns emerge. First, the export crops of wheat, soybeans, and coffee consumed

Table 5: Comparison of the costs of soybean production in Brazil and the United States

	U.S.	Brazil
1. Variable costs		
Machinery	737	672
Labor	457	241
Inputs	964	2299
Other	62	404
Subtotal	2220	3617
2. Fixed costs		
Depreciation	781	427
Interest on capital	293	141
Labor	449	170
Land	2029	937
Other	365	51
Subtotal	3315	1727
3. Total costs per ha	6136	5343
4. Yield (kg/ha)	1900	1920
5. Unit costs (per ton)	3229	2783
6. Port Costs	3368	3882

Note: Units are arbitrary but have been corrected for exchange rates.

almost one-half of all crop credits. Second, the largest fraction was invested in soybean production. Third, very little of the credits was allocated to staple crops such as black beans and manioc.

Soybean production was a major success story. The area harvested increased sixfold in the 1970s, ten times more than any other crop except oranges and wheat; yields increased fivefold. The combination of land, fertilizers, improved seeds, and government-sponsored fiscal credits and incentives produced an internationally competitive export program. Soybeans became one of Brazil's major export crops. From being a small producer at 10% of the international markets, Brazil was able to compete with the United States by 1980. Despite higher costs for such inputs as fertilizers and pesticides, and similar labor costs, Brazil had land costs about one-half of those in the United States, giving it a comparative advantage in costs of production (Table 5).

Most of the soybean production was concentrated in two states: Rio Grande do Sul and Paraná. Unlike that for soybeans, the international market for coffee was highly variable and undependable. Government programs concentrated on replacing coffee fields with soybeans (World Bank, 1982).

Agricultural Modernization and Demographic Change

Just as the industrial sector was modernizing, so was agriculture, as discussed above. A labor-intensive agricultural system was being transformed into an impor-

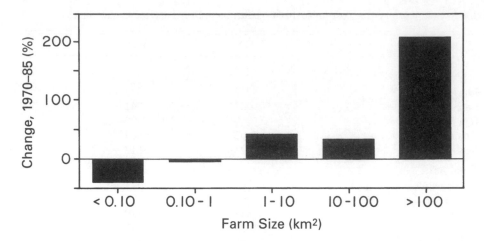

Figure 13. Change in the farm size distribution in Paraná between 1970 and 1985. Units are percent increase or decrease.

tant energy- and machinery-intensive component of the national economy, particularly in Paraná. Land prices rose significantly (World Bank, 1982) as consolidation into larger holdings occurred. Coffee, a labor-intensive crop, was replaced by soybeans and wheat, which utilize machinery. This transformation of land use changed land tenure. Figure 13 shows a loss of small farms in Paraná and an increase in very large, presumably commercial farms.

It has already been noted that the period 1970–80 saw increased migration from rural to urban areas. Part of this migration was a response to 'pull' factors as industrial development created increased opportunities for wages in urban areas. As well, commercialization of agriculture in Paraná shifted the mode of production from labor to machinery, creating a 'push' factor. Demographic data show that large numbers of people left Paraná during this period; the out-migration rate was higher than that of any other state. Undoubtedly many, if not most, of the migrants left for urban areas, but a large number also went to new opportunities in the Rondônia frontier (Hecht and Cockburn, 1989).

Effect on Deforestation in Rondônia State
Much has been written about the large-scale government programs to facilitate the opening of the frontier (e.g., Mahar, 1989; Moran, 1981; Bunker, 1984a; Fearnside, 1990), and the various reasons for them, from the military-oriented view of the need to consolidate the hollow frontier to the need to provide a population safety valve. The long-term drought in the northeast certainly figured prominently. Moreover, much has been made of the massive road-building projects, suggesting them to be the key determinant of change in the region. All of these factors must be considered, but they were largely mechanisms facilitating a transformation that had more fundamental underpinnings. Clearly, migration to the region was largely a response to events and conditions far removed from Rondônia:

changes in land tenure in the south of Brazil and changes in the structure of a rapidly developing national economy, fueled to a large degree by excess petrodollars.

Conclusion

Land-cover conversion rates are not well known. The comparison of current estimates of deforestation in the tropics presented in this paper demonstrates the need for new and objective information. One approach would be to couple historical reconstruction with remote sensing, as shown in Figures 3 and 4. Tabular data, national censuses, or historical documents in which data are reported geographically (e.g., by administrative district) could be mapped in a geographic information system. Remote sensing provides direct observations of some aspects of land-cover change, while historical reconstruction generally relies on indirect estimation from changes in various human-use categories. Nonetheless, the coupling of these two approaches provides a way to 'calibrate' the historical assessment, since the historical trend should overlap with remotely sensed data—both in magnitude and geographically—when both exist.

The advantages of coarse-resolution remote sensing over existing maps are temporal consistency and an explicit definition of actual, rather than estimated, boundaries between cover types. The most straightforward approach would be to derive maps of current land-cover types from remote sensing measurements along a predefined classification system. No single classification will suit all needs. But much could be gained by initiating international efforts to collect the necessary data sets from existing satellite sensors, such as AVHRR, from which various classifications could be made on a case-by-case basis. A remote sensing-based map of current land cover could form the base from which a predisturbance land-cover map could be created by correlating existing natural land cover with physical variables (e.g., temperature, precipitation, edaphic conditions), or through approximation based on simple assumptions of contiguity and spatial clustering. Having such maps of both current and predisturbance land cover, it might then be possible to reconstruct the history of land-cover change with the addition of geographically referenced time series of human use and conversion, such as maps of the expansion of deforestation. Since much of the analysis is spatial, a GIS would be used to organize the data and analysis.

The use of coarse-resolution (1 km or greater) remotely sensed data to map deforestation and other forms of land-cover conversion has frequently been advocated, since it requires fewer data than high-resolution sensors (less than 100 m). In cases where coarse-resolution data sets have been used in the past, however, they have tended to overestimate deforestation. This overestimation bias is partly related to the geometry of clearing, so a single conversion factor cannot readily be developed. Considerable work remains before mixture modeling and other techniques that could derive an accurate estimate from coarse sensors could be used. Meanwhile, the best approach appears to be one based on high-resolution data from such satellite systems as Landsat. The use of photoproducts at scales ranging from 1:250,000 to 1:500,000 provides an efficient, low-cost alternative. These data can be interpreted for areas of deforestation and then encoded digitally in a geographic information system.

The fine-scale dynamics of clearing, abandonment, regrowth, and reclearing are little known and understudied, yet they could be important to analyses of the net flux of carbon. The use of high-spatial-resolution remote sensing data provides detailed information on regrowth and abandonment. There is very little quantitative information on the amount of secondary growth. Some preliminary results from my own work and others' suggest that as much as 20% of the deforested land in the Amazon Basin is in some stage of regrowth at any point in time.

Thematic classification of land cover from remote sensing data follows traditional cartographic approaches. For many aspects of global change research, in the fields of biogeochemistry and water/energy dynamics for instance, there is a need to parameterize land cover beyond basic classes such as forest, savanna, and the like. Here, parameters such as nitrogen status, biomass, NPP, and other land-cover-related factors are required. Functional classifications, of the type mentioned in this paper, could be very important new approaches to classification of land cover. Another approach would be to develop methods to derive direct parameterization of the state of land cover from satellite data. Direct parameterization of canopy chemistry, albedo, surface roughness, NPP, and other variables could be one feasible approach that circumvents the necessity for an a priori system of nomenclature.

Documentation of land-cover change is the first step toward understanding its underlying agents. But it is clear that documentation alone will not yield a complete explanation. Simple correlations between various factors are often facile but not fruitful, and may in fact be misleading or wrong. When it works, correlation between deforestation and some other social factor such as population might help in developing a sampling scheme or for making short-term predictions, but such an approach lacks the ability to explain the phenomenon. Population pressure has frequently been cited as the main driver of land-cover change, yet there is little strong quantitative evidence one way or the other because there is a paucity of hard data on land-cover change.

It could be reasonably argued that if a correlation exists between population and land-cover change, such a relationship could be used as a first-order, short-term predictor of land-cover change, in terms of both extent and spatial distribution. It is not clear, however, that for reasonable short-term predictions complex formulations based on population, economic variables, or social drivers are required. Studies of the Amazon that we are developing now at the University of New Hampshire suggest that the rate and spatial distribution of deforestation are somewhat persistent over short time periods (see Figure 9). These data indicate that, in this case at least, spatial relationships are maintained from one date to the next in generally predictable ways. This, in turn, suggests that one of the highest priorities in deforestation prediction and analysis is an accurate base map of deforestation. Predictions on time horizons beyond one or two decades, however, will require more sophisticated models that consider primary causal factors and incorporate an understanding of the complex social and economic conditions in a particular region.

Future research can be developed in three general areas: (1) acquisition and analysis of new data on existing and predisturbance land cover, using a combination of satellite data, historical reconstructions, and maps; (2) acquisition and analysis of mapped data on the distribution of immediate processes of land-cover change, such as deforestation or agricultural expansion, using satellite data for the contemporary period and tabular census data for historical periods; and (3) development of models of change that might be based on simple extrapolation for short-term analyses and socioeconomic factors for long-term analyses.

References

Aber, J. D., J. M. Melillo, and C. A. McClaugherty. Predicting long-term patterns of mass loss, nitrogen dynamics and soil organic matter formation from initial litter chemistry in forest ecosystems. *Canadian Journal of Botany*, in press.

Aber, J. A., K. J. Nadelhoffer, P. Stuedler, and J. M. Melillo. 1989. Nitrogen saturation in northern forest ecosystems. *BioScience*

Allen, J. C., and D. F. Barnes. 1985. The causes of deforestation in developing countries. *Annals of the Association of American Geographers 75(2)*, 163–184.

Anderson, D. 1986. Declining tree stocks in African countries. *World Development 14(7)*, 853–863.

Andres, L., W. A. Salas, and D. L. Skole. Fourier analysis of multi-temporal AVHRR data applied to global scale land cover classification. *International Journal of Remote Sensing*, in preparation.

Berry, J. K. 1987. Computer assisted map analysis: Potential and pitfalls. *Photogrammetric Engineering and Remote Sensing, 53(10)*, 1405–1410.

Bracken, I., and C. Webster, 1989. Toward a typology of geographical information systems. *International Journal of Geographic Information Systems 3(2)*, 137–152.

Bunker, S. G. 1984a. Modes of extraction, unequal exchange, and the progressive underdevelopment of an extreme periphery: The Brazilian Amazon, 1600–1980. *American Journal of Sociology 89(5)*, 1017–1065.

Bunker, S. G. 1984b. *Underdeveloping the Amazon: Extraction, Unequal Exchange, and the Failure of the Modern State*, University of Illinois Press, Champaign, Illinois.

Burrough, P. A. 1986. *Principles of Geographic Information Systems for Land Resources Assessment*. Oxford University Press, New York, 193 pp.

CNES. 1988. *Spot User's Handbook*. Spot Image Corporation, Reston, Virginia.

Colwell, R. N. 1983. *Manual of Remote Sensing*. American Society of Photogrammetry.

Crutzen, P. J., and M. O. Andreae. 1990. Biomass burning in the tropics: Impact on atmospheric chemistry and biogeochemical cycles. *Science 250*, 1669–1678.

Ehrlich, P.R., and E.O. Wilson. 1991. Biodiversity studies: Science and policy. *Science 253*, 758–762.

Emanuel, W.R., I. Fung, G.G. Killough Jr., B. Moore III, T.-H. Peng. 1985. Modeling the global carbon cycle and changes in atmospheric carbon dioxide levels. In *Atmospheric Carbon Dioxide and the Global Carbon Cycle* (J.R. Trabalka, ed.), DOE/ER- 0239, U.S. Department of Energy, Washington, D.C.

FAO/UNEP. 1981. Tropical Forest Resources Assessment Project. Forest Resources of Tropical America, FAO, Rome, Italy.

Fearnside, P. M. 1990. The rate and extent of deforestation in Brazilian Amazonia. *Environmental Conservation 17*, 213–226.

Fearnside, P. M., A. Tardin, and L.G. Meira. 1990. *Deforestation in the Brazilian Amazon*. National Secretariat of Science and Technology, Brazil.

Field, C., and H. A. Mooney. 1986. The photosynthesis-nitrogen relationship in wild plants. In *On the Economy of Plant Formation and Function* (T.J. Givinish, ed.), Cambridge University Press, New York.

Flint, E. P., and J. F. Richards. 1991. Historical analysis of changes in land use and carbon stock of vegetation in south and southeast Asia. *Canadian Journal of Forest Research 21(1)*, 91–110.

Fung, I. Y., C. J. Tucker, and K. C. Prentice. 1987. Application of advanced very high resolution radiometer vegetation index to study atmosphere–biosphere exchange of CO_2. *Journal of Geophysical Research 92(D3)*, 2999–3015.

Hecht, S. B. 1983. Cattle ranching in the eastern Amazon: Environmental and social implications. In *The Dilemma of Amazonian Development* (E. Moran, ed.), Westview Press, Boulder, Colorado, 155–188.

Hecht, S. B., and A. Cockburn. 1989. *The Fate of the Forest*. Verso, London, 266 pp.

Henderson-Sellers, A. 1987. Effects of change in land use on climate in the humid tropics. In *The Geophysiology of Amazonia, Vegetation and Climate Interactions* (R. E. Dickinson, ed.), John Wiley and Sons, New York.

Henderson-Sellers, A., and V. Gornitz. 1984. Possible climatic impacts of land cover transformation, with particular emphasis on tropical deforestation. *Climatic Change 6*, 231–256.

Holben, B. N., and C. O. Justice. 1981. An examination of spectral band ratioing to reduce the topographic effect of remotely sensed data. *International Journal of Remote Sensing 2*, 115.

Holdridge, L. R. 1947. Determination of world formations from simple climatic data. *Science 105*, 367–368.

Holdridge, L. R. 1967. *Life Zone Ecology*. Tropical Science Center, San Jose, Costa Rica, 206 pp.

Houghton, R. A. 1986. Estimating changes in the carbon content of terrestrial ecosystems from historical data. In *The Changing Carbon Cycle: A Global Analysis* (J. R. Trabalka, and D. E. Reichle, eds.), Springer-Verlag, New York.

Houghton, R. A., and D. L. Skole. 1990. Carbon. In *The Earth As Transformed by Human Action: Global and Regional Changes in the Biosphere over the Past 300 Years* (B. L. Turner II, W. C. Clark, R. W. Kates, J. F. Richards, J. T. Mathews, and W. B. Meyer, eds.), Cambridge University Press, Cambridge, U.K.

Houghton, R. A. , D. L. Skole, and D. S. Lefkowitz. 1991. Changes in the landscape of Latin America between 1850 and 1985: II. A net release of CO_2 to the atmosphere. *Journal of Forest Ecology and Management, 38(3/4)*, 173–200.

IBGE (Instituto Brasileiro de Geografia e Estatistica). 1960. *Censo Agricola de 1960*, VII Recenseamento Geral do Brasil, Instituto Brasileiro de Geografia e Estatistica, Rio de Janeiro, Brazil..

IBGE. 1970a. *Censo Agropecuario*, VIII Recenseamento Geral do Brasil, Instituto Brasileiro de Geografia e Estatistica, Rio de Janeiro, Brazil.

IBGE. 1970b. *Censo Demografico*, VIII Recenseamento Geral do Brasil, Instituto Brasileiro de Geografia e Estatistica, Rio de Janeiro, Brazil.

IBGE. 1970c. *Anuario Estatistico do Brasil – 1970b, Instituto* Brasileiro do Geografia e Estatistica, Rio de Janeiro, Brazil.

IBGE. 1975. *Anuario Estatistico do Brasil—1975*, Instituto Brasileiro do Geografia e Estatistica, Rio de Janeiro, Brazil.

IBGE. 1980a. *Censo Agropecuario*, IX Recenseamento Geral do Brasil, Instituto Brasileiro de Geografia e Estatistica, Rio de Janeiro, Brazil.

IBGE. 1980b. *Censo Demografico*, IX Recenseamento Geral do Brasil, Instituto Brasileiro de Geografia e Estatistica, Rio de Janeiro, Brazil..

IBGE. 1989. *Anuario Estatistico do Brasil—1989*, Instituto Brasileiro do Geografia e Estatistica, Rio de Janeiro, Brazil.

Justice, C. O., J. R. G. Townshend, B. N. Holben, and C. J. Tucker. 1985. Analysis of the phenology of global vegetation using meteorological satellite data. *International Journal of Remote Sensing 6*, 1271–1318.

Kaufman, Y. J., A. Setzer, C. Justice, C. J. Tucker, M. G. Pereira, and I. Fung. 1990. Remote sensing of biomass burning in the tropics. In *Fire in the Tropical Biota* (J.G. Goldammer, ed.), Springer-Verlag, New York.

Kidwell, K. B. 1990. *Global Vegetation Index Users Guide*. National Environmental Satellite Data and Information Service, National Oceanic and Atmospheric Administration, Washington, D.C.

Kidwell, K. B. 1991. *NOAA Polar Orbiter Data Users Guide*. National Environmental Satellite Data and Information Service, National Oceanic and Atmospheric Administration, Washington, D.C.

Lanly, J. P. 1982. *Tropical Forest Resources*. FAO Forestry Paper No. 30, Food and Agriculture Organization of the United Nations, Rome, Italy.

Loveland, T. R., J. W. Merchant, D. Ohlen, and J. F. Brown. 1991. Development of a land-cover characteristics database for the conterminous U.S. *Photogrammetric Engineering and Remote Sensing 11*, 1453–1463.

Maguire, D. J., M. F. Goodchild, and D. W. Rind. 1991. *Geographical Information Systems: Principles and Applications*. John Wiley and Sons, New York.

Mahar, D. J. 1989. *Government Policies and Deforestation in Brazil's Amazon Region*. World Bank, Washington D.C.

Malingreau, J. P. 1986. Global vegetation dynamics: Observations over Asia. *International Journal of Remote Sensing 7(9)*, 1121–1146.

Malingreau, J. P., and C. J. Tucker. 1987. The contribution of AVHRR data for measuring and understanding global processes: Large scale deforestation in the Amazon Basin. In *Proceedings of IGARSS '87 Symposium*, Ann Arbor, Michigan, 443–448.

Malingreau, J. P., and C. J. Tucker. 1988. Large scale deforestation in the southeastern Amazon basin of Brazil. *Ambio 17(1)*, 49–54.

Malingreau, J. P., and C. J. Tucker. 1989. AVHRR for monitoring global tropical deforestation. *International Journal of Remote Sensing 10(4)*, 855–867.

Marden dos Santos, J. 1987. Climate, natural vegetation, and soils in Amazonia: An overview. In *The Geophysiology of Amazonia, Vegetation and Climate Interactions* (R.E. Dickinson, ed.), John Wiley and Sons, New York.

Matthews, E. 1983. Global vegetation and land use: New high resolution data bases for climate studies. *Journal of Climate and Applied Meteorology 22(3)*, 474–487.

Meentemeyer, V. 1978. Macroclimate and lignin control of decomposition. *Ecology 63*, 621–626.

Melillo, J. M., J. D. Aber, and J. F. Mutore. 1982. Nitrogen and lignin control of hardwood leaf litter decomposition dynamics. *Ecology 63(3)*, 621–626.

Melillo, J. M., C. A. Palm, R. A. Houghton, G. M. Woodwell, and N. Myers. 1985. A comparison of two recent estimates of disturbance in tropical forests. *Environmental Conservation 12(1)*, 37–40.

Moran, E. F. 1981. *Developing the Amazon*. Indiana University Press, Bloomington, Indiana, 292 pp.

Mueller-Dombois, D. 1984. Classification and mapping of plant communities: a review with emphasis on tropical vegetation. In *The Role of Terrestrial Vegetation in the Global Carbon Cycle: Measurement by Remote Sensing* (G.M. Woodwell, ed.), SCOPE 23, John Wiley and Sons, Chichester, U.K.

Myers, N. 1980. *Conservation of Tropical Moist Forests*. National Research Council, National Academy of Science, Washington, D.C.

Myers, N. 1989. *Deforestation Rates in Tropical Forests and Their Climatic Implications*. Friends of the Earth, London.

Nelson, R. F., and B. N. Holben. 1986. Identifying deforestation in Brazil using multiresolution satellite data. *International Journal of Remote Sensing 7(3)*, 429–448.

Nelson, R. F., D. Case, N. Horning, V. Anderson, and S. Pillai. 1987a. Continental land cover assessment using Landsat MSS data. *Remote Sensing of the Environment 21*, 61–81.

Nelson, R. F., N. Horning, and T. A. Stone. 1987b. Determining the rate of forest conversion in Mato Grosso, Brazil, using Landsat MSS and AVHRR data. *International Journal of Remote Sensing 8(12)*, 1767–1784.

Olson, J. S., J. A. Watts, and L. J. Allison. 1983. *Carbon in Live Vegetation of Major World Ecosystems*. ORNL-5862, Oak Ridge National Laboratory, Oak Ridge, Tennessee, 164 pp.

Pool, J. C., and S. Stamos. 1987. *The ABCs of International Finance: Understanding the Trade and Debt Crisis*. Lexington Books, Lexington, Massachusetts, 138 pp.

Richards, J. F. 1984. Global patterns of land conversion. *Environment 26(9)*, 6–38.

Richards, J. F., J.S. Olson, and R.M. Rotty. 1983. *Development of a Data Base for Carbon Dioxide Releases Resulting from Conversion of Land to Agricultural Uses*. ORAU/IEA-82-10(M), ORNL/TM-8801, Oak Ridge National Laboratories, Oak Ridge, Tennessee.

Salati, E., and P. B. Vose. 1984. Amazon basin: A system in equilibrium. *Science 225*, 129–138.

Setzer, A. W., and M. C. Pereira. 1990. Amazon biomass burnings in 1987 and an estimate of their tropospheric emissions. *Ambio 20(1)*, 19–22.

Shukla, J., C. Nobre, and P. Sellers. 1990. Amazon deforestation and climate change. *Science 247*, 1322–1325.

Skole, D. L., and C. J. Tucker. 1993. Tropical deforestation and habitat fragmentation in the Amazon: Satellite data from 1978 to 1988. *Science 260*, 1905–1910.

Smith, T. R., S. Menon, J. L. Star, and J. E. Estes. 1987. Requirements and principles for the implementation and construction of large-scale geographic information systems. *International Journal of Geographical Information Systems 1(1)*, 13–32.

Soule, M. E. 1991. Conservation: Tactics for a constant crisis. *Science 253*, 744–750.

Tardin, A. T., and R. Pereira da Cunha. 1990. *Evaluation of Deforestation in the Legal Amazonia Using Landsat-TM Images*. Instituto de Pesquisas Espaciais, Sao Jose dos Campos, Brazil.

Tardin, A. T., D. C.L. Lee, R. J. R. Santos, O. R. Ossis, M. P. S. Barbosa, M. L. Moreira, M. T.

Pereira, D. Silva, and C. P. Santos Filho. 1980. *Subprojecto Desmatamento.* IBDF/CNPq -INPE, Instituto de Pesquisas Espaciais, Sao Jose dos Campos, Brazil.

Townshend, J. R. G., and C. O. Justice. 1988. Selecting the spatial resolution of satellite sensors required for global monitoring of land transformation. *International Journal of Remote Sensing 9(2)*, 187–236.

Townshend, J. R. G., and C. J. Tucker. 1984. Objective assessment of Advanced Very High Resolution Radiometer data for land cover mapping. *International Journal of Remote Sensing 5(2)*, 497–504.

Townshend, J. R. G., C. O. Justice, D. L. Skole, J. P. Malingreau, J. Cihlar, P. Teillet, F. Sadowski, and S. Ruttenberg. The 1 km resolution global dataset: Needs of the International Geosphere Biosphere Program. *International Journal of Remote Sensing*, in press.

Tucker, R., and J. F. Richards. 1983. *Global Deforestation and the Nineteenth Century World Economy.* Duke University Press, Durham, North Carolina.

Tucker, C. J., B. N. Holben, and T. E. Goff. 1984. Intensive forest clearing in Rondônia, Brazil as detected by satellite remote sensing. *Remote Sensing of the Environment 15*, 255–261.

Tucker, C. J., H. E. Dregne, and W. W. Newcomb. 1991. Expansion and contraction of the Sahara desert from 1980 to 1990. *Science 253*, 299–301.

UNESCO (U.N. Educational, Scientific, and Cultural Organization). 1973. *International Classification and Mapping of Vegetation,* UNESCO Ecology and Conservation Series 6, UNESCO, Rome, Italy, 93 pp.

Vitousek, P. M. 1982. Nutrient cycling and nutrient use efficiency. *American Naturalist1 19*, 553–572.

Whittaker, R. H. 1978. *Classification of Plant Communities.* Junk Publishers, The Hague, The Netherlands, 408 pp.

Williams, M. 1990. Global deforestation. In *The Earth As Transformed by Human Action: Global and Regional Changes in the Biosphere over the Past 300 Years* (B.L. Turner II, W. C. Clark, R. W. Kates, J. F. Richards, J. T. Mathews, and W. B. Meyer, eds.), Cambridge University Press, Cambridge, U.K.

Wilson, E. O., and F. M. Peter. 1988. *Biodiversity.* National Academy Press, Washington, D.C.

Wilson, M. F., and A. Henderson-Sellers. 1985. A global archive of land cover and soils data for use in general circulation climate models. *Journal of Climatology 5*, 119–143.

Woodwell, G. M., J. E. Hobbie, R. A. Houghton, J. M. Melillo, B. Moore, A. B. Park, B. J. Peterson, and J. R. Shaver. 1984. Measurement of changes in the vegetation of the earth by satellite imagery. In *The Role of Terrestrial Vegetation in the Global Carbon Cycle: Measurement by Remote Sensing* (G.M. Woodwell, ed.), SCOPE 23, John Wiley and Sons, New York.

Woodwell, G. M., R. A. Houghton, T. A. Stone, and A. B. Park. 1986. Changes in the area forests in Rondônia, Amazon Basin, measured by satellite imagery. In *The Changing Carbon Cycle, a Global Analysis* (J. R. Trabalka, and D. E. Reichle, eds.), Springer-Verlag, New York.

World Bank. 1982. *Brazil: A Review of Agricultural Policies.* A World Bank Country Study, World Bank, Washington, D.C.

WRI (World Resources Institute). 1990. *World Resources: A Guide to the Global Environment.* World Resources Institute, Washington, D.C.

Appendix I

Data Collections Useful for Analysis of Land-Use/Cover Change[1]

John Kelmelis and Fran Rowland

Introduction

Data describing land use and land cover at different times are needed to determine the occurrence of changes in land use/cover. To determine *why* changes have occurred, even more complex data are necessary. These range from information about land use to the social, economic, biological, and physical variables that affect processes on the land surface.

Global change data on the whole are difficult to define; however, they are necessary for research being done in many disciplines and in many organizations, both nationally and internationally. Data that are specifically collected for global change research can be clearly identified. However, a data set that is not currently used for global change research may contain data that will have some use for global change research in the future. Likewise, data useful for researching the causes, status, and effects of changes in land use and cover represent the results of a large portion of the observations and research in the natural and social sciences. Critical variables for land-use/land-cover change research have not yet been well defined; thus identifying the complete set of pertinent data for this research is not an easy task.

The overlap of land-use/land-cover data and the broader category of global change data is large and reflects a volume of data with vague boundaries. A listing of all, or even a large portion, of the data collections that are important to the study of the global change aspects of land-use/cover change would be unmanageably large and is beyond the scope of this document. A sample of the types of data that are available with pointers to sources of those data is somewhat more manageable.

The listing that follows is by no means comprehensive, and would benefit by having additional references from the social science community. The value of this appendix could be significantly increased if data producers would submit information about their data sets to one of the data catalogs or directories listed herein. Data can only become more accessible if they are identified in catalogs or directories that are likely to be queried by potential users. A commitment must be made

---

473

by researchers to preserve data for future use, find suitable archives for such data, and accurately describe the contents of the data base.

Before proceeding to the listings themselves, we provide a brief overview of U.S. data centers and descriptions, with examples, of the various types of data compilations: directories and catalogs, archives, training data sets, and data bases.

Overview of Data Centers and Data Systems

National data centers are a valuable resource for obtaining data bases relevant to land-use/cover change and global change research. Geophysical data are archived within the Department of Defense (DOD), the National Oceanic and Atmospheric Administration (NOAA), the National Aeronautics and Space Administration (NASA), the National Science Foundation (NSF), the National Center for Atmospheric Research (NCAR), the Carbon Dioxide Information Analysis Center (CDIAC) (sponsored by the Department of Energy [DOE]), the Environmental Protection Agency (EPA), and the U.S. Geological Survey (USGS). Some national data centers are:

DOD Environmental Technical Applications Center	704-259-0218
DOE Carbon Dioxide Information Analysis Center	615-574-0390
DOE Energy Information Administration	202-586-8800
EPA National Computer Center	919-541-7862
NASA National Space Science Data Center	301-286-6695
NCAR	303-497-1215
NOAA National Climatic Data Center (NCDC)	704-259-0682
NOAA National Geophysical Data Center (NGDC)	303-497-6764
NOAA National Oceanographic Data Center	202-673-5549
NOAA National Snow and Ice Data Center	303-497-5171
USGS Earth Resources Observation Systems (EROS) Data Center	605-594-6151
USGS Earth Science Information Center	800-USA-MAPS

The World Data Center (WDC) system is an international network of four World Data Centers which were established originally in 1957 by the International Council of Scientific Unions to manage data collected during the International Geophysical Year. WDC-A is located in the United States; WDC-B is located in the USSR; other countries occasionally volunteer to host specific discipline centers (WDC-C); and WDC-D has recently been established in the People's Republic of China. Each WDC consists of several data centers organized by subject. The data centers manage and distribute data bases that are of widespread interest to the international research community. The nine centers composing the U.S.'s WDC-A are:

Glaciology (Snow and Ice), Boulder, CO	303-492-5171
Marine Geology and Geophysics, Boulder, CO	303-497-6487

Meteorology, Asheville, NC	704-259-0682
Oceanography, Washington, DC	202-673-5571
Rockets and Satellites, Greenbelt, MD	301-286-6695
Seismology, Denver, CO	303-236-1500
Solar–Terrestrial Physics, Boulder, CO	303-497-6324
Solid Earth Geophysics, Boulder, CO	303-497-6521

Some federal agencies and others have developed data and information distribution systems to facilitate availability of information to potential users. Some of the major information systems relevant to land-use/cover change and global change research follow:

Department of Commerce National Technical	
Information Service (NTIS)	703-487-4600
NASA Climate Data System	301-286-9760
NASA Oceans Data System	818-354-6980
NASA Pilot Land Data System	301-286-9282
NOAA National Environmental Data Referral Service	202-673-5548
USGS Global Land Information System (GLIS)	800-252-GLIS

Internationally, the U.N. Environment Program (UNEP) is responsible for the development of the Global Environment Monitoring System (GEMS), an environmental data collecting and assessment activity involving 142 countries and hundreds of national and international organizations. The Global Resource Information Database (GRID) was established within the framework of GEMS in 1985. GRID is based on environmental geographic information systems (GIS) which can be used either in conjunction with mainframe computers to study global environmental issues or with microcomputers to study detailed problems at national or local levels. For more information about GEMS and/or GRID, contact UNEP, P.O. Box 30552, Nairobi, Kenya.

INFOTERRA is an international environmental research and referral service established by UNEP. The goal of INFOTERRA is to serve as a link between those who are seeking environmental information and those who have the knowledge and expertise. The INFOTERRA network consists of 137 member countries. All of the INFOTERRA members have established special national focal points (NFPs), which represent their countries in the INFOTERRA system and carry out work on the national level. The U.S. NFP is located at the EPA Headquarters in Washington, D.C. It serves as an agency clearinghouse for international information requests. INFOTERRA produces a publication entitled *The International Directory of Sources* which contains over 5600 registered international sources of information. This directory may be obtained from INFOTERRA/U.S. National Focal Point, U.S. Environmental Protection Agency, 401 M Street, SW (PM211a), Washington, DC 20460; phone: 202-382-1522.

Data Catalogs or Data Directories

A data catalog contains detailed information about whole data sets, typically specific to a discipline, data center, and/or other facility. The catalog contains information needed to identify and retrieve elements of the data set, given the specification of the independent variable range(s), and may contain information extracted from the data sets (e.g., percentage of cloud cover) as well as information to enable ordering (e.g., volume ID, file names). Once data sets of interest are located, the user can obtain them from the archive where the data reside.

A data directory is a collection of uniform descriptions that summarize the contents of a large number of data sets. It provides information suitable for making an initial determination of the existence and contents of each data set. Each directory entry contains brief information for a data set (e.g., type of data, data set name, time and location bounds).

A catalog system is an implementation of one or more directories, plus a guide and/or inventories, integrated with user support mechanisms that provide data access and answers to inquiries. Capabilities may include browsing, data searches, and placing and taking orders. In some cases, electronic linking of individual directories allows users to obtain information about existing data holdings anywhere in the catalog system without having to separately query each individual directory. Such is the case with the Global Change Distributed Information System (GCDIS).

Within the GCDIS, there is one directory exclusively dedicated to global change data base references: the Global Change Master Directory (GCMD). The GCMD also has the capability to electronically link to other directories to allow the user to review additional data holdings. The GCMD is the U.S. coordinating node of an International Directory Network (IDN) that includes nodes in Italy and Japan.

More than 100 national and international data centers representing federal and state governments, academia, private industry, and others contribute data base references to the GCMD. References exist for the following disciplines: geoscience, ocean science, atmospheric science, space physics, solar physics, planetary science, and astronomy. GCMD references of particular interest to land-use/cover change research include glacier, climate, soil, Landsat multispectral scanner system (MSS) and thematic mapper (TM) scenes, Systeme probatoire d'observation de la terre (SPOT) scenes, advanced very high resolution radiometer (AVHRR) images, and more.

Many independent directories contribute data base references to the GCMD. For example, the USGS's Earth Science Data Directory (ESDD) contributes global change data base references to the GCMD through the EROS Data Center, where appropriate data base references are also entered into the Global Land Information System (GLIS). A contributor who submits data base information on an ESDD Input Form (available from ESDD Project Manager, USGS MS 801,

Reston, VA 22092) to the USGS would then effectively have his or her data referenced in three directories, thereby increasing the probability that a potential user would locate the data.

To address the human dimensions of land-use/cover change and global change, references to data bases from the social sciences are also required. The Consortium for International Earth Science Information Network (CIESIN) is developing a directory of such references. The directory currently has more than 80 references to data bases, such as the Agency for Toxic Substances and Disease Registry (ATSDR), the Center for Redevelopment of Industrialized States, the Hispanic Health and Nutrition Examination Survey, and the South African National Cancer Registry.

The CIESIN directory management group is working closely with the GCMD management group to ensure compatability and ease of access to both physical and social science data relevant to global change research. The Inter-University Consortium for Political and Social Research (ICPSR) *Guide to Resources and Services, 1991–92* also has many references to social science data bases in over 130 countries.

Access to the GCMD from the networks indicated can be accomplished at no charge by using the following procedures:

NSI/DECnet (SPAN)
    $SET HOST NSSDCA
    USERNAME: NSSDC
Internet
    TELNET 128.183.36.25
    USERNAME: NSSDC
Omnet
    GOTO NSSDC
Dial-in lines
    Set to 8 bits, no parity, 1 stop bit
    Dial 301-286-9000
    CONNECT 1200 (or 2400 or 300)
    Enter several carriage returns
    ENTER NUMBER
    MD
    CALLING 55201 (or 55202)
    CALL COMPLETE
    Enter several carriage returns
    USERNAME: NSSDC

A prototype PC-based GCMD is currently being evaluated. For more information about any aspect of the GCMD, contact: GCMD User Support Office, NSSDC, Code 933, NASA Goddard Space Flight Center, Greenbelt, MD 20771; phone: 301-794-5186.

Data Archives

Data archives are collections of preserved and maintained data sets and data set documentation such that long-term ability to access and use of such data is provided. An example of an organization responsible for archiving information is the National Archives and Records Administration (NARA).

The NARA Center for Electronic Records appraises, collects, preserves, and provides access to general records in a format designed for computer processing. The center works with contributing agencies to identify and select electronic records likely to have permanent value, including long-term research potential. Some categories of archived data related to land-use/cover change are demographic data, economic and financial statistics, health and social services data, international import and export statistics by commodity, and scientific and technological data. Information about electronic records available from NARA can be obtained by contacting the Reference Staff, National Archives Building, Pennsylvania Ave., Seventh and Ninth Sts., NW, Washington, DC 20408; phone: 202-501-5579.

Of particular interest to land-use/cover change researchers are data archives held at the U.S. Geological Survey's EROS Data Center (EDC). The EDC houses the world's largest collection of space- and aircraft-acquired imagery of the earth, including over 2 million images acquired from Landsat and other satellites and over 8 million aerial photographs. The EDC is also a clearinghouse for information concerning the holdings of foreign Landsat ground reception stations and data acquired by other countries' earth observing satellites. To facilitate access to these data, EDC has developed the GLIS (described above), which is linked electronically to the GCMD.

Although not an archive in the true sense of the word, the World Meteorological Organization (WMO) serves as a facilitator for the development and operation of a comprehensive data collection, exchange, and archiving system for the meteorological services of nations worldwide. The World Weather Watch (WWW) is the basic program of the WMO, and its main elements are the Global Observing System, Global Data Processing System, and the Global Telecommunications System. National meteorological centers (NMCs) in participating nations of WMO/WWW are central data accumulation points. Sample data bases available through WMO include the Global Atmospheric Research Program, INFOCLIMA (International Climatological Data), and the Hydrological Operational Multipurpose Subprogramme. For more information, contact the Director, World Meteorological Center, NOAA, W/NMC, Washington, DC 20233.

The World Health Organization (WHO) is the international health agency of the United Nations and acts as a directing and coordinating authority on international health work. WHO promotes technical cooperation, research, and the generation and transfer of related information. The U.S. component of this organization is the Pan American Health Organization in Washington, D.C.; phone: 202-861-3200. Sample data bases are:

AFRO/CDC—Morbidity/Mortality AFRO
Decade Monitoring of Worldwide Water Supplies and Sanitation
Directory of On-going Research in Cancer Epidemiology
EURODOC—European Region Health and Health Services, 1988-Present
Health Statistics: Population, Live Births, and Mortality (Worldwide)
Joint FAO/WHO Food Contamination Monitoring Programme Data Base

An international organization that works with developing countries is the World Bank, Washington, D.C.; phone: 202-473-0579. The World Bank concentrates on rural and urban development, agriculture, and education, and conducts research programs on topics including economic planning. Data bases are maintained on such subjects as revised external debt, socioeconomic indicators, international business opportunities, and worldwide development projects.

The Agency for International Development (AID) is an international organization that promotes agricultural projects and industrialization as a way of overcoming economic crises. AID also operates a research and documentation center. Contact AID, 320 21st St. NW, Washington, DC; phone: 202-647-5572. Some AID data systems and data bases are the Development Information System, the Economic and Social Data System, and Economic and Social Indicators for 18 Latin American Nations, 1960–1971.

Training Data Sets

Training data sets are collections of data assembled to train or educate potential users of similar data. Land-use/cover change research and other aspects of global change research require combining data in new and innovative ways. Researchers sometimes have difficulty finding high-quality data bases within their own discipline, or even within their own agency or organization. When interdisciplinary efforts require them to combine data from across disciplinary or agency boundaries, the process of locating and using such data is not easy. Training data sets have been developed to stimulate thinking about effective ways to combine different types of data, and also to stimulate creative minds from secondary schools and beyond into areas of multidisciplinary research. Some sample training data sets are included in the listings below.

Data Bases

A data base can be a collection of data sets associated with a system, project, or facility, or a collection of interrelated or independent data items stored together to serve one or more applications.

A national program producing a collection of environmental data bases is the EPA's Environmental Monitoring and Assessment Program (EMAP). EMAP was initiated in 1990 and is being designed to monitor indicators of the condition of U.S. ecological resources. These assessments will require data collection over

479

long periods of time from sampling sites on a national subgrid of a global grid. EMAP will also use remote sensing technology and data from maps, censuses, and other sources to characterize the national landscape features in areas associated with the EMAP sampling grid. For more information, contact EMAP Director, ORD/OMMSQA (RD-680), U.S. EPA, Washington, DC 20460; phone: 202-382-5767.

The International Geosphere–Biosphere Program (IGBP) is producing a collection of data sets of interest to the land-use and land-cover change research community. It is developing a Data and Information System (DIS) and also is using remotely sensed data from the AVHRR to develop a global data set of the land surface. This data set will have a spatial resolution of 1 km and will be generated at least once every ten days for the entire globe. This is an international project involving space agencies and research groups from many countries. Major contributors from the United States are NOAA, NASA, and the USGS. For more information, contact S.I. Rasool, Director, IGBP/DIS, Université de Paris VI, France; phone: 1-44-27-61-68/69/70; Omnet: I.Rasool.

The listing of data bases at the end of this appendix includes many that fit the second definition, i.e., a collection of items stored together to serve various applications. For example, the USGS has data bases on water resources and sea-level issues; the U.S. Department of Commerce (DOC) has census data and county data such as assessor records; various agencies collect environmental data on such land-use-related subjects as agriculture and fertilizer application rates (U.S. Department of Agriculture) and landfills and waste sites. Climate data are available through NOAA, NCAR, and other organizations. In addition, numerous studies and maps provide data and information on various processes and geographic areas.

Data Catalogs and Data Directories

Global Change Research Site Directory

Contents:	Bibliographic references and data sources and composition of physical and biological landscape features for research sites and coastal refuges in the U.S. Southeast.
Coverage:	North Central Florida and Southwest Texas coast
Format:	Digital, maps, tables, text, references
Data Producer:	U.S. Fish and Wildlife Service, National Wetlands Research Center (USFWS-NWRC)
Data Location:	Lafayette, Louisiana
Comments:	Work in progress. Target date of completion, FY 1993.
Point of Contact:	Thomas W. Doyle
	USFWS-NWRC
	1010 Gause Blvd.
	Slidell, LA 70458

                        Phone: 504-646-7355
Submitted By: Thomas W. Doyle

Electronic Products Catalog
Contents: Catalog of recent monographs, periodicals, data products,
                        and videotapes from the U.S. Department of
                        Agriculture's economics agencies.
Coverage: U.S. and world
Format: Digital, text, references
Data Producer: Economic Research Service (ERS); National
                        Agricultural Statistics Service (NASS); World
                        Agricultural Outlook Board
Data Location: ERS-NASS, Washington, DC
Comments: Published quarterly.
Point of Contact: Evelyn Blazer
                        Electronic Products Catalog
                        Room 1232
                        1301 New York Ave., NW
                        Washington, D.C. 20005-4788
                        Phone: 202-219-0305
Submitted By: Barbara Anderson, OIES

Operational Airborne and Satellite Cover Products of the National Operational
Hydrologic Remote Sensing Center (catalog)
Contents: Alphanumeric and graphic remotely sensed hydrology
                        products.
Coverage: Regions covering two-thirds of the U.S. and southern
                        Canada where snow cover is a significant hydrologic variable
Format: Alphanumeric and graphic products
Data Producer: National Operational Hydrologic Remote Sensing
                        Center (NOAA)
Data Location: National Operational Hydrologic Remote Sensing
                        Center, Minneapolis, MN
Point of Contact: Milan Allen
                        National Weather Service, NOAA
                        6301 34th Ave. South
                        Minneapolis, MN 55450
                        Phone: 612-725-3232

Global Network Meta Data File (directory)
Contents: Catalog of data sets used by the Terrestrial Branch of the
                        U.S. Environmental Protection Agency (EPA)
                        Environmental Research Laboratory (ERL)–Corvallis
                        (Oregon). This includes data sets used in global climate

	research. Types of data sets include soils, sunshine, cloudiness, temperature, vegetation, political boundaries, forests, etc.
Coverage:	U.S. and world; some data sets for specific countries
Format:	Digital (ARC/INFO, GRASS, SAS, Spreadsheet), maps, text
Data Producer:	EPA ERL-C
Data Location:	Terrestrial Branch Network, Corvallis, OR
Comments:	Continually being augmented with new research results and data collection. Also releasing a CD-ROM data base of global data sets in cooperation with NGDC.
Point of Contact:	EPA Environmental Research Laboratory 200 SW 35th St. Corvallis, OR 97333
Submitted By:	Peter A. Beedlow, Terrestrial Branch Chief Phone: 503-420-4600

Instituto Nacional de Estadisticas Geograficas y Informacion, Mexico Catalogo

Contents:	Statistics and maps produced by the Mexican government. It lists maps, censuses, aerial photographs, and statistical publications.
Coverage:	Mexico
Format:	Maps, tables
Data Producer:	INEGI, Mexico
Data Location:	INEGI
Point of Contact:	INEGI Balderas Street Mexico City, Mexico There is also an INEGI office in the Mexico City Airport main concourse.
Submitted By:	Diana Liverman

Preliminary and Partial Title List of Holdings, Center for Electronic Records, National Archives and Records Administration (directory)

Contents:	Lists approximately 4500 of 14,000 data sets currently in the custody of the Center for Electronic Records.
Coverage:	See 'Data Archives,' above
Format:	Text, output periodically from a data base that is maintained as an ongoing activity of the Reference Services staff
Data Producer:	Center for Electronic Records, NARA
Data Location:	Center for Electronic Records, Washington, D.C.
Point of Contact:	Reference Services Staff Room 18E National Archives Building 8th and Pennsylvania Aves.

Washington, D.C. 20408
Phone: 202-501-5579
Submitted By: Margaret O. Adams, Assistant Branch Chief, Archival
Services Branch, Center for Electronic Records

Global Land Information System (directory)

Contents: On-line directory, guide, and inventory listings that are
searchable via Internet or dial-up lines. GLIS contains
references to regional, continental, and global land
information including land-use, land-cover, and soils data
cultural and topographic data; and remotely sensed satellite
and aircraft data. Data sets include Landsat MSS and
TM, AVHRR 1-km, continental vegetation indices, and
cartographic data.

Coverage: World and regional
Format: Digital (CCT, 8 mm, CD-ROM, floppy disks) plots, photos
Data Producer: Government agencies
Data Location: Various
Comments: GLIS is accessible via Internet or modem connection.
PC-GLIS software available at no cost to provide enhanced
graphics capabilities.
Point of Contact: User Assistance: 800-252-GLIS. User access to GLIS:
From NSI/DECnet:
$SET HOST GLIS
USERNAME: GLIS
From Internet:
$TELNET glis.cr.usgs.gov
or $TELNET 192.41.204.54
Direct dial:
Set modem to 8 bits, no parity, 1 stop bit
Dial: 605-594-6888
Submitted By: Donna Scholz, USGS EROS Data Center

Cartographic Catalog

Contents: Sources of maps and charts, aerial photographs and imagery,
geodetic control data, map data in digital form, and earth-
science-related books, studies, indexes, software programs,
and reports.
Coverage: Mostly U.S., some international
Format: CD-ROM, microfiche
Data Producer: USGS Earth Science Information Office, compiled from the
indicated sources
Data Location: Earth Science Information Office, Reston, VA
Comments: The Earth Science Information Office is the information

branch of the USGS's National Mapping Program. ESIO gathers descriptions of earth science products held by federal state, and local agencies and private companies. Computerized information systems are used to edit, store, and distribute these descriptions.

Point of Contact: U.S. Geological Survey
Earth Science Information Office
509 National Center
Reston, VA 22092
Phone: 703-648-5914

Submitted By: Walter Wagner
Phone: 703-648-5914

Aerial Photography Summary Record System (APSRS) (catalog)

Contents: Source for over 495,000 aerial photography projects that contain 2,322,090 7.5-minute quadrangles of output data (as of April 1, 1991) held by approximately 580 federal, state, and local government agencies and commercial firms. Each summary record describes a completed U.S. aerial photographic project. It is keyed to a southeast corner latitude-longitude coordinate of a USGS 7.5-minute map.

Coverage: Includes the National Aerial Photography Program (NAPP), Side Looking Airborne Radar (SLAR), and other low- and high-altitude aerial photographic coverage.

Format: Fixed block, fixed record length data base that is processed as a sequential file in CD-ROM and microfiche formats. Aerial photographs are available in a variety of formats.

Data Producer: USGS Earth Science Information Office and EROS Data Center

Data Location: Earth Science Information Office, Reston, VA, and EROS Data Center, Sioux Falls, SD

Comments: The Earth Science Information Office (ESIO) is the information branch of the USGS's National Mapping Program. ESIO acquires information about earth-science-related products held by federal, state, and local governments and private agencies. Computerized information systems are used to edit, store, and disseminate these descriptions. The APSRS is ESIO's automated encyclopedia of aerial photography held by federal and state government agencies and commercial firms.

Point of Contact: U.S. Geological Survey
Earth Science Information Office
509 National Center
Reston, VA 22092

Phone: 703-648-5914

Submitted By: Walter Wagner

Phone: 703-648-5914

NOAA Earth System Data Directory

Contents:	872 descriptions of NOAA data sets, as of 25 October 1991.
Coverage:	Primarily U.S., but some foreign and global
Format:	Digital
Data Producer:	NOAA. Data in NOAA offices are from many different sources.
Data Location:	National Oceanographic Data Center, Washington, DC
Comments:	System is available to the research and general user community at no charge by modem, Internet, and DECnet (SPAN). This system points to the holders of the data sets and in some cases allows direct connection to the data system.
Point of Contact:	Gerald S. Barton
	National Oceanographic Data Center
	NOAA/NESDIS (E/OCx7)
	1825 Connecticut Avenue, NW
	Washington, DC 20235
	Phone: 202-606-4548
Submitted By:	Gerald S. Barton

National Environmental Data Referral Service (NEDRES) (directory)

Contents:	Over 22,000 descriptions of environmental and satellite data sets held by federal, state, and local governments and private, public and academic institutions.
Coverage:	Primarily U.S., but some foreign and global
Format:	Digital. Full text search with powerful tools.
Data Producer:	NOAA
Data Location:	Various
Comments:	System is available to any user using Sprintnet (Telenet) national and international telecommunication systems. The descriptions point to the holders of the data sets.
Point of Contact:	Gerald S. Barton
	National Oceanic Data Center
	NOAA/NESDIS (E/OCx7)
	1825 Connecticut Avenue, NW
	Washington, DC 20235
	Phone: 202-606-4548
Submitted By:	Gerald S. Barton

Earth Science Data Directory (ESDD)

Contents:	References to approximately 3000 earth science and natural

resource data bases concerned with geologic, hydrologic, cartographic, biologic, geographic, sociologic, economic, and demographic sciences. References to data bases that support the protection and management of natural resources are also included. Points of contact and other pertinent information are listed for data bases.

Coverage:	Primarily U.S., but some foreign and global data bases are referenced, particularly those from the Arctic
Format:	The data referenced are stored in many forms, including digital, map, text, etc. The ESDD is maintained on line and is available on a CD-ROM that is published quarterly.
Data Producer:	State, federal, and foreign governments, universities
Data Location:	Producing agencies
Comments:	ESDD is actively seeking new data base references. Data bases are referenced by key words, location, producing agency, etc.
Point of Contact:	ESDD Project Manager U.S. Geological Survey 801 National Center Reston, VA 22092 Phone: 703-648-7112
Submitted By:	Eliot Christian Phone: 703-648-7245

Satellite Data Services Division (SDSD) Electronic Catalog System (ECS)

Contents:	Inventory information about SDSD's AVHRR data sets.
Coverage:	Global AVHRR 4-km data since 1979. Regional 1-km data since 1985, much of it acquired for land-use/cover change studies.
Format:	On-line catalog accessible by telephone
Data Producer:	NOAA, National Environmental Satellite, Data, and Information Service (NESDIS), NCDC, SDSD.
Data Location:	Satellite Data Services Division, Camp Springs, MD
Comments:	NCDC/SDSD has maintained its ECS for several years. SDSD is in the process of rehosting its ECS to a new system, and there will be limited access for about a year. The new ECS will be augmented with a prototype on-line browse system in the same time frame.
Point of Contact:	Satellite Data Services Division Room 100, Princeton Executive Square 5627 Allentown Road Camp Springs, MD 20746 Phone: 301-763-8402 Fax: 301-763-8443

Submitted By: Levin Lauritson

Catalog of Data Bases and Reports

Contents:	References to quality-assured, documented data bases on various topics relating to ecosystems, nutrient cycles, geochemistry, and climate.
Coverage:	Varying, up to global
Format:	Digital, with accompanying statistics, graphics, and text
Data Producer:	Contributors from many institutions in the U.S. and foreign countries
Data Location:	CDIAC, Oak Ridge National Laboratory, TN
Comments:	The numeric data packages described in the catalog are compiled, edited, and produced by CDIAC, based primarily on contributed data; in a few cases, the authors of the data bases are located at CDIAC as well.
Point of Contact:	Thomas A. Boden
	CDIAC, Oak Ridge National Laboratory
	P.O. Box 2008, Mail Stop 6335
	Oak Ridge, TN 37831-6335
	Phone: 615-574-0390
Submitted By:	Robert M. Cushman
	Phone: 615-574-0390

Master Water Data Index (directory)

Contents:	Information about hydrologic sites for which water data are available. The index maintains identification information and a detailed description of each site, including latitude, state county, hydrologic unit, site type, and operating agency.
Coverage:	United States
Format:	Digital
Data Producer:	USGS, National Water Data Exchange
Data Location:	USGS, Reston, VA
Comments:	Original data base created from Water Resources Division (WRD) Catalog of Information of Water Data, WRD MIS Hydrologic Station File, and USGS Water Data Storage and Retrieval System (WATSTORE) and EPA Storage and Retrieval System (STORET).
Point of Contact:	Donald J. Dolnack
	National Water Data Exchange
	U.S. Geological Survey
	Reston, VA 22092
Submitted By:	James S. Burton
	Phone: 703-648-5684

Water Data Sources Directory

Contents:	Information on organizations that collect water data or water-related data. Indicates whether the organization is a member of the USGS's National Water Data Exchange, identifies the type of organization and type of data collected, and maintains information on the geographic area covered by the organization.
Coverage:	U.S. and organizations in other countries including Canada, Australia, Brazil, China, and Thailand
Format:	Digital; also available on diskettes for use on IBM and IBM-compatible computers
Data Producer:	USGS National Water Data Exchange
Data Location:	National Water Data Exchange, Reston, VA
Point of Contact:	J. Wayne Green
	National Water Data Exchange
	U.S. Geological Survey
	Reston, Virginia 22092
Submitted By:	James S. Burton
	Phone: 703-648-5684

Census Catalog and Guide: 1992

Contents:	A cumulative annual description of data products, statistical programs, and services of the Census Bureau. It provides abstracts of the publications, data files, microfiche, maps, and items online. In addition, the catalog offers such features as information about censuses and surveys and telephone contact lists of data specialists at the Census Bureau, the state data centers, and other data processing service centers.
Coverage:	All Census Bureau products, mid-1988 through early 1992; selected statistical products from other federal agencies. (A new edition is issued each summer.)
Format:	Text
Data Producer:	U.S. Bureau of the Census
Data Location:	U.S. Bureau of the Census, Washington, DC
Comments:	Sold by the U.S. Government Printing Office. Publication No. GPO S/N 003-024-08560-7; $17. Order from:
	Government Printing Office
	Superintendent of Documents
	P.O. Box 371954
	Pittsburgh, PA 15250-7954
	Make checks payable to 'Superintendent of Documents.'
Point of Contact:	John McCall
	Data User Services Division
	Bureau of the Census

Washington, DC 20233
Phone: 301-763-1584
Submitted By: John McCall

A Compilation of Inventories of Emissions to the Atmosphere (directory)
Contents: Information about 190 inventories of emissions, emission
                   factors, or supporting information relevant to inventory
                   production. Lists characteristics such as spatial and temporal
                   resolution, geographical region, and epoch.
Coverage: Regional and global
Format: Tables and accompanying text
Data Producer: Global Emissions Inventory Activity, International Global
                   Atmospheric Chemistry Project
Data Location: IGAC Secretariat
                   c/o J. M. Pacyna
                   Norwegian Institute for Air Research
                   P.O. Box 64
                   N 2001 Lillestrom
                   Norway
Point of Contact: T.E. Graedel
                   AT&T Bell Laboratories
                   Room 1D-349
                   Murray Hill, NJ 07974
                   Phone: 908-582-5420
Submitted By: T.E. Graedel

Digitizing the Future (catalog)
Contents: Summary information concerning Defense Mapping Agency
                   digital data and related subjects to all components of the
                   DOD, federal agencies, special program offices, and their
                   contractors.
Coverage: Varies to global
Format: Text
Data Producer: DOD
Data Location: Defense Mapping Agency, Fairfax, VA
Point of Contact: Defense Mapping Agency
                   Customer Assistance Office
                   8613 Lee Highway
                   Fairfax, VA 22031-2137
                   Phone: 800-826-0342

Data Availability at the National Center for Atmospheric Research (NCAR)
(catalog)
Contents: List of data sets available from NCAR. Updated annually.

Coverage:	Varies
Format:	Text
Data Producer:	NCAR
Data Location:	NCAR, Boulder, CO
Comments:	NCAR has 380 data sets. Data can be copied on tape at cost Catalog can be used on line at NCAR (requires NCAR computing project number).
Point of Contact:	NCAR
	Data Support Section
	Scientific Computing Division
	P.O. Box 3000
	Boulder, CO 80307
Submitted By:	Roy Jenne
	Phone: 303-497-1215

Data Available on CD-ROMs from the National Center for Atmospheric Research (NCAR) (catalog)

Contents:	Descriptions of environmental data available on CD-ROMs.
Coverage:	Varies
Format:	Text
Data Producer:	NCAR
Data Location:	NCAR, Boulder, CO
Point of Contact:	NCAR
	Data Support Section
	Scientific Computing Division
	P.O. Box 3000
	Boulder, CO 80307
Submitted By:	Roy Jenne
	Phone: 303-497-1215

Current Global, Land-Surface Data Sets for Use in Climate-Related Studies (at NCAR) (catalog)

Contents:	References to land-use and soil data and information.
Coverage:	Varies
Format:	Text
Data Producer:	NCAR
Data Location:	NCAR
Point of Contact:	NCAR
	Data Support Section
	Scientific Computing Division
	P.O. Box 3000
	Boulder, CO 80307
Submitted By:	Roy Jenne
	Phone: 303-497-1215

Data Archives

Inter-university Consortium for Political and Social Research (ICPSR)

Contents:	World's largest repository of computer-based research and instructional data for the social sciences. Over 29,000 data files containing information on social, cultural, and political phenomena; geography and the environment; health; historical and contemporary census enumerations; law and criminal justice; economics; and education.
Coverage:	Mostly U.S., some international
Format:	Survey data, some digital and coordinate data
Data Producer:	Government and private agencies, leading scholars and researchers
Data Location:	Institute for Social Research, University of Michigan, Ann Arbor, MI
Comments:	ICPSR is a membership-based consortium of 360 colleges and universities around the world. Data are distributed free of charge to researchers at member institutions and at reasonable cost to nonmembers.
Point of Contact:	ICPSR P.O. Box 1248 Ann Arbor, MI 48106 Phone: 313- 763-5010
Submitted By:	Richard Rockwell Phone: 313-764-2570

Center for Electronic Records, National Archives and Records Administration

Contents:	Electronic records created or received by agencies of the federal government, concerning virtually any area or subject in which the government is involved. The center maintains electronic records with continuing value created by the Congress, the courts, the Executive Office of the President, numerous presidential commissions, and nearly 100 bureaus, departments, and other components of executive branch agencies and their contractors.
Coverage:	Data created or collected or received by any federal agency that have been identified as permanently valuable records for retention by the National Archives.
Format:	Federal regulations require that electronic records be transferred to the National Archives in software-independent formats.
Data Producer:	Any federal agency
Data Location:	Center for Electronic Records, Washington, DC

Appendix 1

Point of Contact: Reference Services Staff
Room 18E
National Archives Building
8th and Pennsylvania Aves.
Washington, DC 20408
Phone: 202-501-5579

Submitted By: Margaret O. Adams, Assistant Branch Chief, Archival
Services Branch, Center for Electronic Records

Geography and Map Division, Library of Congress

Contents: Maps of the world at a variety of scales; coverage is strongest
for U.S. and Europe, but includes the best coverage available
for virtually all of the world; includes maps of all types,
ranging from topographic to thematic; collections include ca.
4 million maps and 50,000 atlases.

Coverage: Worldwide, historic as well as current (ca. 1480s to present)

Format: Printed maps and atlases, assorted globes and models

Data Producer: Various

Data Location: Library of Congress, Washington, DC

Comments: Resources available to the public.

Point of Contact: Reference and Bibliography Section
Geography and Map Division
Library of Congress
Washington, DC 20540
Phone: 202-707-6277

Submitted By: Gary L. Fitzpatrick

Scientific Committee on Antarctic Research (SCAR) Library for Geodesy and Cartography

Contents: Approximately 450,000 aerial photographs of Antarctica. The
majority of film is black and white, with some color and
color-infrared photographs. Coverage began in 1946 and
continues through the present. Over 3000 maps and atlases.

Coverage: Antarctica

Format: Photographs, maps, atlases

Data Producer: Primarily USGS

Data Location: USGS EROS Data Center, Sioux Falls, SD

Comments: Ancillary products such as geodetic control data, maps,
browse prints, and miscellaneous reports are available for
inspection at the SCAR library; some cartographic products
are available for sale through:

        Office of International Activities
        U.S. Geological Survey
        515 National Center
        Reston, VA 22092

Point of Contact: Customer Services
USGS EROS Data Center
Sioux Falls, SD 57198
Phone: 605-594-6151
Submitted By: Donna Scholz, USGS EROS Data Center

Satellite Data Services Division (SDSD), National Climatic Data Center (NCDC)
Contents: AVHRR data and derived vegetation index data and cloud-free Antarctic 1-km data set.
Coverage: Worldwide AVHRR 4-km data since 1979. Regional 1-km data since 1985, much of it acquired for land-use/cover change studies. Antarctic data set: Antarctica and surrounding oceans.
Format: Digital data, hard copy imagery
Data Producer: NOAA, NESDIS, NCDC/SDSD
Data Location: Satellite Data Services Division, Camp Springs, MD
Comments: NCDC/SDSD archives AVHRR data and derived products and provides services to a variety of users, including the land-use/cover community. Much of the 1-km data were collected specifically for land use/cover. The Antarctic data were used to construct an Antarctic mosaic.
Point of Contact: Satellite Data Services Division
Room 100, Princeton Executive Square
5627 Allentown Road
Camp Springs, MD 20746
Phone: 301-763-8402
Fax: 301-763-8443
Submitted By: Levin Lauritson

Archive of Environmental Data Bases
Contents: Climatological/meteorological observations, information and derivations, including temperature, precipitation, snowfall, wind, humidity, weather, cloud cover, and pressure.
Coverage: Global, but bulk of the data sets are for the U.S.
Format: Digital, manuscript, publications, microfiche, maps, and tables
Data Producer: NOAA, Federal Aviation Administration, DOD, foreign meteorological services, others
Data Location: NCDC, Asheville, NC
Comments: Period of record for many U.S. stations extends back to the late 1880s. Global data coverage begins in 1967, but some regional data for earlier periods are available. Not all of the observational elements are available in all of the data sets. Approximately 17,000 magnetic tapes, 1.2 million microfiche, and 55 million paper records.

Appendix 1

Point of Contact: National Climatic Data Center
                      Federal Building
                      Asheville, NC 28802
                      Phone: 704-259-0384
Submitted By: Richard M. Davis

Training Data Sets

Global Change Data Base, Africa Data Base
Contents: Vegetation (from satellite and ground study), ecosystems,
                      climate, elevation, land-cover and terrain classes, soils, and
                      other environmental data for Africa. Data are scientifically
                      integrated and peer reviewed, with special attention to
                      educational applications.
Coverage: Africa
Format: Digital
Data Producer: Many original producers. Integrated at the National
                      Geophysical Data Center (NGDC) of NOAA.
Data Location: NGDC, Boulder, CO
Comments: Data are similar to the Global Change Data Base, except that
                      they are solely for Africa. Some additional data available only
                      for Africa are provided, workbooks are being prepared, and
                      special encouragement is given for the development of
                      additional educational materials for the junior high school
                      to graduate level.
Point of Contact: David C. Schoolcraft
                      National Geophysical Data Center
                      325 Broadway
                      Boulder, CO 80303
                      Phone: 303-497-6125
Submitted By: David A. Hastings, NOAA

Joint Educational Initiative—Earth and Space Sciences (JEdI)
Contents: Three CD-ROMs and a teacher's activity book. The disks
                      provide a vast array of data bases, images, and software on a
                      digital medium. The activity book is based on the information
                      on the disks and provides the necessary interface between the
                      scientific and educational communities.
Coverage: Global
Format: Digital
Data Producer: USGS, NASA, NOAA

Data Location: Various
Comments: Leadership of JEdI was transferred to the University of
Maryland in 1991. Information on the project can be obtained
from:
> Robert W. Ridky
> Professor of Geology
> University of Maryland
> Department of Geology
> College Park, MD 20742

Copies of the JEdI materials can be ordered from:
> Nimbus Information Systems
> SR 629, Guildford Farm
> Ruckersville, VA 22968
> Phone: 804-985-1100
> Fax: 804-985-4625

Point of Contact: John Sands, Nimbus Information Systems
Phone: 804-985-1100
Submitted By: Mary Orzech, USGS

International Space Year Global Change Encyclopedia
Contents: A comprehensive set of satellite and other global data with
relevance to studies of global change and the earth system.
Coverage: Global, regional, polar
Format: Packaged on CD-ROMs, accompanied by appropriate soft
ware for access, display, and manipulation
Data Producer: Canadian Centre for Remote Sensing
Data Location: Canadian Centre for Remote Sensing
Comments: This is a joint project being carried out by the Canadian
Centre for Remote Sensing, on behalf of the Canadian Space
Agency, with NOAA and NASA as major contributors, and
the USGS EROS Data Center also contributing to the project.
Point of Contact: Mr. Rejean Simard
2464 Sheffield Road
Ottawa, Ontario, K1A OY7, Canada
Phone: 613-952-2741
Fax: 613-952-7353

Data Bases

Primary Productivity of Vegetation in Different Geographical Zones
Contents: Data on biomass and annual increase of biomass for 7000
sites.

Coverage: Global
Format: Tables, computer tables for a part of the base
Data Producer: Institute of Geography of the USSR Academy of Sciences
Data Location: Institute of Geography of the USSR Academy of Sciences, Moscow
Point of Contact: Department of Biogeography
Institute of Geography
USSR Academy of Sciences
Staromonetny 29
Moscow 109017
USSR
Fax: 095-230-2090
Submitted By: Olga Bykova
Phone: 095-230-2090; 238-9121

Digital Chart of the World
Contents: Worldwide coverage using a topologically structured vector representation of the earth's land surface information on a microcomputer-accessible storage medium.
Coverage: Global
Format: Digital
Data Producer: Defense Mapping Agency in cooperation with the United Kingdom, Canada, and Australia; Environmental Systems Research Institute.
Data Location: Available through the USGS's Earth Science Information Center or call 800-USA-MAPS.
Comments: The objective of this project is to develop, refine, and establish a suite of standards to support future DMA digital data and enhance the utility of digital spatial information in vector format.
Point of Contact: D.M. Danko, Project Manager
Digital Chart of the World Project
Defense Mapping Agency
8613 Lee Highway
Fairfax, VA 22031-2137

Geophysics of North America
Contents: Consolidated collection of land and marine geophysical data for North America on CD-ROM.
Coverage: Northern Hemisphere from approximately the Prime Meridian to the International Date Line
Format: Digital; software allows viewing data as images
Data Producer: NOAA
Data Location: NGDC, Boulder, CO

Point of Contact: Allen Hittelman
                     National Geophysical Data Center
                     NOAA, Code E/GC1; Dept. 720
                     325 Broadway
                     Boulder, CO 80303-3328
                     Phone: 303-497-6591

30-Second U.S. Topography Data for Personal Computers
Contents: The topography data set spans the entire United States
                     (excluding Alaska and Hawaii) and a small portion of the
                     bordering areas. Elevations are given for every 30-second by
                     30-second coordinate cell (approximately one square
                     kilometer).
Coverage: United States (excluding Alaska and Hawaii)
Format: Digital topography data base and access software.
Data Producer: National Geophysical Data Center, NOAA
Data Location: NGDC, Boulder, CO
Point of Contact: Lee Row
                     National Geophysical Data Center
                     NOAA, Code E/GC1
                     325 Broadway
                     Boulder, CO 80303
                     Phone: 303-497-6764

Global Environmental Sensitivity Database
Contents: Data for analysis of the world; being used as input to U.S.
                     Army Construction Engineering Research Lab's global
                     sensitivity analyses with 'GRASS' GIS.
Coverage: Global
Format: Digital/raster
Data Producer: U.S. Army Construction Engineering Research Lab
Data Location: On 3 CD-ROMs available from Rutgers University; phone
                     908-932-9631.
Comments: Extensive digital global data base includes global
                     vegetations, soils, bioproductivity, sensitive species,
                     oceanography, topography, climatology, demographics,
                     economics, cultural concerns, fisheries, and other data.
Point of Contact: Robert C. Lozar
                     USACERL
                     Phone: 217-373-6739

National Transportation Network Databases
Contents: Digital data on major transportation networks in the United
                     States (highways, railroads, waterways). Geographic loca-

497

tions derived from USGS 1:2 million and 1:100,000 DLGs. Enhanced with attribute data on travel times, restrictions, and infrastructure descriptors.

Coverage:	Primarily continental United States
Format:	Digital
Data Producer:	Various (Oak Ridge National Laboratories, Federal Railroad Administration)
Data Location:	Currently with data producers. Establishment of a DOT data clearinghouse is under consideration.
Comments:	Currently under development. No overall funding for data base development has been established, so development activities are largely *ad hoc*. Preliminary data bases for all three networks will be available by the end of FY 1992.
Point of Contact:	Bruce D. Spear Senior Transportation Planner Volpe National Transportation Systems Center Kendall Square Cambridge, MA 02142 Phone: 617-494-2192
Submitted By:	Bruce D. Spear

USGS/NGIC Geomagnetic Observatory Data: 1985–1989
USGS/NGIC Geomagnetic Observatory Data: 1990

Contents:	One-minute geomagnetic field values from the USGS net work of 13 magnetic observatories. Also contains magnetic activity (K-Index) values and other geomagnetic information.
Coverage:	United States, Alaska, Hawaii, Guam, and Puerto Rico
Format:	Digital data on CD-ROM, with access software and documentation
Data Producer:	USGS National Geomagnetic Information Center (NGIC)
Data Location:	NGIC, Golden, CO
Comments:	The NGIC serves as the primary center for definitive geomagnetic data from the nation's magnetic observatory network.
Point of Contact:	National Geomagnetic Information Center U.S. Geological Survey Box 25046, MS 968 DFC Denver, CO 80225-0046
Submitted By:	Donald C. Herzog Phone: 303-236-1365

National Geomagnetic Information Center Data Base

Contents:	Digital one-minute geomagnetic component variation and absolute data reported during the past month to one year.

Some of these data are reported in near real-time (12-minute intervals).

Coverage: Continental United States, Alaska, Hawaii, Guam, Puerto Rico, and Canada

Format: Digital one-minute component data, tables of mean hourly values and K-indices, and magnetograms (graphic)

Data Producer: USGS, Geologic Survey of Canada, and University of Alaska

Data Location: USGS, Denver, CO

Comments: Data are provided to users upon request. Plans are in progress to provide interactive access.

Point of Contact: U.S. Geological Survey
Box 25046, MS 968 DFC
Denver, CO 80225-0046

Submitted By: Lanny Wilson
Phone: 303-236-1370

GEOMAG

Contents: A dial-in service for getting model values of geomagnetic field elements such as declination and total field intensity.

Coverage: U.S. and global

Format: Interactive computer program accessed by modem.

Data Producer: USGS

Data Location: USGS, Denver, CO

Point of Contact: Norman W. Peddie
U.S. Geological Survey
MS 968, Box 25046
Denver Federal Center
Denver, CO 80225-0046

Submitted By: Norman W. Peddie

Global Ecosystem Database

Contents: Global vegetation index data, 1985-88; global climate data, global ecosystem maps, global soil maps

Coverage: Global

Format: Digital images and digital maps on one CD-ROM

Data Producer: NOAA National Geophysical Data Center

Data Location: NGDC, Boulder, CO

Comments: Produced for EPA Global Change Research Program, Environmental Research Laboratory, Corvallis, Oregon

Point of Contact: John Kineman, NGDC
3100 Marine St.
Boulder, CO 80303
Phone: 303-497-6521

Appendix 1

Digital Flood Insurance Rate Maps (FIRMs)

Contents: Thematic flood risk data, prepared in response to National Flood Insurance Program regulations.

Coverage: More than 45 counties have been completed. More than 90 counties to be completed by the end of FY 1992.

Format: USGS DLG-3 optional format, to be available on CD-ROM by the end of the third quarter, FY 1992

Data Producer: Federal Emergency Management Agency (FEMA) (contractor)

Data Location: FEMA, Washington, DC

Comments: New program in FY 1991, scheduled as a ten-year effort to prepare 41,000 digital FIRMs.

Point of Contact: Dan Cotter
FEMA
Insurance Administration—Risk Assessment
500 C St., SW
Washington, D.C. 20472
Phone: 202-646-2757

Submitted By: Dan Cotter

Global Climate Change Research Unit

Contents: Data bases on effects of global climate change (sea-level rise, elevated temperature, carbon dioxide and salinity) on submerged aquatic vegetation, emergent marshes, and bottom land hardwood forests.

Coverage: Global, but primarily U.S.

Format: Text

Data Producer: U.S. Fish and Wildlife Service, National Wetlands Research Center (USFWS-NWRC)

Data Location: Lafayette, Louisiana

Comments: Work in progress. Target date for completion is September 30, 1993.

Point of Contact: William Rizzo
USFWS-NWRC
700 Cajundom Blvd.
Lafayette, LA 70506
Phone: 318-266-8633

Submitted By: William Rizzo

Landsat Habitat Map of Florida

Contents: Based on 1985–89 TM imagery. The classification includes nine upland plant communities, eight wetland plant communities, one aquatic community, and four disturbed communities.

Coverage:	Florida
Format:	Digital
Data Producer:	Florida Camp and Fresh Water Fish Commission
Data Location:	FC&FWFC, Tallahassee, FL
Comments:	Office of Environmental Services, Non-Game Wildlife Section
Point of Contact:	Randy Kautz
	FC&FWFC
	620 S. Meridan Street
	Tallahassee, FL 32399-1600
	Phone: 904-488-6661
Submitted By:	Elijah Ramsey
	Phone: 504-646-7251

National Aerial Photography Program (NAPP)

Contents:	Aerial photographs of the 48 contiguous states. The NAPP program acquires color-infrared or black-and-white aerial photography from flights made at 20,000 feet above mean terrain every five years.
Coverage:	The 48 contiguous United States
Format:	Film transparencies or photographic reproductions of the color-infrared or black-and-white quarter quadrangle photographs are available in 9' × 9' contact print size, as well as in standard enlargement sizes. Custom products are also available. Published indexes show the extent and timing of coverage.
Data Producer:	Agricultural Stabilization and Conservation Service, U.S. Forest Service, National Agricultural Statistics Service, Soil Conservation Service, Bureau of Land Management, USGS, Tennessee Valley Authority
Data Location:	Reston, VA; Sioux Falls, SD; Salt Lake City, UT
Comments:	Coverage of the conterminous United States is scheduled for completion before the end of 1991. In 1992, the program staff began managing a second five-year cycle. Information about NAPP photography can be located within the Aerial Photography Summary Records System, an automated encyclopedia of aerial photography projects held by federal and state agencies and commercial firms.
Point of Contact:	U.S. Geological Survey
	Earth Science Information Center
	509 National Center
	Reston, VA 22092
	Phone: 800-USA-MAPS

U.S. Department of Agriculture
Aerial Photography Field Office
P.O. Box 30010
2222 West 2300 South
Salt Lake City, UT 84130

Photography and Shoreline Map Database

Contents:	Complete inventory of National Ocean Survey photography and shoreline map coverage as overlays on the USGS DLG 1:250,000 map base. Photograph centers and map corners are depicted and supporting data are listed.
Coverage:	United States
Format:	Digital data files displayed through customized Auto Cad licensed software
Data Producer:	National Ocean Survey, Coast and Geodetic Survey
Data Location:	Photogrammetry Branch, Rockville, MD
Comments:	The data files must be used within the data base as currently configured for the DOS-PC environment. The entire operating system is transferrable.
Point of Contact:	Photogrammetry Branch, Support Section 6001 Executive Blvd. Rockville, MD 20852 Phone: 301-443-8601 Fax: 301-443-1009
Submitted By:	Lewis A. Lapine, NOAA

Global Change Data Base

Contents:	Vegetation (from satellite and ground study), ecosystems, climate, elevation/bathymetry, land cover and terrain classes, soils, and other global scientific data. Data are scientifically integrated (not just collected) and peer reviewed, with periodic enhancements.
Coverage:	Global
Format:	Digital. Integrated for use on GIS as well as traditional modeling software.
Data Producer:	Many original producers. Integrated at the NOAA National Geophysical Data Center.
Data Location:	NGDC, Boulder, CO
Comments:	Data are under continual review and improvement by a number of testing and evaluation sites. Integration emphasizes quality control and improved documentation of the data, emphasizing their integrated use. New contributions of global scientific data are eagerly sought. Authorship credit and data exchange can be arranged.

Point of Contact: John J. Kineman (303-497-6900) or
David A. Hastings (303-497-6729)
National Geophysical Data Center
325 Broadway
Boulder, CO 80303

Submitted By: David A. Hastings, NOAA

National Water Data Storage and Retrieval System

Contents: Water data parameters measured or observed either daily or on a continuous basis and numerically reduced to daily values; annual maximum (peak) streamflow (discharge); analysis of water samples; water parameters measured on a schedule less frequent than daily; and inventory data about wells.

Coverage: United States

Format: Digital, tables

Data Producer: USGS

Data Location: USGS, Reston, VA

Comments: This data base resides on the USGS's mainframe computer but will be removed during 1994 and replaced with the National Water Information System–II.

Point of Contact: John Briggs
U.S. Geological Survey
437 National Center
Reston, VA 22092

Submitted By: James S. Burton
Phone: 703-648-5684

Selected Water Resources Abstracts

Contents: Abstracts of current and earlier literature related to water resources. Contains full bibliographic citation and descriptors.

Coverage: Water resources as treated in the life, physical, and social sciences and related legal and engineering aspects

Format: Digital and CD-ROM

Data Producer: Pertinent monographs, journal articles, and reports published worldwide

Data Location: USGS/WRD Data General DIS-II System (Contact: George Knapp, 703-648-6823), DIALOG, Information Services, European Space Agency, and on CD-ROM through:
      NISC (National Information Services Corp.), 301-243-0797
      Silver Platter Information, Inc., 800-343-0064
      Cambridge Scientific Abstracts, 800-843-7751

Comments: Designed to serve the scientific and technical information
                     needs of scientists, engineers, and managers.
Point of Contact: Water Resources Scientific Information Center
                     U.S. Geological Survey
                     425 National Center
                     Reston, VA 22092
Submitted By: Raymand A. Jensen
                     Phone: 703-648-6820

Central Atlantic Regional Ecological Test Site (CARETS)

Contents: 1:100,000 scale land-use and land-cover data of counties in
                     the Chesapeake Bay and Delaware Bay drainage areas.
Coverage: Chesapeake Bay and Delaware Bay drainage areas
Format: Maps on stable base mylar; digital files may now have
                     deteriorated.
Data Producer: USGS
Data Location: USGS Library, Reston, VA
Comments: Data were compiled by USGS in the early 1970s from high-
                     altitude photography. CARETS was a prototype GIS with
                     several layers of digital data.
Point of Contact: Robert Bier
                     USGS
                     National Center, MS 952
                     12201 Sunrise Valley Blvd.
                     Reston, VA 22092
                     Phone: 703-648-5555
Submitted By: Katherine Lins

Aerial Photography

Contents: Metric aerial photographs cover 70% of the U.S. coastline
                     and nearly 900 major U.S. airports. The photography dates
                     back to the 1940s.
Coverage: United States and Trust Territories
Format: Vertical photography, 9" × 9" negative format, black and
                     white, color, and infrared
Data Producer: National Ocean Survey (NOS), Coast and Geodetic Survey
Data Location: Photogrammetry Branch, Rockville, MD
Comments: The photography was obtained for mapping shoreline
                     features and airport facilities and obstructions. Much of the
                     photography is geodetically controlled and tide coordinated.
Point of Contact: Photogrammetry Branch, Support Section
                     6001 Executive Boulevard
                     Rockville, MD 20852
                     Phone: 301-443-8601
                     Fax: 301-443-1009

Submitted By: Lewis A. Lapine, NOAA

Shoreline Maps
Contents: Metric quality maps produced in support of the Nautical
                    Chart Program. These maps cover nearly 100% of the U.S.
                    coastline.
Coverage: United States and Trust Territories. The maps are synoptic in
                    nature.
Format: 1:20,000 scale planimetric maps
Data Producer: National Ocean Survey (NOS), Coast and Geodetic Survey
Data Location: Photogrammetry Branch, Rockville, MD
Comments: The maps depict high- and low-water features as well as near-
                    shore features.
Point of Contact: Photogrammetry Branch, Support Section
                    6001 Executive Boulevard
                    Rockville, MD 20852
                    Phone: 301-443-8601
                    Fax: 301-443-1009
Submitted By: Lewis A. Lapine, NOAA

AVHRR 1 km
Content: AVHRR 1-km satellite data acquired from October 13, 1978,
                    to the present.
Coverage: Global
Format: Digital
Data Producer: USGS EROS Data Center
Data Location: USGS EROS Data Center, Sioux Falls, SD
Comments: Each AVHRR scene is a raster data set of varying length and
                    2048 km width.
Point of Contact: Customer Services
                    USGS EROS Data Center
                    Sioux Falls, SD 57198
                    Phone: 605-594-6151
Submitted By: Donna Scholz, USGS EROS Data Center

Side Looking Airborne Radar (SLAR) Image
Contents: X-band synthetic aperture radar acquired annually by the
                    USGS since 1980.
Coverage: Selected areas in the conterminous U.S. and Alaska
Format: Digital, photographs
Data Producer: USGS
Data Location: USGS EROS Data Center, Sioux Falls, SD
Comments: SLAR photographic products include contact strip images,
                    mosaics, and custom products. Some ancillary products such

as indexes (on paper film, or microfiche) and mission flight logs are available.

Point of Contact: Customer Services
USGS EROS Data Center
Sioux Falls, SD 57198
Phone: 605-594-6151
Submitted By: Donna Scholz, USGS EROS Data Center

Landsat MSS
Contents: Landsat MSS data more than two years old.
Coverage: Global, since July 23, 1972
Format: Digital, photographs
Data Producer: NASA, NOAA
Data Location: USGS EROS Data Center, Sioux Falls, SD
Comments: Each Landsat scene is a raster data set covering an area
185 × 185 km at an approximate resolution of 80 m.
Point of Contact: Customer Services
USGS EROS Data Center
Sioux Falls, SD 57198
Phone: 605-594-6151
Submitted By: Donna Scholz, USGS EROS Data Center

Space Shuttle Handheld and Large Format Camera Photography
Contents: Space shuttle handheld photography, of which 85% is earth-looking views. The rest show satellite deployment, extrave hicular activities, and astronaut activity in the cabin. Shuttle missions began March 6, 1981.
Coverage: Global between 28°N and 28°S latitude
Format: Photographs
Data Producer: NASA
Data Location: USGS EROS Data Center, Sioux Falls, SD
Comments: Most of the photographs are natural color, although a limited amount of black-and-white film has been collected with polarizing filters. A small amount of color-infrared film has been tested on some missions.
Point of Contact: Customer Services
USGS EROS Data Center
Sioux Falls, SD 57198
Phone: 605-594-6151
Submitted By: Donna Scholz, USGS EROS Data Center

Continental AVHRR Greenness Composites
Contents: 10-day and 14-day vegetation index composites derived from AVHRR 1-km satellite data. Each 10-band composite con-

tains vegetation index, geographically registered AVHRR data, and various solar and satellite geometry data.

Coverage:	North America, western USSR, Sahel region of North Africa
Format:	Digital, maps, CD-ROM
Data Producer:	USGS EROS Data Center
Data Location:	USGS EROS Data Center, Sioux Falls, SD
Comments:	These data include both continental and regional composites. Typically, the temporal coverage is for the growing season for the region. Coverage of North America began in 1988, USSR coverage began in 1986, Sahel coverage began in 1988.
Point of Contact:	Customer Services
	USGS EROS Data Center
	Sioux Falls, SD 57198
	Phone: 605-594-6151
Submitted By:	Donna Scholz, USGS EROS Data Center

Global and Regional Topography Data Base

Contents:	Topographic data at several scales and areas of coverage, from the conterminous U.S. to global. Bathymetric data are also available for the U.S. Exclusive Economic Zone and world.
Coverage:	Regional (parts of North America, Europe, the Mediterranean, and the Far East) and global topography. Regional (U.S. Exclusive Economic Zone) and global bathymetry.
Format:	Digital, maps
Data Producer:	Many original producers. Edited, integrated, and enhanced at the NOAA National Geophysical Data Center.
Data Location:	NGDC, Boulder, CO
Comments:	Additional contributions of enhanced data are welcome.
Point of Contact:	Lee W. Row, Topography, 303-497-6764
	Peter W. Sloss, Bathymetry, 303-497-6119
	National Geophysical Data Center
	325 Broadway
	Boulder, CO 80303
Submitted By:	David A. Hastings, NOAA

Appendix II

Participants of the 1991 Global Change Institute

B.L. Turner II (Director)
Graduate School of Geography
Clark University
950 Main Street
Worcester, MA 01610
USA

Diogenes Salas Alves
Departamento de Processamento de Imagens
Instituto de Pesquisas Espaciais
Av. Dos Astronautas 1758
Caixa Postal 515-12201
São Jose Dos Campos
Brazil

Lourdes Arizpe
Instituto Envistigaciones
UNAM
Antropologicas
04510 Mexico
Mexico

Francis Bretherton
Space Science and Engineering Center
University of Wisconsin
Room 349
1225 W Dayton Street
Madison, WI 53706
USA

Stephen Brush
International Agricultural Development
University of California at Davis
Room 141 AOB-IV
Davis, CA 95616
USA

Stanley Buol
Department of Soil Science
North Carolina State University
Box 7619
Raleigh, NC 27695-7619
USA

Olga Bykova
Institute of Geography
USSR Academy of Sciences
Staromonetny per. 29
109017 Moscow
Russia

Ian Douglas
Department of Geography
University of Manchester
Manchester M13 9PL
U.K.

John Eddy
CIESIN
Saginaw Valley State University
2250 Pierce Road
University Center, MI 48710
USA

Michael Fosberg
U.S. Department of Agriculture Forest Service
First Floor Center Wing
201 14th Street, SW FF/ASR
Washington, DC 20250
USA

Kathleen Galvin
Natural Resource Ecology Laboratory
Colorado State University
Grassland Laboratory, Room 144
Fort Collins, CO 80523
USA

Appendix II

Thomas Graedel
Environmental Chemistry Research Laboratory
AT&T Bell Telephone Labs
Room 1D349
600 Mountain Avenue
Murray Hill, NJ 07974-2070
USA

Dean Graetz
Division of Wildlife and Ecology
CSIRO
P.O. Box 84
2602 Lyneham
Australia

Wolf Grossmann
Institute for Ecosystem Studies
Kegelgasse 27
A-1030 Vienna
Austria

Arnulf Grübler
International Institute for Applied Systems
   Analysis
A-2361 Laxenburg
Austria

Richard Houghton
Woods Hole Research Center
P.O. Box 296
Wood Hole, MA 02543
USA

Rattan Lal
Department of Agronomy
Ohio State University
202 Cottman Hall
2021 Coffey Road
Columbus, OH 43210-1086
USA

Diana Liverman
Department of Geography
Pennsylvania State University
325 Walker Building
University Park, PA 16802
USA

Stephen Lonergan
Department of Geography
University of Victoria
Cornett Building
Office B-234
Victoria V8W 2Y2
Canada

John McNeill
History Department
Georgetown University
Washington, DC 20057-1058
USA

Jerry Melillo
The Ecosystems Center
Marine Biological Laboratory
Woods Hole, MA 02543
USA

William Meyer
Graduate School of Geography
Clark University
950 Main Street
Worcester, MA 01610-1477
USA

Shem Migot-Adholla
Agriculture and Rural Development Department
The World Bank
Room N8041
1818 H Street, NW
Washington, DC 20433
USA

Peter Morrisette
P.O. Box 308
Victor, ID 83455
USA

Richard Moss
Woodrow Wilson School
Princeton University
Robertson Hall
Princeton, NJ 08544-1013
USA

Alexi Naumov
Department of Geography
Moscow State University
II9899 Moscow
Russia

Joyce Penner
Atmospheric, Microphysics, and Chemistry
   Department
Lawrence Livermore National Laboratory
P.O. Box 808 (L-262)
Livermore, CA 94550
USA

510

Steve Rayner
Battelle Pacific Northwest Laboratories
901 D Street, SW
Suite 900
Washington, DC 20024
USA

John Richards
Department of History
Duke University
104A W. Duke Building
Durham, NC 27708
USA

William Riebsame
Natural Hazards Center
University of Colorado
Campus Box 482
Boulder, CO 80309-0482
USA

Jennifer Robinson
Applied Landscape Ecology Section
Environmental Research Center Leipzig-Halle
   Inc.
Permoserstrasse 15
04318 Leipzig
Germany

Richard Rockwell
Inter-University Consortium for Political and
   Social Research
P.O. Box 1248
Ann Arbor, MI 48106
USA

Peter Rogers
Division of Applied Sciences
Harvard University
29 Oxford Street
Cambridge, MA 02138
USA

Vernon Ruttan
Department of Agricultural and Applied
   Economics
University of Minnesota
332 Classroom Office Building
1994 Buford Avenue
St. Paul, MN 55108
USA

Franklin Sadowski
Science Information Network
Environmental Research Institute of Michigan
3000 Plymouth Road
Ann Arbor, MI 48108
USA

Colin Sage
Environment Division
Wye College
Wye
Ashford
TN25 5AH Kent
UK

Steven Sanderson
Tropical Conservation and Development
   Program
University of Florida
319 Grinter Hall
Gainesville, FL 32611
USA

David Skole
Institute for the Study of Earth, Oceans, and
   Space
University of New Hampshire
Science and Engineering Building, 4th Floor
Durham, NH 03824
USA

Joel Tarr
Dean's Office for Humanities and Social
   Sciences
Carnegie-Mellon University
Baker Hall 260
Pittsburgh, PA 15213
USA

Evert van Imhoff
Netherlands Interdisciplinary Demographic
   Institute
Lang Houtstraat 19
P.O. Box 11650
2502 AR The Hague
The Netherlands

Michael Williams
School of Geography
University of Oxford
Mansfield Road
OX1 3TB Oxford
UK

Appendix II

Agency Representatives

J. Michael Hall
Office of Global Programs
National Oceanic and Atmospheric
    Administration
1100 Wayne Ave, Suite 1225
Silver Spring, MD 20910
USA

John Kelmelis
U.S. Geological Survey
National Center 521
12201 Sunrise Valley Drive
Reston, VA 22092
USA

Jeffrey Lee
Environmental Research Laboratory
Environmental Protection Agency
200 SW 35th Street
Corvallis, OR 97333
USA

James Sturdevant
Science and Applications Branch
U.S. Geological Survey
Earth Resources Observation Systems
Mundt Federal Building
Sioux Falls, SD 57198
USA

George Njiru
Natural Resource Ecology Laboratory
Colorado State University
Room 125
Fort Collins, CO 80523
USA

Jennifer Olson
CASID
Michigan State University
306 Berkey Hall
East Lansing, MI 48824-1111
USA

Satya Yadav
Department of Agriculture and Applied
    Economics
University of Minnesota
231 Classroom Office Building
1994 Buford Avenue
St. Paul, MN 55108
USA

Stephen Young
Graduate School of Geography
Clark University
950 Main Street
Worcester, MA 01610-1477
USA

Graduate Students

Winifred Hodge
U.S. Army Construction Engineering Research
    Laboratory
University of Illinois
2902 Newmark Drive
Champaign, IL 61826-9005
USA

Index

Index

National Center for Atmospheric Research (NEAR), 399, 474
National Climatic Data Center (NCDC), NOAA, 474
National Computer Center, U.S. EPA, 474
National Environmental Data Referral Service (NEDRES), 475, 485
national focal points (NAPS), 475
National Geomagnetic Information Center Data Base, 498–499
National Geophysical Data Center (NGDC), NOAA, 474
National Institute of Space Research, 110
National Oceanic and Atmospheric Administration (NOAA), CP-3, CP-13, 474
  Earth System Data Directory, 485
  ETOPO5 data set, CP-9
  National Climatic Data Center (NCDC), 474
  National Environmental Data Referral Service (NEDRES), 475, 485
  National Geophysical Data Center (NGDC), 474
  National Oceanographic Data Center, 474
  National Snow and Ice Data Center, 474
  polar orbiting satellites, 448
National Research Council, 9, 72
national resource accounting models, 412
National Science Foundation (NSF), 378, 379, 474
National Snow and Ice Data Center, NOAA, 474
National Space Science Data Center, NASA, 474
National Transportation Network Databases, 497–498
National Water Data Storage and Retrieval System, 503
natural resource accounting models, 426–429
nature myths, 363–364
NEAR (National Center for Atmospheric Research), 399, 474
NCDC (National Climatic Data Center), 474
NDVI, *see* normalized difference vegetation index
NEDRES (National Environmental Data Referral Service), 475, 485
net carbon exchange, modeling, 397–402
net primary production (NPP), 36, 274
  defined, 438
  effects of predicted climate change on, 399–402
  global pattern, CP-14

  modeling, 397–398
  TEM model of, 398–399
'neutral biosphere' assumption, 46
New Forests in Eastern Ontario program, 430
NAPS (national focal points), 475
NGDC (National Geophysical Data Center), 474
NGOs, *see* nongovernmental organizations
NH_3, *see* ammonia
NH_4^+, *see* ammonium ion
$(NH_4)_2SO_4$, *see* ammonium sulfate
nitrate ion (NO_3^-), 192
nitric acid (HNO_3), 192
nitric oxide (NO), 192
  ozone decreases and, 183–184
nitrification, as source of N_2O, 184, 185
nitrogen (N)
  crop requirements, 38
  fertilizer, production of, 310
  limitation, NPP and, 401
  soil system and, 215, 222
  TEM model of, 397
nitrogen dioxide (NO_2), 192
nitrous oxide (N_2O), 183–186
  characteristics of, 178
  increases in, 175, 183–184
  lifetime of, 177
  sources of, 183–186
NMHCs, *see* nonmethane hydrocarbons (NMHCs)
NO_2, *see* nitrogen dioxide
NO_3^-, *see* nitrate ion
NOAA, *see* National Oceanic and Atmospheric Administration
nomadic movement, of animals, 137
nomadic societies, 132, 133
  desertification and, 140
  precipitation and, 137, 138
nonforested land, 97
nongovernmental organizations (NGOs), 336–337
  development assistance by, 345–346
  institutional driving forces and, 331
  international, 36
nonmethane hydrocarbons (NMHCs), 190–191
  importance of, 181
  increases in, 176–177
  lifetimes of, 190–191
  OH and, 190
  oxidation of, 189
  sources of, 191
  surface abundance of, 178, 179
  tropospheric aerosols and, 199

World Meteorological Organization (WMO),
478
World Resources Institute (WRI), 101, 457, 458
World Soil Resources map, CP-8
World Values Survey, 20, 361–363
world views, 375–376

world water balance, 234
World Weather Watch (WWW), 478
World Wildlife Fund, 248
WRI (World Resources Institute), 101, 457, 458

Youssef, N., 273